BIO-NANOPARTICLES

BIO-NANOPARTICLES

Biosynthesis and Sustainable Biotechnological Implications

Edited by

Om V. Singh

WILEY Blackwell

Copyright © 2015 by Wiley-Blackwell. All rights reserved

Published by John Wiley & Sons, Inc., Hoboken, New Jersey
Published simultaneously in Canada

No part of this publication may be reproduced, stored in a retrieval system, or transmitted in any form or by any means, electronic, mechanical, photocopying, recording, scanning, or otherwise, except as permitted under Section 107 or 108 of the 1976 United States Copyright Act, without either the prior written permission of the Publisher, or authorization through payment of the appropriate per-copy fee to the Copyright Clearance Center, Inc., 222 Rosewood Drive, Danvers, MA 01923, (978) 750-8400, fax (978) 750-4470, or on the web at www.copyright.com. Requests to the Publisher for permission should be addressed to the Permissions Department, John Wiley & Sons, Inc., 111 River Street, Hoboken, NJ 07030, (201) 748-6011, fax (201) 748-6008, or online at http://www.wiley.com/go/permissions.

Limit of Liability/Disclaimer of Warranty: While the publisher and author have used their best efforts in preparing this book, they make no representations or warranties with respect to the accuracy or completeness of the contents of this book and specifically disclaim any implied warranties of merchantability or fitness for a particular purpose. No warranty may be created or extended by sales representatives or written sales materials. The advice and strategies contained herein may not be suitable for your situation. You should consult with a professional where appropriate. Neither the publisher nor author shall be liable for any loss of profit or any other commercial damages, including but not limited to special, incidental, consequential, or other damages.

For general information on our other products and services or for technical support, please contact our Customer Care Department within the United States at (800) 762-2974, outside the United States at (317) 572-3993 or fax (317) 572-4002.

Wiley also publishes its books in a variety of electronic formats. Some content that appears in print may not be available in electronic formats. For more information about Wiley products, visit our web site at www.wiley.com.

Library of Congress Cataloging-in-Publication Data

Bio-nanoparticles : biosynthesis and sustainable biotechnological implications / edited by Om V. Singh.
    p. ; cm.
  Includes bibliographical references and index.
  ISBN 978-1-118-67768-1 (cloth)
  I. Singh, Om V., editor.
  [DNLM: 1. Nanoparticles. 2. Bacteria–metabolism. 3. Biotechnology–methods.
4. Industrial Microbiology–methods. 5. Nanotechnology–methods. QT 36.5]
  QR88.3
  579.3′17–dc23

                                                                              2015004421

Set in 9.5/11.5pt Times LT Std by SPi Publisher Services, Pondicherry, India

Printed and bound in Malaysia by Vivar Printing Sdn Bhd

1   2015

*The editor gratefully dedicates this book to
Daisaku Ikeda, Uday V. Singh, and Indu Bala
in appreciation of their encouragement.*

CONTENTS

List of Contributors — xv

Introduction — xvii

1 DIVERSITY OF MICROBES IN SYNTHESIS OF METAL NANOPARTICLES: PROGRESS AND LIMITATIONS — 1
Mahendra Rai, Irena Maliszewska, Avinash Ingle, Indarchand Gupta, and Alka Yadav

- 1.1 Introduction — 1
- 1.2 Synthesis of Nanoparticles by Bacteria — 2
- 1.3 Synthesis of Nanoparticles by Fungi — 9
- 1.4 Synthesis of Nanoparticles by Algae — 12
- 1.5 Applications of Metal Nanoparticles — 16
  - 1.5.1 Nanoparticles as Catalyst — 16
  - 1.5.2 Nanoparticles as Bio-membranes — 17
  - 1.5.3 Nanoparticles in Cancer Treatment — 17
  - 1.5.4 Nanoparticles in Drug Delivery — 17
  - 1.5.5 Nanoparticles for Detection and Destruction of Pesticides — 17
  - 1.5.6 Nanoparticles in Water Treatment — 18
- 1.6 Limitations of Synthesis of Biogenic Nanoparticles — 18
- References — 20

2 ROLE OF FUNGI TOWARD SYNTHESIS OF NANO-OXIDES — 31
Rajesh Ramanathan and Vipul Bansal

- 2.1 Introduction — 31
- 2.2 Fungus-mediated Synthesis of Nanomaterials — 34
  - 2.2.1 Biosynthesis of Binary Nano-oxides using Chemical Precursors — 34
  - 2.2.2 Biosynthesis of Complex Mixed-metal Nano-oxides using Chemical Precursors — 39

	2.2.3	Biosynthesis of Nano-oxides using Natural Precursors Employing Bioleaching Approach	42
	2.2.4	Biosynthesis of Nano-oxides Employing Bio-milling Approach	44
2.3	Outlook		46
References			47

3 MICROBIAL MOLECULAR MECHANISMS IN BIOSYNTHESIS OF NANOPARTICLES 53

Atmakuru Ramesh, Marimuthu Thiripura Sundari, and Perumal Elumalai Thirugnanam

3.1	Introduction		53
3.2	Chemical Synthesis of Metal Nanoparticles		54
	3.2.1	Brust–Schiffrin Synthesis	55
3.3	Green Synthesis		57
3.4	Biosynthesis of Nanoparticles		58
3.5	Mechanisms for Formation or Synthesis of Nanoparticles		61
	3.5.1	Biomineralization using Magnetotactic Bacteria (MTB)	61
	3.5.2	Reduction of Tellurite using Phototroph *Rhodobacter capsulatus*	62
	3.5.3	Formation of AgNPs using Lactic Acid and Bacteria	62
	3.5.4	Microfluidic Cellular Bioreactor for the Generation of Nanoparticles	62
	3.5.5	Proteins and Peptides in the Synthesis of Nanoparticles	65
	3.5.6	NADH-dependent Reduction by Enzymes	65
	3.5.7	Sulfate and Sulfite Reductase	66
	3.5.8	Cyanobacteria	67
	3.5.9	Cysteine Desulfhydrase in *Rhodopseudomonas palustris*	68
	3.5.10	Nitrate and Nitrite reductase	68
3.6	Extracellular Synthesis of Nanoparticles		69
	3.6.1	Bacterial Excretions	69
	3.6.2	Fungal Strains	71
	3.6.3	Yeast: Oxido-reductase Mechanism	72
	3.6.4	Plant Extracts	73
3.7	Conclusion		76
References			78

4 BIOFILMS IN BIO-NANOTECHNOLOGY: OPPORTUNITIES AND CHALLENGES 83

Chun Kiat Ng, Anee Mohanty, and Bin Cao

4.1	Introduction		83
4.2	Microbial Synthesis of Nanomaterials		84
	4.2.1	Overview	84
	4.2.2	Significance of Biofilms in Biosynthesis of Nanomaterials	89
	4.2.3	Synthesis of Nanomaterials using Biofilms	90

4.3	Interaction of Microbial Biofilms with Nanomaterials	90
	4.3.1 Nanomaterials as Anti-biofilm Agents	90
	4.3.2 Nanomaterials as a Tool in Biofilm Studies	92
4.4	Future Perspectives	93
	References	94

5 EXTREMOPHILES AND BIOSYNTHESIS OF NANOPARTICLES: CURRENT AND FUTURE PERSPECTIVES   101

Jingyi Zhang, Jetka Wanner, and Om V. Singh

5.1	Introduction	101
5.2	Synthesis of Nanoparticles	104
	5.2.1 Microorganisms: An Asset in Nanoparticle Biosynthesis	104
	5.2.2 Extremophiles in Nanoparticle Biosynthesis	104
5.3	Mechanism of Nanoparticle Biosynthesis	108
5.4	Fermentative Production of Nanoparticles	111
5.5	Nanoparticle Recovery	114
5.6	Challenges and Future Perspectives	115
5.7	Conclusion	115
	References	116

6 BIOSYNTHESIS OF SIZE-CONTROLLED METAL AND METAL OXIDE NANOPARTICLES BY BACTERIA   123

Chung-Hao Kuo, David A. Kriz, Anton Gudz, and Steven L. Suib

6.1	Introduction	123
6.2	Intracellular Synthesis of Metal Nanoparticles by Bacteria	124
6.3	Extracellular Synthesis of Metal Nanoparticles by Bacteria	129
6.4	Synthesis of Metal Oxide and Sulfide Nanoparticles by Bacteria	131
6.5	Conclusion	135
	References	135

7 METHODS OF NANOPARTICLE BIOSYNTHESIS FOR MEDICAL AND COMMERCIAL APPLICATIONS   141

Shilpi Mishra, Saurabh Dixit, and Shivani Soni

7.1	Introduction	141
7.2	Biosynthesis of Nanoparticles using Bacteria	144
	7.2.1 Synthesis of Silver Nanoparticles by Bacteria	144
	7.2.2 Synthesis of Gold Nanoparticles by Bacteria	145
	7.2.3 Synthesis of other Metallic Nanoparticles by Bacteria	145
7.3	Biosynthesis of Nanoparticles using Actinomycete	146
7.4	Biosynthesis of Nanoparticles using Fungi	147

7.5	Biosynthesis of Nanoparticles using Plants	148
7.6	Conclusions	149
References		149

8 MICROBIAL SYNTHESIS OF NANOPARTICLES: AN OVERVIEW   155
Sneha Singh, Ambarish Sharan Vidyarthi, and Abhimanyu Dev

8.1	Introduction		156
8.2	Nanoparticles Synthesis Inspired by Microorganisms		157
	8.2.1	Bacteria in NPs Synthesis	162
	8.2.2	Fungi in NPs Synthesis	167
	8.2.3	Actinomycetes in NPs Synthesis	170
	8.2.4	Yeast in NPs Synthesis	171
	8.2.5	Virus in NPs Synthesis	173
8.3	Mechanisms of Nanoparticles Synthesis		174
8.4	Purification and Characterization of Nanoparticles		176
8.5	Conclusion		177
References			179

9 MICROBIAL DIVERSITY OF NANOPARTICLE BIOSYNTHESIS   187
Raveendran Sindhu, Ashok Pandey, and Parameswaran Binod

9.1	Introduction		187
9.2	Microbial-mediated Nanoparticles		187
	9.2.1	Gold	188
	9.2.2	Silver	190
	9.2.3	Selenium	191
	9.2.4	Silica	192
	9.2.5	Cadmium	192
	9.2.6	Palladium	193
	9.2.7	Zinc	193
	9.2.8	Lead	194
	9.2.9	Iron	195
	9.2.10	Copper	195
	9.2.11	Cerium	196
	9.2.12	Microbial Quantum Dots	196
	9.2.13	Cadmium Telluride	197
	9.2.14	Iron Sulfide-greigite	198
9.3	Native and Engineered Microbes for Nanoparticle Synthesis		198
9.4	Commercial Aspects of Microbial Nanoparticle Synthesis		199
9.5	Conclusion		200
References			200

10 SUSTAINABLE SYNTHESIS OF PALLADIUM(0) NANOCATALYSTS AND THEIR POTENTIAL FOR ORGANOHALOGEN COMPOUNDS DETOXIFICATION 205
Michael Bunge and Katrin Mackenzie

    10.1 Introduction 205
    10.2 Chemically Generated Palladium Nanocatalysts for Hydrodechlorination: Current Methods and Materials 206
          10.2.1 Pd Catalysts 206
          10.2.2 Data Analysis 207
          10.2.3 Pd as Dehalogenation Catalyst 207
          10.2.4 Intrinsic Potential *vs*. Performance 208
          10.2.5 Concepts for Pd Protection 210
    10.3 Bio-supported Synthesis of Palladium Nanocatalysts 211
          10.3.1 Background 211
    10.4 Current Approaches for Synthesis of Palladium Catalysts in the Presence of Microorganisms 212
          10.4.1 Pd(II)-Tolerant Microorganisms for Future Biotechnological Approaches 213
          10.4.2 Controlling Size and Morphology during Bio-Synthesis 214
          10.4.3 Putative and Documented Mechanisms of Biosynthesis of Palladium Nanoparticles 215
          10.4.4 Isolation of Nanocatalysts from the Cell Matrix and Stabilization 216
    10.5 Bio-Palladium(0)-nanocatalyst Mediated Transformation of Organohalogen Pollutants 217
    10.6 Conclusions 218
    References 219

11 ENVIRONMENTAL PROCESSING OF Zn CONTAINING WASTES AND GENERATION OF NANOSIZED VALUE-ADDED PRODUCTS 225
Abhilash and B.D. Pandey

    11.1 Introduction 225
          11.1.1 World Status of Zinc Production 226
          11.1.2 Environmental Impact of the Process Wastes Generated 226
          11.1.3 Production Status in India 227
          11.1.4 Recent Attempts at Processing Low-Grade Ores and Tailings 228
    11.2 Physical/Chemical/Hydrothermal Processing 229
          11.2.1 Extraction of Pb-Zn from Tailings for Utilization and Production in China 229
          11.2.2 Vegetation Program on Pb-Zn Tailings 229
          11.2.3 Recovering Valuable Metals from Tailings and Residues 229

	11.2.4	Extraction of Vanadium, Lead and Zinc from Mining Dump in Zambia	230
	11.2.5	Recovery of Zinc from Blast Furnace and other Dust/Secondary Resources	230
	11.2.6	Treatment and Recycling of Goethite Waste	231
	11.2.7	Other Hydrometallurgical Treatments of Zinc-based Industrial Wastes and Residues	231
11.3	Biohydrometallurgical Processing: International Scenario		233
	11.3.1	Bioleaching of Zn from Copper Mining Residues by *Aspergillus niger*	233
	11.3.2	Bioleaching of Zinc from Steel Plant Waste using *Acidithiobacillus ferrooxidans*	234
	11.3.3	Bacterial Leaching of Zinc from Chat (Chert) Pile Rock and Copper from Tailings Pond Sediment	234
	11.3.4	Dissolution of Zn from Zinc Mine Tailings	234
	11.3.5	Microbial Diversity in Zinc Mines	234
	11.3.6	Chromosomal Resistance Mechanisms of *A. ferrooxidans* on Zinc	235
	11.3.7	Bioleaching of Zinc Sulfides by *Acidithiobacillus ferrooxidans*	235
	11.3.8	Bioleaching of High-Sphalerite Material	235
	11.3.9	Bioleaching of Low-Grade ZnS Concentrate and Complex Sulfides (Pb-Zn) using Thermophilic Species	236
	11.3.10	Improvement of Stains for Bio-processing of Sphalerite	236
	11.3.11	Tank Bioleaching of ZnS and Zn Polymetallic Concentrates	237
	11.3.12	Large-Scale Development for Zinc Concentrate Bioleaching	237
	11.3.13	Scale-up Studies for Bioleaching of Low-Grade Sphalerite Ore	238
	11.3.14	Zinc Resistance Mechanism in Bacteria	238
11.4	Biohydrometallurgical Processing: Indian Scenario		238
	11.4.1	Electro-Bioleaching of Sphalerite Flotation Concentrate	239
	11.4.2	Bioleaching of Zinc Sulfide Concentrate	239
	11.4.3	Bioleaching of Moore Cake and Sphalarite Tailings	239
11.5	Synthesis of Nanoparticles		240
11.6	Applications of Zinc-based Value-added Products/Nanomaterials		244
	11.6.1	Hydro-Gel for Bio-applications	244
	11.6.2	Sensors	244
	11.6.3	Biomedical Applications	245
	11.6.4	Antibacterial Properties	245
	11.6.5	Zeolites in biomedical applications	246
	11.6.6	Textiles	246

	11.6.7 Prospects of Zinc Recovery from Tailings and Biosynthesis of Zinc-based Nano-materials	246
11.7	Conclusions and Future Directions	247
References		248

12 INTERACTION BETWEEN NANOPARTICLES AND PLANTS: INCREASING EVIDENCE OF PHYTOTOXICITY   255
Rajeshwari Sinha and S.K. Khare

12.1	Introduction	255
12.2	Plant–Nanoparticle Interactions	256
12.3	Effect of Nanoparticles on Plants	256
	12.3.1 Monocot Plants	257
	12.3.2 Dicot Plants	257
12.4	Mechanisms of Nanoparticle-induced Phytotoxicity	257
	12.4.1 Endocytosis	257
	12.4.2 Transfer through Ion Channels Post-ionization	262
	12.4.3 Aquaporin Mediated	262
	12.4.4 Carrier Proteins Mediated	262
	12.4.5 Via Organic Matter	262
	12.4.6 Complex Formation with Root Exudates	262
	12.4.7 Foliar Uptake	263
12.5	Effect on Physiological Parameters	263
	12.5.1 Loss of Hydraulic Conductivity	263
	12.5.2 Genotoxic Effects	263
	12.5.3 Absorption and Accumulation	263
	12.5.4 Generation of Reactive Oxygen Species (ROS)	264
	12.5.5 Biotransformation of NPs	264
12.6	Genectic and Molecular Basis of NP Phytotoxicity	266
12.7	Conclusions and Future Perspectives	266
References		267

13 CYTOTOXICOLOGY OF NANOCOMPOSITES   273
Horacio Bach

13.1	Introduction	273
13.2	Cellular Toxicity	274
	13.2.1 Mechanisms of Cellular Toxicity	274
	13.2.2 Effect of Glutathione (GSH) in Oxidative Stress	276
	13.2.3 Damage to Cellular Biomolecules	277
13.3	Nanoparticle Fabrication	281
	13.3.1 Physico-chemical Characteristics of NPs	282
	13.3.2 Cellular Uptake	284
	13.3.3 Factors Affecting the Internalization of NPs	287

13.4	Immunological Response	289
	13.4.1 Cytokine Production	289
	13.4.2 Cytotoxicity, Necrosis, Apoptosis, and Cell Death	290
13.5	Factors to Consider to Reduce the Cytotoxic Effects of NP	292
13.6	Conclusions and Future Directions	293
	References	294

14 NANOTECHNOLOGY: OVERVIEW OF REGULATIONS AND IMPLEMENTATIONS — 303

Om V. Singh and Thomas Colonna

14.1	Introduction	303
14.2	Scope of Nanotechnology	305
14.3	Safety Concerns Related to Nanotechnology	310
14.4	Barriers to the Desired Regulatory Framework	311
	14.4.1 Regulatory Framework in the United States	312
	14.4.2 Global Efforts toward Regulation of Nanotechnology	315
14.5	Biosynthesis of Microbial Bio-nanoparticles: An Alternative Production Method	317
14.6	Conclusion	325
	References	326

Name index — 331

Subject index — 333

LIST OF CONTRIBUTORS

Abhilash, CSIR-National Metallurgical Laboratory (NML), Jamshedpur, Jharkand, India

Horacio Bach, Department of Medicine, Division of Infectious Diseases, University of British Columbia, Vancouver, BC, Canada

Vipul Bansal, NanoBiotechnology Research Laboratory (NBRL), School of Applied Sciences, RMIT University, Melbourne, VIC, Australia

Parameswaran Binod, CSIR-National Institute for Interdisciplinary Science and Technology, Pappanamcode, Trivandrum, Kerala, India

Michael Bunge, Institute of Applied Microbiology, Justus Liebig University of Giessen, Giessen, Germany

Bin Cao, School of Civil and Environmental Engineering, Nanyang Technological University, Singapore; Singapore Centre on Environmental Life Sciences Engineering, Nanyang Technological University, Singapore

Thomas Colonna, Center for Biotechnology Education, Zanvyl Krieger School of Arts and Sciences, The Johns Hopkins University, Rockville, MD, USA

Abhimanyu Dev, Department of Pharmaceutical Sciences and Technology, Birla Institute of Technology, Mesra, Ranchi, Jharkand, India

Saurabh Dixit, Center for Nanobiotechnology Research, Alabama State University, Montgomery, AL, USA

Anton Gudz, Department of Chemistry, University of Connecticut, Storrs, CT, USA

Indarchand Gupta, Department of Biotechnology, SGB Amravati University, Amravati, Maharashtra, India

Avinash Ingle, Department of Biotechnology, SGB Amravati University, Amravati, Maharashtra, India

S.K. Khare, Enzyme and Microbial Biochemistry Laboratory, Department of Chemistry, Indian Institute of Technology, Delhi, New Delhi, India

David A. Kriz, Department of Chemistry, University of Connecticut, Storrs, CT, USA

Chung-Hao Kuo, Department of Chemistry, University of Connecticut, Storrs, CT, USA

Katrin Mackenzie, Department of Environmental Engineering, Helmholtz Centre for Environmental Research – UFZ, Leipzig, Germany

Irena Maliszewska, Division of Medicinal Chemistry and Microbiology, Faculty of Chemistry, Wrocław University of Technology, Wrocław, Wybrzeże Wyspiańskiego, Poland

Shilpi Mishra, Department of Biological Sciences, Alabama State University, Montgomery, AL, USA

Anee Mohanty, School of Civil and Environmental Engineering, Nanyang Technological University, Singapore; Singapore Centre on Environmental Life Sciences Engineering, Nanyang Technological University, Singapore

Chun Kiat Ng, Singapore Centre on Environmental Life Sciences Engineering, Nanyang Technological University, Singapore; Interdisciplinary Graduate School, Nanyang Technological University, Singapore

Ashok Pandey, CSIR-National Institute for Interdisciplinary Science and Technology, Pappanamcode, Trivandrum, Kerala, India

B.D. Pandey, CSIR-National Metallurgical Laboratory (NML), Jamshedpur, Jharkand, India

Mahendra Rai, Department of Biotechnology, SGB Amravati University, Amravati, Maharashtra, India; Institute of Chemistry, Biological Chemistry Laboratory, Universidade Estadual de Campinas, Campinas, SP, Brazil

Rajesh Ramanathan, NanoBiotechnology Research Laboratory (NBRL), School of Applied Sciences, RMIT University, Melbourne, VIC, Australia

Atmakuru Ramesh, International Institute of Biotechnology and Toxicology (IIBAT), Padappai, Tamil Nadu, India

Raveendran Sindhu, CSIR-National Institute for Interdisciplinary Science and Technology, Pappanamcode, Trivandrum, Kerala, India

Om V. Singh, Division of Biological and Health Sciences, University of Pittsburgh, Bradford, PA, USA

Sneha Singh, Department of Bio-Engineering, Birla Institute of Technology, Mesra, Ranchi, Jharkand, India

Rajeshwari Sinha, Enzyme and Microbial Biochemistry Laboratory, Department of Chemistry, Indian Institute of Technology-Delhi, New Delhi, India

Shivani Soni, Department of Biological Sciences, Alabama State University, Montgomery, AL, USA

Steven L. Suib, Department of Chemistry, University of Connecticut, Storrs, CT, USA

Marimuthu Thiripura Sundari, International Institute of Biotechnology and Toxicology (IIBAT), Padappai, Tamil Nadu, India

Perumal Elumalai Thirugnanam, International Institute of Biotechnology and Toxicology (IIBAT), Padappai, Tamil Nadu, India

Ambarish Sharan Vidyarthi, Institute of Engineering and Technology, Lucknow, Uttar Pradesh, India

Jetka Wanner, Division of Biological and Health Sciences, University of Pittsburgh, Bradford, PA, USA

Alka Yadav, Department of Biotechnology, SGB Amravati University, Amravati, Maharashtra, India

Jingyi Zhang, Division of Biological and Health Sciences, University of Pittsburgh, Bradford, PA, USA

INTRODUCTION

Nanoparticles are the building blocks of nanotechnology; they are defined as particles having more than one dimension measuring 100 nm or less. Nanostructured materials are being promoted as better built, longer lasting, cleaner, safer, and smarter products for use in communications, medicine, transportation, agriculture, and other industries. Applications such as molecular recognition, biomolecule-nanocrystal conjugates as fluorescence labels for biological cells, and DNA-mediated groupings of nanocrystals are widespread, intriguing researchers from both biological and engineering fields. The diversity of nanotechnology covers fields from biology to materials science and physics to chemistry.

The controlled size, shape, composition, crystallinity, and structure-dependent properties of nanoparticles govern the unique properties of nanotechnology. The controlled biosynthesis of nanoparticles is of high scientific and technological interest; the microorganisms grab target ions from their environment and turn them into the element metal through enzymatic mechanisms generated via intra- or extracellular activities.

Nanoparticles are known to enable cleaner and safer applications of various technologies. However, the current physiochemical methods for creating them (chemical vapor deposition, solgel technique, hydrothermal synthesis, precipitation, and micro-emulsion) are hazardous, environmentally unfriendly, cumbersome, and expensive, and require conditions of high temperature, pH, and/or pressure levels for synthesis. Due to current manufacturing practices, the United States Environmental Protection Agency (USEPA) issued a Significant New Use Rule (SNUR) in 2008 for 56 chemicals, including two nanoparticles (73 F.R. 65743): siloxane-modified silica nanoparticles, PMN No. P-05-673; and siloxane-modified alumina nanoparticles, PMN No. P-05-687.

The comparatively insignificant efforts toward safer and cleaner biosynthesis of nanoparticles may have green biotechnological implications. It would benefit human society to learn from microorganisms, as they have the potential to assist us in dealing with emerging diseases due to their tremendous metabolic strategies and ability to alter the physical and chemical forms of ionic molecules. The products of microbial metabolism (enzymes and proteins) are referred to as primary and secondary metabolic products, and they have proven their importance to biotechnology.

This book continues to bridge the technology gap and focuses on exploring microbial diversity and the respective mechanisms regulating the biosynthesis of metal nanoparticles. **Chapter 1** (Rai et al.) presents a wide array of microorganisms that are employed as biological agents for the biosynthesis of nanoparticles of different shapes and sizes. The possible mechanisms and applications of synthesized metal nanoparticles is described, and details of the biochemical aspects of major events that occur during biosynthesis of metal nanoparticles are discussed. Inspired by the concept of microbial tolerance at high metal ion concentration, Ramanathan and Bansal explore the role of fungi in the synthesis of nano-oxide in **Chapter 2**, presenting the facts of biosynthesis of binary nano-oxide and mixed-metal nano-oxides using chemical precursors and natural precursors employing bioleaching and bio-milling approaches. In **Chapter 3**, Ramesh et al. elaborate on the microbial molecular mechanisms used in biosynthesis of nanoparticles. Ng et al. discuss unique approaches to microbial synthesis of nanomaterials in **Chapter 4**, with a special emphasis on molecular mechanisms and the emerging use of biofilms, the application of nanomaterials in biofilm control, and future prospects for microbial biofilms in biotechnology.

Extremophiles are the most mysterious category of life on planet Earth and perhaps also on other planets. Even though nature offers abundant opportunities to life forms that can consume or produce sufficient energy for their survival, normal survival may not be possible in environments that experience extreme conditions (e.g., temperature, pressure, pH, salinity, geological scale/barriers, radiation, chemical extremes, lack of nutrition, osmotic barriers, or polyextremity). Due to extraordinary properties, certain organisms (mostly bacteria, archaea, and a few eukaryotes) can thrive in such extreme habitats; they are called extremophiles. In **Chapter 5**, Zhang et al. discuss the role of extremophiles in the biosynthesis of nanoparticles. In view of the tremendous industrial potential of producing nanoparticles via extremophiles, this chapter also sheds light on specifics of fermentation media and the recovery of nanoparticles from the microbiological process using standard microorganisms. Discussions on limitations and challenges further outline the future of extremophile-mediated nanotechnology.

Most metallic nanoparticles are of size 0.1–1000 nm and have different shapes, including triangular, spherical, rod, and other irregular shapes. Because of their extremely small size and specific shape allowing them to bind with molecules of interest, nanoparticles have unusual characteristics that make them valuable. In **Chapter 6**, Kuo et al. elaborate on the biosynthesis of size-controlled metal and metal oxide nanoparticles by bacteria. Further, in **Chapter 7**, Mishra et al. discuss the methodologies applied to biosynthesize a variety of nanoparticles with medical and commercial significance. In **Chapters 8** and **9**, Singh et al. and Sindhu et al. provide overviews of microbial diversity in nanoparticle biosynthesis with insights into characterization and purification from biological medium.

The unique properties of nanoparticles have attracted attention from scientists to harness their great functionality. In **Chapter 10**, Bunge and Mackenzie compare the sustainable synthesis of palladium nanoparticles using chemical and biological methods, exploring their potential for detoxification of organohalogen compounds from contaminated waters. In **Chapter 11**, Abhilash and Pandey present a unique approach for the generation of nano-sized value-added products of commercial significance using environmental wastes. According to the authors, continuous depletion of high-grade resources necessitates the conversion of lower-grade ores into value-added products of commercial significance, and they claim to have developed a green process for the production of nanomaterials.

The increasing use of nanoparticles poses challenging issues of safety for the environment and human society. The stability of nanoparticles is one of the prime reasons for their broad potential; however, medical professionals, ecologists, and environmentalists are concerned about their safety. The release of nanoparticles has made it necessary to study nanotoxicity, which is being increasingly evidenced in microbial, human, and animal systems. In **Chapter 12**, Sinha and Khare present the facts about the involvement of plants in nanotechnology, along with comprehensive analyses of available information pertaining to the newly discovered domain of phytonanotoxicology. In **Chapter 13**, Bach discusses aspects of the toxicity of nanoparticles at a cellular level, the mechanisms of cytotoxicity, and the effects of the physicochemical characteristics of nanoparticles. The role of nanoparticles has also been defined in eliciting oxidative and nitrosative stress, including apoptosis and immunological responses.

The science of nanotechnology has great promise as a source of novel products with beneficial applications. It exploits the fact that quantum effects and higher surface-area-to-mass ratios give materials different properties on the nanoscale. However, like many new forms of technology, nanotechnology implemented with engineered nanoscale materials (ENMs) has its potential dangers. A legal regulatory framework is being sought to control the ENMs and deal with the risks they pose to human health and the ecosystem. Microbial nanoparticles produced using enzymes and proteins from biological sources present a potential solution to these problems. In **Chapter 14**, Singh and Colonna discuss the safety issues and the legal framework of regulatory policies in the United States and worldwide related to the nanotechnology field.

This book, *Bio-nanoparticles: Biosynthesis and Sustainable Biotechnological Implications*, is a collection of outstanding articles elucidating several broad-ranging areas of progress and challenges in the utilization of microorganisms as sustainable resources in nanotechnology. This book will contribute to research efforts in the scientific community and commercially significant work for corporate businesses. The expectations are to establish long-term safe and sustainable forms of nanotechnology through microbial implementation of nanoparticle biosynthesis with minimum impact on the ecosystem.

We hope readers will find these articles interesting and informative for their research pursuits. It has been my pleasure to put together this book with Wiley-Blackwell Press. I would like to thank all of the contributing authors for sharing their quality research and ideas with the scientific community through this book.

1

DIVERSITY OF MICROBES IN SYNTHESIS OF METAL NANOPARTICLES: PROGRESS AND LIMITATIONS

Mahendra Rai

Department of Biotechnology, SGB Amravati University, Amravati Maharashtra, India; and Institute of Chemistry, Biological Chemistry Laboratory, Universidade Estadual de Campinas, Campinas, SP, Brazil

Irena Maliszewska

Division of Medicinal Chemistry and Microbiology, Faculty of Chemistry, Wroclaw University of Technology, Wroclaw, Wybrzeże Wyspiańskiego, Poland

Avinash Ingle, Indarchand Gupta, and Alka Yadav

Department of Biotechnology, SGB Amravati University, Amravati, Maharashtra, India

1.1. INTRODUCTION

Nanotechnology is a widely emerging field involving interdisciplinary subjects such as biology, physics, chemistry, and medicine (Bankar et al., 2010; Zhang, 2011; Rai and Ingle, 2012). Nanotechnology involves the synthesis of nanoparticles using the top-down and bottom-up approach (Kasthuri et al., 2008; Bankar et al., 2010; Nagajyothi and Lee, 2011). However, due to the growing environmental concern and the adverse

---

Bio-Nanoparticles: Biosynthesis and Sustainable Biotechnological Implications, First Edition.
Edited by Om V. Singh.
© 2015 John Wiley & Sons, Inc. Published 2015 by John Wiley & Sons, Inc.

effects of physical and chemical synthesis, most researchers are looking to the biological protocols for nanoparticle synthesis (Rai et al., 2008). The biological method of synthesis involves a wide diversity of biological entities that could be harnessed for the synthesis of metal nanoparticles (Sharma et al., 2009; Vaseeharan et al., 2010; Zhang et al., 2011a; Gupta et al., 2012; Rajesh et al., 2012). These biological agents emerge as an environmently friendly, clean, non-toxic agent for the synthesis of metal nanoparticles (Sastry et al., 2003; Bhattacharya and Gupta, 2005; Riddin et al., 2006; Duran et al., 2007; Ingle et al., 2008; Kumar and Yadav, 2009; Vaseeharan et al., 2010; Thakkar et al., 2011; Zhang et al., 2011b; Rajesh et al., 2012).

A wide array of microorganisms such as bacteria, fungi, yeast, algae, and actinomycetes are majorly employed as biological agents for the synthesis process (Kumar and Yadav, 2009; Satyavathi et al., 2010). The synthesis of metal nanoparticles employs both intracellular and extracellular methods (Sharma et al., 2009; Mallikarjuna et al., 2011). Some examples of these microbial agents include bacteria (Husseiny et al., 2007; Shahverdi et al., 2007, 2009), fungi (Kumar et al., 2007; Parikh et al., 2008; Gajbhiye et al., 2009), actinomycetes (Ahmad et al., 2003al Golinska et al., 2014), lichens (Shahi and Patra, 2003), and algae (Singaravelu et al., 2007; Chakraborty et al., 2009). These diverse groups of biological agents have many advantages over physical and chemical methods such as easy and simple scale-up, easy downstream processing, simpler biomass handling and recovery, and economic viability (Rai et al., 2009a; Thakkar et al., 2011; Renugadevi and Aswini, 2012). These different biological agents such as bacteria, fungi, yeast, algae, and acitnomycetes therefore demonstrate immense biodiversity in the synthesis of nanoparticles and lead to green nanotechnology (Vaseeharan et al., 2010; Singh et al., 2011, 2013; Thakkar et al., 2011).

The present review also deals with the diversity of microbes involved in the synthesis of metal nanoparticles. The possible mechanisms and different applications for the synthesis of metal nanoparticles are also discussed.

1.2. SYNTHESIS OF NANOPARTICLES BY BACTERIA

Although it is known that bacteria have the ability to produce various inorganic nanoparticles (e.g., metal, calcium, gypsum, silicon), research in this area is usually focused on the formation of metals and metals sulfide/oxide (Fig. 1.1).

Different bacteria from different habitats and nutritional modes have been studied for the synthesis of metallic nanocrystals, as summarized in Table 1.1. Some of the earliest reports on the reduction and accumulation of inorganic particles in bacteria can be traced back to the 1960s, where zinc sulfide was described in sulfate-reducing bacteria (Temple and Le-Roux, 1964). Later studies in this area date back to the 1980s, when Beveridge and Murray (1980) described how the incubation of gold chloride with *Bacillus subtilis* resulted in the production of octahedral gold nanoparticles with a dimension of 5–25 nm within the bacterial cell. It is believed that organophosphate compounds secreted by the bacterium play an important role in the formation of these nanostructures (Southam and Beveridge, 1996). Another example

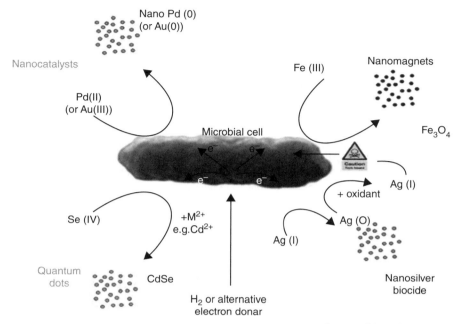

Figure 1.1. Mechanisms of microbial fabrication of nanobiominerals, catalyzed by enzymatic reductive biotransformations of redox active metals, driven by a suitable electron donor such as hydrogen. In some cases, for example transformations of Fe(III) minerals and Se(IV), redox mediators such as AQDS (anthraquinone-2,6 disulfonate) are utilized to increase the kinetics of metal reduction and hence nanobiomineral formation. Source: Lloyd, J.R., Byrne, J.M., Coker, V.S. 2011. Biotechnological synthesis of functional nanomaterials. *Current Opinion in Biotechnology* 22: 509–515. Copyright © 2011, Elsevier. *See insert for color representation of the figure.*

of bacterial reduction and precipitation of gold was described by Kashefi and coworkers (2001). These authors demonstrated that iron-reducing anaerobic bacteria *Shewanella algae* can reduce gold ions in the presence of H_2 gas, which results in the formation of 10–20 nm gold nanoparticles. It was further hypothesized that specific hydrogenase might be involved in the reduction of gold ions when hydrogen was used as an electron donor.

Karthikeyan and Beveridge (2002) observed microbial reduction of gold ions by *Pseudomonas aeruginosa*, resulting in intracellular accumulation of gold nanoparticles with a particle diameter of 20 nm. In another study, it was shown that bacterium *Rhodopseudomonas capsulate* was capable of reducing gold ions into gold nanoparticles (He et al., 2007). When the biomass of *R. capsulate* was incubated with gold ions at neutral pH, gold nanoparticles of 10–20 nm size were formed. Further, when the same reaction was carried out under acidic pH conditions, triangular gold particles

TABLE 1.1. List of Different Metallic Nanoparticles Synthesized by Bacteria

Metallic material	Bacteria (reference)
Au^0	*Bacillus subtilis* (Beveridge and Murray, 1980); *Shewanella algae* (Kashefi et al., 2001); *Rhodopseudomonas capsulate* (Kashefi et al., 2001; He et al., 2007, 2008); *Pseudomonas aeruginosa* (Karthikeyan and Beveridge, 2002); *Lactobacilli* strains (Nair and Pradeep, 2002); *Thermomonospora* sp. (Ahmad et al., 2003b); *Rhodococcus* sp. (Ahmad et al., 2003a); *Ralstonia metallidurans* (Reith et al., 2006); *Actinobacter* sp. (Bharde et al., 2007); *Streptomyces viridogens* strain HM10 (Balagurunathan et al., 2011); *Streptomyces griseus* (Derakhshan et al., 2012); *Streptomyces hygroscopicus* (Sadhasivam et al., 2012); *Streptomyces* sp. ERI-3 (Zonooz et al., 2012)
Ag^0	*Pseudomonas stutzeri* A259 (Klaus et al., 1996; Joerger et al., 2000); *Corynebacterium* sp. SH09 (Zhang et al., 2005); *Enterobacteriaceae* (*Klebsiella pneumoniae*, *E. coli* and *Enterobacter cloacae*) (Shahverdi et al., 2007); *Morganella spp.* (Parikh et al., 2008); *Bacillus licheniformis* (Kalishwaralal et al., 2008); *Lactobacillus fermentum* (De-Gusseme et al., 2010); *Morganella psychrotolerans* (Ramanathan et al., 2011); *Escherichia coli* AUCAS 112 (Kathiresan et al., 2010); *Idiomarina* sp. PR58-8 (Seshadri et al., 2012)
Fe_3S_4	*M. magnetotacticum* (Mann et al., 1984; Philipse and Maas, 2002); *Magnetospiryllum* (Farina et al., 1990); Sulfate-reducing bacteria (Mann et al., 1990); *M. gryphiswaldense* (Lang et al., 2006); *Acinetobacter* sp. (Bharde et al., 2008)
Fe_3O_4, Fe_2O_3	Magnetotactic bacteria (Blakemore, 1975; Mann et al., 1984); *Geobacter metallireducens* (Vali et al., 2004); *Actinobacter* sp. (Bharde et al., 2005)
Pt^0	*Shewanella algae* (Konishi et al., 2007)
Pd^0	*Desulfovibrio desulfuricans* (Yong et al., 2002a,b)
Cu^0	*Serratia* sp. (Hasan et al., 2008); *E. coli* (Singh et al., 2010)
Co_3O_4	Marine cobalt-resistant bacterial strain (Kumar et al., 2008)
CdS	*Clostridium thermoaceticum* (Cunningham and Lundie, 1993); *R. palustris* (Bai et al., 2009)
ZnS	Sulfate-reducing bacteria (Labrenz et al., 2000)
Se^0	*Thauera selenatis* (DeMoll-Decker and Macy, 1993; Bledsoe et al., 1999; Sabaty et al., 2001); *Rhizobium selenitireducens* strain B1 (Hunter and Kuykendall, 2007; Hunter et al., 2007); *E. coli* (Avazeri et al., 1997); *Clostridium pasteurianum* (Yanke et al., 1995); *Bacillus selenitireducens* (Afkar et al., 2003); *Pseudomonas stutzeri* (Lortie et al., 1992); *Wolinella succinogenes* (Tomei et al., 1992); *Enterobacter cloacae* (Losi and Frankenberger, 1997); *Pseudomonas aeruginosa* (Yadav et al., 2008); *Pseudomonas alkaphila* (Zhang et al., 2011a)
Te^0	*Sulfurospirillum barnesii*, *B. selenireducens* (Baesman et al., 2007)
Ti^0	*Lactobacillus* sp. (Prasad et al., 2007), *Bacillus* sp. (Prakash et al., 2009)
UO_2	*Micrococcus lactilyticus* (Woolfolk and Whiteley, 1962); *Alteromonas putrefaciens* (Myers and Nealson, 1988); *G. metallireducens* GS-15 (Lovley et al., 1991); *S. oneidensis* MR-1 (Marshall et al., 2006); *Desulfosporosinus* sp. (O'Loughlinej et al., 2003)

with an edge length of c. 500 nm were obtained as well as the spherical gold nanoparticles (He et al., 2008).

The ability of bacterium *Ralstonia metallidurans* to precipitate colloidal gold nanoparticles from aqueous gold chloride solution has recently been reported, but the exact mechanism of this process is not yet clear (Reith et al., 2006). In a series of reports, Sastry's group screened different bacterial strains for the biosynthesis of gold and silver nanoparticles with control over morphologies and size distribution. It was shown that alkalothermophylic actinomycete, *Thermomonospora* sp., synthesized spherical gold nanoparticles of size 8 nm with a narrow size distribution (Ahmad et al., 2003*b*). In other case, an alkalotolerant actinomycete, *Rhodococcus* sp., was reported for the formation of nearly monodisperse 10 nm gold nanoparticles (Ahmad et al., 2003*a*). Further study demonstrated that the size and shape of gold nanoparticles could be controlled by adjusting the reaction parameters. Gold nanoparticles with variable size and shape were obtained using a strain of actinomycetes, *Actinobacter* sp. (Bharde et al., 2007). When this strain reacted with gold ions in the absence of molecular oxygen, gold nanoparticles of triangular and hexagonal shapes were produced along with some spherical particles. It was consequently concluded that protease enzyme secreted by *Actinobacter* sp. was responsible for the reduction of gold ions. Moreover, it was described how molecular oxygen slows down the reduction of gold ions in an unknown manner, possibly by inhibiting the action of protease secreted by this strain. Other bacterial systems able to reduce gold ions to make nanoparticles include: *Streptomyces viridogens* strain HM10 (Balagurunathan et al., 2011), *Streptomyces griseus* (Derakhshan et al., 2012), *Streptomyces hygroscopicus* (Sadhasivam et al., 2012), and *Streptomyces* sp. ERI-3 (Zonooz et al., 2012). The reduction appears to be initiated via electron transfer from NADH-dependent enzymes as an electron carrier.

Some bacteria have been reported for the formation of more than one metal nanoparticle and bimetallic alloy. Nair and Pradeep (2002) synthesized nanoparticles of gold, silver, and their alloys using different *Lactobacillus* strains, lactic acid-producing bacteria, and the active bacterial component of buttermilk. It is well known that ionic silver is highly toxic to most microbial cells. Nonetheless, several bacterial strains have been reported to be silver resistant. *Pseudomonas stutzeri* A259, isolated from a silver mine in Utah (USA), was the first bacterial strain with reductive potential to form silver crystals (Klaus et al., 1996; Joerger et al., 2000). This bacterium produced a small number of crystalline α-form silver sulfide acanthite (Ag_2S), crystallite particles with the composition of silver and sulfur at a ratio of 2:1. The resistance mechanism of *P. stutzeri* towards silver ions is poorly understood however; several groups postulated that efflux cellular pumps might be involved in the formation of silver nanoparticles (Silver, 2003; Ramanathan et al., 2011). Parikh and co-workers (2008) used *Morganella* sp., a silver-resistant bacterium for the synthesis of silver nanoparticles, and established a direct correlation between the silver-resistance machinery of this bacteria and silver nanoparticles biosynthesis. Three silver-resistant homologue genes (*silE*, *silP*, and *silS*) were recognized in *Morganella* sp. The presence of these genes suggested that this organism has a unique mechanism for protection against the toxicity of silver ions that involves the formation of silver nanoparticles. Similarly, the intracellular production of silver nanoparticles by the highly silver-tolerant marine bacterium, *Idiomarina* sp. PR58-8, was described by Seshadri et al. (2012). The strain of *Escherichia coli* AUCAS 112 isolated from mangrove sediments is also capable of reducing the silver ions at a faster rate (Kathiresan et al., 2010).

Zhang and co-workers (2005) demonstrated that dried cells of *Corynebacterium* sp. SH09 produced silver nanoparticles at 60 °C in 72 h on the cell wall in the size range of 10–15 nm with diamine silver complex $[Ag(NH_3)_2]^+$. The culture supernatants of *Enterobacteriaceae* (*Klebsiella pneumoniae*, *E. coli*, and *Enterobacter cloacae*) also formed silver nanoparticles by reducing Ag^+ to Ag^0. With the addition of piperitone, silver ions reduction was partially inhibited, which suggested the involvement of nitro-reductase in this process (Shahverdi et al., 2007). Similarly, the formation of silver nanoparticles by *Bacillus licheniformis* (Kalishwaralal et al., 2008), *Lactobacillus fermentum* (De-Gusseme et al., 2010), and psychro-tolerant bacteria *Morganella psychrotolerans* (Ramanathan et al., 2011) was described.

Attempts to synthesize metallic nanoparticles such as iron, platinum, palladium, copper, selenium, and uranium have only been made recently. Most of the work in this area has been oriented towards synthesis of magnetite nanoparticles. For instance, a simple and green approach for the synthesis of iron sulfide has been shown in bacteria such as *Magnetospiryllum* (Farina et al., 1990), *M. magnetotacticum* (Mann et al., 1984; Philipse and Maas, 2002), sulfate-reducing bacteria (Mann et al., 1990), *M. gryphiswaldense* (Lang and Schüler, 2006), and *Acinetobacter* sp. (Bharde et al., 2008). Nanoparticles formed by these strains showed predominant morphologies of octahedral prism, cubo-octahedral, and hexagonal prism in the size range 2–120 nm. The results obtained indicated that bacterial sulfate reductases are responsible for the process. The possibility of using bacteria for the formation of iron oxide nanoparticles has also been studied. Magnetotactic bacteria (Blakemore, 1975; Mann et al., 1984), iron-reducing bacteria *Geobacter metallireducens* (Vali et al., 2004), and *Actinobacter* sp. (Bharde et al., 2005) were able to form magnetite (Fe_3O_4) or maghemite (γ-Fe_2O_3).

The process of magnetic nanoparticles mineralization can be divided into four steps: (1) vesicle formation and iron transport from outside of the bacterial membrane into the cell; (2) magnetosomes alignment in a chain; (3) initiation of crystallization; and (4) crystal maturation (Fig. 1.2; Faramarzi and Sadighi, 2013).

Platinum and palladium nanoparticles have also been explored as they are interesting materials with a wide range of applications. Konishi et al. (2007) demonstrated platinum nanoparticle synthesis by the resting cells of *Shewanella algae* in the presence of N_2–CO_2 gas mixture. *S. algae* were able to reduce $PtCl_6^{-2}$ ions to platinum nanoparticles during 60 minutes of reaction time at neutral pH and 25 °C. It should be noted that the underlying biochemical process as to how these organisms are able to synthesize platinum nanoparticles is still unclear and remains an open challenge. The first report for the synthesis of palladium nanoparticles appeared in early 2000 using a sulfate-reducing bacterium *Desulfovibrio desulfuricans* (Yong et al., 2002a,b). The formation of palladium nanoparticles of approximately 50 nm in size in the presence of molecular hydrogen or formate as an electron donor when reacted with aqueous palladium ions at pH 2–7 was demonstrated.

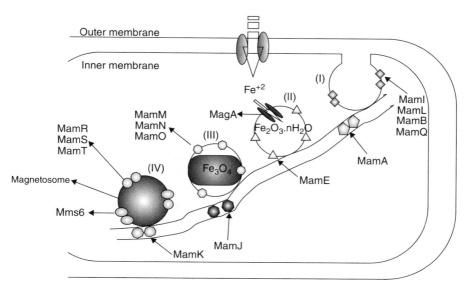

Figure 1.2. Magnetosome bio-mineralization in magnetotactic bacteria (MTB). (I) MamI, MamL, MamB, and MamQ proteins initiate the membrane invagination and form a vesicular membrane around the magnetosome structure. (II) The protease-independent function of MamE recruits other proteins such as MamK, MamJ, and MamA to align magnetosomes in a chain. (III) Iron uptake occurs via MagA, a transmembrane protein, and initiation of magnetic crystal bio-mineralization occurs through MamM, MamN, and MamO proteins. (IV) Finally, MamR, MamS, MamT, MamP, MamC, MamD, MamF, MamG, the protease-dependent function of MamE, and Mms6, a membrane tightly bounded by GTP-ase, regulate crystal growth and determine morphology of the produced magnetic nanoparticles. Source: Faramarzi, M.A., Sadighi, A. 2013. Insights into biogenic and chemical production of inorganic nanomaterials and nanostructures. *Advances in Colloid and Interface Science* 189–190: 1–20. Copyright © 2013, Elsevier. *See insert for color representation of the figure.*

Gram-negative rod, such as *Serratia* sp. (Hasan et al., 2008) isolated from insect gut and *Escherichia coli* (Singh et al., 2010) when challenged with aqueous copper precursors, were able to synthesize a mixture of copper and copper oxide quasi-spherical nanoparticles of size range 10–40 nm. It was postulated that two proteins/peptides of molecular weight 25 kDa and 52 kDa might be involved in the reduction and stabilization of these nanoparticles. Kumar et al. (2008) reported the extracellular synthesis of ferromagnetic Co_3O_4 nanoparticles by a marine cobalt-resistant bacterium isolated from the Arabian Sea.

Different bacterial strains have also been explored for the synthesis of semiconductors (so-called quantum dots), such as CdS and ZnS. Production of CdS nanocrystals was observed in the case of *Clostridium thermoaceticum* (Cunningham and Lundi, 1993), *K. pneumoniae* (Smith et al., 1998), *E. coli* (Sweeney et al., 2004), and *R. palustris* (Bai et al., 2009). It is believed that bacterial cysteine desulfhydrase is responsible for the formation of CdS nanoparticles. Labrenz et al. (2000) has demonstrated that sphalerite crystals (ZnS) are produced within natural biofilms, which are dominated by the sulfate-reducing bacteria.

Selenium and tellurium respiration studies have fascinated scientists as these ions are rarely in contact with microbes in their natural environment. Some studies have suggested that SeO_3^{2-} reduction may involve the periplasmic nitrite reductase in *Thauera selenatis* (DeMoll-Decker and Macy, 1993; Bledsoe et al., 1999; Sabaty et al., 2001) and *Rhizobium selenitireducens* strain B1 (Hunter and Kuykendall, 2007; Hunter et al., 2007), nitrate reductase in *E. coli* (Avazeri et al., 1997), hydrogenase I in *Clostridium pasteurianum* (Yanke et al., 1995), and arsenate reductase in *Bacillus selenitireducens* (Afkar et al., 2003) or some of the non-enzymatic reactions (Tomei et al., 1992). Most research on the biogenesis of selenium nanoparticles are based on anaerobic systems. However, there are also a few reports in the literature on the aerobic formation of these nanostructures by bacteria such as: *Pseudomonas stutzeri* (Lortie et al., 1992), *Enterobacter cloacae* (Losi and Frankenberger, 1997), *Pseudomonas aeruginosa* (Yadav et al., 2008), *Bacillus* sp. (Prakash et al., 2009), and *Pseudomonas alkaphila* (Zhang et al., 2011a). Another metalloid semiconductor, tellurium (which belongs to the chalcogen family), has been reduced from tellurite to elemental tellurium by two anaerobic bacteria, *Sulfurospirillum barnesii* and *B. selenireducens*. Interestingly, the two different species yielded a different morphology of tellurium nanoparticles. *S. barnesii* formed extremely small nanospheres of diameter less than 50 nm that coalesced to form large aggregates, while *B. selenitireducens* initially formed nanorods of approximately 10 nm in diameter by 200 nm length, which clustered together forming large rosettes (c. 1000 nm) made from individual shards of 100 nm width by 1000 nm length (Baesman et al., 2007). Titanium nanoparticles of spherical aggregates of 40–60 nm were also produced extracellularly using the culture filtrate of *Lactobacillus* sp. at room temperature (Prasad et al., 2007).

Another interesting metal, especially in its nanoparticulate form, is uranium. In the same way as for gold, uranium is soluble in the oxidized form, U(VI); the reduced form of uranium, U(IV), is insoluble however. Among the first reports of U(IV) synthesis, Woolfolk and Whiteley (1962) found that the cell-free extracts of *Micrococcus lactilyticus* reduced U(VI) to UIV). *Alteromonas putrefaciens* grown in the presence of hydrogen as an electron donor and U(VI) as an electron acceptor also reduced U(VI) to U(IV) (Myers and Nealson, 1988). In line with these observations, Lovley et al. (1991) demonstrated that *G. metallireducens* GS-15, when grown anaerobically in the presence of acetate and U(VI) as electron donor and electron acceptor, reduced soluble uranium U(VI) to insoluble U(IV) oxidizing acetate to CO_2. Marshall and co-workers (2006) found out that c-type cytochrome (MtrC) on the outer membrane and extracellular of dissimilatory metal-reducing bacterium *S. oneidensis* MR-1 was involved in the reduction of U(VI) predominantly with extracellular polymeric substance as UO2-EPS in cell suspension and intracellularly in periplasm. *Desulfosporosinus* sp., a gram-positive sulfate-reducing bacterium isolated from sediments when incubated with mobile hexavalent uranium U(VI) reduced to tetravalent uranium U(IV) which precipitated uraninite. These uraninite (UO_2) crystals coated on the cell surface of *Desulfosporosinus* sp. were within the size range 1.7 ± 0.6 nm (O'Loughlinej et al., 2003).

1.3. SYNTHESIS OF NANOPARTICLES BY FUNGI

Fungi have also been used for the synthesis of different kinds of metal nanoparticles (Basavaraja et al., 2008; Bawaskar et al., 2010; Raheman et al., 2011). Rai et al. (2009b) proposed the term "myconanotechnology" to describe research carried out on nanoparticles synthesized by fungi, the integrated discipline of mycology and nanotechnology. Many fungal species have so far been exploited for the synthesis of metal nanoparticles, including endophytic fungi.

Ahmad et al. (2003c) exploited *Fusarium oxysporum* for the synthesis of silver nanoparticles. They reported that when exposed to the fungus *F. oxysporum*, aqueous silver ions were reduced in solution thereby leading to the formation of an extremely stable silver hydrosol. The silver nanoparticles were in the range of 5–15 nm in dimensions and stabilized in solution by proteins secreted by the fungus. Exposure of *F. oxysporum* to an aqueous solution of K_2ZrF_6, resulting in the protein-mediated extracellular hydrolysis of zirconium hexafluoride anions at room temperature, leading to the formation of crystalline zirconia nanoparticles, was reported by Bansal et al. (2004). They concluded that as *F. oxysporum* is a plant pathogen which is not exposed to such ions during its life cycle, it secretes proteins capable of hydrolyzing ZrF_6^2.

Bansal et al. (2005) reported the synthesis of silica and titania nanoparticles by *F. oxysporum* when the fungal cell filtrate was challenged with their respective salts, that is, K_2SiF_6 and K_2TiF_6. The resulting nanoparticles formed were in the range of 5–15 nm and had an average size of 9.8 ± 0.2 nm. Duran et al. (2005) studied the production of metal nanoparticles by several strains of *F. oxysporum*. They found that when aqueous silver ions were exposed to the cell filtrate of *F. oxysporum*, they reduced to the formation of silver nanoparticles. The resulting silver nanoparticles were in the range of 20–50 nm. They hypothesized that the reduction of the metal ions occurs by a nitrate-dependent reductase enzyme and a shuttle quinone extracellular process. *F. oxysporum* f. sp. *lycopersici*, causing wilt in tomato, was exploited for intracellular and extracellular production of platinum nanoparticles (Riddin et al., 2006). Bharde et al. (2006) reported the synthesis of magnetite nanoparticles from *F. oxysporum*. *F. acuminatum* isolated from infected ginger was successfully exploited for the mycosynthesis of silver nanoparticles by Ingle et al. (2008). Ingle and coworkers (2008) also proposed the hypothetical mechanism for the synthesis of silver nanoparticles. According to them, NADH-dependent nitrate reductase played an important role in the synthesis of nanoparticles (Fig. 1.3).

They reported the formation of polydispersed, spherical nanoparticles in the range of 4–50 nm with average diameter of 13 nm. Ingle et al. (2009) used *F. solani* isolated from infected onion for the synthesis of silver nanoparticles; again, polydispersed, spherical nanoparticles were synthesized in the size range of 5–35 nm.

Soil-born fungus *Aspergillus fumigatus* is also reported to produce the silver nanoparticles extracellularly when the cell extract was challenged with aqueous silver ions (Bhainsa and D'Souza, 2006). Gade et al. (2008) reported the biosynthesis of silver nanoparticles from *A. niger* isolated from soil, and also suggested the mechanism for the action of silver nanoparticles on *E. coli*. Similarly, Jaidev and Narsimha

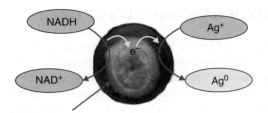

Fungal biomass secrets NADH–
dependant nitrate reductase

Figure 1.3. Possible mechanism of enzymatic reduction of the silver ions. Source: Ingle, A., Gade, A., Pierrat, S., Sonnichsen, C., Rai, M.K. Mycosynthesis of silver nanoparticles using the fungus *Fusarium acuminatum* and its activity against some human pathogenic bacteria. *Current Nanoscience* 4: 141–144. Copyright © 2008, Bentham Science Publishers.

(2010) reported the synthesis of silver nanoparticles using *A. niger*. Vigneshwaran et al. (2007) reported the potential of *A. flavus* for the intracellular production of silver nanoparticles; when treated with aqueous silver ions, the silver nanoparticles were synthesized in the cell wall. Moharrer and co-workers (2012) reported the extracellular synthesis of silver nanoparticles using *A. flavus* which was isolated from soil of Ahar copper mines. Extracellular biosynthesis of silver nanoparticles was studied using the fungus *Cladosporium cladosporioides* (Balaji et al., 2009). The transmission electron microscope (TEM) image showed polydispersed and spherical particles with size ranges of 10–100 nm.

Mukherjee et al. (2001) were the first to report the use of fungal systems for the synthesis of silver and gold nanoparticles. They observed for *Verticillium* sp. that the reduction of the metal ions occurred intracellularly, leading to the formation of gold and silver nanoparticles in the size range of 2–20 nm. Sastry et al. (2003) also used fungal systems and actinomycetes for the synthesis of silver nanoparticles and gold nanocrystals by intra- and extracellular processes, respectively. They reported a number of silver nanoparticles on the surface of the mycelial wall of *Verticillium*. The mycelia also showed an individual *Verticillium* cell with silver particles clearly bound to the surface of the cytoplasmic membrane, confirming the intracellular formation of the silver nanoparticles. TEM analysis showed the formation of the silver nanoparticles in *Verticillium* cells of average diameter 25 ± 12 nm. In case of *Thermomonospora* sp. when exposed to aqueous gold ions, the metal ions were reduced extracellularly, yielding gold nanoparticles with a much-improved polydispersity in the range of 7–12 nm. Bharde et al. (2006) reported the formation of nanoparticulate magnetite at room temperature extracellularly, after the treatment of cell filtrate of fungus *Verticillium* sp. with mixtures of ferric and ferrous salts. Extracellular hydrolysis of the anionic iron complexes by cationic proteins secreted by the fungi results in the room-temperature synthesis of crystalline magnetite particles that exhibit the signature of a ferromagnetic transition with a negligible amount of spontaneous magnetization at low temperature. TEM analysis revealed a number of cubo-octahedrally shaped iron oxide particles ranging in size from 100 to 400 nm.

Chen et al. (2003) studied the extracellular formation of silver nanoparticles using *Phoma* sp. 3.2883. Another species of *Phoma* used for the synthesis of silver nanoparticles was *P. glomerata* (Birla et al., 2009). In this study, the authors reported extracellular synthesis of silver nanoparticles when the fungal cell filtrate was treated with aqueous silver ions (silver nitrate, 1 mM) and incubated at room temperature. The silver nanoparticles synthesized were found to be in the range of 60–80 nm when the colloidal solution of these silver nanoparticles was analyzed using scanning electron microscopy (SEM).

An endophytic fungus *Colletotrichum* sp., growing in the leaves of Geranium, produced gold nanoparticles when exposed to chloroaurate ions. These particles are predominantly decahedral and icosahedral in shape, ranging in size from 20 to 40 nm (Shivshankar et al., 2003). Lichen fungi (*Usnea longissima*) have shown synthesis of bioactive nanoparticles (usnic acid) in specified medium (Shahi and Patra, 2003).

Ahmad et al. (2005) reported the ability of fungus *Trichothecium* sp. for the synthesis of gold nanoparticles both intra- and extracellularly. When kept in a stationary condition, the fungal biomass resulted in rapid extracellular synthesis of gold nanoparticles with spherical, rod-like, and triangular morphology; when the biomass was shaken on a rotary shaker it resulted in the intracellular synthesis of nanoparticles. The authors therefore concluded that the enzymes and proteins which are released in the stationary phase are not released under shaking conditions, resulting in the intracellular and extracellular synthesis of gold nanoparticles.

The agriculturally important fungus *Trichoderma asperellum* was used to depict the capacity of extracellular synthesis of silver nanoparticles. The silver nanoparticles were in the range of 13–18 nm with stability of up to 6 months due to the binding of proteins as capping legands. Synthesis of silver nanoparticles by yeast strain MKY_3 has been reported, leading to particles ranging 2–5 nm in size. The extracellular synthesis of nanoparticles was useful, as nanoparticles could be obtained in large quantities (Kowshik et al., 2003).

Das and Marsili (2010) reviewed the role of bacteria and fungi in the synthesis of metal nanoparticles, and also suggested the possible mechanism of the synthesis of nanoparticles using microorganisms (Fig. 1.4). According to them, natural processes such as biomineralization may be mimicked to design efficient nanoparticle synthesis techniques. Biomineralization processes exploit biomolecular templates that interact with the inorganic material throughout its formation, resulting in the synthesis of particles of defined shape and size.

Many successful attempts have been made for the synthesis of metal nanoparticles using different fungi such as *A. foetidus*-MTCC-8876 (Roy et al., 2013), *Penicillium citrinum* (Honary et al., 2013), and *Epicoccum nigrum* (Qian et al., 2013) for the synthesis of silver nanoparticles and *Phanerochaete chrysosporium* (Sanghi et al., 2011) and *F. oxysporum* f. sp. *cubense* JT1 (Thakker et al., 2013) for gold nanoparticles synthesis.

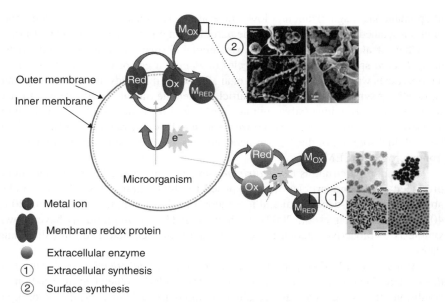

- ● Metal ion
- ● Membrane redox protein
- ● Extracellular enzyme
- ① Extracellular synthesis
- ② Surface synthesis

Figure 1.4. Biomineralization process for nanoparticle synthesis. Source: Das, S.K., Marsili, E. A green chemical approach for the synthesis of gold nanoparticles: characterization and mechanistic aspect. *Reviews in Environmental Science and Biotechnology* 9: 199–204. Copyright © 2010, Springer. *See insert for color representation of the figure.*

1.4. SYNTHESIS OF NANOPARTICLES BY ALGAE

The synthesis of metal nanoparticles using algae is relatively unexplored, but it is a more biocompatible method than the other biological methods. The process of synthesis of metal nanoparticle using algal extract was found to be a rapid and non-toxic process (Jena et al., 2013). Many scientists therefore became interested in the use of algal systems for successful synthesis of metal nanoparticles. Merin et al. (2010) exploited four different micro algae – *Chaetoceros calcitrans*, *Chlorella salina*, *Isochrysis galbana*, and *Tetraselmis gracilis* – for the synthesis of silver nanoparticles. Synthesized nanoparticles were also checked for their antibacterial potential against human pathogens (*Klebsiella* sp., *Proteus vulgaricus*, *Pseudomonas aeruginosa*, and *E. coli*). The silver nanoparticles produced by the above mentioned algal species showed significant activity with all four bacteria (Merin et al., 2010). In recent years, Kalabegishvili et al. (2012) reported the reduction of chloroaurate ($HAuCl_4$) into gold nanoparticles when expose to blue-green algae *Spirulina platensis*. Similarly, Mahdieh et al. (2012) used the same algae (*Spirulina platensis*) for the synthesis of silver nanoparticles by treating an aqueous solution of silver ions with a live biomass of *S. platensis*. The transmission electron studies carried out for nanoparticles showed the production of silver nanoparticles with an average size of ~ 12 nm.

Jena and co-workers (2013) described how not only the extract but also whole cells of algal biomass can be used for the synthesis of silver nanoparticles. In their study, they used fresh extract and whole cell of microalga *Chlorococcum humicola*. The *in vivo* and *in vitro* formation of nanoparticles were characterized using different analytical methods including ultraviolet–visible (UV-vis) spectroscopy, scanning electron microscopy (SEM), Fourier transform infrared spectroscopy (FTIR), X-ray diffraction (XRD), and transmission electron microscopy (TEM), confirming the synthesis of nanoparticles of 16 nm in size. These biosynthesized silver nanoparticles were found to be significantly active against *E. coli*.

There are also some other reports suggesting the potential of crude extracts of some microalgae such as *Kappaphycus alvarezii* and *Gelidiella acerosa* for the production of metallic nanoparticles, which could be used for medicinal purposes (Rajasulochana et al., 2010; Vivek et al., 2011). These investigations imply that the vast types of algae most commonly found can be efficiently used for the synthesis of nanoparticles in an environmentally friendly approach.

Cyanobacteria, also called blue-green algae, are a well-known group of aquatic and photosynthetic microorganisms. They are also the source organisms for the synthesis of several types of nanoparticles. Very few studies have yet been performed on cyanobacteria-mediated synthesis of nanoparticles. However, these bacteria are a good source of the enzymes responsible for the catalysis reaction that leads to the reduction of ions to their corresponding nanoparticles (Mubarak-Ali et al., 2011). Various cyanobacteria have been found to be capable of synthesizing nanoparticles. Lengke et al. (2006a) investigated the synthesis of silver nanoparticles by using *Plectonema boryanum*. The investigators proposed the involvement of the cyanobacterial metabolites in the synthesis of silver nanoparticles from its salt.

Cyanobacteria are thought to synthesize silver nanoparticles intracellularly. During this synthesis, the silver ions become reduced to neutral silver with the help of the protein molecule present inside the cell, and the nanoparticle becomes released outside the dead cell. The released nanoparticles are capped with the protein (Mubarak-Ali et al., 2011). Recently, Sudha et al. (2013) reported the synthesis of silver nanoparticles by using *Aphanothece* sp., *Oscillatoria* sp., *Microcoleus* sp., *Aphanocapsa* sp., *Phormidium* sp., *Lyngbya* sp., *Gleocapsa* sp., *Synechococcus* sp., and *Spirulina* sp. The silver nanoparticles synthesized were found to be spherical in shape, without aggregation, and with good antibacterial activity.

From a synthesis point of view, it is obvious that some components of algal biomass or the cell are used to reduce the ions to the nanoparticles. In this respect, X-ray absorption spectroscopy (XAS) results for the *Chlorella-vulgaris*-mediated synthesis of gold nanoparticles showed that the reduction of gold ions to gold nanoparticles was coordinated with the conversion of sulfhydryl residues into sulfur atoms (Watkins et al., 1987). On the other hand, the carbonyl groups ($C\equiv O$) of cellulosic materials of *Sargassumnatans* cell wall were reported to be the main functional group in gold precipitation (Kuyucak and Volesky, 1989). However, the cell wall components such as hydroxyl (OH^-) groups of saccharides and carboxylate anion (COO^-) of aminoacids residues also interacts with gold (Lin et al., 2005). Collectively, it can be said that various molecules or their components play an important role in the formation of various types of nanoparticles, especially metal nanoparticles.

Brayner et al. (2007) used commonly occurring cyanobacteria such as *Anabaena*, *Calothrix*, and *Leptolyngbya* for intracellular synthesis of gold (Au), silver (Ag), palladium (Pd), and platinum (Pt) nanoparticles. The study also indicated the role of the intracellular nitrogenase enzyme in the reduction of corresponding metal ions and which, after release in the surrounding medium, were stabilized by polysaccharides produced by the same algae. Likewise, *Lyngbya majuscula* and *Spirulina subsalsa* have also been exploited for the synthesis of gold nanoparticles, both intracellularly and extracellularly (Chakraborty

TABLE 1.2. Synthesis of Metal Nanoparticles using Different Algae

Serial No.	Name of fungi	Type of nanoparticle synthesized	Size (nm)	Shape	Reference
1	*Plectonema boryanum*	Gold	10–6000	Octahedral platelets	Lengke et al., 2006a
2	*Plectonema boryanum*	Platinum	<300	Spherical	Lengke et al., 2006b
3	*Plectonema boryanum*	Paladium	<30	Spherical and elongate	Lengke et al., 2007
4	*Sargassum wightii*	Gold	8–12	Spherical	Singaravelu et al., 2007
5	*Spirulina platensis*	Silver, Gold, Au core and Ag shell	7–16, 6–10, 17–25	All spherical	Govindraju et al., 2008
6	*Chaetoceros calcitrans* *Chlorella salina* *Isochrysis galbana* *Tetraselmis gracilis*	Silver	Not mentioned	Not mentioned	Merin et al., 2010
7	*Oscillatoria willei* NTDM01	Silver	100–200	Not mentioned	Mubarak-Ali et al., 2011
9	*Oscillatoria willei*	Silver	100–200	Spherical	Mubarak-Ali et al., 2011
10	*Spirulina platensis*	Silver	11.6 (avg)	Mostly spherical	Mahdieh et al., 2012
11	*Aphanothece* sp. *Oscillatoria* sp. *Microcoleus* sp. *Apanocapsa* sp. *Pormidium* sp. *Lyngbya* sp. *Gleocapsa* sp. *Synecoccus* sp. *Spirulins* sp.	Silver	40–80	Spherical	Sudha et al., 2013

et al., 2009). In an interesting study, Mubarak-Ali et al. (2012) reported the synthesis of pigment-stabilized CdS nanoparticles. They first extracted C-phyotoerythrin (C-PE) from a marine cyanobacterium *Phormidium tenue* NTDM05, and later used these C-PE to synthesize CdS nanoparticles. All of these studies signify that cyanobacteria are promising biological agents for the synthesis of stable nanoparticles. However, as very few reports have described their use in the synthesis of nanoparticles, there is a great need for further research. Some of the algal species used for the synthesis of metal nanoparticles to date are listed in Table 1.2.

After summarizing the above mentioned studies on the mechanism for nanoparticle synthesis using different algae, we propose a general mechanism for the intracellular synthesis of metal nanoparticles. In this mechanism different compounds or metabolites such as proteins or nitrogenase enzyme, present intracellularly in the algal mycelium itself, are responsible for the reduction of metal ions to its nanoparticulate form (Fig. 1.5).

Figure 1.6 depicts the general mechanism for the extracellular synthesis of metal nanoparticles, based on the hypothesis given by Mahdieh et al. (2012). In this two-step mechanism, the aqueous metal ion first adheres to the surface of algal cells due to the electrostatic attraction between the positively charged metal ions

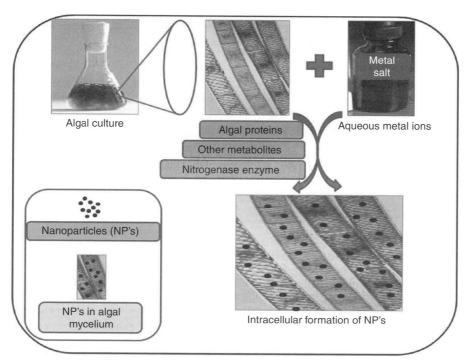

Figure 1.5. General mechanism for the intracellular synthesis of metal nanoparticles using algae. *See insert for color representation of the figure.*

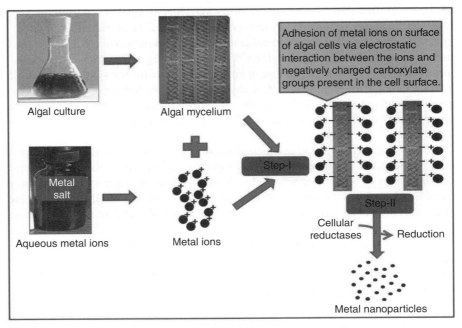

Figure 1.6. General mechanism for the extracellular synthesis of metal nanoparticles using algae. *See insert for color representation of the figure.*

and negatively charged carboxylate ions present on the cell surface of algae. The ions are then reduced to metal nanoparticles as a result of the secretion of reductase enzyme (cellular reductases) by algal cells.

1.5. APPLICATIONS OF METAL NANOPARTICLES

Metal nanoparticles offer applications in many disciplines including agriculture, catalysis, and biomedical biosensors.

1.5.1. Nanoparticles as Catalyst

The application of nanoparticles as a catalyst is a rapidly expanding field in nanotechnology. Due to their distinctive properties, nanoparticles are an ideal material for catalyst. Platinum and gold bimetallic nanoparticles have been used as electrocatalyst for polyelectrolyte fuel cells for the conversion of exhaust heat to energy (Toshima, 2013). Titanium and silver heterostructures have also demonstrated electrochemical properties, and could be used as photocatalysts (Zhang et al., 2013). Similarly, gold nanoparticles also demonstrate properties of photocatalysis (Kawamura et al., 2013).

1.5.2. Nanoparticles as Bio-membranes

Nanotube membranes with molecular dimension <1 nm could be used as channels for the separation of molecules and ions between solutions. These nanotube membranes could separate nano-sized molecules based on their size while membranes of dimension 20–60 nm could be used to separate proteins (Gupta et al., 2012).

1.5.3. Nanoparticles in Cancer Treatment

Gold nanoparticles have shown potential in the treatment of cancer (Bhattacharya and Mukherjee, 2008; Chauhan et al., 2011). Vascular endothelial growth factor (VEGF) acts as a potential angiogenic factor and blood vessel permeabilizing agent after ligand binding to VEGF receptors (VEGFRs) on endothelial cells. Blocking the interaction of the VEGF with its receptors could be a possible way to inhibit angiogenesis. Quantum dots are luminescent crystals that allow specific drugs such as proteins, oligonucleotides, and siRNA (small interfering RNA) to penetrate targeted cancer cells in the central nervous system; they are therefore utilized for imaging in biological crystals. However, the toxicity issues of quantum dots is a major obstacle in its medical application to humans (Dikpati et al., 2012).

1.5.4. Nanoparticles in Drug Delivery

Metal nanoparticles with magnetic properties work as an effective molecular carrier for gene separation and also show promising application in drug delivery (Bava et al., 2013). For drug delivery, magnetic nanoparticles are injected into the drug molecule which is to be delivered; these particles are then guided towards the chosen site under a localized magnetic field. These magnetic carriers can carry large doses of drugs (Lu et al., 2007; Perez-Martinez et al., 2012). Similarly, silica-coated nanoparticles are also used in drug delivery due to their high stability, surface properties, and compatibility. Silica nanoparticles are also used in biological applications such as artificial implants (Dikpati et al., 2012; Perez-Martinez et al., 2012).

1.5.5. Nanoparticles for Detection and Destruction of Pesticides

Pesticides are hazardous to both human beings and the environment, contaminating drinking and surface water. The unique properties of nanoparticles allow their use in the detection and destruction of pesticides. The large-surface-area-to-volume-ratio property of nanoparticles plays a crucial role in the catalytic reactions used to degrade pesticides (Aragay et al., 2012). The optical properties of nanoparticles are related to their size and surface-induced changes in electronic structure, which helps in the detection of pesticides. For the destruction of pesticides, a photocatalytic oxidation method employing titanium nanoparticles is used (Aragay et al., 2012).

1.5.6. Nanoparticles in Water Treatment

Polyethylenimine-derived (PEI) nanoparticles and dendrimers have a number of applications such as gene delivery, catalysis, and electronics (Dikpati et al., 2012; Perez-Martinez et al., 2012). Magnetite nanoparticles are used in wastewater treatment and the removal of heavy metals from water. Das et al. (2009) reported the use of nanogold-bioconjugate in water hygiene management. The authors described the single-step removal of some model organophosphorus pesticide from water along with some microorganisms. Magnetite nanoparticles can be harnessed as adsorbents for separating and removing contaminants in water by applying an external magnetic field (Carlos et al., 2013).

1.6. LIMITATIONS OF SYNTHESIS OF BIOGENIC NANOPARTICLES

Nanotechnology in general and nanomaterials in particular have the potential to revolutionize different sectors such as pharmacy and medicine, electronics and agriculture and allied sectors such as food processing, packaging, and storage with modern tools (Rai and Ingle, 2012). There are different approaches for the synthesis of nanoparticles and each approach – whether physical, chemical, or biological – has its own limitations. In this chapter we have discussed some of the major limitations of biological (biogenic) synthesis of nanoparticles.

In nanotechnology, the synthesis of monodisperse nanoparticles is always a challenge for researchers and is one of the major limitations of biogenic approaches. It was reported that biogenic synthesis of nanoparticles failed to synthesize monodisperse nanoparticles, while chemical and physical methods are found to be significant for the synthesis of monodisperse nanoparticles (Tran et al., 2013). Another major limitation of the biogenic approach regards the mechanism of the synthesis of metal nanoparticles. There are many different mechanisms proposed for the elucidation of biogenic synthesis of metal nanoparticles using different biological systems; a survey of the literature suggests that many scientists have proposed a hypothetical mechanism based on their studies (Duran et al., 2005; Huang et al., 2007; Kumar et al., 2007; Li et al., 2007; Ingle et al., 2008). Some studies proposed that NADH-dependent nitrate reductase is responsible for the reduction of aqueous silver ions (Duran et al., 2005; Kumar et al., 2007; Ingle et al., 2008). Duran et al. (2005) further suggested that anthraquinone pigments and their derivatives synthesized by certain fungi were responsible for the reduction of metal ions into their respective nanoparticles. Gade et al. (2011) proposed a three-step mechanism for the synthesis of silver nanorods using fungus *Phoma sorghina* and reported that anthraquinone pigments played a key role in the formation of silver nanorods. Similarly, the exact mechanism for the formation of nanoparticles using bacteria and plants is not yet known; many investigators have suggested possible mechanisms however (Chandran et al., 2006; Huang et al., 2007; Li et al., 2007).

Fourier transform infrared spectroscopy (FTIR) studies carried out for the characterization of synthesized nanoparticles showed the presence of capped proteins

(Ingle et al., 2008, 2009; Birla et al., 2009; Gajbhiye et al., 2009; Bawaskar et al., 2010; Gade et al., 2010). Jain et al. (2010) studied the sodium dodecyl sulfate polyacrylamide gel electrophoresis (SDS-PAGE) profiles of the extracellular proteins secreted by *Aspergillus flavus*, which has been used for the synthesis of silver nanoparticles. They reported the presence of two proteins having a molecular weight of 32 and 35 kDa. Further, they proposed a two-step mechanism for synthesis and stabilization of nanoparticles. The first step involves a 32 kDa protein which may be a reductase and is responsible for the reduction of silver ions into silver nanoparticles. The second step involves 35 kDa proteins which bind with nanoparticles and confer stability. This is the only study which determines the exact protein involved in the capping of nanoparticles, playing a key role in their stabilization. Due to the lack of other publications on this topic, we believe that further extensive studies are needed to determine the exact proteins responsible for nanoparticle capping and their stabilization.

CONCLUSION

In summary, there are three types of methods applied for the synthesis of nanoparticles, viz. physical, chemical and biological. Physical methods are tedious and time consuming, whereas many chemical methods make use of toxic chemicals for the purpose. On the other hand, biological methods of synthesis are quite safe and has the capability to produce nanoparticles with good biocompatibility.

Among biological methods, various types of microorganisms including bacteria, fungi and algae can be used. Biological systems, especially those using bacteria, have the potential to offer cheap and scalable, green, synthetic production of the latest generation of nanomaterials; it should however, be emphasized that, in many bacterial mediated synthesis methods, the actual biochemical mechanism remains poorly investigated. Likewise, many studies are upholding the use of fungi and algae for the size controlled and stable nanoparticle synthesis.

The metal nanoparticles are getting recognition for treatment of dreadful diseases like cancer, AIDS and tuberculosis. Moreover, they have potential to help in cleaning the environment by destructions of hazardous material like pesticide.

Although the biogenic nanoparticles have many advantages, toxicity issues associated with them are also important. The reports showing the toxic effects of the biogenic nanoparticles are emerging with slower pace. However, there is need of exhaustive study concerning the hazardous effects of nanoparticles.

ACKNOWLEDGEMENTS

We gratefully acknowledge financial support from the NCN, grant no. NN 507 5150 38. MR also thanks FAPESP (No. 2012/03731-3) for financial assistance to visit the Institute of Chemistry, Universidade Estadual de Campinas, Campinas, SP, Brazil.

REFERENCES

Afkar, E., Lisak, J., Saltikov, C., Basu, P., Oremland, R.S., Stolz, J.F. 2003. The respiratory arsenate reductase from *Bacillus selenitireducens* strain MLS10. *FEMS Microbiology Letters* 226: 107–112.

Ahmad, A., Senapati, S., Khan, M. I., Kumar, R., Ramani, R., Shrinivas, V., Sastry, M. 2003a. Intracellular synthesis of gold nanoparticles by a novel alkalotolerent actinomycete, *Rhodococcus* species. *Nanotechnology* 14: 824–828.

Ahmad, A., Senapati, S., Khan, M.I., Kumar, R., Sastry, M. 2003b. Extracellular biosynthesis of monodisperse gold nanoparticles by a novel extremophilic actinomycete, *Thermomonospora* sp. *Langmuir*, 19: 3550–3553.

Ahmad, A., Mukherjee, P., Mandal, D., Senapati, S., Khan, M.I., Kumar, R., Sastry, M. 2003c. Extracellular biosynthesis of silver nanoparticles using the fungus *Fusarium oxysporum*. *Colloids and Surfaces B: Biointerfaces* 28: 313–318.

Ahmad, A., Senapati, S., Khan, M.I., Kumar, R., Sastry, M. 2005. Extra-/intracellular biosynthesis of gold nanoparticles by an alkalotolerant fungus, *Trichothecium* sp. *Journal of Biomedical Nanotechnology* 1(1): 47–53.

Aragay, G., Pino, F., Merkoci, A. 2012. Nanomaterials for sensing and destroying pesticides. *Chemical Reviews* 112: 5317–5338.

Avazeri, C., Turner, R.J., Pommier, J., Weiner, J.H., Giordano, G., Vermeglio, A. 1997. Tellurite and selenate reductase activity of nitrate reductases from *Escherichia coli*: correlation with tellurite resistance. *Microbiology* 143: 1181–1189.

Baesman, S.M., Bullen, T.D., Dewald, J., Zhang, D., Curran, S., Islam, F.S., Beveridge, T.J., Oremland, R.S. 2007. Formation of tellurium nanocrystals during anaerobic growth of bacteria that use the oxyanions as respiratory electron acceptors. *Applied and Environmental Microbiology* 73: 2135–2143.

Bai, H.J., Zhang, Z.M., Guo, Y., Yang, G.E. 2009. Biosynthesis of cadmium sulfide nanoparticles by photosynthetic bacteria *Rhodopseudomonas palustris*. *Colloids and Surfaces B: Biointerfaces* 70:142–146.

Balagurunathan, R., Radhakrishnan, M., Rajendran, R.B., Velmurugan, D. 2011. Biosynthesis of gold nanoparticles by actinomycete *Streptomyces viridogens* strain HM10. *Indian Journal of Biochemistry and Biophysics* 48: 331–335.

Balaji, D.S., Basavaraja, S., Deshpande, R., Mahesh, D.B., Prabhakar, B.K., Venkataraman, A. 2009. Extracellular biosynthesis of functionalized silver nanoparticles by strains of Cladosporium cladosporioides fungus. *Colloids Surface B: Biointerfaces* 68(1): 88–92.

Bankar, A., Joshi, B., Kumar, A.R., Zinjarde, S. 2010. Banana peel extract mediated novel route for the synthesis of silver nanoparticles. Colloids and Surfaces A: *Physicochemical and Engineering Aspects* 368: 58–63.

Bansal, V., Rautaray, D., Ahmad, A., Sastry, M. 2004. Biosynthesis of zirconia nanoparticles using the fungus *Fusarium oxysporum*. *Journal of Material Chemistry* 14: 3303–3305.

Bansal, V., Rautaray, D., Bharde, A., Ahire, K., Sanyal, A., Ahmad, A., Sastry, M. 2005. Fungus-mediated biosynthesis of silica and titania particles. *Journal of Material Chemistry* 15: 2583–2589.

Basavaraja, S., Balaji, S.D., Lagashetty, A., Rajasab, A.H., Venkataraman, A. 2008. Extracellular biosynthesis of silver nanoparticles using the fungus *Fusarium semitectum*. *Materials Research Bulletin* 43(5): 1164–1170.

REFERENCES

Bava, A., Cappellini, F., Pedretti, E., Rossi, F., Caruso, E., Vismara, E., Chiriva-Internati, M., Bernardini, G., Gornati, R. 2013. Heparin and Carboxymethyl-chitosan metal nanoparticles: An evaluation on their cytotoxicity. *BioMed Research International* 2013: Article ID 314091, doi: http://dx.doi.org/10.1155/2013/314091.

Bawaskar, M., Gaikwad, S., Ingle, A., Rathod, D., Gade, A., Duran, N., Marcato, P.D, Rai, M. 2010. A new report on mycosynthesis of silver nanoparticles by *Fusarium culmorum*. *Current Nanoscience* 6: 376–380.

Beveridge, T.J., Murray, R.G. 1980. Sites of metal deposition in the cell wall of *Bacillus subtilis*. *Journal of Bacteriology* 141: 876–887.

Bhainsa, K.C., D'Souza, S.F. 2006. Extracellular biosynthesis of silver nanoparticles using the fungus *Aspergillus fumigatus*. *Colloids Surface B: Biointerfaces* 47: 160–164.

Bharde, A., Wani, A., Shouche, Y., Joy, P.A., Prasad, B.L.V., Sastry, M. 2005. Bacterial aerobic synthesis of nanocrystalline magnetite. *Journal of the American Chemical Society* 127: 9326–9327.

Bharde, A., Rautaray, D., Bansal, V., Ahmad, A., Sarkar, I., Yusuf, S.M., Sanyal, M., Sastry, M. 2006. Extracellular biosynthesis of magnetite using fungi. *Small* 2(1): 135–141.

Bharde, A., Kulkarni, A., Rao, M., Prabhune, A., Sastry, M.J. 2007. Bacterial enzyme mediated biosynthesis of gold nanoparticles. *Nanoscience and Nanotechnology* 7: 4369–4377.

Bharde, A.A., Parikh, R.Y., Baidakova, M., Jouen, S., Hannoyer, B., Enoki, T., Prasad, B.L.V., Shouche, Y.S., Ogale, S., Sastry, M. 2008. Bacteria-mediated precursor-dependent biosynthesis of super paramagnetic iron oxide and iron sulfide nanoparticles. *Langmuir* 24: 5787–5794.

Bhattacharya, D., Gupta, R.K. 2005. Nanotechnology and potential of microorganisms. *Critical Review in Biotechnology* 24(4): 199–204.

Bhattacharya, R., Mukherjee, P. 2008. Biological properties of naked nanoparticles. *Advances in Drug Delivery Reviews* 60: 1289–1306.

Birla, S.S., Tiwari, V.V., Gade, A.K., Ingle, A.P., Yadav, A.P., Rai, M.K. 2009. Fabrication of silver nanoparticles by *Phoma glomerata* and its combined effect against *Escherichia coli*, *Pseudomonas aeruginosa* and *Staphylococcus aureus*. *Letters in Applied Microbiology* 48: 173–179.

Blakemore, R. 1975. Magnetotactic bacteria. *Science* 190: 377–379.

Bledsoe, T.L., Cantafio, A.W., Macy, J.M. 1999. Fermented whey- an inexpensive feed source for a laboratory-scale selenium-bioremediation reactor system inoculated with *Thauera selenatis*. *Applied Microbiology and Biotechnology* 51: 682–685.

Brayner, R., Barberousse, H., Hemadi, M., Djedjat, C., Yéprémian, C., Coradin, T., Livage, J., Fiévet, F., Couté, A. 2007. Cyanobacteria as bioreactors for the synthesis of Au, Ag, Pd, and Pt nanoparticles via an enzyme-mediated route. *Journal of Nanoscience and Nanotechnology* 7(8): 2696–2708.

Carlos, L., Einschlag, F.S.G., Gonzalez, M.C., Martire, D.O. 2013. Applications of magnetite nanoparticles for heavy metal removal from wastewater. In *Waste Water: Treatment Technologies and Recent Analytical Developments* (Einschlag, F.S.G., Carlos, L. eds). Intech Open Science Publisher, Croatia, pp. 63–77.

Chakraborty, N., Banerjee, A., Lahiri, S., Panda, A., Ghosh, A.N., Pal, R. 2009. Biorecovery of gold using cynobacteria and eukaryotic alga with special reference to nanogold formation – a novel phenomenon. *Journal of Applied Phycology* 21(1): 145–152.

Chandran, S. P., Ahmad, A., Chaudhary, M., Pasricha, R., Sastry, M. 2006. Synthesis of gold nanotriangles and silver nanoparticles using Aloe vera plant extract. *Biotechnology Program* 22(2): 577–583.

Chauhan, A., Zubair, S., Tufail, S., Sherwani, A., Sajid, M., Raman, S.C., Azam, A., Owais, M. 2011. Fungus-mediated biological synthesis of gold nanoparticles: potential in detection of liver cancer. *International Journal of Nanomedicine* 6: 2305–2319.

Chen, J.C., Lin, Z.H., Ma, X.X. 2003. Evidence of the production of silver nanoparticles via pretreatment of *Phoma* sp.3.2883 with silver nitrate. *Letters in Applied Microbiology* 37: 105–108.

Cunningham, D.P., Lundie, L.L. 1993. Precipitation of cadmium by *Clostridium thernoaceticum*. *Applied and Environmental Microbiology* 59: 7–14.

Das, S.K., Marsili, E. 2010. A green chemical approach for the synthesis of gold nanoparticles: characterization and mechanistic aspect. *Reviews in Environmental Science and Biotechnology* 9: 199–204.

Das, S.K., Das, A.R., Guha, A.K. 2009. Gold nanoparticles: microbial synthesis and application in water hygiene management. *Langmuir* 25(14): 8192–8199.

De-Gusseme, B., Sintubin, L., Baert, L., Thibo, E., Hennebel, T., Vermeulen, G., Uyttendaele, M., Verstraete, W., Boon, N. 2010. Biogenic silver for disinfection of water contaminated with viruses. *Applied and Environmental Microbiology* 76: 1082–1087.

DeMoll-Decker, H., Macy, J.M. 1993. The periplasmic nitrite reductase of *Thauera selenatis* may catalyze the reduction of selenite to elemental selenium. *Archives in Microbiology* 160: 241–247.

Derakhshan, F.K., Dehnad, A., Salouti, M. 2012. Extracellular biosynthesis of gold nanoparticles by metal resistance bacteria: *Streptomyces griseus*. *Synthesis and Reactivity in Inorganic, Metal-Organic, and Nano-Metal Chemistry* 42: 868–871.

Dikpati, A., Madgulkar, A.R., Kshirsagar, S.J., Bhalekar, M.R., Chahal, A.S. 2012. Targeted drug delivery to CNS using nanoparticles. *Journal of Advanced Pharmaceutical Sciences* 2(1): 179–191.

Duran, N., Marcato, P.D., Alves, O.L., D'Souza, G., Esposito, E. 2005. Mechanistic aspects of biosynthesis of silver nanoparticles by several *Fusarium oxysporum* strains. *Nanobiotechnology* 3: 8–14.

Duran, N., Marcarto, P.D., De Souza, G.I.H., Alves, O.L., Esposito, E. 2007. Antibacterial effect of silver nanoparticles produced by fungal process on textile fabrics and their effluent treatment. *Journal of Biomedical Nanotechnology* 3: 203–208.

Faramarzi, M.A., Sadighi, A. 2013. Insights into biogenic and chemical production of inorganic nanomaterials and nanostructures. *Advances in Colloid and Interface Science* 189–190: 1–20.

Farina, M., Esquivel, D.M.S., de Barros, H.G.P.L. 1990. Magnetic iron-sulphur crystals from a magnetotactic microorganism. *Nature* 343: 256–258.

Gade, A.K., Bonde, P.P., Ingle, A.P., Marcato, P., Duran, N., Rai, M.K. 2008. Exploitation of *Aspergillus niger* for synthesis of silver nanoparticles. *Journal of Biobased Material Bioenergy* 2:1–5.

Gade, A., Gaikwad, S., Tiwari, V., Yadav, A., Ingle, A., Rai, M. 2010. Biofabrication of silver nanoparticles by *Opuntia ficus-indica*: In vitro antibacterial activity and study of the mechanism involved in the synthesis. *Current Nanoscience* 6(4): 370–375.

Gade, A., Rai, M., Kulkarni, S. 2011. *Phoma sorghina*, a phytopathogen mediated synthesis of unique silver rods. *International Journal of Green Nanotechnology*, 3: 153–159.

Gajbhiye, M., Kesharwani, J., Ingle, A., Gade, A., Rai, M. 2009. Fungus-mediated synthesis of silver nanoparticles and their activity against pathogenic fungi in combination with fluconazole. *Nanomedicine: Nanotechnology, Biology, and Medicine* 5: 382–386.

Golinska, P., Wypij, M., Ingle, A.P., Gupta, I., Dahm, H., Rai, M. 2014. Biogenic synthesis of metal nanoparticles from actinomycetes: Biomedical applications and cytotoxicity. *Applied Microbiology and Biotechnology* 98: 8083–8097.

Govindraju, K., Basha, S.K., Kumar, V.G., Singaravelu, G. 2008. Silver, gold and bimetallic nanoparticles production using single-cell protein (*Spirulina platensis*) Geitler. *Journal of Material Science* 43: 5115–5122.

Gupta, S., Sharma, K., Sharma, R. 2012. Myconanotechnology and applications of nanoparticles in biology. *Recent Research in Science and Technology* 4(8): 36–38.

Hasan, S., Singh, S., Parikh, R.Y., Dharne, M.S., Patole, M.S., Prasad, B.L.V., Shouche, Y.S. 2008. Bacterial Synthesis of copper/copper oxidenanoparticles. *Journal for Nanoscience and Nanotechnology* 8: 3191–3196.

He, S., Guo, Z., Zhang, Y., Zhang, S., Wang, J., Gu, N. 2007. Biosynthesis of gold nanoparticles using the bacteria *Rhodopseudomonas capsulate*. *Material Letters* 61: 3984–3987.

He, S., Zhang, Y., Guo, Z., Gu, N. 2008. Biological synthesis of gold nanowires using extract of *Rhodopseudomonas capsulate*. *Biotechnology Progress* 24: 476–480.

Honary, S., Barabadi, H., Gharaei-Fathabad, E., Naghibi, F. 2013. Green synthesis of silver nanoparticles induced by the fungus *Penicillium citrinum*. *Tropical Journal of Pharmaceutical Research* 12(1): 7–11.

Huang, J., Chen, C., He, N., Hong, J., Lu, Y., Qingbiao, L., Shao, W., Sun, D., Wang, X.H., Wang, Y., Yiang, X. 2007. Biosynthesis of silver and gold nanoparticles by novel sundried *Cinnamomum camphora* leaf. *Nanotechnology* 18: 105–106.

Hunter, W.J., Kuykendall, L.D. 2007. Reduction of selenite to elemental red selenium by *Rhizobium* sp. strain B1. *Current Microbiology* 55: 344–349.

Hunter, W.J., Kuykendall, L.D., Manter, D.K. 2007. *Rhizobium selenireducens* sp. nov.: a selenite reducing a-Proteobacteria isolated from a bioreactor. *Current Microbiology* 55: 455–460.

Husseiny, M.I., El-Aziz, M.A., Badr, Y., Mahmoud, M.A. 2007. Biosynthesis of gold nanoparticles using *Pseudomonas aeruginosa*. *Spectrochimica Acta Part A: Molecular and Biomolecular Spectroscopy* 67: 1003–1006.

Ingle, A., Gade, A., Pierrat, S., Sonnichsen, C., Rai, M.K. 2008. Mycosynthesis of silver nanoparticles using the fungus *Fusarium acuminatum* and its activity against some human pathogenic bacteria. *Current Nanoscience* 4: 141–144.

Ingle, A., Rai, M., Gade, A., Bawaskar, M. 2009. *Fusarium solani*: a novel biological agent for the extracellular synthesis of silver nanoparticles. *Journal of Nanoparticle Research* 11: 2079–2085.

Jaidev, L.R., Narasimha, G. 2010. Fungal mediated biosynthesis of silver nanoparticles, characterization and antimicrobial activity. *Colloids and Surfaces B: Biointerfaces* 81: 430–433.

Jain, N., Bhargava, A., Majumdar, S., Tarafdar, J.C., Panwar, J. 2010. Extracellular biosynthesis and characterization of silver nanoparticles using *Aspergillus flavus* NJP08: A mechanism perspective. *Nanoscale* 3: 635–641.

Jena, J., Pradhan, N., Dash, B.P., Sukla, L.B., Panda, P.K. 2013. Biosynthesis and Characterization of silver nanoparticles using microalga *Chlorococcum humicola* and its antibacterial activity. *International Journal of Nanomaterials and Biostructures* 3(1): 1–8.

Joerger, R., Klaus, T., Granquist, C.G. 2000. Biologically produced silver-carbon composite materials for optically thin film coatings. *Advanced Materials* 12: 407–409.

Kalabegishvili, T., Kirkesali, E., Frontasyeva, M.V., Pavlov, S.S., Zinicovscaia, I., Faanhof, A. 2012. Synthesis of gold nanoparticles by blue-green algae *Spirulina platensis*. *Proceedings of the International Conference of Nanomaterials: Applications and Properties* 1(2): 6–9.

Kalishwaralal, K., Deepak, V., Ramakumarpandian, S., Nellaiah, H., Sangiliyandi, G. 2008. Extracellular biosynthesis of silver nanoparticles by the culture supernatant of *Bacillus licheniformis*. *Material Letters* 62: 4411–4416.

Karthikeyan, S., Beveridge, T.J. 2002. *Pseudomonas aeruginosa* biofilms react with and precipitate toxic soluble gold. *Environmental Microbiology* 4: 667–675.

Kashefi, K., Tor, J.M., Nevin, K.P., Lovley, D.R. 2001. Reductive precipitation of gold by dissimilatory Fe(III)-reducing *Bacteria* and *Archaea*. *Applied and Environmental Microbiology* 67: 3275–3279.

Kasthuri, J., Kanthiravan, K., Rajendiran, N. 2008. Phyllanthin-assisted biosynthesis of silver and gold nanoparticles: a novel biological approach. *Journal of Nanoparticle Research* 15: 1075–1085.

Kathiresan, K., Manivannan, S., Nabeel, M.A., Dhivya, B. 2009. Studies on silver nanoparticles synthesized by a marine fungus, *Penicillium fellutanum* isolated from coastal mangrove sediment. *Colloids Surface B: Biointerfaces* 71: 133–137.

Kathiresan, K., Nabeel, M.A., SriMahibala, N., Asmathunisha, N., Saravanakumar, K. 2010. Analysis of antimicrobial silver nanoparticles synthesized by coastal strains of *Escherichia coli* and *Aspergillus niger*. *Canadian Journal of Microbiology* 56: 1050–1059.

Kawamura, G., Nogami, M., Matsuda, A. 2013. Shape-controlled metal nanoparticles and their assemblies with optical functionalities. *Journal of Nanomaterials* 2013: Article ID 631350.

Klaus, T., Joerger, R., Olsson, E., Granqvist, C.G. 1996. Silver-based crystalline nanoparticles, microbially fabricated. *Proceedings of the National Academy of Sciences, USA* 96: 13611–13614.

Konishi, Y., Ohno, K., Saitoh, N., Nomura, T., Nagamine, S., Hishida, H., Takahashi, Y., Uruga, T. 2007. Bioreductive deposition of platinum nanoparticles on the bacterium *Shewanella algae*. *Journal of Biotechnology* 128: 648–653.

Kowshik, M., Ashtaputre, S., Kharrazi, S., Vogel, W., Urban, J., Kulkarni, S.K., Paknikar, K.M. 2003. Extracellular synthesis of silver nanoparticles by silver-tolerant yeast strain MKY3. *Nanotechnology* 14: 95.

Kumar, S.A., Abyaneh, M.K., Gosavi, S.W., Kulkarni, S.K., Pasricha, R., Ahmad, A., Khan, M.I. 2007. Nitrate reductase-mediated synthesis of silver nanoparticles from $AgNO_3$. *Biotechnology Letters* 29: 439–445.

Kumar, U., Shete, A., Harle, A.S., Kasyutich, O., Schwarzacher, W., Pundle, A., Poddar, P. 2008. Extracellular bacterial synthesis of protein-functionalized ferromagnetic Co_3O_4 nanocrystals and imaging of self-organization of bacterial cells under stress after exposure to metal ions. *Chemistry of Materials* 20: 1484–1491.

Kumar, V., Yadav, S.K. 2009. Plant-mediated synthesis of silver and gold nanoparticles and their applications. *Journal of Chemical Technology and Biotechnology* 84: 151–157.

Kuyucak, N., Volesky, B. 1989.The mechanism of gold biosorption. *Biorecovery* 1: 219–235.

Labrenz, M., Druschel, G.K., Thomsen-Ebert, T., Gilbert, B., Welch, S.A., Kemner, K.M., Graham, A., Logan, G.H, Summons, R.E., De Stasio, G., Bond, P.L., Lai, B., Kelly, S.D., Jillian, F., Banfield, J.F. 2000. Formation of sphalerite (ZnS) deposits in natural biofilms of sulfate-reducing bacteria. *Science* 90, 1744–1747.

Lang, C., Schűler, D. 2006. Biogenic nanoparticles: production, characterization, and application of bacterial magnetosomes. *Journal of Physics: Condensed Matter* 18: S2815–S2828.

Lengke, M., Ravel, B., Feet, M.E., Wanger, G., Gordon, R.A., Southam, G. 2006a. Mechanisms of gold bioaccumulation by filamentous cynobacteria from gold (III)-chloride complex. *Environmental Science and Technology* 40: 6304–6309.

Lengke, M.F., Fleet, ME., Southam, G. 2006b. Synthesis of platinum nanoparticles by reaction of filamentous cyanobacteria with platinum (IV) chloride complex. *Langmuir* 22(17): 7318–7323.

Lengke, M.F., Fleet, M.E., Southam, G. 2007. Synthesis of palladium nanoparticles by reaction of filamentous cyanobacterial biomass with a palladium (II) chloride complex. *Langmuir* 23(17): 8982–8987.

Li, S., Qui, L., Shen, Y., Xie, A., Yu, X., Zhang, L., Zhang, Q. 2007. Green synthesis of silver nanoparticles using Capsicum annum L. extract. *Green Chemistry* 9: 852–858.

Lin, Z., Wu, J., Xue, R., Yang, Y. 2005. Spectroscopic characterization of Au^{3+} biosorption by waste biomass of *Saccharomyces cerevisiae*. *Spectrochimica Acta Part A: Molecular and Biomolecular Spectroscopy* 61: 761–765.

Lloyd, J.R., Byrne, J.M., Coker, V.S. 2011. Biotechnological synthesis of functional nanomaterials. *Current Opinion in Biotechnology* 22: 509–515.

Lortie, L., Gould, W.D., Rajan, S., Meeready, R.G.L., Cheng, K.J. 1992. Reduction of elemental selenium by a *Pseudomonas stutzeri* isolate. *Applied and Environmental Microbiology* 58: 4042–4044.

Losi, M., Frankenberger, W.T. 1997. Reduction of selenium by *Enterobacter cloacae* SLD1a-1: isolation and growth of bacteria and its expulsion of selenium particles. *Applied and Environmental Microbiology* 63: 3079–3084.

Lovley, D.R., Phillips, E.J.P., Gorby, Y.A., Landa, E.R. 1991. Microbial reduction of uranium. *Nature* 350: 413–416.

Lu, A.H., Salabas, E.L., Schuth, F. 2007. Magnetic nanoparticles: synthesis, protection, functionalization and application. *Angewandte Chemie International Edition* 46: 1222–1244.

Mahdieh, M., Zolanvari, A., Azimee, A.S., Mahdieh, M. 2012. Green biosynthesis of silver nanoparticles by *Spirulina platensis*. *Scientia Iranica F* 19(3): 926–929.

Mallikarjuna, K., Narasimha, G., Dilip, G.R., Praveen, B., Shreedhar, B., Shree-Lakshmi, C., Reddy, B.V.S., Deva-Prasad Raju, B. 2011. Green synthesis of silver nanoparticles using *Ocimum* leaf extract and their characterization. *Digest Journal of Nanomaterials and Biostructures* 6(1): 181–186.

Mann, S., Frankel, R.B., Blakemore, R.P. 1984. Structure, morphology and crystal growth of bacterial magnetite. *Nature* 310: 405–407.

Mann, S., Sparks, N.H.C., Frankel, R.B., Bazylinski, D.A., Jannasch, H.W. 1990. Biomineralization of ferrimagneticgreigite (Fe_3S_4) and iron pyrite (FeS_2) in a magnetotactic bacterium. *Nature* 343: 258–261.

Marshall, M.J., Beliaev, A.S., Dohnalkova, A.C., Kennedy, D.W., Shi, L., Wang, Z., Boyanov, M.I., Lai, B., Kemner, K.M., McLean, J.S., Reed, S.B., Culley, D.E., Bailey, V.L., Simonson, C.J., Saffarini, D.A., Romine, M.F., Zachara, J.M., Fredrickson, J.K. 2006. *c*-Type cytochrome-dependent formation of U(IV) nanoparticles by *Shewanella oneidensis*. *PLOS Biology* 4: 1324–1333.

Merin, D.D., Prakash, S., Bhimba, B.V. 2010. Antibacterial screening of silver nanoparticles synthesized by marinemicro algae. *Asian Pacific Journal of Tropical Medicine* 3(10): 797–799.

Moharrer, S., Mohammadi, B., Gharamohammadi, R.A., Yargoli, M. 2012. Biological synthesis of silver nanoparticles by *Aspergillus flavus*, isolated from soil of Ahar copper mine. *Indian Journal of Science and Technology* 5: 2443–2444.

Mubarak-Ali, D., Sasikala, M., Gunasekaran, M., Thajuddin, N. 2011. Biosynthesis and characterization of silver nanoparticles using marine cyanobacterium, *Oscillatoria willei* ntdm01. *Digest Journal of Nanomaterials and Biostructures* 6(2): 385–390.

Mubarak-Ali, D., Gopinath,V., Rameshbabu, N., Thajuddina, N. 2012. Synthesis and characterization of CdS nanoparticles using C-phycoerythrin from the marine cyanobacteria. *Materials Letters* 74: 8–11.

Mukherjee, P., Ahmad, A., Mandal, D., Senapati, S., Sainkar, S.R., Khan, M.I., Ramani, R., Parischa, R., Ajayakumar, P.V., Alam, M., Sastry, M., Kumar, R. 2001. Bioreduction of AuCl4 ions by the fungus, *Verticillium* sp. And surface trapping of gold nanoparticles formed. *Angewandte Chemie International Edition* 40: 3585–3588.

Myers, C.R., Nealson, K.H. 1988. Bacterial manganese reduction and growth with manganese oxide as the sole electron acceptor. *Science* 240: 1319–1321.

Nagajyothi, P.C., Lee, K.D. 2011. Synthesis of plant mediated silver nanoparticles using *Dioscoreabatatas* rhizome extract and evaluation of their antimicrobial activities. *Journal of Nanomaterials* 2011: Article ID 573429.

Nair, B., Pradeep, T. 2002. Coalescence of nanoclusters and the formation of submicron crystallites assisted by *Lactobacillus* strains. *Crystal Growth and Designs* 4: 295–298.

O'Loughlinej, E.J., Kelly, S.D., Cook, R.E., Csencsits, R., Kemner, K.M. 2003. Reduction of uranium (VI) by mixed iron(II)/iron(III) hydroxide (green rust): formation of UO_2 nanoparticles. *Environmental Science and Technology* 37: 721–727.

Parikh, R.Y., Singh, S., Prasad, B.L.V., Patole, M.S., Sastry.M., Shouche, Y.S. 2008. Extracellular synthesis of crystalline silver nanoparticles and molecular evidence of silver resistance from *Morganella sp.*: towards understanding biochemical synthesis mechanism. *ChemBioChem* 9: 1415–1422.

Perez-Martinez, F.C., Carrion, B., Cena, V. 2012. The use of nanoparticles for gene therapy in the nervous system. *Journal of Alzheimer's Disease* 31: 697–710.

Philipse, A.P., Maas, D. 2002. Magnetic colloids from magnetotactic bacteria: chain formation and colloidal stability. *Langmuir* 18: 9977–9984.

Prakash, T.N., Sharma, N., Prakash, R., Raina, K.K., Fellowes, J., Pearce, C.I., Lloyd, J.R., Pattrick, R.A.D. 2009. Aerobic microbial manufacture of nanoscale selenium: exploiting nature's bio-nanomineralization potential. *Biotechnology Letters* 31: 1857–1862.

Prasad, K., Jha, A.K., Kulkarni, A.R. 2007. Lactobacillus assisted synthesis of titanium nanoparticles. *Nanoscale Research Letters* 2: 248–250.

Qian, Y., Yu, H., He, D., Yang, H., Wang, W., Wan, X., Wang, L. 2013. Biosynthesis of silver nanoparticles by the endophytic fungus *Epicoccum nigrum* and their activity against pathogenic fungi. *Bioprocess and Biosystems Engineering* 36(11): 1613–1619.

Raheman, F., Deshmukh, S., Ingle, A., Gade, A., Rai, M. 2011. Silver nanoparticles: Novel antimicrobial agent synthesized from an endophytic fungus *Pestalotia* sp. isolated from leaves of *Syzygium cumini* (L). *NanoBiomedical Engineering* 3: 174–178.

Rai, M., Ingle, A. 2012. Role of nanotechnology in agriculture with special reference to management of insect-pest. *Applied Microbiology and Biotechnology* 94(2): 287–293.

Rai, M., Yadav, A., Gade, A. 2008. Current trends in Phytosynthesis of metal nanoparticles. *Critical Reviews in Biotechnology* 28(4): 277–284.

Rai, M., Yadav, A., Gade, A. 2009a. Silver nanoparticles: as a new generation of antimicrobials. *Biotechnology Advances* 27: 76–83.

Rai, M., Yadav, A., Bridge, P., Gade, A. 2009b. Myconanotechnology: A new and emerging science. In: *Applied Mycology* (M.K. Rai, P.D. Bridge, eds). CAB International, New York, 14: 258–267.

Rajasulochana, P., Dhamotharan, R., Murugakoothan P., Murugesan, S., Krishnamoorthy, P. 2010. Biosynthesis and characterization of gold nanoparticles using the alga *Kappaphycus alvarezii*. *International Journal of Nanoscience* 9(5): 511.

Rajesh, S., Raja, D.P., Rathi, J.M., Sahayaraj, K. 2012. Biosynthesis of silver nanoparticles using Ulvafasciata (Delile) ethyl acetate extract and its activity against *Xanthomonas compestris pv. malvacearum*. *Journal of Biopesticides* 5(sppli): 119–128.

Ramanathan, R., O'Mullane, A.P., Parikh, R.Y., Smooker, P.M., Bhargava, S.K., Bansal, V. 2011. Bacterial kinetics-controlled shape-directed biosynthesis of silver nanoplates using *Morganella psychrotolerans*. *Langmuir* 27: 714–719.

Reith, F., Rogers, S.L., McPhail, D.C., Webb, D. 2006. Biomineralization of gold: biofilms on bacterioform gold. *Science* 313: 233–236.

Renugadevi, K., Aswini, R.V. 2012. Microwave irradiation assisted synthesis of silver nanoparticles using *Azadirachta indica* leaf extract as a reducing agent and *in vitro* evaluation of its antibacterial and anticancer activity. *International Journal of Nanomaterials and Biostructures* 2(2): 5–10.

Riddin, T.L., Gericke, M., Whiteley, C.G. 2006. Analysis of the inter- and extracellular formation of platinum nanoparticles by *Fusarium oxysporum*f sp. *lycopersici* using response surface methodology. *Nanotechnology* 17: 3482–3489.

Roy, S., Mukherjee, T., Chakraborty, S., Das, T.K. 2013. Biosynthesis, characterization & antifungal activity of silver nanoparticles synthesized by the fungus *Aspergillus foetidus*MTCC-8876. *Digest Journal of Nanomaterials and Biostructures* 8(1): 197–205.

Sabaty, M., Avazeri, C., Pignol, D., Vermeglio, A. 2001. Characterization of the reduction of selenate and tellurite by nitrate reductases. *Applied and Environmental Microbiology* 67: 5122–5126.

Sadhasivam, S., Shanmugam, P., Veerapandian, M., Subbiah, R., Yun, K. 2012. Biogenic synthesis of multidimensional gold nanoparticles assisted by *Streptomyces hygroscopicus* and its electrochemical and antibacterial properties. *Biometals* 25: 351–360.

Sanghi, R., Verma, P., Puri, S. 2011. Enzymatic formation of gold nanoparticles using *Phanerochaete chrysosporium*. *Advances in Chemical Engineering and Science* 1: 154–162.

Sastry, M., Ahmad, A., Khan, M.I., Kumar, R. 2003. Biosynthesis of metal nanoparticles using fungi and actinomycetes. *Current Science* 85(2): 162–170.

Satyavathi, R., Krishna, M.B., Rao, S.V., Saritha, R., Rao, D.N. 2010. Biosynthesis of silver nanoparticles using *Coriandrum sativum* leaf extract and their application in non-linear optics. *Advanced Science Letters* 3: 1–6.

Seshadri, S., Prakash, A., Kowshik, M. 2012. Biosynthesis of silver nanoparticles by marine bacterium, *Idiomarina* sp. PR58-8. *Bulletin of Material Science* 35: 1201–1205.

Shahi, S.K., Patra, M. 2003. Microbially synthesized bioactive nanoparticles and their formulation active against human pathogenic fungi. *Reviews on Advance Materials Science* 5: 501–509.

Shahverdi, A.R., Fakhimi, A., Shahverdi, H.R., Minaian, S. 2007. Synthesis and effect of silver nanoparticles on the antibacterial activity of different antibiotics against *Staphylococcus aureus* and *Escherichia coli*. *Nanomedicine* 3: 168–171.

Shahverdi, A.R., Minaeian, S., Shahverdi, H.R., Jamalifar, H., Shahverdi, N., Wong, C.W., NurYasumira, A.A. 2009. Rapid biosynthesis of silver nanoparticles using culture supernatant of bacteria with microwave irradiation. *E-Journal of Chemistry* 6(1): 61–70.

Sharma, V.K., Yngard, R.A., Lin, Y. 2009. Silver nanoparticles: Green synthesis and their antimicrobial activities. *Advances in Colloid and Interface Science* 145: 83–96.

Shivshankar, S., Ahmad, A., Pasricha R., Sastry, M. 2003. Bioreduction of chloroaurate ions by geranium leaves and its endophytic fungus yields gold nanoparticles of different shapes. *Journal of Material Chemistry* 13: 1822–1826.

Silver, S. 2003. Bacterial resistance: molecular biology and uses misuses of silver compounds. *FEMS Microbiology Review* 27: 341–353.

Singaravelu, G., Arockiamary, J.S., Ganesh-Kumar, V., Govindraju, K. 2007. A novel extracellular synthesis of monodiaperse gold nanoparticles using marine alga, *Sargassum wightii* Greville. *Colloids and Surfaces B: Biointerfaces* 57: 97–101.

Singh, M., Kalaivani, R., Manikandan, S., Sangeetha, N., Kumaraguru, A.K. 2013. Facile green synthesis of variable metallic gold nanoparticle using *Padinagymnospora*, a brown marine macroalga. *Applied Nanoscience* 3: 145–151.

Singh, R.P., Shukla, V.K., Yadav, R.S., Sharma, P.K., Singh, P. K., Pandey, A.C. 2011. Biological approach of zinc oxide nanoparticles formation and its characterization. *Advance Materials Letters* 2(4): 313–317.

Singh, V., Patil, R., Anand, A., Milani, P., Gade, W. 2010. Biological synthesis of copper oxide nanoparticles using *Escherichia coli*. *Current Nanoscience* 6: 365–369.

Smith, P.R., Holmes, J.D., Richardson, D.J., Russell, D.A., Sodeau, J.R. 1998. Photophysical and photochemical characterization of bacterial semiconductor cadmium sulphide particles. *Journal of the Chemical Society, Faraday Transactions* 94: 1235–1241.

Southam, G., Beveridge, T.J. 1996. The occurence of sulfur and phosphorus within bacterially derived crystalline and pseudocrystalline octahedral gold formed *in vitro*. *Geochimica et Cosmochimica Acta* 60: 4369–4376.

Sudha, S.S., Rajamanickam, K., Rengaramanujam, J. 2013. Microalgae mediated synthesis of silver nanoparticles and their antibacterial activity against pathogenic bacteria. *Indian Journal of Experimental Biology* 52: 393–399.

Sweeney, R.Y., Mao, C., Gao, X., Burt, J.L., Belcher, A.M., Georgiou, G., Iverson, B.L. 2004. Bacterial biosynthesis of cadmium sulphide nanocrystals. *Chemistry and Biology* 11: 1553–1559.

Temple, K.L., Le-Roux, N.W. 1964. Syngenosis of sulphide ores: desorption of adsorbed metal ions and their precipitation as sulphides. *Economic Geology* 59: 647–655.

Thakker, J.N., Dalwadi, P., Dhandhukia, P.C. 2013. Biosynthesis of gold nanoparticles using *Fusarium oxysporum* f. sp. *cubense*JT1, a plant pathogenic fungus. *ISRN Biotechnology*, 2013: Article ID 515091 (5 pages).

Thakkar, K.N., Mhatre, S.S., Parikh, R.Y. 2011. Biological synthesis of metallic nanoparticles. *Nanomedicine* 6(2): 257–262.

Tomei, F.A., Barton, L.L., Lemanski, C.L., Zocco, T.G. 1992. Reduction of selenate and selenite to elemental selenium by *Wolinella succinogenes*. *Canadian Journal of Microbiology* 38: 1328–1333.

Toshima, N. 2013. Metal nanoparticles for energy conversion. *Pure and Applied Chemistry* 85(2): 437–451.

Tran, Q.H., Nguyen, V.Q., Le, A.T. 2013. Silver nanoparticles: synthesis, properties, toxicology, applications and perspectives. *Advances in Natural Sciences: Nanoscience and Nanotechnology* 4: 033001 (20 pp.).

Vali, H., Weiss, B., Li, Y.L., Sears, S.K., Kim, S.S., Kirschvink, J.L., Zhang, C. 2004. Formation of tabular single domain magnetite induced by *Geobacter metallireducens* GS-15. *Proceedings of the National Academy of Sciences USA* 101: 16121–16126.

Vaseeharan, B., Ramasamy, P., Chen, J.C. 2010. Antibacterial activity of silver nanoparticles (AgNPs) synthesized by tea leaf extracts against pathogenic *Vibrio harveyi* and its protective efficacy on juvenile *Feneropenaeus indicus*. *Letters in Applied Microbiology* 50(4): 352–356.

Vigneshwaran, N., Ashtaputre, N.M., Varadarajan, P.V., Nachane, R.P., Paralikar, K.M., Balasubramanya, R.H. 2007. Biological synthesis of silver nanoparticles using the fungus *Aspergillus flavus*. *Materials Letters* 61(6): 1413–1418.

Vivek, M., Senthil Kumar, P., Teffi, S., Sudha, S. 2011. Biogenic silver nanoparticles by *Gelidiella acerosa* extract and their antifungal effects. *Avicenna Journal of Medical Biotechnology* 3(3): 143.

Watkins II, J.W., Elder, R.C., Greene, B., Darnall, D. 1987. Determination of gold binding in an algal biomass using EXAFS and XANES spectroscopies. *Inorganic Chemistry* 26(7): 1147–1151.

Woolfolk, C.A., Whiteley, H.R. 1962. Reduction of inorganic compounds with molecular hydrogen by *Micrococcus lactilyticus*. I. Stoichiometry with compounds of arsenic, selenium, tellurium, transition and other elements. *Journal of Bacteriology* 84: 647–658.

Yadav, V., Sharma, N., Prakash, R., Raina, K.K., Bharadwaj, L.M., Tejo-Prakash, N. 2008. Generation of selenium containing nano-structures by soil bacterium, *Pseudomonas aeruginosa*. *Biotechnology* 7: 299–304.

Yanke, L.J., Bryant, R.D., Laishley, E.J. 1995. Hydrogenase (I) of *Clostridium pasteurianum* functions a novel selenite reductase. *Anaerobe* 1: 61–67.

Yong, P., Farr, J.P.G., Harris, I.R., Macaskie, L.E. 2002a. Palladium recovery by immobilized cells of *Desulfovibrio desulfuricans* using hydrogen as the electron donor in a novel electrobioreactor. *Biotechnology Letters* 24: 205–212.

Yong, P., Rowson, N.A., Farr, J.P.G., Harris, I.R., Macaskie, L.E. 2002b. Bioreduction and biocrystallization of palladium by *Desulfovibrio desulfuricans* NCIMB 8307. *Biotechnology and Bioengineering* 80: 369–379.

Zhang, G., Duan, H., Lu, B., Xu, Z. 2013. Electrospining directly synthesized metal nanoparticles decorated on both sidewalls of TiO_2 nanotubes and their applications. *Nanoscale* 5(13): 5801–5808.

Zhang, H., Li, Q., Lu, Y., Sun, D., Lin, X., Deng, X., He, N., Zheng, S. 2005. Biosorption and bioreduction of diamine silver complex by *Corynebacterium*. *Journal of Chemical Technology and Biotechnology* 80: 285–290.

Zhang, X. 2011. Application of microorganisms in biosynthesis of nanomaterials- a review. *Wei Sheng Wu Xue Bao* 51(3): 297–304.

Zhang, W., Chen, Z., Liu, H., Zhang, L., Gao, P., Li, D. 2011a. Biosynthesis and structural characteristics of selenium nanoparticles by *Pseudomonas alcaliphila*. *Colloids and Surfaces B: Biointerfaces* 88: 196–201.

Zhang, X., Yan, S., Tyagi, R.D., Surampalli, R.Y. 2011b. Synthesis of nanoparticles by microorganisms and their application in enhancing microbiological reaction rates. *Chemosphere* 82(11): 489–494.

Zonooz, N.F., Salouti, M., Shapouri, R., Nasseryan, J. 2012. Biosynthesis of gold nanoparticles by *Streptomyces* sp. ERI-3 supernatant and process optimization for enhanced production. *Journal of Cluster Science* 23: 375–382.

ROLE OF FUNGI TOWARD SYNTHESIS OF NANO-OXIDES

Rajesh Ramanathan and Vipul Bansal

NanoBiotechnology Research Laboratory (NBRL), School of Applied Sciences, RMIT University, Melbourne, VIC, Australia

2.1. INTRODUCTION

Nanotechnology has been a subject of immense importance as it has laid the foundations to revolutionize modern technological concepts that have given birth to wide-ranging applications spanning the fields of electronics (Ramanathan et al., 2013), sensing (Plowman et al., 2009; Sawant et al., 2009), catalysis (Shephard et al., 1997; Choi et al., 2002; Bansal et al., 2009), biosensors (Nassif et al., 2004; Tian et al., 2004; Kang et al., 2010), optics (Huang et al., 1997), and drug delivery (Kang et al., 2010). The application of nanomaterials is vast, but realizing the true potential of nanotechnology requires that efficient methods to fabricate nanomaterials be developed before these materials may be used for real-world applications. During the initial phase of research in the field of nanomaterial synthesis, physical routes such as vapor condensation (Swihart, 2003), spray pyrolysis (Burda et al., 2005), photo-irradiation (Sakamoto et al., 2009), and thermal decomposition (Sun et al., 2000) were routinely used for the synthesis of nanomaterials (Fig. 2.1). These methods employ a 'top-down' approach, which typically involves reducing the size of the bulk material to nano-dimensions through externally controlled engineering tools. With advancements in the field, wet chemical methods (bottom-up approach) for the synthesis of nanomaterials were developed that have received a much wider acceptance than physical methods (Fig. 2.1; Bansal et al., 2012). This is because

---

Bio-Nanoparticles: Biosynthesis and Sustainable Biotechnological Implications, First Edition.
Edited by Om V. Singh.
© 2015 John Wiley & Sons, Inc. Published 2015 by John Wiley & Sons, Inc.

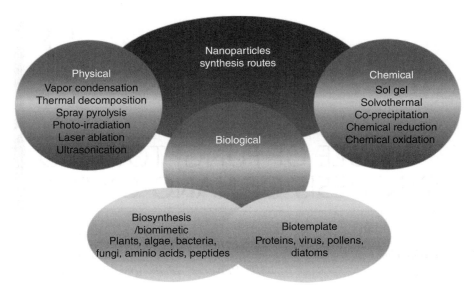

Figure 2.1. Various approaches for the synthesis of nanoparticles.

chemical routes provide excellent control over the shape and size of the nanoparticles, which is highly important as these parameters govern their nanoscale properties (Burda et al., 2005). Typically, chemical processes involve the reduction or oxidation of metal ions, which is followed by the assembly of the obtained atomic/molecular nanocrystals by utilizing different capping agents, solvents, and templates. Although solution-based nanomaterial synthesis routes have enjoyed a long history and have typically dominated the nanosphere, concern regarding the negative impact on the environment has been heightened due to the use of toxic chemicals, reaction conditions involving extremes of temperature, pressure, and pH, as well as the release of harmful by-products during chemical synthesis of nanomaterials (Thakkar et al., 2010; Bansal et al., 2011, 2012). Alternative strategies have therefore been investigated for synthesizing nanomaterials which employ 'green', environmently friendly fabrication routes.

Unsurprisingly, Mother Nature has developed a large repertoire of functional assemblages that are daunting for us to emulate in our laboratories (Bansal et al., 2011, 2012). A classic example includes the assimilation of inorganic materials in hard tissues that are in fact biocomposites containing structural biomolecules, in addition to some 60 different kinds of minerals, that perform a myriad of vital biological functions (Mann et al., 1989; Mann, 1997, 2001). Additionally, some of the biological organisms exhibit the ability to synthesize inorganic materials, both intra- and extracellularly. Examples of such biological entities include magnetotactic bacteria that synthesize magnetite (Frankel et al., 1979; Stolz et al., 1986; Schüler,

INTRODUCTION

2002), diatoms that synthesize silica (Drum and Gordon, 2003; Hildebrand, 2008), and S-layer bacteria that synthesize gypsum and calcium carbonate as surface layers (Schultze-Lam et al., 1992). The synthesis of these inorganic nanostructures in nature is often correlated with the role of these materials in performing different biological functions. However, studying these minerals from the material science perspective may also yield significant potential to transform the fields of physical, biological, and materials sciences. This has led to the development of a new route, commonly referred to as 'biological synthesis', for fabricating nanomaterials either by using biological entities directly or employing biological entities as a underlying template for growing nanostructured materials (Fig. 2.1).

The biological approach, also referred to as biosynthesis approach, typically involves employing whole living organisms for the synthesis of bioinorganic materials. This approach of nanomaterial synthesis is still largely in the discovery phase, where a variety of nanomaterials are synthesized using different organisms such as fungi, bacteria, algae, and plants (Thakkar et al., 2010; Bansal et al., 2011, 2012). These organisms might also include those that are known to create specific functional materials in natural habitats, for example, magnetotactic bacteria for the synthesis of magnetite (Frankel et al., 1979; Stolz et al., 1986; Schüler, 2002) and diatoms for silica synthesis (Drum and Gordon, 2003; Hildebrand, 2008). Although the use of diatoms or magnetic bacteria to synthesize nanomaterials is interesting, it is not however very appealing from a fundamental perspective as these organisms are already known to create inorganic materials during their natural growth. This led to the development of another interesting aspect in biological synthesis, where microorganisms were employed for the synthesis of those inorganic materials which are not known to be encountered by these organisms in their natural growth environments (Bansal et al., 2012). The concept of using microorganisms for deliberate synthesis of new nanomaterials stems from their potential ability to remediate toxic metals while tolerating high metal ion concentration through the development of specific resistance machineries (Beveridge and Murray, 1980; Mehra and Winge, 1991; Silver, 1996). Some of the earlier reports on the accumulation of inorganic particles in the natural habitats of microorganisms were gold in Precambrian algal blooms (Rouch et al., 1995), cadmium sulfide (CdS) in bacteria (Van de Meene et al., 2002) and yeast (Reese et al., 1988), and magnetite in bacteria (Temple and Le Roux, 1964). Inspired by these observations, the past decade has seen several bacteria, yeast, and fungi being employed for the deliberate synthesis of nanomaterials. Examples include bacteria-based synthesis of metals (Southam and Beveridge, 1996; Klaus-Joerger et al., 1999; Ramanathan et al., 2011*a*), oxides (Roh et al., 2001; Philipse and Maas, 2002), and sulfides (Dameron et al., 1989; Cunningham and Lundie, 1993; Smith et al., 1998; Watson et al., 2000); plant-based synthesis of metals (Shankar et al., 2004), oxides, and sulfides; and algae-based synthesis of metals (Ahmad et al., 2003*a*). Several extensive reviews in the area are available that provide a comprehensive overview of some of these approaches (Sastry et al., 2005; Narayanan and Sakthivel, 2010; Thakkar et al., 2010; Bansal et al., 2012).

2.2. FUNGUS-MEDIATED SYNTHESIS OF NANOMATERIALS

The extracellular biological synthesis of nanomaterials involving eukaryotic microorganisms such as fungi offers several advantages over other biological entities, including issues associated with scale-up and commercial implications. This is mainly because fungi: (1) can be cultured under controlled environments with relative ease; (2) are more resistant to mutations than bacteria and may therefore retain nanoparticle synthesis ability for extended populations; and (3) are well-known to produce abundant amounts of extracellular enzymes and proteins that have now been demonstrated to be responsible for the conversion of metal ions to nanoparticles (Bansal et al., 2011). These properties make fungi preferable candidates for commercial nano-product development based on biofermentation techniques. Given the ability of fungi to produce copious amounts of extracellular enzymes and their role in metal reduction (Ahmad et al., 2003b), this route using fungal organisms offers the potential for scale-up to synthesize nanomaterials using a eco-friendly and economically viable 'green' approach. There are several reports on employing fungal microorganisms as efficient biological entities for the extracellular biosynthesis of various metal, oxide, and sulfide nanoparticles, which exhibit interesting properties leading to exciting new applications. In this chapter, we focus on providing an overview of the ability of fungi to hydrolyze metal precursors to synthesize oxide nanomaterials. In this context, it is notable that the capability of fungi to leach silicates and carbonates from their complex ores was reported as early as in 1980s (Kiel and Schwartz, 1980; Klages et al., 1981). However, the use of fungi in deliberate synthesis of oxide-based nanomaterials was not explored for the first time until the early 2000s (Bansal et al., 2004).

For clarity, in this chapter the fungus-mediated biological synthesis of oxide nanoparticles will be discussed in the context of the following four major aspects: binary nano-oxides using chemical precursors; complex mixed-metal nano-oxides using chemical precursors; nano-oxides using natural precursors employing bioleaching approach; and nano-oxides employing bio-milling approach.

2.2.1. Biosynthesis of Binary Nano-oxides using Chemical Precursors

Zirconia (zirconium oxide or ZrO_2), a technologically promising binary oxide, is of significant interest due to its physiochemical properties such as hardness, shock wear, strong alkali and acid resistance, low frictional resistance, and high melting temperature (Heuer, 1987). Due to its use as an abrasive, a resistant coating for cutting tools, and for the development of high-friction engine components, this oxide is also known as 'ceramic steel' (Garvie et al., 1975). Given the significant interest of zirconia, various chemical synthesis methods including sol-gel (Xu et al., 2003), aqueous precipitation (Southon et al., 2002), and hydrothermal methods (Noh et al., 2003) have been employed for its synthesis. These routes are generally not energy-efficient and employ harsh reaction conditions. Conversely, with the hope that synthesis of zirconia nano-oxide using an eco-friendly biosynthesis approach may

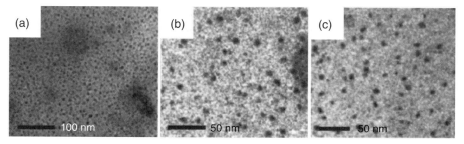

Figure 2.2. Transmission electron microscopy images of (a) zirconia; (b) silica; and (c) titania nanoparticles biosynthesized using the fungus F. oxysporum. Source: Images reprinted from Bansal, V.; Rautaray, D.; Ahmad, A.; Sastry, M., Biosynthesis of zirconia nanoparticles using the fungus Fusarium oxysporum. Journal of Materials Chemistry 2004, 14(22), 3303–3305 and Bansal, V.; Rautaray, D.; Bharde, A.; Ahire, K.; Sanyal, A.; Ahmad, A.; Sastry, M., Fungus-mediated biosynthesis of silica and titania particles. Journal of Materials Chemistry 2005, 15(26), 2583–2589. Reproduced by permission of The Royal Society of Chemistry.

overcome the limitations of chemical routes, zirconia was investigated as the first oxide-based nanomaterial to be synthesized by fungi (Bansal et al., 2004).

The biosynthesis of this technologically important material was first demonstrated using fungus *Fusarium oxysporum*, where the fungus was exposed to anionic hexafluorozirconate (ZrF_6^{2-}) complex under physiological conditions in aqueous environments. This resulted in the extracellular manifestation of fairly regular, quasi-spherical, crystalline zirconia nanoparticles with an average diameter of 7.3 ± 2.0 nm (Fig. 2.2; Bansal et al., 2004). In addition to employing this fungus for nano-zirconia synthesis, biomolecular analysis revealed the role of two low-molecular-weight (c. 24 and 28 kDa) cationic proteins in hydrolyzing the ZrF_6^{2-} ions to form ZrO_2 nanoparticles. The discovery that *F. oxysporum* could synthesize nano-zirconia was particularly intriguing as this fungus was not previously known to encounter such metal ions during its growth cycle in natural habitats. The rationale of using the plant pathogenic fungus *F. oxysporum* for the synthesis of nano-zirconia was based on the fact that plant pathogenic fungi produce a vast array of extracellular hydrolases to degrade host plant components in their natural habitats. The use of such fungi that produce copious amount of hydrolases may therefore lead to the hydrolysis of hexafluorozirconate precursor into nano-zirconia.

The regenerative capability of these biological systems (e.g., ZrF_6^{2-} ions were not toxic to the fungus *F. oxysporum* for concentrations of <1 mM), coupled with the observation that such fungi (*F. oxysporum*) were capable of hydrolyzing metal complexes that they did not encounter during their regular growth cycle, was a significant breakthrough. It showed enormous promise, particularly for developing large-scale eco-friendly routes for the synthesis of metal oxide nanoparticles.

The ability of fungus *F. oxysporum* to synthesize zirconia provided motivation to further expand the field of fungal species for the synthesis of different oxide nanomaterials. Another widely known nano-oxide is silica (silicon dioxide or SiO_2), which has proven applications as molecular sieves, resins, catalysts, and in biomedicine (Iler, 1979). Due to the wide applicability of silica, chemical synthesis routes for nano-silica synthesis are well established, but these routes often require harsh reaction conditions such as high temperatures, pressures, and extremes of pH (Ramanathan et al., 2011b). In contrast, biological entities such as diatoms, cyanobacteria and plants not only synthesize silica nanoparticles, but also have the ability to assemble individual silica nanoparticles at the nanoscale to form exquisite hierarchical silica morphologies under mild physiological conditions (Drum and Gordon, 2003; Bradbury, 2004; Lopez et al., 2005; Hildebrand, 2008). Although several reports on biological and bio-inspired methods for silica synthesis are available in the literature, they predominantly focus on the use of diatoms, sponges, and the proteins/polypeptides produced by these organisms, all of which are familiar to silicate environments in their natural habitats during an important process called 'biosilicification' (Perry et al., 2000; Drum and Gordon, 2003; Patwardhan et al., 2003; Lopez et al., 2005; Hildebrand, 2008; Ramanathan et al., 2011b). However, the use of fungi which are generally not known to encounter such ions during their life cycle provides a unique model to synthesize silica-based nanomaterials in an efficient manner.

Similar to that observed in the case of nano-zirconia, the fungus *F. oxysporum* showed the ability to hydrolyze hexafluorosilicate (SiF_6^{2-}) precursor to form quasi-spherical crystalline silica nanoparticles with an average diameter of 9.8 ± 2 nm (Fig. 2.2; Bansal et al., 2005a). In addition, two low-molecular-weight cationic proteins (c. 21 and 24 kDa) were found to be responsible for the hydrolysis of the precursor salt. The molecular weight of one of these proteins (c. 24 kDa) was also similar to that observed in the case of zirconia, suggesting the importance of similar proteins in the hydrolysis process. It is unsurprising that the proteins involved in the hydrolysis are cationic in nature, as the proteins involved in the biosilicification process, namely silicateins and silaffins, are also cationic proteins and peptides (Brutchey and Morse, 2008; Hildebrand, 2008). It is worth noting that although several biological routes have synthesized silica nanoparticles, the ability to form complex silica morphologies such as those exhibited by diatoms in natural habitats is still a considerable challenge to emulate under laboratory settings (Lopez et al., 2005). Only recently, it was elucidated that the use of ionic liquids in place of water as a solvent allows the supersaturated salt solutions in the marine environment where these diatoms grow to be mimicked, which leads to significant control over the silica morphology (Ramanathan et al., 2011b). To the best knowledge of the authors, this is the only study where the silica structures synthesized using a biomimetic approach in the laboratory resemble the diatomaceous silica structures observed in nature.

In addition to the ability of fungus *F. oxysporum* to synthesize silica and zirconia nanoparticles, the fungus was also able to hydrolyze an anionic hexafluorotitanate (TiF_6^{2-}) precursor to synthesize quasi-spherical brookite titania (titanium dioxide or TiO_2) nanoparticles with an average diameter of 10.2 ± 1 nm (Fig. 2.2; Bansal et al., 2005a). Similar to zirconia and silica, nano-titania is also of significant importance with the applications of this material ranging from photocatalysis to paints and cosmetics (Armstrong et al., 2004; Pearson et al., 2005).

Interestingly, the proteins involved in the hydrolysis were similar to those observed in the case of silica, further strengthening the important role played by these low-molecular-weight extracellular cationic proteins produced by fungi in synthesis of nano-oxides (Bansal et al., 2005a).

Note that the fungus *F. oxysporum* utilizes 21, 24 and/or 28 kDa proteins for the synthesis of silica, titania, and zirconia nanoparticles. It is likely that the proteins involved in all three cases are similar, and the differences observed in molecular weights might be due to the different level of post-translational modifications. This process requires further understanding to elucidate the role of fungal metabolic pathways towards nano-oxide synthesis. An interesting observation from the fungus-mediated nano-oxide synthesis route was the ability of *F. oxysporum* to stabilize titania nanoparticles in the brookite phase, which is the rarest polymorph of titania and is considered to be the most difficult to synthesize under laboratory conditions (Deng et al., 2008). This observation suggested the potential of biological organisms to fabricate difficult-to-synthesize nanomaterials via an energy-efficient process that are otherwise challenging for chemical synthesis routes.

The ability of *F. oxysporum* to biosynthesize the commonly known technologically important oxides such as silica, titania, and zirconia prompted further study of this versatile organism for the synthesis of bismuth oxide (Bi_2O_3) as an important optoelectronic material (Roa et al., 1969). One important property of this material is that it exists in the monoclinic phase (α-Bi_2O_3) at room temperature, but is transformed into the cubic phase (δ-Bi_2O_3) at higher temperatures. This transformed phase shows extremely high ionic conductivity, making it a potential candidate for fuel cell electrolyte and sensor applications (Skorodumova et al., 2005). Challenging the fungus *F. oxysporum* with a bismuth nitrate [$Bi(NO_3)_3$] precursor resulted in sub-10 nm, quasi-spherical Bi_2O_3 nanoparticles with mixed monoclinic and tetragonal phases (Uddin et al., 2008). The stabilization of both monoclinic and tetragonal phases at room temperature is in itself significant, as the presence of metastable β-Bi_2O_3 tetragonal phase is generally central to higher temperatures (Fan et al., 2005).

These observations direct us towards biosynthesis using fungi, which not only has the capability to synthesize nanoparticles but also to stabilize the energy-intensive phases of nano-oxides that can otherwise only be synthesized at extreme reaction conditions using chemical synthesis routes. Other common interesting observations from all of the abovementioned studies on *F. oxysporum*-mediated nano-oxides synthesis are the ability of this fungus to produce sub-10-nm oxide nanoparticles and the high aqueous- as well as organic solvent-dispersibility of these nano-oxides. This high dispersibility of bio-oxides is most likely due to the nanoparticle surface-bound amphiphilic biomacromolecules and proteins that can reorient themselves in different solvents, rendering these particles highly dispersible in a range of solvents.

In addition to the abovementioned oxide nanoparticles that mainly find applications in catalysis and electronics, ceria (cerium oxide or CeO_2) is another important metal oxide. It not only exhibits catalytic and electronic capabilities, but was also recently identified as a potential agent to inhibit cellular aging and mimic the activity of antioxidant enzyme superoxide dismutase (Das et al., 2007). Given its potential applicability in biological systems, chemical synthesis routes for the synthesis of ceria restricted the use of these materials in a biological setting. Recently, the biosynthesis of ceria nanoparticles was explored in *Humicola* sp., where the exposure of cerium nitrate [$Ce(NO_3)_3$] to this fungus resulted in formation of ceria with an average diameter of 16 nm (Khan and Ahmad, 2013a). These nanoparticles showed a mixed oxidation state of cerium (Ce^{4+} and Ce^{3+}) that has the potential to be used for free radical quenching. However, the underlying bimolecular understanding of the biosynthesis process is not currently clear.

The use of fungi for biosynthesis of nanomaterials is quite fascinating with research in this area primarily dominated by nanotechnologists and materials scientists. However, another aspect encompassing the synthesis of nanomaterials is understanding the process from a biological perspective. This research focus stems from the ability of fungi to survive in high concentrations of heavy metal ions such as Ni, Co, Fe, Mg, Mn, Cu, Al, Cr, and Zn (Valix et al., 2001).

One such study that outlined the biological outlook is the ability of several filamentous fungi to contribute significantly to Mn oxidation in soil by depositing Mn(IV) oxides (Thomson et al., 2005). To understand the process, a Mn-depositing *Acremonium*-like hypomycete fungal strain KR21-2 was isolated from a manganese deposit occurring on the wall of a storage bottle containing Mn(III, IV) oxide-coated stream-bed pebbles and stream water. On exposure of this fungus to manganese(II) sulfate, the initiation of the oxidation process occurred only after 42 hours. After 72 hours, densely aggregated δ-MnO_2 crystals associated with the fungal hyphae were formed (Miyata et al., 2006). This process took significantly longer reaction time than the previous studies, when fungus *F. oxysporum* was able to synthesize zirconia, silica, titania, and bismuth oxide nanoparticles within 24 hours of reaction.

Additionally, a laccase-like 61 kDa Mn(II)-oxidizing monomer enzyme was identified (multicopper oxidase) that showed the ability to convert Mn^{2+} ions to MnO_2. The use of this purified enzyme for the synthesis of MnO_2 nanoparticles only yielded sub-micron quasi-spherical nanoparticles with erose surfaces. However, this reaction occurred within 30 minutes of incubation with Mn^{2+} ions, highlighting the efficiency of the enzyme in this oxidation reaction. Further analysis of purified enzyme-medidated synthesis revealed the presence of Mn(II) and Mn_2O_3 impurity along with the predominant δ-MnO_2 phase in these particles, which is in contrast to the pure phase (δ-MnO_2) nanoparticles obtained during a fungus-mediated biosynthesis approach (Miyata et al., 2006). This difference was attributed to the complex cell components, exudates, and the difference in the rate of reaction in two cases.

At the time of writing, to the best of the authors' knowledge only these six binary nano-oxides (zirconia, titania, silica, bismuth oxide, ceria, and manganese oxide) have been successfully fabricated using fungus-mediated biosynthesis routes. Although the ability of fungi toward the synthesis of a large number of other metal oxides remains to be seen, the above-discussed studies clearly outline the unique ability of fungi to synthesize technologically important binary nano-oxides.

2.2.2. Biosynthesis of Complex Mixed-metal Nano-oxides using Chemical Precursors

Section 2.2.1 described the ability of fungi to synthesize simple binary oxides, highlighting the stabilizing of energy-intensive phases of nano-oxides at room temperature. In addition to these properties, some fungal species show the ability to synthesize complex mixed-metal oxides which are otherwise difficult to synthesize by chemical routes. One such material of significant medical importance is magnetite nanoparticles that have applications as a magnetic resonance imaging (MRI) contrast agent, in magnetic memory storage and biomedicine (Frankel et al., 1979).

Given the high importance of magnetic nanomaterials, it is not surprising that several biosynthesis routes have been explored for the synthesis of magnetite (Fe_3O_4/FeO. Fe_2O_3) and greigite (Fe_3S_4) by magnetotactic bacteria and iron-reducing bacteria (Stolz et al., 1986; Philipse and Maas, 2002; Schüler 2002). However, the formation of magnetite typically occurs by dissimilatory Fe(III) reduction under strict anaerobic conditions, and particles are typically formed intracellularly (within the cells) in these bacterial species. Such intracellularly formed nanoparticles offer low value, as their use requires complex nanoparticle harvesting steps that has substantial implications for large-scale commercial applications.

An elegant fungi-mediated biosynthesis approach, that explored challenging *F. oxysporum* and *Verticillium* sp. with aqueous solution containing a mixture of ferricyanide $[Fe(CN)_6]^{3+}$ ions and ferrocyanide $[Fe(CN)_6]^{4+}$ ions in appropriate stoichiometric ratios, yielded phase-pure magnetite nanoparticles without any significant $\gamma\text{-}Fe_2O_3$ impurity (Bharde et al., 2006). This was the first time that a cationic precursor salt was used for the biosynthesis of a nano-oxide using fungi-mediated approach. Notably, the cyanide complexes were not found to be toxic to these fungi. This is consistent with previous studies that have reported the capability of fungal metabolic pathways to utilize and degrade cyanide species, thereby leading to bioremediation of cyanide toxicity from the environment.

The magnetite nanoparticles synthesized by *F. oxysporum* were quasi-spherical in shape and in the size range 20–50 nm, while *Verticillium* sp. synthesized 100–400 nm cubo-octahedral assemblies of magnetite nanoparticles that were composed of smaller 10–40 nm particles (Fig. 2.3; Bharde et al., 2006). This suggests a difference in nanoparticle assembly in the case of *Verticillium* in comparison to that synthesized by magnetic bacteria. Bimolecular studies of *Verticillium* revealed the over-expression of two extracellular proteins corresponding to *c*. 13 kDa and 55 kDa in the presence of iron-cyanide precursor complexes. Furthermore, employment of these proteins for *in vitro* nanoparticle synthesis showed the positive ability of the 55 kDa protein but the inability of the 13 kDa protein for the hydrolysis of iron-cyanide precursors (Bharde et al., 2006). This highlighted the role of the latter protein of capping the nanoparticles, which stabilizes the particles in aqueous solutions. These particles also revealed interesting magnetic properties, demonstrating the properties of a ferrimagnetic transition with a negligible amount of spontaneous magnetization at low temperature and the presence of a strong antiferromagnetic exchange interaction within the nanoparticles.

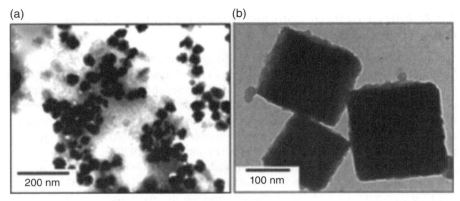

Figure 2.3. Transmission electron microscopy images of magnetite nanoparticle synthesized using (a) *Fusarium oxysporum* and (b) *Verticillium* sp. Source: Bharde, A.; Rautaray, D.; Bansal, V.; Ahmad, A.; Sarkar, I.; Yusuf, Seikh M.; Sanyal, M.; Sastry, M., Extracellular Biosynthesis of Magnetite using Fungi. *Small*, 2(1), 135–141. Copyright © 2006, Wiley-VCH Verlag GmbH & Co. KGaA.

Another important mixed-metal oxide, barium titanate (BT or $BaTiO_3$), is of significant importance for ferroelectric applications in non-volatile memories, thin-film capacitors, pyroelectric detectors, infrared imaging, and micro-electromechanical devices (Shaw et al., 2000). The nanoparticulate form of this pervoskite oxide has been reported for its application as a dielectric candidate for integration in the microelectronics industry (Shaw et al., 2000). However, with growing interest in miniaturizing electronic devices, it is a challenge to synthesize sub-10-nm BT particles of tetragonal symmetry at room temperature, as chemical synthesis routes almost always yield cubic and paraelectric phases at room temperature below a critical particle size. The fungus *F. oxysporum* showed the ability to synthesize 4–5-nm barium titanate under ambient conditions while stabilizing these particles in a tetragonal phase at room temperature, thus exhibiting a well-defined ferroelectric-paraelectric transition at room temperature (Bansal et al., 2006a). The ability of this fungus to synthesize a mixed-metal complex nano-oxide and not a mixture of barium oxide and titania nanoparticles is intriguing, as is the ability of this fungus to hydrolyze chemical precursors to form various binary oxides as described in the previous section.

The ability to synthesize the tetragonal phase of BT nanoparticles prompted the study of the ferroelectric behavior of such biogenic materials using a range of complementary techniques such as high-temperature X-ray diffraction (XRD), differential scanning calorimetry (DSC), dielectric measurements, and Kelvin probe microscopy. DSC measurements revealed an exothermic/endothermic (ferroelectric/paraelectric) transition at 114 °C, which is marginally lower than the phase-transition temperature of 128 °C in bulk BT (Ma et al., 1997). Further measurements using DSC revealed the behavior of this biogenic material as a 'ferroelectric relaxor', thereby enabling the use of BT for piezoelectric applications (Fig. 2.4). Due to the outstanding polarizability displayed by individual biogenic BT nanoparticles, they could also be successfully applied for electrical 'write–read' memory applications (Bansal et al., 2006a).

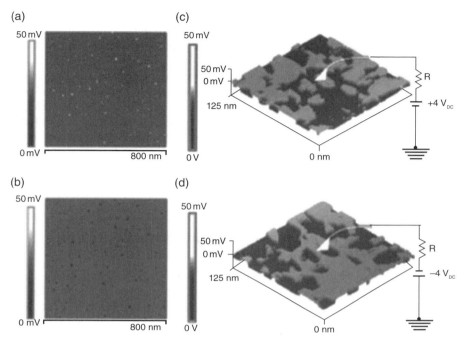

Figure 2.4. (a, b) Low- and (c, d) higher-magnification surface potential microscopy (SPM) images of BaTiO$_3$ nanoparticles synthesized using *Fusarium oxysporum*. SPM images from ferroelectric BaTiO$_3$ nanoparticles obtained in potential mode after application of +4V (a, c) and −4V (b, d) external DC bias voltages, where the reversal in image contrast on reversal of bias voltage is observed. Source: Bansal, V.; Poddar, P.; Ahmad, A.; Sastry, M., Room-Temperature Biosynthesis of Ferroelectric Barium Titanate Nanoparticles. *Journal of American Chemical Society*, 128(36), 11958–11963. Copyright © 2006, American Chemical Society. *See insert for color representation of the figure.*

Although theoretical studies suggested the destabilization of tetragonal phase while converting to high-temperature cubic phase below 30 nm size range, the biogenic BT nanoparticles could retain tetragonal phase even at sizes of <10 nm (Bansal et al., 2006a). This interesting observation was attributed to the charged biomolecules associated with biogenic BT particles that might avoid the stabilization of cubic phase by balancing for excess surface energy in ultrafine particles, resulting in the room-temperature tetragonal phase. These observations provide new opportunities of biogenic BT nanoparticles for ultra-high-density non-volatile ferroelectric memories and ultra-small capacitors, revolutionizing the electronics industry.

Another important example of a fungus-mediated approach is the ability of alkalotolerant and thermophilic fungus *Humicola* sp. for the synthesis and stabilization of a delafossite mineral of copper and aluminum oxide (copper aluminate; Ahmad et al., 2007). This material is particularly important due to its p-type conductivity at room temperature and its selective and reversible electric resistivity response to ozone gas at room temperature that can be used to develop transparent ozone sensors

(Kawazoe et al., 1997). The challenge of stabilizing the Cu^{1+} state in the reaction environment and reduction of Al^{3+} at lower temperatures during chemical synthesis is a well-known limitation to fabricate this material. The ability of *Humicola* sp. to synthesize 5–6-nm β-$CuAlO_2$ nanoparticles at 50 °C further affirms the unique valence-controlled stoichiometric nanosynthesis capability of the fungal biosynthesis processes (Ahmad et al., 2007). More importantly, the material formed was found to be free of any impurities corresponding to CuO, Cu_2O or Al_2O_3, which is similar to the fungus-mediated synthesis of barium titanate that was free from BaO, TiO_2 or $BaCO_3$ impurity. The ability of the fungus in retaining the stoichiometry and concurrently limiting nanoparticle growth was attributed to the distinct capping proteins (four proteins in the range *c.* 20–97 kDa). The important properties of the material for photoluminescence, microwave, and radio-frequency absorption were also demonstrated, thereby showing the broad applicability of these materials.

2.2.3. Biosynthesis of Nano-oxides using Natural Precursors Employing Bioleaching Approach

The previous section outlined the synthesis of nano-oxides using a biosynthesis approach, where fungi were challenged with appropriate chemical precursors that led to the formation of a diverse range of binary and complex metal oxide nanoparticles. However, to make the synthesis of nanoparticles entirely eco-friendly, significant efforts were made to utilize naturally available materials (e.g., white sand and zircon sand) and agro-industrial by-products (e.g., rice husk) as an alternative to chemical precursors during the fungus-mediated biosynthesis approach. The biosynthesis approach involving natural precursors is similar to a bioleaching process in which different microorganisms leach out different metal components from their natural ores, either in their natural habitats or under industrial settings. The employment of fungi for bioleaching has gained particular importance and is a prominent method for metal recovery. For example, iron-oxidizer *Gallionella* sp. was found to be present in copper mines and *Bacillus coagulans* was reported from the waste dunes of bauxite mines in India (Natarajan, 1999). Fungi such as *Aspergillus niger* and *Penicillium* sp. have also been found to be among the most effective organisms for metal recovery (Mulligan and Kamali, 2003; Bharde et al., 2006).

In the context of nanomaterial biosynthesis, the use of bioleaching typically employs a 'top-down' approach of nanomaterial synthesis in contrast to the 'bottom-up' approach that applies to the biosynthesis approaches described in the previous section. Typically, a bottom-up approach involves hydrolysis of oxide precursor ions which nucleate together to form oxide nanoparticles, whereas a top-down approach involves the conversion of various oxides or oxide-composites present in natural forms to their nanoparticulate form. The bioleaching approach employed for nano-oxide synthesis using different precursors is described in the following sections.

Silica bioleaching from sand. The ability of fungus *F. oxysporum* to hydrolyze hexaflorosilicate ions to form silica nanoparticles extracellularly was discussed in Section 2.2.1 (Bansal et al., 2005*a*). The same fungus also showed the ability to leach silica from white sand to form monodispersed 2–5 nm highly porous, quasi-spherical nano-crystallites of silica (Bansal et al., 2005b). The leaching process was significantly rapid, with crystalline silica particles being formed within 24 hours of the reaction. Interestingly, the calcination of these porous particles led to hollow silica nanocrystals, indicating that the bioleached silica was in fact a protein-silica organo-nanocomposite material (Bansal et al., 2005b). The mechanistic aspect of silica nanoparticle bioleaching revealed four distinct processes: (1) the dissolving of the silicates present in sand by the fungus in the form of silicic acid; (2) silicic acid uptake by the fungus; (3) hydrolysis of silicic acid intracellularly (within the fungal biomass), leading to the formation of crystalline silica nanoparticles; and (4) leaching out of silica nanoparticles from within the fungus to the extracellular solution. The monodispersity observed in the leached biosilica nanoparticles was attributed to the stabilization of the particles by fungal proteins that prevented their aggregation both within the fungal biomass and in the solution. This first report of a completely green synthesis route for nanomaterial biosynthesis, where even the precursor material used was a naturally available material, has fueled the use of fungi for bioleaching applications.

The high selectivity and specificity associated with molecules and processes in biological organisms is a well-known phenomenon. In a follow-up study, the same fungus (*F. oxysporum*) was employed for the selective and specific leaching of a metal oxide from a complex mixed-metal oxide sand precursor (Bansal et al., 2007). With the already known ability of *F. oxysporum* to synthesize simple and complex metal oxide nanoparticles, as well as bioleach silica nanoparticles from white sand, the selective leaching capability of this organism was further demonstrated by exposing this fungus to naturally available zircon sand ($ZrSi_xO_y$) which contains both zirconium and silicon components. On exposure to zircon sand, *F. oxysporum* exhibited selective leaching of silica nanoparticles from the zircon sand without allowing the leaching of zircon component (Bansal et al., 2007). These bioleached silica nanoparticles were quasi-spherical in shape, 2–10 nm in diameter and of cristobalite polymorphic form. The role of cationic proteins in the hydrolysis of precursors to form silica and zirconia nanoparticles was described in Section 2.2.1. However, in the case of zircon sand where a mixed system of SiO_2–ZrO_2 coexists, it was believed that the electrostatic interactions between the cationic proteins and the anionic substrates (SiO_2: pI ~ 2; ZrO_2: pI ~ 4) played an important role in providing the specificity of leaching silica from a complex system. This enabled the fungal enzymes to specifically target the low-dielectric silica due to higher affinity and simultaneously enrich the high-dielectric zirconia content in zircon sand. Hence, bioleaching not only provided a simple approach to synthesize important nanomaterials (silica) but also helped to enhance the quality of raw material for other processes (i.e., high-quality zircon with a smaller silica component).

Silica bioleaching from glass. Bioleaching of silica from different sands by *F. oxysporum* is a simple and elegant approach to selectively leach technologically important nanomaterials. Additionally, the synthesis of organic acids by this fungus allows the slow and controlled leaching of essential elements and dynamically caps them to form nanoparticles, all of which is performed at room temperature (Kulkarni et al., 2008). These studies provided a gateway to employ another fungal species (*Humicola* sp.) for leaching silicates from borosilicate glass as a precursor material. The incubation of *Humicola* sp. with borosilicate glass enabled the leaching of sub-10 nm monodispersed silicate nanoparticles (Kulkarni et al., 2008). However, detailed analysis revealed the presence of B–O and B–O–Si bonds in the nanomaterial, suggesting that the particles leached were in fact borosilicate type and not pristine silica. In addition, the processed glass surface also exhibited morphological modifications, further confirming the leaching of silicate materials from the surface.

Silica bioleaching and biotransformation from rice husk. The previous section outlined the process of bioleaching by fungus *F. oxysporum* from silicate precursors. Additionally, the same fungus was not only able to leach hydrated amorphous silica from plants that are accumulated predominantly in the form of phytoliths, but also transformed them into crystalline materials (Bansal et al., 2006b). Plants, including dicots, monocots, conifers, and sphenophytes, have been reported to contain high amounts of hydrated silica, with some plant families containing up to 50–70% of siliceous materials (Lanning et al., 1958). On exposure to rice husk, the fungus *F. oxysporum* was not only able to bioleach the amorphous silica present into the surrounding aqueous environment, it was also found to be capable of biotransforming amorphous plant silica into high-value crystalline quartz (silica) nanoparticles of 2–6 nm diameter at room temperature. The choice of rice husk stemmed from its widespread availability, as it is a waste material disposed of by rice mills as an agro-industrial by-product. The calcination of the silica nanoparticles leached from rice husk resulted in interesting nanoscale morphologies such as orthorhombic quartz plates and cube-like structures (Fig. 2.5; Bansal et al., 2006b). Rigorous experimentation confirmed that the bioleaching and biotransformation of silica occurred simultaneously and the process was fairly rapid, yielding about 90, 95, and 97% of the total silica available in the rice husk within 8, 12, and 24 hours, respectively. However, the nanoparticles synthesized during this process indicated a heterogeneous distribution of micrometer-sized particles of inconsistent morphologies.

As observed previously, fungal proteins were found to be strongly anchored to these silica nanoparticles bioleached from rice husk. The involvement of two low-molecular-weight cationic proteins (*c.* 15–20 kDa) and one higher-molecular-weight cationic protein (*c.* 200 kDa) were found to be critical for *in vitro* bioleaching of silica from rice husk. However, the use of these proteins in their purified form in the absence of fungal biomass only yielded amorphous silica particles, suggesting the important role of the fungal metabolic pathways in imparting crystallinity to the bioleached amorphous silica nanoparticles.

2.2.4. Biosynthesis of Nano-oxides Employing Bio-milling Approach

Bio-milling is a top-down process for synthesizing nanomaterials, where a bulk material is reduced to its nano-dimensions. In several ways, this approach is similar to that observed in the case of bioleaching, but is classified separately for clarity and

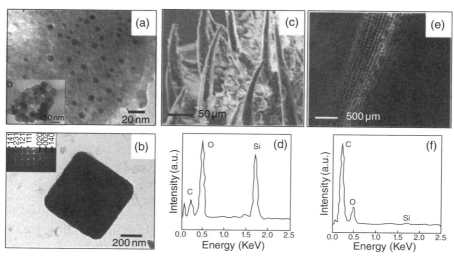

Figure 2.5. Transmission electron micrographs of (a) SiO$_2$ nanoparticles bioleached from rice husk after their biotransformation into quartz nanoparticles by *Fusarium oxysporum* and (b) single crystal quartz plates obtained after calcination of bioleached silica at 400 °C for 2 hours. Scanning electron micrographs and corresponding EDX patterns from rice husk (c, d) before and (e, f) after exposure to *Fusarium oxysporum*. Inset in (a) shows amorphous silica particles formed via rice husk bioleaching using cationic fungal proteins. Inset in (b) shows selected area electron diffraction pattern from the quartz plate shown in the main figure. Source: Bansal, V.; Ahmad, A.; Sastry, M., Fungus-mediated biotransformation of amorphous silica in rice husk to nanocrystalline silica. *Journal of American Chemical Society*, 128(43), 14059–14066. Copyright © 2006, American Chemical Society.

due to its similarity to a physical ball-milling approach, where a bulk material is crushed to its nano-form by hard balls. The first report on the bio-milling approach proposed the use of an alkalotolerant and thermophilic fungus *Humicola* sp., where the fungus was believed to chemically break down 250–400 nm bismuth manganese oxide (BiMnO$_3$) particles into 35–65 nm nanoparticles while maintaining their crystalline structure and the phase purity (Mazumder et al., 2007). However, at a later stage, a corrigendum for the work submitted indicated that bio-milled BiMnO$_3$ nanoparticles did not retain their original phase and crystallinity during the bio-milling process; instead, the fungus *Humicola* sp. converted the BiMnO$_3$ into BiOCl nanoparticles (Mazumder et al., 2008).

More recently, the same fungus was utilized to transform anatase TiO$_2$ particles to the metastable brookite phase TiO$_2$ nanoparticles using a bio-milling approach (Khan and Ahmad, 2013b). In the context of TiO$_2$, the property of the material largely depends on its crystalline phase including anatase, brookite, and rutile (Armstrong et al., 2004; Deng et al., 2008; Pearson et al., 2011) and therefore conversion of larger anatase particles into smaller brookite nanoparticles seems to provide a facile strategy to control nano-oxide crystallinity. Note also that when a hexafluorotitanate precursor was initially used, a different fungus *F. oxysporum* also led to the synthesis of brookite titania nanoparticles, which is considered to be the most difficult phase of titania to be synthesize by chemical routes (Bansal et al., 2005a). The ability of two different fungi to perform the same non-standard task clearly shows that the fungi-based biological synthesis routes are not merely an addition to the existing long list of chemical synthesis routes. Instead, biology seems to offer certain niche opportunities when it comes to nanomaterial synthesis.

2.3. OUTLOOK

The ability of synthesizing technologically important nano-oxides using a unique and facile fungus-mediated route has been discussed; fungi are observed not only to synthesize simple nano-oxides, but also chemically difficult-to-synthesize complex mixed-metal oxides. In addition to synthesizing nanomaterials from chemical precursors, these fungi also show an ability to leach nanomaterials from natural materials and biological waste by-products, which makes the approach completely eco-friendly. The role of plant pathogenic fungi *F. oxysporum* in the synthesis of nano-oxides is particularly interesting, as it has shown the ability to synthesize both simple and complex oxides while demonstrating selective bioleaching capabilities. Moreover, the biomolecular studies also reveal the involvement of low-molecular-weight cationic proteins and peptides in nano-oxide synthesis that seem to play a critical role in the synthesis of oxide-based nanomaterials. Further, studies on biotransformation of amorphous silica in rice husk to silica nano-crystallites confirm the active role played by the fungal metabolism in nano-oxide synthesis.

Although several studies have outlined the ability of biological entities for the synthesis of nanomaterials, the applicability of such materials mostly remains elusive. Only some of the nano-oxides synthesized using fungi, such as barium titanate and copper aluminate, have been utilized for applications such as those in magnetic memory devices and photoluminescence, respectively. Also interesting is the observation that fungus-mediated biosynthesis processes yields metastable oxide phases such as brookite titania and tetragonal barium titanate that are typically not stable at room temperature, especially in the nano-size regime. This further outlines the importance of such biological routes in synthesizing oxide-based nanomaterials.

Although the field of nanoparticle biosynthesis has been extensively studied, there are a number of questions that require serious attention before this approach can compete with the existing physical and chemical approaches. One major question that requires considerable focus is to determine the exact mechanistic aspect of nanoparticle biosynthesis by these organisms. Likewise, surface chemistry of biosynthesized nanoparticles still requires substantial understanding. An understanding of the important questions of the mechanism and surface chemistry may enable this approach to be used at the commercial scale for the large-scale synthesis of inorganic nanomaterials, which will have implications for the future of industrial processes.

REFERENCES

Ahmad, A., Senapati, S., Khan, M.I., Kumar, R., Sastry, M. 2003a. Extracellular biosynthesis of monodisperse gold nanoparticles by a novel extremophilic actinomycete, *Thermomonospora sp*. *Langmuir* 19(8): 3550–3553.

Ahmad, A., Mukherjee, P., Senapati, S., Mandal, D., Khan, M.I., Kumar, R., Sastry, M. 2003b. Extracellular biosynthesis of silver nanoparticles using the fungus *Fusarium oxysporum*. *Colloids and Surfaces B* 28(4): 313–318.

Ahmad, A., Jagadale, T., Dhas, V., Khan, S., Patil, S., Pasricha, R., Ravi, V., Ogale, S. 2007. Fungus-based synthesis of chemically difficult-to-synthesize multifunctional nanoparticles of $CuAlO_2$. *Advanced Materials* 19(20): 3295–3299.

Armstrong, A.R., Armstrong, G., Canales, J., Bruce, P.G. 2004. TiO_2-B Nanowires. *Angewandte Chemie International Edition* 43(17): 2286–2288.

Bansal, V., Rautaray, D., Ahmad, A., Sastry, M. 2004. Biosynthesis of zirconia nanoparticles using the fungus Fusarium oxysporum. *Journal of Materials Chemistry* 14(22): 3303–3305.

Bansal, V., Rautaray, D., Bharde, A., Ahire, K., Sanyal, A., Ahmad, A., Sastry, M. 2005a. Fungus-mediated biosynthesis of silica and titania particles. *Journal of Materials Chemistry* 15(26): 2583–2589.

Bansal, V., Sanyal, A., Rautaray, D., Ahmad, A., Sastry, M. 2005b. Bioleaching of sand by the fungus Fusarium oxysporum as a means of producing extracellular silica nanoparticles. *Advanced Materials* 17(7): 889–892.

Bansal, V., Poddar, P., Ahmad, A., Sastry, M. 2006a. Room-temperature biosynthesis of ferroelectric barium titanate nanoparticles. *Journal of American Chemical Society* 128(36): 11958–11963.

Bansal, V., Ahmad, A., Sastry, M. 2006b. Fungus-mediated biotransformation of amorphous silica in rice husk to nanocrystalline silica. *Journal of American Chemical Society* 128(43): 14059–14066.

Bansal, V., Syed, A., Bhargava, S.K., Ahmad, A., Sastry, M. 2007. Zirconia enrichment in zircon sand by selective fungus-mediated bioleaching of silica. *Langmuir* 23(9): 4993–4998.

Bansal, V., O'Mullane, A.P., Bhargava, S.K. 2009. Galvanic replacement mediated synthesis of hollow Pt nanocatalysts: Significance of residual Ag for the H_2 evolution reaction. *Electrochemical Communications* 11(8): 1639–1642.

Bansal, V., Ramanathan, R., Bhargava, S.K. 2011. Fungus-mediated biological approaches towards 'green' synthesis of oxide nanomaterials. *Australian Journal of Chemistry* 64(3): 279–293.

Bansal, V., Bharde, A., Ramanathan, R., Bhargava, S.K. 2012. Inorganic materials using 'unusual' microorganisms. *Advances in Colloid and Interface Science* 179–182: 150–168.

Beveridge, T.J., Murray, R.G. 1980. Sites of metal deposition in the cell wall of *Bacillus subtilis*. *Journal of Bacteriology* 141(2): 876–887.

Bradbury, J. 2004. Nature's nanotechnologists: unveiling the secrets of diatoms. *PLoS Biology* 2(10): e306.

Brutchey, R.L., Morse, D.E. 2008. Silicatein and the translation of its molecular mechanism of biosilicification into low temperature nanomaterial synthesis. *Chemical Reviews* 108(11): 4915–4934.

Burda, C., Chen, X., Narayanan, R., El-Sayed, M.A. 2005. Chemistry and properties of nanocrystals of different shapes. *Chemical Reviews* 105(4): 1025–1102.

Choi, H.C., Kundaria, S., Wang, D., Javey, A., Wang, Q., Rolandi, M., Dai, H. 2002. Efficient formation of iron nanoparticle catalysts on silicon oxide by hydroxylamine for carbon nanotube synthesis and electronics. *Nano Letters* 3(2): 157–161.

Cunningham, D.P., Lundie, L.L. Jr. 1993. Precipitation of cadmium by *Clostridium thermoaceticum*. *Applied and Environmental Microbiology* 59(1): 7–14.

Dameron, C.T., Reese, R.N., Mehra, R.K., Kortan, A.R., Carroll, P.J., Steigerwald, M.L., Brus, L.E., Winge, D.R., Biosynthesis of cadmium sulphide quantum semiconductor crystallites. *Nature* 338(6216): 596–597.

Das, M., Patil, S., Bhargava, N., Kang, J.-F., Riedel, L.M., Seal, S., Hickman, J.J. 2007. Autocatalytic ceria nanoparticles offer neuroprotection to adult rat spinal cord neurons. *Biomaterials* 28(10): 1918–1925.

Deng, Q., Wei, M., Ding, X., Jiang, L., Ye, B., Wei, K. 2008. Brookite-type TiO_2 nanotubes. *Chemical Communications* 31: 3657–3659.

Drum, R., Gordon, R. 2003. Star Trek replicators and diatom nanotechnology. *Trends in Biotechnology* 21(8): 325–328.

Fan, H.T., Teng, X.M., Pan, S.S., Ye, C., Li, G.H., Zhang, L.D. 2005. Optical properties of delta-Bi_2O_3 thin films grown by reactive sputtering. *Applied Physics Letters* 87(23): 231916–231923.

Frankel, R., Blakemore, R., Wolfe, R. 1979. Magnetite in freshwater magnetotactic bacteria. *Science* 203(4387): 1355–1357.

Garvie, R.C., Hannink, R.H., Pascoe, R.T. 1975. Ceramic steel? *Nature* 258(5537): 703–704.

Heuer, A.H. 1987. Transformation toughening in ZrO_2 containing ceramics. *Journal of the American Ceramic Society* 70(10): 689–698.

Hildebrand, M. 2008. Diatoms, biomineralization processes, and genomics. *Chemical Reviews* 108(11): 4855–4874.

Huang, H.H., Yan, F.Q., Kek, Y.M., Chew, C.H., Xu, G.Q., Ji, W., Oh, P.S., Tang, S.H. 1997. Synthesis, characterization, and nonlinear optical properties of copper nanoparticles. *Langmuir* 13(2): 172–175.

Iler, R.K. 1979. *The Chemistry of Silica*. John Wiley & Sons, New York.

Kang, B., Mackey, M.A., El-Sayed, M.A. 2010. Nuclear targeting of gold nanoparticles in cancer cells induces DNA damage, causing cytokinesis arrest and apoptosis. *Journal of American Chemical Society* 132(5): 1517–1519.

Kawazoe, H., Yasukawa, M., Hyodo, H., Kurita, M., Yanagi, H., Hosono, H. 1997. P-type electrical conduction in transparent thin films of $CuAlO_2$. *Nature* 389(6654): 939–942.

Khan, S.A., Ahmad, A. 2013a. Fungus mediated synthesis of biomedically important cerium oxide nanoparticles. *Materials Research Bulletin* 48(10): 4134–4138.

Khan, S.A., Ahmad, A. 2013b. Phase, size and shape transformation by fungal biotransformation of bulk TiO_2. *Chemical Engineering Journal* 230: 367–371.

Kiel, H., Schwartz, W. 1980. Leaching of a silicate and carbonate copper ore with heterotrophic fungi and bacteria, producing organic acids. *Zeitschrift für Allgemeine Mikrobiologie* 20(10): 627–636.

Klages, D., Meyer, I., Schwartz, W., Naïveke, R. 1981. Leaching of ores with heterotrophic microorganisms. Development of a screening method. *Zeitschrift für Allgemeine Mikrobiologie* 21(10): 729–737.

REFERENCES

Klaus-Joerger, T., Joerger, R., Olsson, E., Granqvist, C.-G. 1999. Silver-based crystalline nanoparticles, microbially fabricated. *Proceedings of the National Academy of Sciences, USA* 96(24): 13611–13614.

Kulkarni, S., Syed, A., Singh, S., Gaikwad, A., Patil, K., Vijayamohanan, K., Ahmad, A., Ogale, S. 2008. Silicate nanoparticles by bioleaching of glass and modification of the glass surface. *Journal of Non-Crystalline Solids* 354(29): 3433–3437.

Lanning, F.C., Ponnaiya, B.W.X., Crumpton, C.F. 1958. The chemical nature of silica in plants. *Plant Physiology* 33(5): 339–343.

Lopez, P., Gautier, C., Livage, J., Coradin, T. 2005. Mimicking biogenic silica nanostructures formation. *Current Nanoscience* 1(1): 73–83.

Ma, Y., Vileno, E., Suib, S.L., Dutta, P.K. 1997. Synthesis of tetragonal $BaTiO_3$ by microwave heating and conventional heating. *Chemistry of Materials* 9(12): 3023–3031.

Mann, S. 1997. Biomineralization: the form (id) able part of bioinorganic chemistry! *Journal of the Chemical Society, Dalton Transactions* 21: 3953–3962.

Mann, S. 2001. *Biomineralization. Principles and Concepts in Bioinorganic Materials Chemistry*. Oxford University Press, Oxford.

Mann, S., Webb, J., Williams, R.J.P. 1989. *Biomineralization. Chemical and Biochemical Perspectives*. Wiley-VCH, Weinheim, Germany.

Mazumder, B., Uddin, I., Khan, S., Ravi, V., Selvraj, K., Poddar, P., Ahmad, A. 2007. Bio-milling technique for the size reduction of chemically synthesized $BiMnO_3$ nanoplates. *Journal of Materials Chemistry* 17(37): 3910–3914.

Mazumder, B., Uddin, I., Khan, S., Ravi, V., Selvraj, K., Poddar, P., Ahmad, A. 2008. Bio-milling technique for the size reduction of chemically synthesized $BiMnO_3$ nanoplates. *Journal of Materials Chemistry* 18(48): 5998–6001.

Mehra, R.K., Winge, D.R. 1991. Metal ion resistance in fungi: Molecular mechanisms and their regulated expression. *Journal of Cellular Biochemistry* 45(1): 30–40.

Miyata, N., Tani, Y., Maruo, K., Tsuno, H., Sakata, M., Iwahori, K. 2006. Manganese (IV) oxide production by *Acremonium sp.* strain KR21-2 and extracellular Mn (II) oxidase activity. *Applied and Environmental Microbiology* 72(10): 6467–6473.

Mulligan, C.N., Kamali, M. 2003. Bioleaching of copper and other metals from low-grade oxidized mining ores by *Aspergillus niger*. *Journal of Chemical Technology and Biotechnology* 78(5): 497–503.

Narayanan, K.B., Sakthivel, N. 2010. Biological synthesis of metal nanoparticles by microbes. *Advances in Colloid and Interface Science* 156(1–2): 1–13.

Nassif, N., Roux, C., Coradin, T., Bouvet, O.M.M., Livage, J. 2004. Bacteria quorum sensing in silica matrices. *Journal of Materials Chemistry* 14(14): 2264–2268.

Natarajan, K.A. 1999. *Biogeochemistry of rivers in tropical South and Southeast Asia* (Ittekkot, V., Subramanian, V., Annadurai, S., eds), Heft 82, Geologisch-Paläontologisches Institute und Institute für Biogeochemie und Meereschemie, University of Hamburg, Hamburg, Germany.

Noh, H.-J., Seo, D.-S., Kim, H., Lee, J.-K. 2003. Synthesis and crystallization of anisotropic shaped ZrO_2 nanocrystalline powders by hydrothermal process. *Materials Letters* 57(16–17): 2425–2431.

Patwardhan, S., Mukherjee, N., Steinitz-Kannan, M., Clarson, S. 2003. Bioinspired synthesis of new silica structures. *Chemical Communications* 2003(10): 1122–1123.

Pearson, A., Jani, H., Kalantar-zadeh, K., Bhargava, S.K., Bansal, V. 2011. Gold nanoparticle-decorated keggin ions/TiO$_2$ photococatalyst for improved solar light photocatalysis. *Langmuir* 27(11): 6661–6667.

Perry, C., Keeling-Tucker, T. 2000. Biosilicification: the role of the organic matrix in structure control. *Journal of Biological Inorganic Chemistry* 5(5): 537–550.

Philipse, A.P., Maas, D. 2002. Magnetic colloids from magnetotactic bacteria: chain formation and colloidal stability. *Langmuir* 18(25): 9977–9984.

Plowman, B., Ippolito, S.J., Bansal, V., Sabri, Y.M., O'Mullane, A.P., Bhargava, S.K. 2009. Gold nanospikes formed through a simple electrochemical route with high electrocatalytic and surface enhanced Raman scattering activity. *Chemical Communications* 33: 5039–5041.

Rao, C.N.R., Rao, G.V.S., Ramdas, S. 1969. Phase transformations and electrical properties of bismuth sesquioxide. *Journal of Physical Chemistry* 73(3): 672–675.

Ramanathan, R., O'Mullane, A.P., Parikh, R.Y., Smooker, P.M., Bhargava, S.K., Bansal, V. 2011*a*. Bacterial kinetics-controlled shape-directed biosynthesis of silver nanoplates using *Morganella psychrotolerans*. *Langmuir* 27(2): 714–719.

Ramanathan, R., Campbell, J.L., Soni, S.K., Bhargava, S.K., Bansal, V. 2011*b*. Cationic amino acids specific biomimetic silicification in ionic liquid: a quest to understand the formation of 3-D structures in diatoms. *PloS one* 6(3): e17707.

Ramanathan, R., Kandjani, A.E., Walia, S., Balendhran, S., Bhargava, S.K., Kalantar-Zadeh, K., Bansal, V. 2013. 3-D nanorod arrays of metal-organic KTCNQ semiconductor on textile for flexible organic electronics. *RSC Advances* 3: 17654–17658.

Reese, R. N., Mehra, R.K., Tarbet, E.B., Winge, D.R. 1988. Studies on the gamma-glutamyl Cu-binding peptide from *Schizosaccharomyces pombe*. *Journal of Biological Chemistry* 263(9): 4186–4192.

Roh, Y., Lauf, R.J., McMillan, A.D., Zhang, C., Rawn, C.J., Bai, J., Phelps, T.J. 2001. Microbial synthesis and the characterization of metal-substituted magnetites. *Solid State Communications* 118(10): 529–534.

Rouch, D.A., Lee, B.T.O., Morby, A.P. 1995. Understanding cellular responses to toxic agents: a model for mechanism-choice in bacterial metal resistance. *Journal of Industrial Microbiology and Biotechnology* 14(2): 132–141.

Sastry, M. 2006. Extracellular biosynthesis of magnetite using fungi. *Small* 2(1): 135–141.

Sastry, M., Ahmad, A., Khan, M.I., Kumar, R. 2005. *Microbial Nanoparticle Production*. Wiley-VCH Verlag GmbH & Co. KGaA, pp. 126–135.

Sawant, P.D., Sabri, Y.M., Ippolito, S.J., Bansal, V., Bhargava, S.K. 2009. In-depth nano-scale analysis of complex interactions of Hg with gold nanostructures using AFM-based power spectrum density method. *Physical Chemistry Chemical Physics* 11(14): 2374–2378.

Schüler, D. 2002. The biomineralization of magnetosomes in *Magnetospirillum gryphiswaldense*. *International Microbiology* 5(4): 209–214.

Schultze-Lam, S., Harauz, G., Beveridge, T. 1992. Participation of a cyanobacterial S layer in fine-grain mineral formation. *Journal of Bacteriology* 174(24): 7971–7981.

Shankar, S.S., Rai, A., Ankamwar, B., Singh, A., Ahmad, A., Sastry, M. 2004. Biological synthesis of triangular gold nanoprisms. *Nature Materials* 3(7): 482–488.

Shaw, T.M., Trolier-McKinstry, S., McIntyre, P.C. 2000. The properties of ferroelectric films at small dimensions. *Annual Review of Materials Science* 30(1): 263–298.

Shephard, D.S., Maschmeyer, T., Johnson, B.F.G., Thomas, J.M., Sankar, G., Ozkaya, D., Zhou, W., Oldroyd, R.D., Bell, R.G. 1997. Bimetallic nanoparticle catalysts anchored inside mesoporous silica. *Angewandte Chemie International Edition* 36(20): 2242–2245.

REFERENCES

Silver, S. 1996. Bacterial resistances to toxic metal ions – a review. *Gene* 179(1): 9–19.

Skorodumova, N.V., Jonsson, A.K., Herranen, M., Stromme, M., Niklasson, G.A., Johansson, B., Simak, S.I. 2005. Random conductivity of delta-Bi_2O_3 films. *Applied Physics Letters* 86(24): 241910–241913.

Smith, P.R., Holmes, J.D., Richardson, D.J., Russell, D.A., Sodeau, J.R. 1998. Photophysical and photochemical characterisation of bacterial semiconductor cadmium sulfide particles. *Journal of the Chemical Society, Faraday Transactions* 94(9): 1235–1241.

Southam, G., Beveridge, T.J. 1996. The occurrence of sulfur and phosphorus within bacterially derived crystalline and pseudocrystalline octahedral gold formed in vitro. *Geochimica et Cosmochimica Acta* 60(22): 4369–4376.

Southon, P.D., Bartlett, J.R., Woolfrey, J.L., Ben-Nissan, B. 2002. Formation and characterization of an aqueous zirconium hydroxide colloid. *Chemistry of Materials* 14(10): 4313–4319.

Stolz, J.F., Chang, S.-B.R., Kirschvink, J.L. 1986. Magnetotactic bacteria and single-domain magnetite in hemipelagic sediments. *Nature* 321(6073): 849–851.

Swihart, M.T. 2003. Vapor-phase synthesis of nanoparticles. *Current Opinion in Colloid and Interface Science* 8(1): 127–133.

Sun, S., Murray, C.B., Weller, D., Folks, L., Moser, A. 2000. Monodisperse FePt nanoparticles and ferromagnetic FePt nanocrystal superlattices. *Science* 287: 1989–1992.

Temple, K.L., Le Roux, N.W. 1964. Syngenesis of sulfide ores, desorption of adsorbed metal ions and their precipitation as sulfides. *Economic Geology* 59(4): 647–655.Sakamoto, M., Fujistuka, M., Majima, T. 2009. Light as a construction tool of metal nanoparticles: Synthesis and mechanism. *Journal of Photochemistry and Photobiology C* 10(1): 33–56.

Thakkar, K.N., Mhatre, S.S., Parikh, R.Y. 2010. Biological synthesis of metallic nanoparticles. *Nanomedicine: Nanotechnology, Biology and Medicine* 6(2): 257–262.

Thompson, I.A., Huber, D.M., Guest, C.A., Schulze, D.G. 2005. Fungal manganese oxidation in a reduced soil. *Environmental Microbiology* 7(9): 1480–1487.

Tian, L., Yam, L., Wang, J., Tat, H., Uhrich, K.E. 2004. Core crosslinkable polymeric micelles from PEG-lipid amphiphiles as drug carriers. *Journal of Materials Chemistry* 14(14): 2317–2324.

Uddin, I., Adyanthaya, S., Syed, A., Selvaraj, K., Ahmad, A., Poddar, P. 2008. Structure and microbial synthesis of sub-10nm Bi2O3 nanocrystals. *Journal of Nanoscience and Nanotechnology* 8(8): 3909–3913.

Valix, M., Tang, J.Y., Malik, R. 2001. Heavy metal tolerance of fungi. *Minerals Engineering* 14(5): 499–505.

Van De Meene, A.M.L., Pickett-Heaps, J.D. 2002. Valve morphogenesis in the centric diatom *Proboscia Alata* Sundstrom. *Journal of Phycology* 38(2): 351–363.

Watson, J.H.P., Cressey, B.A., Roberts, A.P., Ellwood, D.C., Charnock, J.M., Soper, A.K. 2000. Structural and magnetic studies on heavy-metal-adsorbing iron sulphide nanoparticles produced by sulphate-reducing bacteria. *Journal of Magnetism and Magnetic Materials* 214(1–2): 13–30.

Xu, H., Qin, D.-H., Yang, Z., Li, H.-L. 2003. Fabrication and characterization of highly ordered zirconia nanowire arrays by sol-gel template method. *Materials Chemistry and Physics* 80(2): 524–528.

3

MICROBIAL MOLECULAR MECHANISMS IN BIOSYNTHESIS OF NANOPARTICLES

Atmakuru Ramesh, Marimuthu Thiripura Sundari, and Perumal Elumalai Thirugnanam

International Institute of Biotechnology and Toxicology (IIBAT), Padappai, Tamil Nadu, India

3.1. INTRODUCTION

Microscopic analysis of elements was performed several decades ago, defining accurate dimensions and size. The size of an element refers to the internuclear distance between the atoms in the element or the bond length. The C–C length in graphite is 142.6 pm (picometers; 1 pm $=1\times10^{-12}$ m), the bond length in Ni–Ni is 249.2 pm. The size of neutral atoms depends upon the way in which the measurement is made, the geometry and the environment. Theoretically, this indicates that the smallest plausible size of a Ni particle is 249.2 pm or 0.249 nm. Recent trends and innovations in material science have led to the synthesis of several transition metals of nanoparticle size. Synthesis techniques are mainly grouped within three classifications: chemical, green, and biological.

The widely used chemical synthesis procedures are designed to take place during the liquid phase. These can be performed using a range of precursors and synthesis conditions with varying temperature, time, and concentration of reactants. Variation of

---

Bio-Nanoparticles: Biosynthesis and Sustainable Biotechnological Implications, First Edition.
Edited by Om V. Singh.
© 2015 John Wiley & Sons, Inc. Published 2015 by John Wiley & Sons, Inc.

these parameters leads to different size selection and geometries of the resulting particles.

The liquid-phase chemical synthesis reaction mechanism involves: (1) dissolving the transition metal salts in aqueous solutions; (2) dissociation into cations and anions; and (3) formation of neutral particles (or nanotization).

The first experimental evidence that supports the theory that a soccer ball-shaped nanoparticle commonly called a buckyball is the result of a breakdown of larger structures rather than being built atom-by-atom from the ground up.

The aqueous precipitation of a chemical process involves the surface controlling agents shortly after the formation of precipitates which interfere with the nucleating and growing particle, thereby hindering the agglomeration facilitating the formation of fine and controlled nanoparticles. Dispersions of nanoparticles intrinsically in aqueous phase are thermodynamically metastable due to their very high surface area. This represents a positive contribution to the free enthalpy of the system. If the activation energies are not sufficiently high, evolution of the nanoparticle dispersion occurs causing an increase in nanoparticle size.

There are several chemical methods involving the reduction of metal ions to nanoparticles. Overall, the mechanism of formation involves dissociation, nucleation and Oswald ripening. The majority of these chemical reduction processes involves the use of large quantities of hazardous chemicals which are a nuisance to human health and the environment. New techniques which can overcome this problem are under investigation. The concept of fabricating nanoparticles by a simple chemical-free and eco-friendly process is an interesting and important topic for exploration.

3.2. CHEMICAL SYNTHESIS OF METAL NANOPARTICLES

Nanoparticles have been fabricated by chemical synthesis under specific conditions. During the precipitation process from liquid phases, surface controlling agents have been added during or shortly after the formation of precipitates. This process is based on the hypothesis that molecules able to interact with the particle surface are influenced by nucleation and growth through the interfacial free energy to avoid agglomeration and to control size. The facile one-pot synthesis of fluorescent SiNPs (Zhong et al., 2013) followed the bottom-up method via *in situ* growth under microwave irradiation using hydrophilic molecules such as 3-aminopropyl trimethoxysilane $C_6H_{17}NO_3Si$ as the silicon source in the presence of trisodium citrate dehydrate $C_6H_5Na_3O_7$. The prepared SiNPs of *c.* 2.2 nm are water dispersible, a simple method which can be executed within 10 minutes under the defined microwave irradiation conditions. The mechanism for the preparation of SiNPs is depicted in Figure 3.1. The precursor $C_6H_{17}NO_3Si$ molecules readily undergo reduction by trisodium citrate ($C_6H_5Na_3O_7$ as reduction reagent) under microwave irradiation through an oxidation–reduction reaction, forming crystal nuclei in the first step (step 1: A → B). Subsequently, the Oswald ripening stage is initiated (steps 2 and 3), resulting in the growth of small-size nanocrystals (step 2: B → C). The growth continues, eventually producing more stable and larger-size silicon nanocrystals in the third step (step 3: C → D).

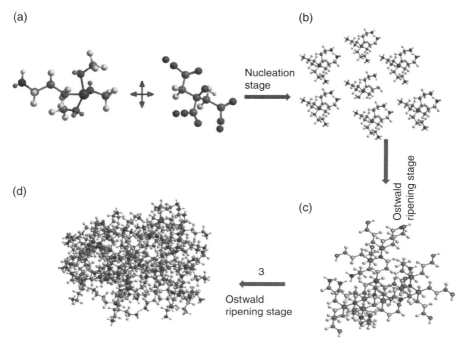

Figure 3.1. Bottom-up -one-pot synthesis of SiNPs. Source: Redrawn from Yiling Zhong, Fei Peng, Feng Bao, Siyi Wang, Xiaoyuan Ji, Liu Yang, Yuanyuan Su, Shuit-Tong Lee, Yao He. Large-Scale Aqueous Synthesis of Fluorescent and Biocompatible Silicon Nanoparticles and Their Use as Highly Photostable Biological Probes. *Journal of the American Chemical Society* 135(22): 8350–8356. Copyright © 2013, American Chemical Society. *See insert for color representation of the figure.*

3.2.1. Brust–Schiffrin Synthesis

The Brust–Schiffrin method (BSM; Brust et al., 1994; Chen et al., 2000; Shon et al., 2001) involves the precursor chloroauric acid (Au^{3+}) which is phase-transferred into toluene using a phase transfer catalyst followed by the addition of dodecanethiol to reduce Au^{3+} to Au^{1+}. The second reducing agent sodium borohydride is allowed to initiate the reduction process in the organic phase. The first-stage phase transfer of chloroauric acid into toluene uses tetraoctylammonium bromide (TOAB), a phase transfer catalyst (PTC), and the reaction is described by Equation (3.1):

$$H^+AuCl_4^-(aq)+(R_8)_4N^+Br^-(org) \to (R_8)_4N^+AuX_4^-(org)+HX(aq) \quad (3.1)$$

where X represents both Cl and Br as the extent of substitution of Cl⁻ by Br⁻ is not known.

The reduction of Au^{3+} to Au^{1+} state by alkanethiol (RSH) in the BSM is described by Equation (3.2) (Brust et al., 1994; Chen et al., 2000; Shon et al., 2001):

$$(R_8)_4 N^+ AuX_4^- + 3RSH \rightarrow -[(AuSR)_n]- + RSSR + (R_8)_4 N^+ + 4X^- + 3H^+ \quad (3.2)$$

The organic phase at this stage consists of TOAB, dialkyl disulfide (RSSR), Au^{1+} in the form of $-[(AuSR)]_n-$ polymer, and either AuX_4^- complexed with TOA ion or excess RSH. The second-phase reduction of Au^{3+} and Au^{1+} to Au^0 by sodium borohydride is described by Equations (3.3) and (3.4):

$$-[(AuSR)_n]- + BH_4^- + RSH + RSSR \xrightarrow{K_2} Au_x(Sr)_y \quad (3.3)$$

$$(R_8)_4 N^+ AuX_4^- + BH_4^- + RSH + RSSR \xrightarrow{K_3} Au_x(SR)_y \quad (3.4)$$

The reduction of AuX_4^- by alkanethiol was subsequently found (Goulet and Lennox, 2010; Li et al., 2011) to be produced through the intermediate AuX_2^- via the reaction described by Equation 3.5:

$$(R_8)_4 N^+ AuX_4^- + 2RSH \xrightarrow{K_1} (R_8)_4 N^+ AuX_2^- + RSSR + 2HX \quad (3.5)$$

Using the theory of colloids, the continuous nucleation–growth passivation-based model was tested. The new mechanisms (see Fig. 3.2) showed that slow and sustained nucleation with particle growth contained by capping of the particle surface

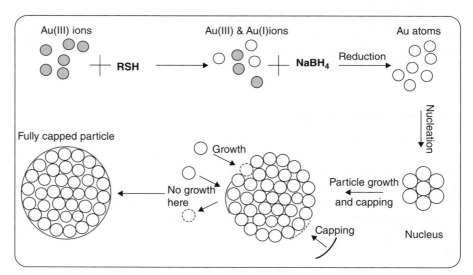

Figure 3.2. Continuous nucleation–growth-capping mechanism. Source: Siva Rama Krishna Perala, Sanjeev Kumar. On the Mechanism of Metal Nanoparticle Synthesis in the Brust-Schiriffin Method. Langmuir 29(31): 9863–9873. Copyright © 2013, American Chemical Society.

also leads to the formation of uniformly sized nanoparticles. The mechanism is quite different from the classical LaMer mechanism, sequential nucleation–growth-capping, and the thermodynamic mechanism. In a constanting synthesis environment, the continuous nucleation–growth-capping mechanism leads to complete capping of particles (no more growth) at the same size while new particles are continuously born, in principle leading to the synthesis of more monodisperse particles. The key features are size tuning by varying the amount of capping agent instead of reducing agent.

The relative role of the different mechanisms, such as nucleation, growth, coagulation, capping, and ripening of nanoparticles, all influence the particle size distribution. Methods of externally modulating these mechanisms to exercise the desired control on mean size and polydispersity are depicted in Figure 3.2.

3.3. GREEN SYNTHESIS

Green synthesis involves the use of plant material or biomass for the production of nanoparticles. Green synthesis techniques utilize relatively non-toxic chemicals to synthesize nanomaterials, and include the use of non-toxic solvents such as water, biological extracts and systems, and microwave-assisted synthesis. The size of nanoparticles (typically <30 nm) mainly depends on their interfacial properties such as dissolution, oxidation, adsorption/desorption, electron transfer, redox cycles, Fenton reactions, and surface acido-basicity.

Plant-extract-mediated synthesis has been experiencing a growth in interest. Biosynthesis of nanoparticles is a bottom-up approach. The plant biomass reduces the metal ions, resulting in the formation of nanoparticles. The reason that plant extracts work so well in the synthesis of nanoparticles is that they act as reducing agents as well as capping agents. When compared to the use of whole plant extracts and plant tissue, the use of plant extracts for making nanoparticles is simpler. Several plant extracts such as alfalfa, aloe vera, amla, Capsicum annuum, geranium, coriandrum, tea, and neem were used for green synthesis. The reducing agents involved in the reduction process include the various water-soluble natural products (e.g., alkaloids, phenolic compounds, terpenoids, flavonoids, and co-enzymes). Researchers focused mainly on the synthesis of silver (Ag) and gold (Au) nanoparticles due to their medical and environmental applications.

A simple, green, one-pot synthesis of gold nanoparticles (AuNPs) was achieved through the reaction of an aqueous mixture of potassium tetrachloroaurate (III) and the macrocycle cucurbit[7]uril in the presence of sodium hydroxide at room temperature, without introducing any kind of traditional reducing agents and/or external energy (Premkumar and Geckeler, 2010). The reaction is depicted in Figure 3.3. Cucurbit[7]uril (CB[7]) facilitated the reduction during the incubation under room-temperature conditions, resulting in CB[7] protecting the Au nanoparticles. Cucurbiturils are macrocyclic molecules composed of glycoluril ($-C_4H_2N_4O_2-$) monomers linked by methylene bridges ($-CH_2-$). The oxygen atoms are located along the edges of the band and are tilted inwards, forming a partly enclosed cavity. Cucurbiturils are

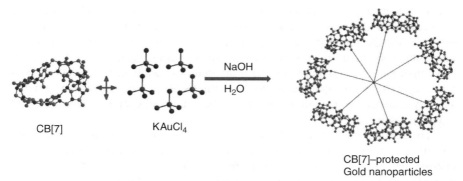

Figure 3.3. Synthesis of gold nanoparticles from cucurbit[7]uril using potassium tetrachloroaurate (III) as a metal precursor. Source: Thathan Premkumar, Kurt E. Geckeler. Cucurbit[7]uril as a Tool in the Green Synthesis of Gold Nanoparticles. *Chemistry–An Asian Journal* 5(12): 2468–2476. Copyright © 2010, John Wiley & Sons, Inc. *See insert for color representation of the figure.*

efficient host molecules in molecular recognition and have a particularly high affinity for positively charged or cationic compounds. High association constants with positively charged molecules are attributed to the carbonyl groups that line each end of the cavity, and can interact with cations in a similar fashion to crown ethers.

Plant constituents such as geraniol possess reducing properties, and reduce the metal ion to nanoparticles of a uniform size and shape in the range of 1–10 nm with an average size of 6 nm. During glycolysis, plants produce a large number of H^+ ions along with NAD, which acts as a strong redoxing agent which is responsible for the formation of nanoparticles. Water-soluble antioxidative agents such as ascorbic acids are also responsible for the reduction.

3.4. BIOSYNTHESIS OF NANOPARTICLES

The biogenesis of nanoparticles within the cell and periplasmic space of a microorganism has been a topic of interest for the past two decades. The biological route to inorganic materials synthesis is not merely an addition to the existing synthesis routes of nanomaterials. The advantage of nanoparticles is their shorter wavelength which renders them transparent, a property very useful for application in cosmetics, coatings, and packaging. Further, metallic nanoparticles can be attached to single strands of DNA non-destructively, providing applications in medicine (mainly in targeted drug delivery). Nanoparticles can traverse the vasculature and localize any target organ. The main issue is the creation and stabilization of these nanoparticles at <20 nm size. The complexity involved in the procedures and conditions have limited the research in this field. Nevertheless, unrelenting efforts have been made in exploring the use of those microorganisms, which are typically not known to be encountered in their natural environment, due to their eco-friendliness, low-energy intensiveness, and economically viable synthesis (Bansal et al., 2012).

Unicellular organisms such as bacteria and algae are also capable of synthesizing inorganic materials, both intra- and extracellularly (Simkiss and Wilbur, 1989). When immobilized by the enzymes of the live cells, the metal ions undergo reduction and subsequent nanotization; using bioresources such as enzymes, proteins, and plant extracts produces intra- and/or extracellular nanoparticles.

The microorganisms have developed a defensive mechanism, protecting them from the toxic nature of heavy metal residues (Faramarzi and Sadighi, 2013). Through multiple mechanisms involving enzymatic oxidation, reduction, sorption, and chelation with peptides or polysaccharides on the cell, the bacterial cell facilitates the nucleation and growth through extracellular or intracellular production of nanoparticles. The inward and outward transportation of heavy metals through the membrane of the microorganisms and the resistance following the transport and passive mechanisms leads to extracellular precipitation, a characteristic of prokaryotes (Krumov et al., 2009; Moghaddam, 2010).

The general bio-defensive protection mechanisms within the live cell tend to reduce metal ions to non-toxic metal salts and metal nanoparticles. These mechanisms include:

1. metal ion excretion across permeable membranes (Bruins et al., 2000);
2. enzymatic oxidation or reduction (Prakash et al., 2010);
3. accumulation of metallic nanparticles outside the plasma membrane (Pugazhenthiran et al., 2009);
4. binding at peptides (e.g., phytochelatin, a peptide with the general structure of (γ–GluCys) n–Gly, where $n =2$–5) that prevents DNA disruption and cell cycle damage due to metal toxicity (Krumov et al., 2007);
5. efflux pump systems, for example chemiosmotic, proton gradient and ATP-dependent pumps (Nies, 1999); and
6. precipitation in the form of carbonates, phosphates, and sulfides and volatilization through methylation and/or ethylation (De et al., 2008).

The biological templates employed for inorganic, transition, or semiconductor nanomaterials production includes bacterial spores (Jain et al., 2009), flagella (Deplanche et al., 2008), S-layer (Das and Marsili, 2010), the fifth repeated peptide unit of silaffin polypeptide R5 (which contains, 19-amino acids residues; Wetherbee, 2002), and viruses (Merzlyak and Lee, 2006), as well as fungi (Section 3.6.2), yeast (Section 3.6.3) and plant extracts (Section 3.6.4). The interactions of metal ions with the biological moieties such as cells, membranes, organelles, DNA, and proteins direct the protein corona, particle wrapping, cellular uptake, and biocatalytic processes, resulting the formation of both intracellular and extracellular nanoparticles. See Table 3.1 for a list of widely used species for the synthesis of nanomaterials.

TABLE 3.1. Species used for the Synthesis of NPs

Nanoparticles synthesized	Species used
Bacteria	
Au	*Bacillus subtilis, Shewanella algae, Rhodopseudomonas capsulata, Pseudomonas aeruginosa, Lactobacilli* strains, *Ralstonia metallidurans, Shewanella oneidensis, Pseudomonas* capsulate, *Stenotrophomonas maltophilia*
Ag	*Pseudomon sastutzeri* AG259, *Bacillus licheniformis, Bacillus subtilis, Lactobacilli* strains, *Fermentum* LMG 8900, *Farciminis* LMG 9189, *Garvieae* LMG8162, *Brevis* LMG 11437, *Parabuchneri* LMG11772, *Rhamnosus* LMG 18243, *Plantarum* LMG 24830/LMG24832, *Mucosae, Lactobacillus fructivorans, Pedicoccus pentosaceus* LMG 9445, *Morganella morganii* Subgroup (A, B, C, D, E, F, G1, G2) *Morganella morganii* RP42, *Morganella psychrotolerans*
Pt	*Shewanella algae*
Pd	*Desulfovibrio desulfuricans*
Se	*Sulfurospirillum barnesii, Bacillus selenitireducens, Selnihalanaero bactershriftii, Pseudomonas alkaliphila*
Te	*Sulfurospirillum barnesii, Bacillus selenitireducens*
Cu	*Serratia* sp., *Escherichia coli*
Au–Ag	*Lactobacilli* strains
Actinomycetes	
Au	*Thermomonospora* sp., *Rhodococcus* sp., *Actinobacter* sp.
Yeast	
Ag	Yeast strain MKY3
CdS	*Candida glabrata, Schizosaccharomyces pombe*
Fungi	
Ag	*Verticillium* sp., *Fusarium oxysporum, Fusarium semitactum, Fusarium acuminatum, Trichoderma asperellum, Phoma* sp.3.2883, *Aspergillus flavus, Penicillium* sp., *Phomaglomerata, Fusarium solani, Coriolus versicolor, Cladosporium cladosporioides*
Au	*Verticillium* sp., *Fusarium oxysporum, Colletotrichum* sp., *Rhizopus oryzae, Colletotrichum* sp., *Trichothecium* sp., *Verticilium luteoalbum, Helminthosporum solani*
Ag–Au	*Fusarium semitactum*
Magnetite	*Verticillium* sp., *Fusarium oxysporum*
Zirconia, CdSeQD, CdS, Si, Pt, Ti	*Fusarium oxysporum*
Plant extracts	
Ag	*Polyalthialongifolia, Pelagonium graveolens, Capsicum annum, Jatropha curcas,* Mesophytes (*Bryophyllum* sp.), Mesophytes (*Cyperus* sp.), Hydrophytes (*Bryophyllum* sp.), *Opuntia ficus, Moringa oleifera, Crossandra infundibuliformis, Desmodium trifolium, Datura metel, Melia azedarach, Eclipta prostrate, Eucalyptus* hybrid, *Piper longum, Cinnamomum zeylanicum, Syzygium cumini, Ocimum sanctum, Musa paradisiaca*

MECHANISMS FOR FORMATION OR SYNTHESIS OF NANOPARTICLES

Nanoparticles synthesized	Species used
Au	*Dioscorea bulbifera, Psidium guajava, Alfalfa* sprouts
Au–Ag	*Cinnamomum camphora, Phyllanthus amarus, Chenopodium album*
Cyanaobacteria	
Au	*Plectonema boryanum* UTEX 485
Ag	*Plectonema boryanum* UTEX 485
Pt	*Plectonema boryanum* UTEX 485
CdS	*Escherichia coli, Clostridium thermoaceticum*

3.5. MECHANISMS FOR FORMATION OR SYNTHESIS OF NANOPARTICLES

Pseudomonas stutzeri AG259 was the first bacteria strain with reductive potential to form silver nanoparticles (AgNPs). The biosynthesis of AgNPs through Ag^+ ions when grown in an $AgNO_3$-enriched medium follows a naturally occurring protective pathway against the toxicity of the silver ions. The intracellularly produced AgNPs particle size ranged from 35 to 46 nm (Klaus et al., 1999). The reduction of bismuth nitrate to intracellular elemental bismuth nanoparticles of <150 nm using a non-pigmented *Serratia marcescens* bacterium isolated from the Caspian Sea also follows a similar procedure (Nazari et al., 2012).

3.5.1. Biomineralization using Magnetotactic Bacteria (MTB)

Magnetotactic bacteria (MTB) consist of fastidious and mostly aquatic prokaryotes with different morphologies including coccal, vibroid, helical, rod-shape, and multicellular forms able to produce magnetic nanoparticles. The nano-sized magnetic particles including magnetite (Fe_3O_4), an iron sulfide, gregite (Fe_3S_4), or a combination of gregite and iron pyrite (FeS_2) are covered by a biological lipid bilayer associated with some proteins critical for magnetic nanoparticle production (Arakaki et al., 2003; Bazylinski and Frankel, 2004; Mohanpuria et al., 2008).

The reduction process of metal ions through magnetotactic bacteria follows a different mechanism. The process of magnetic nanoparticle production can be described in four steps: (1) vesicle formation and iron transport from outside of the bacterial membrane into the cell; (2) accumulation of ferrous/ferric ions in the cell and the vesicles; (3) control of iron oxidation–reduction process; and (4) crystal maturation and morphology regulation.

The crucial role of MamI, MamL, MamB, and MamQ proteins in initiation of the membrane invagination and biogenesis of lipoprotein membrane around the magnetosomes has been confirmed by *Magnetospirillum magnetotacticum* AMB-1. The protease-independent function of MamE recruits other proteins such as MamK, MamJ, and MamA to align magnetosomes in a chain. The mechanism was designed over four stages:

- Stage I: MamI, MamL, MamB, and MamQ proteins initiate the membrane invagination and form a vesicular membrane around the magnetosome structure.

- Stage II: The protease-independent function of MamE recruits other proteins such as MamK, MamJ, and MamA to align magnetosomes in a chain.
- Stage III: Iron uptake occurs via MagA, a transmembrane protein, and initiation of magnetic crystal bio-mineralization occurs through MamM, MamN, and MamO proteins.
- Stage IV: MamR, MamS, MamT, MamP, MamC, MamD, MamF, MamG, protease-dependent function of MamE, and Mms6, a membrane tightly bounded GTP-ase, regulate crystal growth.

3.5.2. Reduction of Tellurite using Phototroph *Rhodobacter capsulatus*

The Tellurite (TeO_3^{2-}) processing reduction mechanism in the cells of the facultative phototroph *Rhodobacter capsulatus* shows how tellurite enters cells as a response to a ΔpH-dependent mechanism (Borsetti et al., 2003). This drives an acetate permease (ActP) transport mechanism, allowing the toxic metalloid to reach the cytosolic space (Borghese and Zannoni, 2010). Inside the cell, tellurite is rapidly reduced to metallic Te^0 by a membrane-bound nitrate reductase (Nar) (Sabaty et al., 2001) and/or the thioredoxin (Trx)-glutathione (GSH) system (Painter-type reaction; Zannoni et al., 2008); conversely, tellurite may also enhance the activity of superoxide dismutase (SOD; Borsetti et al., 2005) which is necessary to inactivate reactive oxygen species (ROS) that are produced in parallel to tellurite reduction (Fenton's reaction).The scheme also shows crystals of tellurium by *in vitro* suspended membrane vesicles (chromatophores) as the membrane-bound thiol disulfide oxidoreductase (DsbB), which functions as an electron conduit between the reduced quinone pool (QH2) and the toxic metalloid that displaces DsbA, the orthodox redox partner of DsbB in the absence of tellurite (Borsetti et al., 2007). The mechanism is depicted in Figure 3.4 (Turner et al., 2012).

3.5.3. Formation of AgNPs using Lactic Acid and Bacteria

A different mechanism which involves a cyclic ring of monosaccharides, such as glucose as a reducing agent and RH as a protonated anionic functional group, are shown to be located on the bacterial cell wall. The role of pH values in the reduction of Ag^+ to AgNPs using lactic acid and bacteria and the mechanism is depicted in Figure 3.5 (Sintubin et al., 2009). By raising the pH values, the dissociation of protons occurs and results in the formation of negatively charged Ag^+ absorption sites, while glucose rings turned into their open-ring forms. The oxidation of the aldehyde functional group results in the reduction of Ag^+ ions to AgNPs.

3.5.4. Microfluidic Cellular Bioreactor for the Generation of Nanoparticles

Figure 3.6 depicts the microdroplet generation model using a microfluidic device (Lee et al., 2012). The system involves the following steps: (1) microdroplets are produced

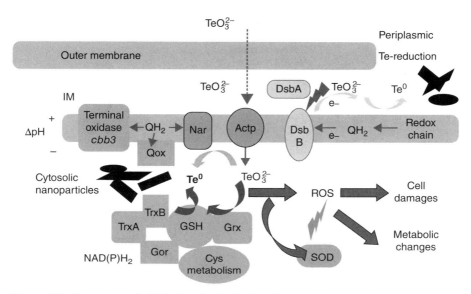

Figure 3.4. Processing of tellurite by the facultative phototroph *R. capsulatus*. Cys: cysteine; Grx: glutharedoxin; NAD(P)H$_2$: Gor; NAD(P)H$_2$: glutathione oxidoreductase; redox chain: respiratory electron transport chain; Qox: quinol oxidase; cbb$_3$: cytochrome c oxidase. Source: Raymond J. Turner, Roberto Borghese and Davide Zannoni. Microbial processing of tellurium as a tool in biotechnology. *Biotechnology Advances* 30(5): 954–963. Copyright © 2012, Elsevier. *See insert for color representation of the figure.*

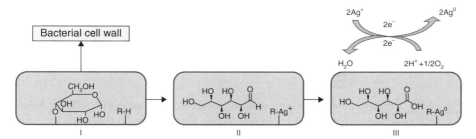

Figure 3.5. The synthesis of silver nanoparticles using a bacterial cell wall. Source: Redrawn from Liesje Sintubin, Wim De Windt, Jan Dick, Jan Mast, David van der Ha, Willy Verstarete, Nico Boon. 2009. Lactic acid bacteria as reducing and capping agent for the fast and efficient production of silver nanoparticles. *Applied Microbiology and Biotechnology* 84(4): 741–7491. Copyright © 2009, Springer. *See insert for color representation of the figure.*

with the mixture of cell extracts and N-Isopropylacrylamide (NIPAM) monomer in a microfluidic device; (2) the polymerized NIPAM monomers serve as an artificial membrane; (3) different types of precursor solutions are dispersed in the artificial cellular bioreactors; and finally (4) the precursors are transferred into the cells and NPs are

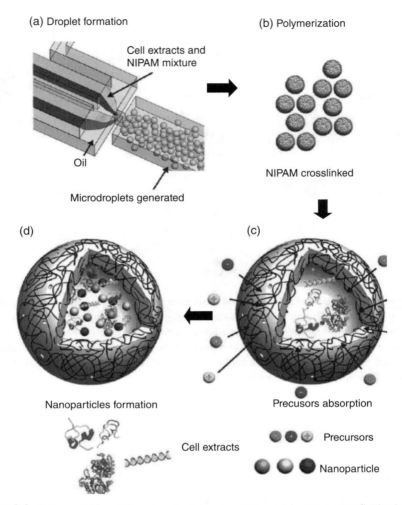

Figure 3.6. Representation of the microdroplet-generation model using a microfluidic device. Source: Kyoung G. Lee, Jongin Hong, Kye Won Wang, Nam Su Heo, Do Hyun Kim, Sang Yup Lee, Seok Jae Lee, Tae Jung Park. In Vitro Biosynthesis of Metal Nanoparticles in Microdroplets. *ACS Nano* 6(8): 6998–7008. Copyright © 2012, American Chemical Society. *See insert for color representation of the figure.*

subsequently formed in the artificial cellular bioreactors. The hydrogel polymer, recombinant *Escherichia coli* bacteria cell extracts, and a microdroplet-based microfluidic device to fabricate artificial cellular bioreactors were used as part of a bioreactor system to facilitate the synthesis of different metal nanoparticles including quantum dots (QD), transition metals, magnetic particles, and noble crystals. The combination of cell extracts, microdroplet-based microfluidic device, and hydrogel were able to produce a huge quantity of artificial cellular bioreactors of uniform size and shape. The method was successfully applied for the synthesis of artificial cellular bioreactors to produce various nanoparticles, including quantum dots, iron, and gold, and was demonstrated by simply dipping the reactors into the metal precursor solutions.

3.5.5. Proteins and Peptides in the Synthesis of Nanoparticles

Dipeptide and tripeptide mediated self-assembly of colloidal gold to produce AuNPs in the presence of $HAuCl_4$ as a reductive agent highlighted the effects of reductive groups of polar amino acids in the production of nanoparticles (Fig. 3.7; Bhattacharjee et al., 2005; Faramarzi and Sadighi, 2013). Gold nanoparticle-tripeptide (GNP-tripeptide) conjugates were prepared by a peptide *in situ* redox technique using a newly designed tripeptide, which has a C-terminus tyrosine residue which reduced Au^{3+} to Au, and the terminally located free amino group was bound to the gold nanoparticle (GNP) surface resulting in highly stable Au colloids.

3.5.6. NADH-dependent Reduction by Enzymes

In general, the microbial resistance against metal ions involves biosorption, bioaccumulation, extra-cellular complexation, efflux system, and alteration of solubility and toxicity via reduction or oxidation. An oxidoreductase is an enzyme that catalyzes the transfer of electrons from one molecule, the reductant (also called the electron donor), to another the oxidant (also called the electron acceptor). This group of enzymes usually utilizes NADP or NAD^+ as cofactors. The NADH-dependent reductase, which is responsible for the biosynthesis, gains electrons from NADH and oxidizes it to NAD^+. The *in vitro* synthesis of silver hydrosol of 10–25 nm from *Fusarium oxysporum* with capping peptide phytochelatin through enzymatic route using NADPH-dependent nitrate reductase (44 kDa) has been successfully demonstrated. The mechanistic

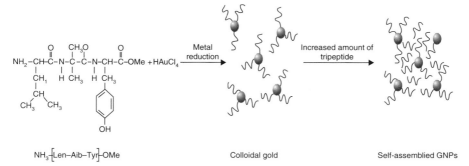

Figure 3.7. Production of gold nanoparticles induced by a tripeptide consisting of leucine (Leu), α-aminoisobutyric acid (Aib), and tyrosine (Tyr) at the C-terminus of the peptide, which is able to reduce Au^{3+} to Au^0, capped by methylene oxide (OMe). Source: Mohammad Ali Faramarzi, Armin Sadighi. Insights into biogenic and chemical production of inorganic nanomaterials and nanostructures. *Advances in Colloid and Interface Science* 189–190: 1–20. Copyright © 2013, Elsevier.

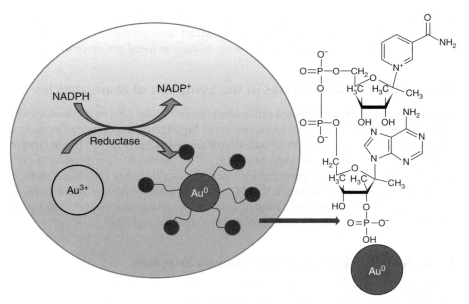

Figure 3.8. The synthesis of gold nanoparticles by *Stenotrophomonas maltophilia* through enzymatic reduction. Source: Redrawn from Yogesh Nangia, Nishima Wangoo, Nisha Goyal, G Shekhawat, C Raman Suri. 2009. A novel bacterial isolate *Stenotrophomonas maltophilia* as living factory for synthesis of gold nanoparticles. *Microbial Cell Factories* 8: 39.

aspect explains the involvement of enzymes, the quinine derivative of napthoquinones and anthraquinones as redox centers, in the reduction of silver nanoparticles (Narayanan and Sakthivel, 2010a).

The biosynthesis of GNPs via charge capping in *Stenotrophomonas maltophilia* gram-negative bacterium involves a NADPH-dependent reductase enzyme that converts Au^{3+} to Au^0 through electron shuttle enzymatic metal reduction processes. The bio-reduction of Au^{3+} to Au^0 by *Stenotrophomonas maltophilia* through enzymatic reduction is depicted in Figure 3.8 (Nangia et al., 2009).

3.5.7. Sulfate and Sulfite Reductase

The *Rhodobacter sphaeroides* bacteria-mediated synthesis of ZnS nanoparticles involves a series of reductase mechanisms such as sulfate permease, sulfurylase, and sulfite reductase. Initially the soluble sulfate enters to immobilized beads via diffusion, and is later carried into the interior membrane of the *R. sphaeroides* cell, reducing sulfite by ATP sulfurylase and phosphoadenosinephosphosulfate reductase and next sulfite to sulfide by sulfite reductase. The sulfide then reacts with O-acetyl serine through O-acetylserinethiolyase forming cysteine (Holmes et al., 1997), cysteine produces S^{2-} by a cysteine desulfhydrase in the presence of zinc, the resulting S^{2-} reacts with the soluble zinc salt, and the ZnS nanoparticles are synthesized and discharged from immobilized cells.

In the synthetic process, the particle size is controlled by the culture time of *R. sphaeroides*, and simultaneously the immobilized beads act to separate the ZnS nanoparticles from the *Rhodobacter sphaeroides* (Bai et al., 2006).

3.5.8. Cyanobacteria

The biosynthesis mechanisms of gold by cyanobacteria (*Plectonemaboryanum* UTEX 485) from gold(III)-chloride initially results in the precipitation of amorphous gold(I)-sulfide at the cell walls, before depositing metallic gold in the form of octahedral(III) platelets near cell surfaces and in solutions (Lengke et al., 2006). The mechanism was hypothesized involving the sulfate-reducing bacterial to reduce the gold(I) thiosulfate complex to elemental gold following three possible mechanisms involving iron sulfide, localized reducing conditions and metabolism (Lengke and Southam, 2006), and the interaction of *P. boryanum* UTEX485 with $Au(S_2O_3)_2^{-3}$, all promoting the accumulation of nanoparticles at membrane vesicles.

The silver nanoparticles were successfully synthesized by *Plectonemaboryanum* UTEX 485 at 100 °C. The interaction of $AgNO_3$ solution with cyanobacterium promoted the precipitation of grayish-black silver nanoparticles in solution state within 28 days (Lengke et al., 2007). The bacteria utilized nitrate as the major source for the growth of silver nanoparticles, generation of metabolic energy, and redox balancing. The presence of silver nanoparticles in the cells indicates entry of Ag^+ and NO_3^- into the cyanobacteria cells and their dissemination. The general mechanism is given in Equation (3.6). The mechanism of nitrate reduction by cyanobacterial metabolic processes first to nitrite and then to ammonium is described in Equations (3.7) and (3.8):

$$AgNO_3 \rightarrow Ag^+ + NO_3^- \quad (3.6)$$

$$NO_3^- + 2H^+ + 2e^- \rightarrow NO_2^- + H_2O \quad (3.7)$$

$$NO_2^- + 8H^+ + 6e^- \rightarrow NH_4^+ + 2H_2O \quad (3.8)$$

The resulting ammonium is fixed as the amide group of glutamine. The spherical silver nanoparticles were observed at the tested conditions of 25–100 °C both intracellularly (<10 nm) and extracellularly (1–200 nm). The biosynthesis of silver nanoparticles by bacteria, either extracellular or intracellular, using *Pseudomonas stutzeri* AG259, *Escherichia coli*, *Vibrio cholerae*, *Pseudomonas aeruginosa*, *Salmonella typhus*, *Phoma* sp. 3.2883, *Aspergillus flavus* NJP08, and *Staphylococcus aureus* has been demonstrated by reacting the cells with silver (I) nitrate ($AgNO_3$). The reduction mechanism was facilitated by the interaction of deoxyribonucleic acid (DNA) or sulfur-containing proteins. The bioreduction of silver (I) nitrate transition using fungi (*Fusarium oxysporum*) was speculated to be facilitated by nitrate-dependent reductase, and for other fungi (e.g., *Verticillium* sp. and *Aspergillus fumigates*) the involvement of the carboxylate groups of the cell walls was hypothesized. Silver nanoparticles (2–20 nm) inside live alfalfa plants were due to silver metal uptake from agar media. The formation of spherical silver nanoparticles with germanium leaf extract suggested reduction by plant terpenoids.

3.5.9. Cysteine Desulfhydrase in *Rhodopseudomonas palustris*

The synthesis of cadmium sulfide (CdS) nanoparticles from the cysteine desulfhydrase in *Rhodopseudomonas palustris* follows a simple route. *Rhodopseudomonas palustris* reduces the CdS particles in the quantum size regime. The cysteine-desulfhydrase-producing S^{2-} in the *R. palustris* located in the cytoplasm was found to be responsible for the formation of CdS nanocrystals, while protein secreted by *R. palustris* stabilized the cadmium sulfide nanoparticles which were ultimately excreted through the cell wall (Bai et al., 2009). This bioreduction of metal ions involved the interaction of the positively charged metal ions with the negatively charged sulfhydryl groups on the cysteine or imidazole rings on the histidine residue of metallothioneins via passive processes. See Figure 3.9 for a depiction of the process.

3.5.10. Nitrate and Nitrite reductase

Fusarium Oxysporum. The nitrate-dependent reductase of soil-bound microorganisms also facilitates rapid reduction of metal ions, resulting in the formation of extracellular nanoparticles. In *Fusarium oxysporum*, the nitrate-dependent reductase enzyme is conjugated with an electron donor (quinine), which reduces the metal ion and brings it to elemental form.

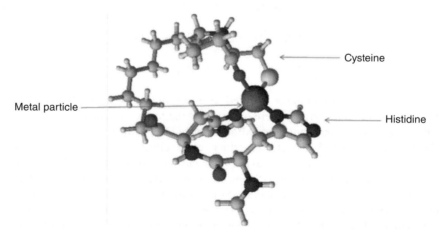

Figure 3.9. The interaction of metal ions with sulfhydryl groups on cysteine/imidazole rings on the histidine residues of metallothioneins. Source: Tamsyn Louise Riddin. 2008. Investigating the enzymatic mechanism of platinum nanoparticle synthesis in sulfate-reducing bacteria. Master of thesis. Rhodes University. *See insert for color representation of the figure.*

Enterobacteriaceae. The centrifuge extracts of *Enterobacteriaceae* (*Klebsiella pneumonia*, *E. coli*, and *Enterobacter cloacae*) culture also facilitates the rapid synthesis of silver nanoparticles by reducing Ag^+ to Ag^0. The presence of nitro reductase enzymes in the reduction process was identified by inhibition through the addition of piperitone (Narayanan and Sakthivel, 2010*a*).

3.6. EXTRACELLULAR SYNTHESIS OF NANOPARTICLES

3.6.1. Bacterial Excretions

The biochemical reduction of sodium selenite was performed using reducing agents such as proteins excreted by *Bacillus subtilis* for the extracellular synthesis of Se nanoparticles (Wang et al., 2010). Figure 3.10 depicts the formation of m-Se/protein composites and t-Se nanowires. The author proposed that in the first step, SeO_3^{2-} are firstly trapped on the surface of the proteins excreted from the bacteria via electrostatic interaction between SeO_3^{2-} and positively charged groups in the protein/enzymes; further reduction leads to the formation of Se nuclei. In the second step, growth occurs by further reduction of SeO_3^{2-} and the accumulation of newly formed Se atoms onto these nuclei. The first two steps are nucleation and growth steps. The flexible linkage of proteins and large numbers of biomolecules in the solutions leads to the disordered growth of Se in the third step, and in the fourth step the primary and larger particles are rearranged into more compact/dense spherical nanoparticles. In the fifth step, the spherical colloids split into irregular colloids

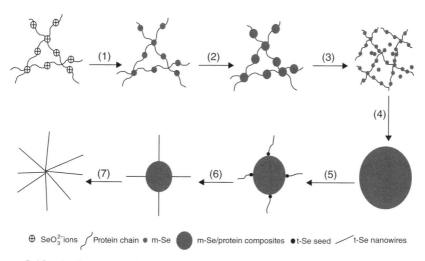

Figure 3.10. The formation of t-Se nanoparticles using the bacterial strain of *Bacillus subtilis*. Source: Tingting Wang, Liangbao Yang, Buchang Zhang, Jinhuai Liu. Extracellular biosynthesis and transformation of selenium nanoparticles and application in H_2O_2 biosensor. *Colloids and Surfaces B: Biointerfaces* 80: 94–102. Copyright © 2010, Elsevier. *See insert for color representation of the figure.*

along with protein chains and t-Se seeds. In the sixth step, further t-Se were grown from the seeds with the dissolution of m-Se colloidal particles. Finally, the formation of actinomorphic nanowires was obtained. Here, the excreted proteins not only inhibit the further growth of spherical monoclinic Selenium (m-Se) nanospheres but also form the m-Se/protein composites and highly anisotrophic one-dimensional t-Se nanowires.

Extracellular spherical silver nanoparticles (AgNPs) were synthesized by Prakash et al. (2011) in the range of 10–30 nm using the *Bacillus cereus* bacteria culture. Different known concentrations of silver nitrate solutions were added to the bacteria. During the synthesis, the molecular oxygen is released by both the oxidase and reductase processes for transformation by the tautomerization of quinines for the conversion of silver ions to silver oxide. The NADH acts as an electron carrier and transfers the electrons to receptors.

The synthesis of (Se^0) selenium nanospheres under aerobic conditions by a bacterial strain *Bacillus cereus*, isolated from coalmine soil, was studied by Dhanjal and Cameotra (2010). The aerobic reduction of selenate (SeO$_4^{2-}$) and selenite (SeO$_3^{2-}$) to Se^0 was carried out by a *Bacillus cereus*, as depicted in Figure 3.11. The extracellular nanospheres of selenium were observed by atomic force microscopic imaging to be of average size 150–200 nm. The reduction seems to be initiated by means of electron transfer from the NADPH/NADH by NADPH/NADH-dependent reductase as an electron carrier. The results show: (1) selenite reduction at 0 hours; (2) formation of red elemental selenium in membrane fraction after 3–4 hours of incubation; and (3) prolonged incubation for 12 hours also resulted in formation of red elemental selenium in soluble fraction also.

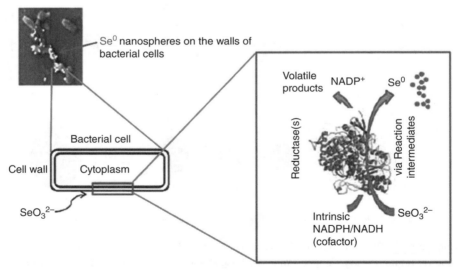

Figure 3.11. The formation of Se nanoparticles using a bacterial strain. Source: Soniya Dhanjal, Swaranjit Singh Cameotra. 2010. Aerobic biogenesis of selenium nanospheres by *Bacillus cereus* isolated from coalmine soil. *Microbial Cell Factories* 9: 52.

3.6.2. Fungal Strains

The production (Mukherjee et al., 2001) of AgNPs using *Verticillium*, a soil-borne fungus, takes place under two separate steps. In the first step, Ag^+ ions are trapped on the surface of the fungal cell wall through electrostatic interactions between the positive charge of Ag^+ ions and the negative charge of the enzymes carboxylic groups. In the second step the reductase enzymes, which are located on the surface of the cell wall, reduce Ag^+ ions to AgNPs and generate silver nuclei. Further reduction results in the accumulation of silver ions on the nuclei (Nithya and Raghunathan, 2009; Arya, 2010). The same observation was reported when using the strain of *Fusarium oxysporum* (reductase), responsible for the reduction of Ag^+ ions and the formation of extracellular silver nanoparticles. Ag (0) reduction in this case was due to a conjugation between the electron shuttle and the reductase participation, as shown in Figure 3.12 (Duran et al., 2005).

The cell free filtrate of fungus *Aspergillus flavus* NJP08 facilitated the synthesis of silver naoparticles. The SDS-PAGE (sodium dodecyl sulfate polyacrylamide gel electrophoresis) profile of the extracellular proteins showed the presence of two intense bands at 32 and 35 kDa, responsible for the synthesis of silver nanoparticles from aqueous silver ions and stability of the nanoparticles (Jain et al., 2011).

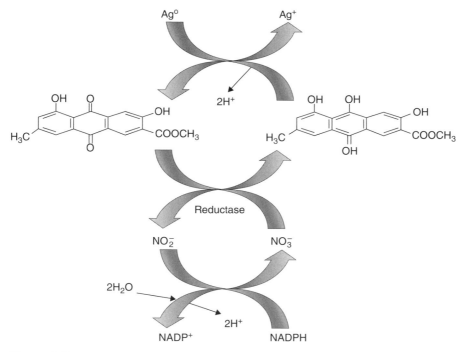

Figure 3.12. The two-step mechanism for the synthesis of silver nanoparticles. Source: Redrawn from Nelson Duran, Priscyla D Marcato, Oswaldo L Alves, Gabriel IH DeSouza, Elisa Esposito. 2005. Mechanistic aspects of biosynthesis of silver nanoparticles by several *Fusarium oxysporum* strains. *Journal of Nanobiotechnology* 3: 8.

In the formation of zirconia nanoparticles, the fungus-secreted proteins were found capable of hydrolyzing metal halide precursors under acidic conditions. The proteins involved in the reduction of metal nanoparticles were cationic proteins with molecular weights of around 21–24 kDa (Bharde et al., 2006). In the case of $[Fe(CN)_6]_3$ and $[Fe(CN)_6]_4$, the cationic protein of 55 kDa of *Verticillium* sp. was found responsible for extracellular synthesis (Rai et al., 2009).

3.6.3. Yeast: Oxido-reductase Mechanism

The transformation of cadmium, silver, titanium and antimony metal hydroxides $(M(OH)_2)$ in the yeast takes place at two distinct levels. The first is on the cell membrane immediately after the addition of the precursor $M(OH)_2$ solution, which triggers tautomerization of quinones and makes molecular oxygen available for the transformation. Subsequently, the oxygenases (mainly the monooxygenases or cytochrome P450 oxidases) or mixed-function oxidase break down or reduce the metal ions to nanoparticles. An oxygenase is any enzyme that oxidizes a substrate by transferring the oxygen from molecular oxygen O_2 (as in air) to it. The oxygenases form a class of oxidoreductases which facilitate the reduction mechanism.

When challenged with Cd^+, yeast responded by activating phytochelation (PC) synthase and Cd^+ is then complexed to form PC–Cd. A sulfide is then added to this complex in the membrane to form intracellular $Cd–S^{-2}$ complex (Krumov et al., 2007; Mohanpuria et al., 2008). The silica transporter proteins (SIT) located at specific silica-deposited vesicles (SDV) of the diatom *Cylindrotheca fusiformis* (Arya, 2010) facilitates the formation of silaffin, a bio-silica protein which is enriched in charged amino acids and is responsible for the polymerization of Si in these vesicles.

The extracellular biosynthesis of cadmium telluride CdTe QDs (Bao et al., 2010) was carried out using yeast cells (*Saccharomyces cerevisiae*). The precursors were Cd and Te salts. During the synthesis, yeast cells secreted the protein ligands to act as caps on the nanoparticles. The protein-capped CdTe QDs play a role in the extracellular growth. The mechanism represents the extracellular growth pathway for the biosynthesis of protein-capped CdTeQDs and an endocytosis pathway. In the first step of the endocytosis pathway, QDs are initially adhered to the cell membrane via electrostatic interactions; second, they become embedded by deforming the membrane and endocytose into the cell; in the third step they are released from the endosome into the cytoplasm by disruption of the endosomal membrane; and finally, in the fourth step they enter the nucleus via a nuclear translocation.

The extracellular method using isolates of yeast *Yarrowia lipolytica* when reacted with 3, 4-dihydroxy-L-phenylalanine (L-DOPA) and chloroauric acid facilitated the reduction of silver and gold to silver and gold nanostructures (Fig. 3.13). Here, the melanin present in the yeast isolate acted as an electron exchanger either in oxidizing or reducing metals. The conversion of the hydroxyl groups to the quinone groups produces reducing agents for the conversion of metal ions into elemental nanostructures (Apte et al., 2013).

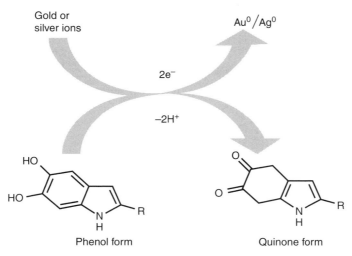

Figure 3.13. The formation of silver and gold nanoparticles by L-DOPA-induced melanin derived from yeasts. Source: Redrawn from Mugdha Apte, Gauri Girme, Ashok Bankar, Ameeta Ravi Kumar and Smita Zinjarde. 2013. 3, 4-dihydroxy-L-phenylalanine-derived melanin from *Yarrowia lipolytica* mediates the synthesis of silver and gold nanostructures. *Journal of Nanobiotechnology* 11: 2.

3.6.4. Plant Extracts

The synthesis of spherical gold nanoparticles (AuNP) using the aqueous extract of *Cissus quadrangularis* (CQE) by microwave irradiation obtained by the reduction of hydrogen tetrachloroaurate (HAuCl$_4$) and by CQE revealed a sharp surface plasmon resonance (SPR) peak at 530 nm, confirming the presence of AuNP with an average size of 12.0 ± 3.2 nm as measured through transmission electron microscopy (TEM; Bhuvanasree et al., 2013).

Silver nanoparticles were prepared by using different plant extracts from xerophytes (*Bryophyllum* sp.), mesophytes (*Cyperus* sp.), and hydrophytes (*Hydrilla* sp.). The suggested mechanism for xerophytes (such as *Bryophyllum* sp.) is depicted in Figure 3.14. Pyruvate is produced initially by malic acid through oxidative decarboxylation in the presence of NAD$^+$/NADP$^+$-dependent malic enzyme. The pyruvate is subsequently converted into phosphoenol pyruvate (PEP) in the presence of PEP carboxykinase, which is used in the glycolytic pathway. The anthraquinone moiety emodin also undergoes redial tautomerization, facilitating the reduction of silver ions (Jha et al., 2009).

The mechanism proposed for mesophytes (*Cyperus* sp.) is depicted in Figure 3.15, which shows the production of silver nanoparticles from the tautomerization of quinone moieties such as benzoquinones (e.g., cyperquinone (type I), dietchequinone (type II), and remirin (type III)) through redial tautomerization (Jha et al., 2009).

When hydrophytes (*Potamogeton* sp. or *Hydrilla* sp.) are used, the antioxidant ascorbate is oxidized in antioxidative reactions and the enzyme dehydroascorbate

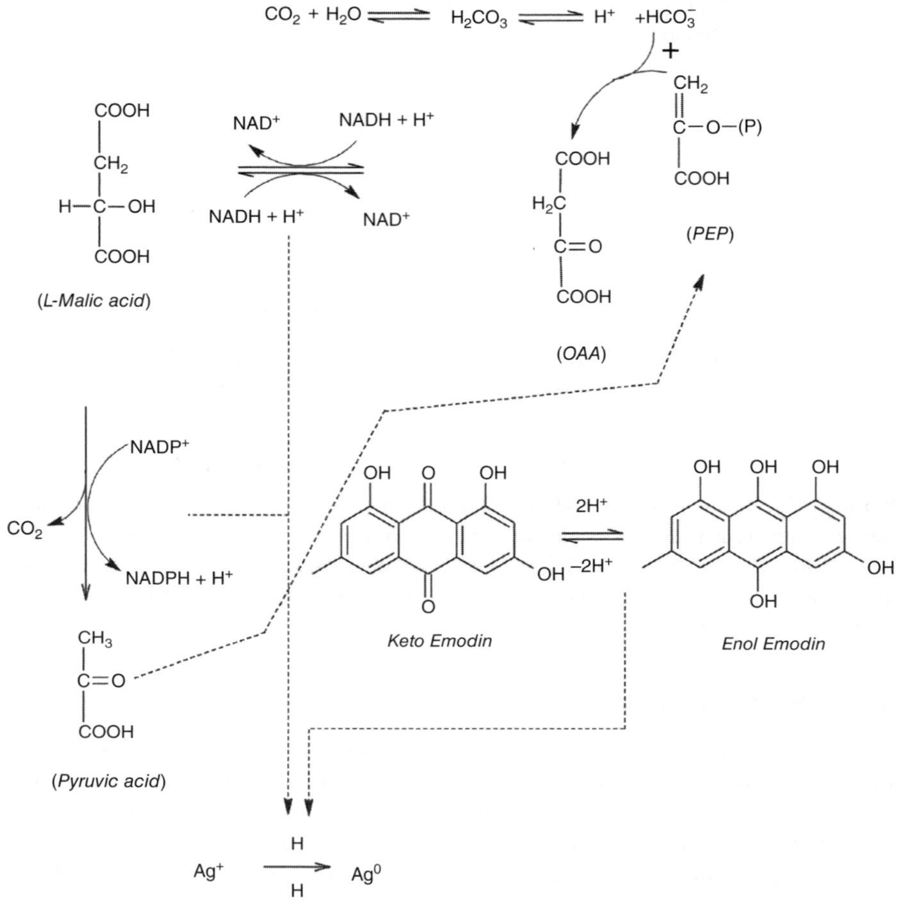

Figure 3.14. The synthesis of silver nanoparticles using xerophytes. Source: Redrawn from Anal K Jha, K Prasad, Kamlesh Prasad, A R Kulkarni. Plant system. Nature's nanofactory. *Colloids and Surfaces B: Biointerfaces* 73(2): 219–223. Copyright © 2009, Elsevier.

(DHA) reductase catalyzes the re-reduction of DHA to ascorbate. Under alkaline conditions, protocatechuic acid is formed by the oxidation of catechol through protocatechaldehyde. This facilitates the participation of hydrogen in the reduction of silver ions, resulting in the synthesis of silver nanoparticles as shown in the proposed mechanism in Figure 3.16 (Jha et al., 2009).

Mukherjee et al. (2012) demonstrated the biological route for the synthesis of biocompatible gold nanoparticles (AuNPs) from chloroauric acid ($HAuCl_4$) using a water extract of *Eclipta alba*. The very fast formation of AuNPs using *Eclipta* extract is depicted in Figure 3.17 (Mukherjee et al., 2012).

A new synthesis method of silver and gold using the leaf extract of *Coleus amboinicus* Lour (Narayanan and Sakthivel, 2010b, 2011) has been reported; the

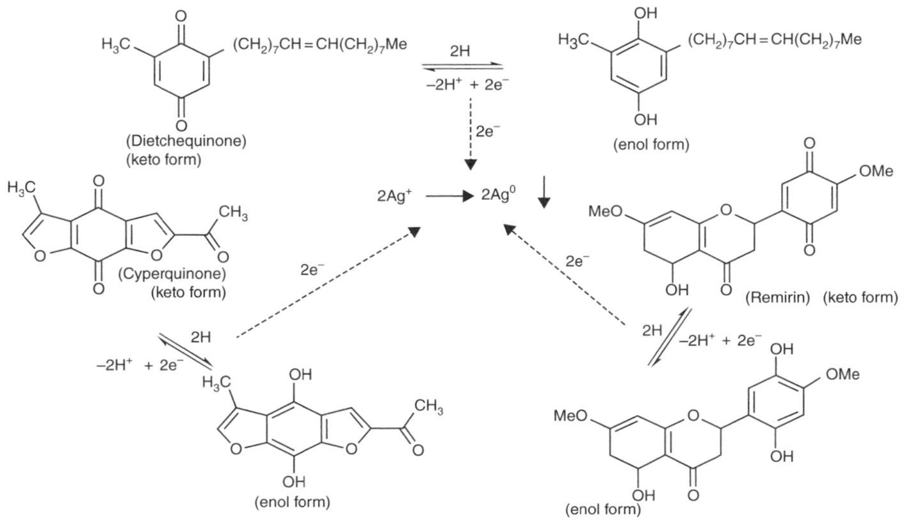

EXTRACELLULAR SYNTHESIS OF NANOPARTICLES 75

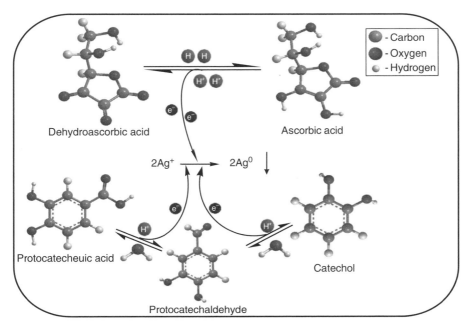

Figure 3.15. The synthesis of silver nanoparticles using mesophytes. Source: Redrawn from Anal K Jha, K Prasad, Kamlesh Prasad, A R Kulkarni. Plant system. Nature's nanofactory. *Colloids and Surfaces B: Biointerfaces* 73(2): 219–223. Copyright © 2009, Elsevier.

Figure 3.16. The synthesis of silver nanoparticles using hydrophytes. Source: Redrawn from Anal K Jha, K Prasad, Kamlesh Prasad, A R Kulkarni. Plant system. Nature's nanofactory. *Colloids and Surfaces B: Biointerfaces* 73(2): 219–223. Copyright © 2009, Elsevier.

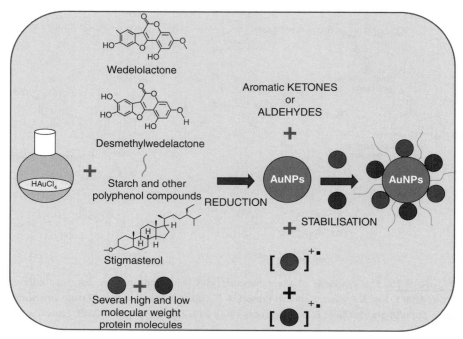

Figure 3.17. The biocompatibility of AuNPs observed by *in vitro* cell culture assays. Source: Priyabrata Mukherjee, Absar Ahmad, Deendayal, Mandal Satyajyoti Senapati, Sudhakar R Sainkar, Mohammad I Khan, R Ramani, Renu Parischa, P V Ajayakumar, Mansoor Alam, Murali Sastry, Rajiv Kumar. Bioreduction of AuCl4- Ions by the Fungus, *Verticillium* sp. and Surface Trapping of the Gold Nanoparticles Formed. *Angewandte Chemie International Edition* 40(19): 3585–3588. Copyright © 2001, John Wiley & Sons, Inc.

biosynthesis of silver nanoparticles in plants is shown in Figure 3.18 (Duran et al., 2011). Phytochemicals such as carvacrol, caryophyllene, patchoulane, and flavonoids found in the extract reduce metal ions in the solution to nanoparticles. Because the reagents used are naturally derived (Fig. 3.18a–d), this new method is both green and sustainable. Duran et al. (2011) discuss the influence of leaf extract on the size and shape of the silver nanoparticles. An increase in the concentration of leaf extract leads to a change in the shape of the nanoparticles from isotrophic nanostructures such as trianglar, decahedral, and hexagonal to isotrophic spherical nanoparticles.

3.7. CONCLUSION

The synthesis of nanoparticles by microbial microorganisms is not new; however, understanding the mechanisms of how these bacteria, fungi, yeasts, and plant extracts are involved in the process is a difficult task. The development of advanced imaging tools has helped researchers to understand these mechanisms and the involvement of reducing agents and enzymes in the microbial metabolism involved in the synthesis of

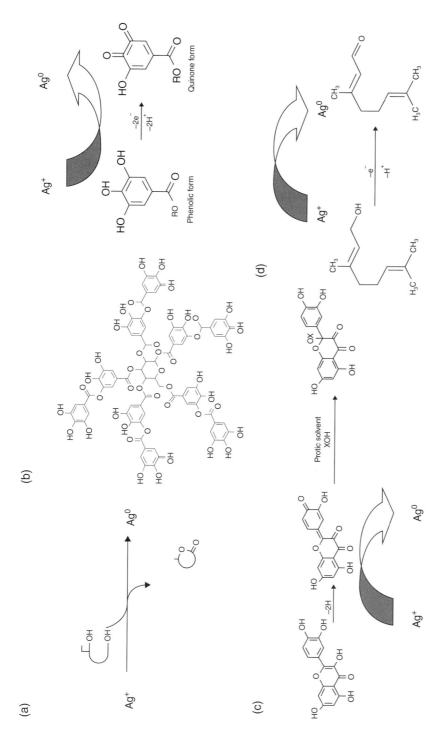

Figure 3.18. The extracellular synthesis of silver nanoparticles using leaf extract. Reprinted from Nelson Duran, Priscyla D Marcato, Marcela Duran, Alka Yadav, Aniket Gade, Mahendra Rai. Mechanistic aspects in the biogenic synthesis of extracellular metal nanoparticles by peptides, bacteria, fungi, and plants. *Applied Microbiology and Biotechnology* 90: 1609–1624. Copyright © 2011, Springer.

nanoparticles. Synthesized NPs by microorganisms are thought to be clean, efficient, and non-toxic and the process can be classified as "green chemistry." With greater understanding of microbial molecular mechanisms of NP synthesis, research is booming in fields such as biomedical applications, targeted drug delivery, and clinical diagnostics, and achieving the target of low-cost public monitoring programs.

REFERENCES

Apte, M., Girme, G., Bankar, A., Kumar, A.R., Zinjarde, S. 2013. 3, 4-dihydroxy-L-phenylalanine-derived melanin from Yarrowialipolytica mediates the synthesis of silver and gold nanostructures. *Journal of Nanobiotechnology* 11: 2.

Arakaki, A., Webb, J., Matsunaga, T. 2003. A novel protein tightly bound to bacterial magnetic particles in *Magnetospirillum magneticum* strain AMB-1. *The Journal of Biological Chemistry* 278: 8745–8750.

Arya, V. 2010. Living systems: eco-friendly nanofactories. *Digest Journal of Nanomaterials and Biostructures* 5(1): 9–21.

Bai, H.-J., Zhang, Z.-M., Gong, J. 2006. Biological synthesis of semiconductor zinc sulfide nanoparticles by immobilized *Rhodobacter sphaeroides*. *Biotechnology Letters* 28: 1135–1139.

Bai, H.-J., Zhang, Z.M., Guo, Y., Yang, G.E. 2009. Biosynthesis of cadmium sulfide nanoparticles by photosynthetic bacteria *Rhodopseudomonas palustris*. *Colloids and Surfaces B: Biointerfaces* 70(1): 142–146.

Bansal, B., Bharde, A., Ramanathan, R., Bhargava, S.K. 2012. Inorganic materials using 'unusual' microorganisms. *Advances in Colloid and Interface Science*, 179–182(1): 150–168.

Bao, H., Hao, N., Yang, Y., Zhao, D. 2010. Biosynthesis of biocompatible cadmium telluride quantum dots using yeast cells. *Nano Research* 3: 481–489.

Bazylinski, D.A., Frankel, R.B. 2004. Magnetosome formation in prokaryotes. *Nature Reviews Microbiology* 2: 217–230.

Bharde, A., Rautray, D., Bansal, V., Sarkar, I., Yusaf, S.M., Sanyal, M., Sastry, M. 2006. Extracellular biosynthesis of magnetite using fungi. *Small* 2(1): 135–141.

Bhattacharjee, R.R., Das, A.K., Haldar, D., Si, S., Banerjee, A., Mandal, T.K. 2005. Peptide-assisted synthesis of gold nanoparticles and their self-assembly. *Journal of Nanoscience and Nanotechnology*, 5(7): 1141–1147.

Bhuvanasree, S.R., Harini, D., Rajaram, A., Rajaram, R. 2013. Rapid synthesis of gold nanoparticles with *Cissus quadrangularis* extract using microwave irradiation. *Spectrochimica Acta Part A: Molecular and Biomolecular Spectroscopy* 106: 190–196.

Borghese, R., Zannoni, D. 2010. Acetate permease (ActP) is responsible for tellurite (TeO_3^{2-}) uptake and resistance in cells of the facultative phototroph *Rhodobacter capsulatus*. *Applied and Environmental Microbiology* 76(3): 942–944.

Borsetti, F., Toninello, A., Zannoni, D. 2003.Tellurite uptake by cells of the facultative phototroph *Rhodobacter capsulatus* is a ΔpH dependent process. *FEBS Letters* 554(3): 315–318.

Borsetti, F., Tremaroli, V., Michelacci, F., Borghese, R., Winterstein, C., Daldal, F., Zannoni, D. 2005. Tellurite effects on *Rhodobacter capsulatus* cell viability and superoxide dismutase activity under oxidative stress conditions. *Research in Microbiology* 156(7): 807–813.

Borsetti, F., Francia, F., Turner, R.J., Zannoni, D. 2007. The thiol:disulfideoxidoreductase DsbB mediates the oxidizing effects of the toxic metalloid tellurite (TeO$_3^{2-}$) on the plasma membrane redox system of the facultative phototroph *Rhodobacter capsulatus*. *Journal of Bacteriology* 189(3): 851–859.

Bruins, M.R., Kapil, S., Oehme, F.W. 2000. Microbial resistance to metals in the environment. *Ecotoxicology and Environmental Safety* 45(3): 198–207.

Brust, M., Walker, M., Bethell, D., Schiffrin, D.J., Whymam, R. 1994. Synthesis of thiol-derivatised gold nanoparticles in a two phase liquid- liquid system. *Journal of the Chemical Society, Chemical Communications* 7: 801–802.

Chen, S., Templeton, A.C., Murray, R.W. 2000. Monolayer-proteced cluster growth dynamics. *Langmuir* 16(7): 3543–3548.

Das, S.K., Marsili, E. 2010. A green chemical approach for the synthesis of gold nanoparticles: characterization and mechanistic aspect. *Reviews in Environmental Science and Biotechnology* 9(3): 199–204.

De, J., Ramaiah, N., Vardanyan, L. 2008. Detoxification of toxic heavy metals by marine bacteria highly resistant to mercury. *Marine Biotechnology* 10(4): 471–477.

Deplanche, K., Woods, R.D., Mikheenko, I.P., Sockett, R.E., Macaskie, L.E. 2008. Manufacture of stable palladium and gold nanoparticles on native and genetically engineered flagella scaffolds. *Biotechnology and Bioengineering* 101(5): 873–880.

Dhanjal, S., Cameotra, S.S. 2010. Aerobic biogenesis of selenium nanospheres by *Bacillus cereus* isolated from coalmine soil. *Microbial Cell Factories* 9: 52.

Duran, N., Marcato, P.D., Alves, O.L., DeSouza, G.I.H., Esposito, E. 2005. Mechanistic aspects of biosynthesis of silver nanoparticles by several *Fusarium oxysporum* strains. *Journal of Nanobiotechnology* 3: 8.

Duran, N., Marcato, P.D., Duran, M., Yadav, A., Gade, A., Rai, M. 2011. Mechanistic aspects in the biogenic synthesis of extracellular metal nanoparticles by peptides, bacteria, fungi, and plants. *Applied Microbiology and Biotechnology* 90: 1609–1624.

Faramarzi, M.A., Sadighi, A. 2013. Insights into biogenic and chemical production of inorganic nanomaterials and nanostructures *Advances in Colloid and Interface Science* 189–190: 1–20.

Goulet, P.G.J., Lennox, R.B. 2010. New insights into Brust-Schiffrin metal nanoparticle synthesis. *Journal of American Chemical Society* 132(28): 9582–9584.

Holmes, J.D., Richardson, D.J., Saed, S., Evans-Gowing, R., Russell, D.A., Sodeau, J.R. 1997.Cadmium-specific formation of metal sulfide 'Q-particle' by *Klebsiella pneumoniae*. *Microbiology* 143: 2521–2530.

Jain, D., Daima, H.K., Kachhwaha, S., Kothari, S.L. 2009. Synthesis of plant-mediated silver nanoparticles using papaya fruit extract and evaluation of their anti microbial activities. *Digest Journal of Nanomaterials and Biostructures* 4(4): 723–727.

Jain, N., Bhargava, A., Majumdar, S., Tarafdar, J.C., Panwar, J. 2011. Extracellular biosynthesis and characterization of silver nanoparticles using *Aspergillus flavus* NJP08: A mechanism perspective. *Nanoscale* 3: 635–641.

Jha, A.K., Prasad, K., Prasad, K., Kulkarni, A.R. 2009. Plant system. nature's nanofactory. *Colloids and Surfaces B: Biointerfaces* 73(2): 219–223.

Klaus, T., Joerger, R., Olsson, E., Granqvist, C.-G. 1999. Silver-based crystalline nanoparticles, microbially fabricated. *Proceedings of the National Academy of Sciences of the USA* 96(24): 13611–13614.

Krumov, N., Oder, S., Perner-Nochta, I., Angelov, A., Posten, C. 2007. Accumulation of CdS nanoparticles by yeasts in a fed-batch bioprocess. *Journal of Biotechnology* 132(4): 481–486.

Krumov, N., Perner-Nochta, I., Oder, S., Gotcheva, V., Angelov, A., Posten, C. 2009. Production of inorganic nanoparticles by microorganisms. *Chemical Engineering and Technology* 32(7): 1026–1035.

Lee, K.G., Hong, J., Wang, K.W., Heo, N.S., Kim, D.H., Lee, S.Y., Lee, S.J., Park, T.J. 2012. In vitro biosynthesis of metal nanoparticles in microdroplets. *ACS Nano* 6(8): 6998–7008.

Lengke, M.F., Southam, G. 2006. Bioaccumulation of gold by sulfate-reducing bacteria cultured in the presence of gold(I)-thiosulfate complex. *Geochimica et Cosmochimica Acta* 70: 3646–3661.

Lengke, M.F., Ravel, B., Fleet, M.E., Wanger, G., Gordon, R.A., Southam, G. 2006. Mechanisms of gold bioaccumulation by filamentous cyanobacteria from gold(III)-chloride complex. *Environmental Science and Technology* 40(20): 6304–6309.

Lengke, M.F., Fleet, M.E., Southam, G. 2007. Biosynthesis of silver Nanoparticles by filamentous Cyanobacteria from a Silver (I) nitrate complex. *Langmuir* 23(5): 2694–2699.

Li, Y., Zaluzhna, O., Xu, B., Gao, Y., Modest, J.M., Tong, J.Y. 2011. Mechanistic insights into the Brust-Schiffrin two phase synthesis of organo-chalcogenate-protected metal nanoparticles. *Journal of American Chemical Society* 133(7): 2092–2095.

Merzlyak, A., Lee, S.-W. 2006. Phage as templates for hybrid materials and mediators for nanomaterial synthesis. *Current Opinion in Chemical Biology* 10: 246–252.

Moghaddam, K.M. 2010. An introduction to microbial metal nanoparticle preparation method. *The Journal of Young Investigations*, 19(19): 1–6.

Mohanpuria, M., Rana, N.K., Yadav, S.K. 2008. Biosynthesis of nanoparticles: technological concepts and future applications. *Journal of Nanoparticle Research* 10(3): 507–517.

Mukherjee, P., Ahmad, A., Mandal, D., Senapati, S., Sainkar, S.R., Khan, M.I., Ramani, R., Parischa, R., Ajayakumar, P.V., Alam, M., Sastry, M., Kumar, R. 2001. Bioreduction of $AuCl_4^-$ ions by the fungus, *Verticillium* sp. and surface trapping of the gold nanoparticles formed. *Angewandte Chemie International Edition* 40(19): 3585–3588.

Mukherjee, S., Sushma, V., Patra, S., Barui, A.K., Bhadra, M.P., Sreedhar, B., Patra, C.R. 2012. Green chemistry approach for the synthesis and stabilization of biocompatible gold nanoparticles and their potential applications in cancer therapy. *Nanotechnology* 23(45).

Nangia, Y., Wangoo, N., Goyal, N., Shekhawat, G., Suri, C.R. 2009. A novel bacterial isolate *Stenotrophomonas maltophilia* as living factory for synthesis of gold nanoparticles. *Microbial Cell Factories* 8: 39.

Narayanan, K.B., Sakthivel, N. 2010*a*. Biological synthesis of metal nanoparticles by microbes. *Advances in Colloid and Interface Science* 156: 1–13.

Narayanan, K.B., Sakthivel, N. 2010*b*. Phytosynthesis of gold nanoparticles using leaf extract of *Coleus amboinicus* Lour. *Materials Characterization* 61(11): 1232–1238.

Narayanan, K.B., Sakthivel, N. 2011. Extracellular synthesis of silver nanoparticles using the leaf extract of *Coleus amboinicus* Lour. *Material Research Bulletin* 46(10): 1708–1713.

Nazari, P., Faramarzi, M.A., Sepehrizadeh, Z., Mofid, M.R., Bazaz, R.D., Shahverdi, A.R. 2012. Biosynthesis of bismuth nanoparticles using *Serratia marcescens* isolated from the Caspian Sea and their characterisation. *IET Nanobiotechnology* 6(2): 58–62.

Nies, D.H. 1999. Microbial heavy-metal resistance. *Applied Microbiology and Biotechnology* 51: 730–750.

Nithya, R., Raghunathan, R. 2009. Synthesis of silver nanoparticles using Pleurotus sajor caju and its antimicrobial study. *Digest Journal of Nanomaterials and Biostructures* 4(4): 623–629.

Perala, S.R.K., Kumar, S. 2013. On the mechanism of metal nanoparticle synthesis in the Brust-Schiriffrin method. *Langmuir* 29(31): 9863–9873.

Prakash, A., Sharma, S., Ahmad, N., Ghosh, A., Sinha, P. 2010. Bacteria mediated extracellular synthesis of metallic nanoparticles. *International Research Journal of Biotechnology* 1(5): 71–79.

Prakash, A., Sharma, S., Ahmad, N., Ghosh, A., Sinha, P. 2011. Synthesis of AgNPs by *Bacillus Cereus* Bacteria and their antimicrobial potential. *Journal of Biomaterials and Nanobiotechnology* 2: 156–162.

Premkumar, T., Geckeler, K.E. 2010. Cucurbit [7] uril as a tool in the green synthesis of gold nanoparticles. *Chemistry – An Asian Journal* 5(12): 2468–2476.

Pugazhenthiran, N., Anandan, S., Kathiravan, G., Prakash, N.K.U., Crawford, S., Ashokkumar, M. 2009. Microbial synthesis of silver nanoparticles by *Bacillus* sp. *Journal of Nanoparticle Research* 11: 1811–1815.

Rai, M., Yadav, A., Bridge, P., Gade, A., 2009. Myconanotechnology: a new and emerging science. In *Applied Mycology* (eds M. Rai and P.D. Bridge). CAB International.

Riddin, T.L. 2008. Investigating the enzymatic mechanism of platinum nanoparticle synthesis in sulfate-reducing bacteria. MSc thesis. Rhodes University.

Sabaty, M., Avazeri, C., Pignol, D., Vermeglio, A. 2001. Characterization of the reduction of selenate and tellurite by nitrate reductases. *Applied and Environmental Microbiology* 67(11): 5122–5126.

Shon, Y.-S., Mazzitelli, C., Murray, R.W. 2001. Unsymmetrical disulfides and thiol mixtures produce different mixed monolayer-protected gold clusters. *Langmuir* 17(25): 7735–7741.

Simkiss, K., Wilbur, K.M. 1989. *Biomineralization: Cell Biology and Mineral Deposition*. Academic Press, New York, 337 pp.

Sintubin, L., De Windt, W., Dick, J., Mast, J., van der Ha, D., Verstarete, W., Boon, N. 2009. Lactic acid bacteria as reducing and capping agent for the fast and efficient production of silver nanoparticles. *Applied Microbiology and Biotechnology* 84(4): 741–749.

Turner, R.J., Borghese, R., Zannoni, D. 2012. Microbial processing of tellurium as a tool in biotechnology. *Biotechnology Advances* 30(5): 954–963.

Wang, T., Yang, L., Zhang, B., Liu, J. 2010. Extracellular biosynthesis and transformation of selenium nanoparticles and application in H_2O_2 biosensor. *Colloids and Surfaces B: Biointerfaces* 80(1): 94–102.

Wetherbee, R. 2002. Biomineralization. The diatom glasshouse. *Science* 298(5593): 547.

Zannoni, D., Borsetti, F., Harrison, J.J., Turner, R.J. 2008. The bacterial response to the chalcogen metalloids Se and Te. *Advances in Microbial Physiology* 53: 1–71.

Zhong, Y., Peng, F., Bao, F., Wang, S., Ji, X., Yang, L., Su, Y., Lee, S.-T., He, Y. 2013. Large-scale aqueous synthesis of fluorescent and biocompatible silicon nanoparticles and their use as highly photostable biological probes. *Journal of the American Chemical Society* 135(22): 8350–8356.

4

BIOFILMS IN BIO-NANOTECHNOLOGY: OPPORTUNITIES AND CHALLENGES

Chun Kiat Ng

Singapore Centre on Environmental Life Sciences Engineering, Nanyang Technological University Singapore and Interdisciplinary Graduate School, Nanyang Technological University, Singapore

Anee Mohanty and Bin Cao

School of Civil and Environmental Engineering, Nanyang Technological University Singapore and Singapore Centre on Environmental Life Sciences Engineering, Nanyang Technological University, Singapore

4.1. INTRODUCTION

Microorganisms in most natural, engineered, and medical settings are found to be growing in biofilm mode, where the microbial cells form surface-associated communities. Advanced imaging and molecular techniques have revealed that biofilms are well-organized structures made up of cells encased in a biofilm matrix formed by self-produced extracellular polymeric substances (EPS) including polysaccharide, proteins, lipids, and nucleic acids. Extensive studies have demonstrated a wide range of advantages of the presence of an EPS matrix for the biofilm mode of life (Flemming

---

Bio-Nanoparticles: Biosynthesis and Sustainable Biotechnological Implications, First Edition.
Edited by Om V. Singh.
© 2015 John Wiley & Sons, Inc. Published 2015 by John Wiley & Sons, Inc.

and Wingender, 2010a; Flemming, 2011). For example, cells growing in biofilms exhibit a high resistance to various physicochemical stresses such as starvation, radiation, desiccation, and exposure to toxic chemicals including biocides. Because of these advantages, biofilms have been increasingly applied to a wide range of bioprocesses, including the efficient removal of toxic contaminants from industrial wastewater and biofilm-based biocatalysis for chemical synthesis (Singh et al., 2006; Bishop, 2007; Cao et al., 2010; Martin and Nerenberg, 2012). On the other hand, since biofilms are structurally stable, they are difficult to remove. Hence, in many engineered and medical systems such as membrane-based processes and medical devices, biofilms may have detrimental impacts.

Bio-nanotechnology is an emerging field of science at the interface between biotechnology and nanotechnology, where the interaction between biomolecules or organisms and nanostructures takes place, promoting the technological advancement in both biotechnology and nanotechnology. Various biological systems have been applied to nanotechnology, for example, biosynthesis of nanomaterials, where biomolecules or cells are employed to synthesize nanomaterials. In addition, nanotechnology has tremendous potential in biotechnological applications such as drug delivery, biosensing, and antimicrobial therapy.

In this chapter, we review: (1) microbial synthesis of nanomaterials, with a special emphasis on molecular mechanisms and the emerging use of biofilms in this application; and (2) the application of nanomaterials in biofilm control. Further, the available opportunities and challenges of microbial biofilms in bio-nanotechnology are discussed.

4.2. MICROBIAL SYNTHESIS OF NANOMATERIALS

4.2.1. Overview

There are many instances where nature acts as a guide to human advancement in technology. In the search for a more environmentally benign approach to synthesize nanomaterials, researchers drew inspiration from the way microorganisms interact with metal(loid)s in the natural environment. Microorganisms have the potential to change the oxidation state of metal(loid)s, and these microbial processes have provided access to novel applications including biosynthesis of metal nanomaterials (Gadd, 2010; Lloyd et al., 2011). A wide variety of microorganisms including bacteria, yeast, and viruses have been exploited for the synthesis of metal nanomaterials (Li et al., 2003; Bigall et al., 2008; Suresh et al., 2010; Jain et al., 2011; Kumar and Mamidyala, 2011; Sathish et al., 2011; Jung et al., 2012).

A brief summary of recent work on bacterial synthesis of metal(loid) nanomaterials is shown in Table 4.1. Various bacteria, mostly non-pathogenic environmental bacteria, have been employed to produce metal(loid) nanomaterials. Among them, metal-reducing bacteria from *Shewanella* and *Geobacter* genera are of great interest because they have a relatively well-characterized electron transport chain that can transport electrons across cell membranes and is capable of reducing metal(loid)s at

TABLE 4.1. Summary of Recent Work on Bacterial Synthesis of Metal(loid) Nanomaterials

Metal (loid)	Microorganism	Size (nm)	Shape	Application	Reference
Au	Rhodopseudomonas capsulata	10–20	Spherical	Optoelectronics	He et al. (2007)
	Shewanella algae	10–20	Irregular	Recovery of noble metals	Konishi et al. (2007a)
	S. oneidensis	7–17	Spherical	Biomedical	Suresh et al. (2011)
	Desulfovibrio desulfuricans	20	Irregular	Catalyst	Deplanche et al. (2008)
	Geobacillus stearothermophilus	12	Spherical	Drug delivery and biosensor	Mohammed Fayaz et al. (2011)
Pd	Rhodobacter sphaeroides	3–10	Irregular	Catalyst	Redwood et al. (2008)
	Geobacter sulfurreducens	11–17	Irregular	Electronics	Yates et al. (2013)
	Cupriavidus necator	3–30	Spherical	Generation of hydrogen	Bunge et al. (2010)
	Pseudomonas putida	3–30	Spherical	Dechlorination	Bunge et al. (2010)
	Clostridium butyricum	<50	Spherical	Dechlorination	Hennebel et al. (2011)
	Escherichia coli	5	Spherical	Recovery of palladium	Deplanche et al. (2010)
	D. desulfuricans	5	Irregular	Catalyst	Bennett et al. (2013)
Au/Pd	S. oneidensis	–	–	Catalyst	Heugebaert et al. (2012)
	E. coli	16	Quasi-spherical	Catalyst	Deplanche et al. (2012)
	S. oneidensis	–	–	Dehalogenation	De Corte et al. (2012)

(continued)

TABLE 4.1. (Continued)

Metal (loid)	Microorganism	Size (nm)	Shape	Application	Reference
Ag	S. oneidensis	40	Spherical	Antimicrobial	Ng et al. (2013c)
	Bacillus licheniformis	40–50	–	Solar energy	Kalimuthu et al. (2008)
	E. coli	50	Spherical	Bio-labelling	Gurunathan et al. (2009)
	G. stearothermophilus	25	Spherical	Drug delivery	Mohammed Fayaz et al. (2011)
	P. putida MnB6	–	–	Disinfection	Meerburg et al. (2012)
Pt	E. coli	14	Irregular	Asymmetric catalysis	Bennett et al. (2012)
	E. coli	4.5	Spherical	Catalyst	Attard et al. (2013)
	S. algae	5	Spherical	Catalyst	Konishi et al. (2007b)
CdTe	E. coli	2.0–3.2	Spherical	Quantum dots	Bao et al. (2010)
CdS	E. coli	2–5	Spherical	Semiconductor	Sweeney et al. (2004)
Se	Bacillus sp. MSh-1	80–220	Spherical	Anti-Leishmania	Beheshti et al. (2013)
	E. coli K-12	10–90	Spherical	Photonics	Dobias et al. (2011)
	B. cereus	150–200	Spherical	Medical application	Dhanjal and Cameotra (2010)
	P. aeruginosa	<10	Spherical	Anti-oxidative drug therapy	Yadav et al. (2008)
As-S	Shewanella sp. HN-41	40–70	Tubular	Optoelectronics	Jiang et al. (2009)
	S. oneidensis MR-1	30–60	Tubular	Semiconductor	Jiang et al. (2009)
	S. putrefaciens CN-32	30–60	Tubular	Metal remediation	Jiang et al. (2009)

the cell exterior, giving a relatively narrow size distribution and minimizing post-production extraction and purification (Narayanan and Sakthivel, 2010; Kalathil et al., 2011a; Sintubin et al., 2012). Strict anaerobic conditions are often required for *Geobacter* species, while *Shewanella* can grow well in both aerobic and anaerobic conditions, rendering *Shewanella* an important bacterium in bio-nanotechnology. *S. oneidensis* is one often-used metal-reducing bacterium in biosynthesis of metal(loid) nanomaterials.

Although a wide range of bacteria have been used for the synthesis of metal(loid) nanomaterials, the molecular mechanisms underlying the biosynthesis processes are largely unknown. Based on previous literature, the conclusions we can draw include: (1) both cell-mediated and biologically influenced redox reactions are often involved; (2) dominating mechanisms vary in different bacteria; and (3) primary molecular sites responsible for the biosynthesis in a certain bacterium can be very different for different metal(loid)s.

In the biosynthesis of nanomaterials by bacteria, both cell-mediated and biologically influenced reactions have been reported. In organisms such as *E. coli* and *P. aeruginosa*, the conversion of soluble metal(loid)s to their insoluble nanosized structures is a general detoxification mechanism where surface proteins, polysaccharides, and periplasmic proteins are often involved and there are no specific biomolecular sites responsible for the biosynthesis of nanomaterials.

In *S. oneidensis*, *c*-type cytochromes, a class of redox proteins, have been shown to play a central role in metal reduction (Fredrickson et al., 2008). The outer membrane of *S. oneidensis* contains *c*-type cytochromes including two best-characterized outer membrane *c*-type cytochromes (OMCs) MtrC and OmcA (Fig. 4.1; Shi et al., 2006).

Extensive studies have suggested the important role of OMCs in efficient reduction of various metal(loid) s including Fe(III), Cr(VI), Tc(VII), and U(VI) (Marshall et al., 2006; Belchik et al., 2011). However, whether OMCs influence the size and activity including antimicrobial, toxicological, catalytic, and optical features of the extracellular nanoparticles produced by *S. oneidensis* remains unknown. In a recent study, Ng *et al.* (2013a) investigated the influence of OMCs, that is, MtrC and OmcA, on the size and activity of the extracellular nanoparticles using the production of Ag nanoparticles and Ag_2S nanoparticles by *S. oneidensis* MR-1 as a model system. Ag and Ag_2S nanoparticles were produced using cell suspensions of the wild-type (WT) and the mutant lacking OMCs to reduce $AgNO_3$ or $AgNO_3$ and $Na_2S_2O_3$, respectively. Although both the WT and the mutant produced extracellular nanoparticles, the size of the nanoparticles produced by the mutant lacking OMCs was found to be significantly lower than those produced by the WT. Ag nanoparticles have been demonstrated to exhibit biocidal activity towards a broad range of bacteria and have been termed the 'new generation' of antimicrobial agents (Morones et al., 2005; Sintubin et al., 2012). Ag_2S nanoparticles have a documented catalytic property in reducing methylviologen (MV^{2+}) (Kryukov et al., 2004). Ng *et al.* (2013a) demonstrated that the Ag and Ag_2S nanoparticles produced by the mutant exhibited higher antibacterial activity and catalytic activity, respectively. This study suggests that, in bacterial synthesis of nanomaterials, it might be possible to 'tailor' the size

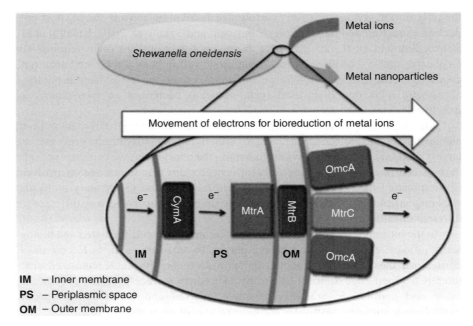

Figure 4.1. Illustration of the metal-reducing (MTR) pathway in S. oneidensis. The MTR pathway consists of a number of electron carrier proteins including CymA, MtrA, MtrC, and OmcA. The electrons obtained from bacterial metabolism are transferred across the periplasmic space to the outer membrane, where MtrC and OmcA can use the electrons to reduce metal(loid)s. *See insert for color representation of the figure.*

and activity of the extracellular biogenic nanostructures via a controlled expression of surface proteins. This is the first report exploring the possibility of using genetic manipulation to influence or even control the biosynthesis of nanomaterials.

The primary molecular sites responsible for the biosynthesis of different metal(loid)s using the same bacterium can be very different. For example, OMCs in *S. oneidensis* are essential for reductive formation of UO_2 nanoparticles (Marshall et al., 2006) but do not play an essential role in biosynthesis Pd nanoparticles, while hydrogenases facilitate the reductive formation of Pd nanoparticles (Ng et al., 2013b).

During the past decade with rapid development in nanotechnology, the design and synthesis of bimetallic nanomaterials have attracted considerable attention. Bimetallic nanomaterials often show multiple functions and superior performance in catalysis (Liu et al., 2012). For example, the addition of Au to Pd can greatly enhance the catalytic activity, selectivity, and stability of Pd catalysts (Chen et al., 2005). Bacterial synthesis of bimetallic nanomaterials is attracting increasing interest. One recent example is the synthesis of Au/Pd nanomaterials using *E. coli* and *S. oneidensis* (De Corte et al., 2012; Deplanche et al., 2012; Heugebaert et al., 2012). Biogenic Au/Pd bimetallic nanomaterials have been demonstrated to exhibit excellent catalytic activities. With sustainable development in nanotechnology, more research on biogenic nanostructures containing two or more metal(loid)s is expected.

4.2.2. Significance of Biofilms in Biosynthesis of Nanomaterials

The bioreductive generation of metallic nanomaterials using the ability of certain bacterial cells to reduce metal precursors to zero-valent state enzymatically with the presence of an electron donor is usually inhibited by the presence of oxygen. Highly heterogeneous microenvironments in biofilms likely provide anoxic conditions in the biofilm matrix that are favorable for the generation of nanomaterials even under bulk aerobic conditions. Figure 4.2 shows a representative depth-resolved profile of dissolved oxygen in a *S. oneidensis* MR-1 biofilm. Although the biofilm was growing with an oxygen-saturated medium, anoxic zones with low or depleted oxygen in the biofilm could be observed, suggesting that the inherently heterogeneous microenvironments of microbial biofilms render biofilms a promising biological system, capable of providing favorable conditions for the reductive generation of metallic nanomaterials.

The presence of a self-produced EPS matrix as a major biofilm component is an important feature of a biofilm. A role of EPS in extracellular reduction of metal(loid)s has been proposed (Marshall et al., 2006; Cao et al., 2011). For example, the formation of uraninite (UO_2) nanoparticles in uranium reduction by *S. oneidensis* was demonstrated to be modulated by EPS (Marshall et al., 2006). Further, the extracellular nanoparticles associated with EPS have been found to be an important fraction of biogenic nanoparticles. In a recent study, Ng et al. (2013a) demonstrated that, in biosynthesis of Ag_2S using *S. oneidensis*, the Ag_2S nanoparticles extracted from the cell-free supernatant fraction exhibited significantly higher catalytic activity than the Ag_2S nanoparticles associated with the cells. This finding highlights the

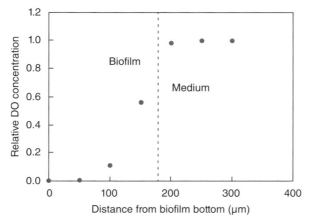

Figure 4.2. A representative depth-resolved profile of dissolved oxygen concentration in a *S. oneidensis* biofilm growing in air-saturated media.

need for novel methods that can potentially retain extracellular nanomaterials effectively in biosynthesis processes. The EPS matrix renders the use of biofilms a promising strategy to achieve this because the *in situ* generated nanomaterials in biofilms can be readily immobilized in the biofilm matrix.

Taken together, the inherently heterogeneous microenvironments favorable for bioreduction of metal(loid)s and the cohesive self-produced EPS matrix potentially applicable to effectively retain extracellular nanomaterials highlight the significance of biofilms as a promising biological system in biosynthesis of nanomaterials.

4.2.3. Synthesis of Nanomaterials using Biofilms

The use of biofilms for the synthesis of nanomaterials is an emerging topic in biotechnology and nanotechnology. Although the power of biofilms in biosynthesis of nanomaterials has not been fully explored, a variety of nanomaterials have been synthesized using biofilms, mostly electrochemically active biofilms (EABs). EABs are biofilms formed by electrochemically active microorganisms capable of exchanging electrons with solid electrodes and catalyzing redox transformation of metal(loid)s in the presence of appropriate electron donors (Borole et al., 2011; Babauta et al., 2012).

Recently, EABs were employed to synthesize Ag nanoparticles by reducing $AgNO_3$ with acetate as the electron donor (Kalathil et al., 2011*b*). Compared to conventional microbial approaches employing planktonic cells, EAB-based production was a lot faster (2h compared to 4–96h) and resulted in much smaller nanoparticles (*c.* 4nm compared to 10–50nm). Au nanoparticles as well as $Au@TiO_2$ and $Ag@TiO_2$ nanocomposites have also been synthesized using EABs (Kalathil et al., 2012; Khan et al., 2013). In addition, EABs can also be used to synthesize nanoparticles indirectly. For example, Au nanoparticles could be synthesized in the cathode of a microbial fuel cell with an EAB growing on the anode (Kalathil et al., 2013).

4.3. INTERACTION OF MICROBIAL BIOFILMS WITH NANOMATERIALS

4.3.1. Nanomaterials as Anti-biofilm Agents

Biofilms are physiologically very different from their planktonic counterparts in many aspects, one of which is their resistance to antimicrobial agents. It has been reported that biofilms are 100–1000 times more resistant to antibiotics than planktonic cells (Gristina et al., 1987; Prosser et al., 1987). As a result, commonly used biocides in hospitals and medical settings are ineffective to eradicate biofilms (Smith and Hunter, 2008). The riddle with biofilm resistance has been a topic of interest over decades (Lewis, 2001; Mah and O'Toole, 2001*b*; Davies, 2003). Diffusion limitation within biofilms due to the presence of complex matrix structure could be one protective mechanism for cells in biofilms against biocides (Stewart, 1996; Billings et al., 2013; Tseng et al., 2013). The negatively charged polysaccharides in biofilm matrix can also bind to positively charged antibiotics such as aminoglycosides

(Shigeta et al., 1997). Matrix also retards the movement of small molecules such as defensins and other antimicrobial peptides, and thus synergistically works with various antibiotic-degrading enzymes such as β-lactamase to increases the chance of their degradation (Giwercman et al., 1991). Slow growth rates of cells in biofilms could be another mechanism attributed to antibiotic resistance because certain antibiotics such as penicillin target dividing cells and preferentially kill fast-growing cells (Mah and O'Toole, 2001a; Stewart and Costerton, 2001).

Because of their inherently resistant nature, biofilms can have an enormous detrimental impact in medical and engineering settings. For example, more than half of the chronic infections in immune-compromised patients are related to biofilm formation (Donlan, 2001). In addition, biofilm formation in industrial settings causes biofouling in membrane systems and biocorrosion in water and/or oil pipelines and storage tanks, resulting in the loss of billions of dollars every year (Costerton et al., 1987). Various strategies preventing biofilm formation or disrupting established biofilms have been reported to control biofilms. Among them, bactericidal activities of nanomaterials have been exploited to combat biofilms.

Various nanomaterials have been shown to exhibit antimicrobial activities (Li et al., 2008; Huh and Kwon, 2011). The small size and high surface area to volume ratio are the unique properties that allow nanomaterials to closely interact with microorganisms (Morones et al., 2005). Although the exact mechanism of the antibacterial effect of most nanomaterials is still not conclusive, studies have shown that nanomaterials may damage cell membranes, deactivate membrane-bound enzymes, or generate reactive oxygen species (ROS; Fig. 4.3; Zhang et al., 2007; Li et al., 2010; Marambio-Jones and Hoek, 2010).

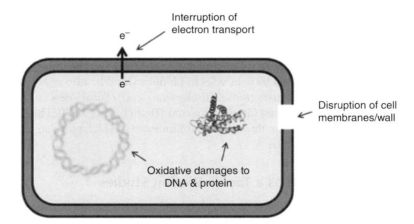

Figure 4.3. The multiple-sites-targeted action of nanomaterials against bacteria. Certain nanomaterials can enter cell membranes and physically disrupt them. Some nanomaterials can generate ROS and oxidatively damage DNA, proteins, electron transport chains, and cell membranes. Certain nanomaterials can also release toxic ions that can enter cells and cause stress responses and damages. *See insert for color representation of the figure.*

Nanomaterials with a broad spectrum of antimicrobial activities have been coated on surfaces of medical devices to prevent biofilm formation. Silver nanoparticles have been used to coat catheters, which showed an excellent inhibition of biofilm formation of six different bacteria (Roe et al., 2008). In another recent study, a silver nanoparticle coating on titanium implants showed a significant reduction in biofilm formation (Secinti et al., 2011). Cousins et al. (2007) modified the polystyrene surfaces by attaching spherical silica nanoparticles ranging from 4 to 21 nm in size and demonstrated that the modified surfaces could inhibit the growth and attachment of *C. albicans*. Similarly, surface modification with bactericidal nanomaterials has also been explored for biofouling control in membrane processes. Hybrid organic/inorganic reverse osmosis (RO) membranes with surface-coated titanium dioxide (TiO_2) nanoparticles was fabricated to minimize biofouling (Kwak et al., 2001).

In addition, incorporation of bactericidal nanomaterials is another approach to provide biomaterials with anti-biofilm properties. For example, incorporation of nanomaterials such as nanosilver and ZnO nanoparticles into resin-based dental composites has been used to control oral biofilms (Ahn et al., 2009; Aydin Sevinc and Hanley, 2010). Silver nanoparticles have been incorporated into polysulfone ultrafiltration membranes to effectively kill bacteria and to inactivate bacteriophages, preventing microbial adhesion and other types of fouling (Zodrow et al., 2009). A thin-film layer of nanocomposite membrane containing acid-modified multi-walled carbon nanotubes (MWNTs) and silver nanoparticles has been demonstrated to possess bactericidal activities against *Pseudomonas aeruginosa* PAO1 and a high permeability of water (Kim et al., 2012). Recently, a layer-by-layer assembly method was used to fabricate novel silver nanocomposite nanofiltration (NF) and forward osmosis (FO) membranes, which exhibited a better performance compared to conventional membranes as well as an excellent antibacterial activity against both Gram-positive and Gram-negative bacteria (Liu et al., 2013).

Nanomaterials can also be exploited to deliver anti-biofilm agents. Nitric oxide (NO) can disrupt cell–cell communications, inducing biofilm dispersal (McDougald et al., 2012). NO-releasing silica nanoparticles have been tested for their efficacy as anti-biofilm agents and more than 99% of cells in biofilms could be effectively killed in biofilms of *Pseudomonas aeruginosa, Escherichia coli, Staphylococcus aureus, Staphylococcus epidermidis*, and *Candida albicans* (Hetrick et al., 2009). The nanoparticles facilitate NO penetration through the biofilm matrix, which plays an important role in killing the biofilm cells.

4.3.2. Nanomaterials as a Tool in Biofilm Studies

In addition to the use as anti-biofilm agents, nanomaterials can also be used as a tool to understand biofilm structure and properties. For example, amphiphilic surface-engineered CdSe/ZnS core/shell quantum dots (QDs) have been employed to explore the hydrophilic/hydrophobic balance in bacterial biofilms and revealed a patterned distribution of hydrophobic microdomains in a biofilm matrix. The presence of hydrophobic microdomains in highly hydrated EPS networks in biofilms suggests that biofilm matrix may protect biofilm cells by accumulating toxic hydrophobic xenobiotics outside the cells.

Understanding and controlling the mechanisms underlying the emergent properties of biofilms has become a major focus of contemporary fundamental microbiology. Mechanical properties are one important aspect of biofilms that help us understand biofilm structural stability, assembly, and disassembly. Various approaches widely used in materials sciences have been implemented to study biofilm mechanical properties, and it was found that biofilms are viscoelastic liquids or solids depending on the applied stress and the timescale (Galy et al., 2012). Such studies provide a better understanding of macroscopic mechanical response of biofilms to external stress. Biofilms are highly heterogeneous and dynamic systems (Flemming and Wingender, 2010*b*); the mechanical properties are therefore also expected to be highly heterogeneous in both spatial and temporal scales. Fluorescent nanoparticles have been reported to quantify diffusion properties of biofilms with a spatial and temporal resolution (Habimana et al., 2011; Peulen and Wilkinson, 2011; Golmohamadi et al., 2013). Other biofilm properties such as biofilm porosity and matrix hydrophobicity can be estimated based on diffusion measurements.

Nanomaterials have found tremendous biomedical applications, although to date not much work has been done to apply them in microbial studies. More and more biofilm biologists recognize the importance of nanotechnology-enabled strategies in elucidating emergent biofilm properties. Specifically, the nanotechnology-enabled approaches that allow biofilm properties to be mapped with a spatio-temporal resolution will attract increasing interest from biofilm research communities.

4.4. FUTURE PERSPECTIVES

Biofilms in bio-nanotechnology, specifically the use of biofilms for nanomaterials production and the use of nanomaterials as anti-biofilm agents, are an emerging topic that provide great opportunities for researchers in both microbiology and nanotechnology. Biofilms have been demonstrated to be a superior microbial factory for the production of nanomaterials, especially metal(loid) nanomaterials. The *in-situ*-formed nanomaterials are immobilized in biofilm matrix, forming biofilm–nanomaterial hybrids which can be exploited as potential multifunctional catalytic systems with a nanocatalytic activity of the *in-situ*-formed nanomaterials and a biocatalytic activity of the microbial cells. An integrated effort from material scientists, chemical engineers, organic synthesis chemists, and microbiologists is needed to further explore the biofilm–nanomaterial hybrid catalytic systems and their applications in chemical synthesis. In addition, the impacts of biofilm matrix components on the catalytic activity of nanomaterials that are produced and immobilized in the biofilm matrix, as well as the impacts of such nanomaterials on the microbial activities in the biofilms, are also important topics to be explored.

The capabilities of biofilms in producing and immobilizing metal(loid)s nanomaterials render biofilm-based processes a promising approach for the recovery of precious metals from industrial wastes. For example, biofilms can be used to reductively immobilize precious metals such as Pt, Pd, Rh, and Au from acid leaches of electronic wastes. Further studies are required to develop and apply fed-batch or continuous biofilm-based bioprocesses for precious metal recovery from electronic wastes.

The application of nanomaterials in biofilm control has been extensively studied. However, most previously reported work has been focused on the killing and growth-inhibiting effects of nanomaterials. Since antibiotic resistance has become a global issue, antivirulence drugs that disarm pathogens by targeting the production of virulence traits such as bacterial toxins are preferable antimicrobial agents and hold great promise in combating microbial infections (Marra, 2004; Cegelski et al., 2008; Rasko and Sperandio, 2010). The development of novel nanomaterials that do not exhibit bactericidal activities but can effectively attenuate bacterial virulence will therefore become an interesting research focus in bio-nanotechnology. In addition, the use of nanomaterials for drug delivery, targeting various diseases, has been extensively reported. Little has been done in drug delivery targeting microbial biofilms. The development of nanotechnology and biofilm biology provides an opportunity for biofilm-targeted drug delivery as an emerging anti-biofilm strategy.

ACKNOWLEDGEMENTS

This work was supported by the National Research Foundation and Ministry of Education Singapore under its Research Centre of Excellence Program, Singapore Centre on Environmental Life Sciences Engineering (SCELSE) (M4330005.C70), a Start-up Grant from College of Engineering (M4080847.030), and a Seed Money Project Grant (M4081178.030.500000) from the Sustainable Earth Office, Nanyang Technological University, Singapore.

REFERENCES

Ahn, S.-J., Lee, S.-J., Kook, J.-K., Lim, B.-S. 2009. Experimental antimicrobial orthodontic adhesives using nanofillers and silver nanoparticles. *Dental Materials* 25: 206–213.

Attard, G.A., Bennett, J.A., Mikheenko, I., Jenkins, P., Guan, S., Macaskie, L.E. et al. 2013. Semi-hydrogenation of alkynes at single crystal, nanoparticle and biogenic nanoparticle surfaces: the role of defects in Lindlar-type catalysts and the origin of their selectivity. *Faraday Discussions* 162: 57–75.

Aydin Sevinc, B., Hanley, L. 2010. Antibacterial activity of dental composites containing zinc oxide nanoparticles. *Journal of Biomedical Materials Research B: Applied Biomaterials* 94: 22–31.

Babauta, J., Renslow, R., Lewandowski, Z., Beyenal, H. 2012. Electrochemically active biofilms: facts and fiction. A review. *Biofouling* 28: 789–812.

Bao, H., Lu, Z., Cui, X., Qiao, Y., Guo, J., Anderson, J.M., Li, C.M. 2010. Extracellular microbial synthesis of biocompatible CdTe quantum dots. *Acta Biomaterialia* 6: 3534–3541.

Beheshti, N., Soflaei, S., Shakibaie, M., Yazdi, M.H., Ghaffarifar, F., Dalimi, A., Shahverdi, A.R. 2013. Efficacy of biogenic selenium nanoparticles against Leishmania major: In vitro and in vivo studies. *Journal of Trace Elements in Medicine and Biology* 27: 203–207.

Belchik, S.M., Kennedy, D.W., Dohnalkova, A.C., Wang, Y., Sevinc, P.C., Wu, H. et al. 2011. Extracellular reduction of hexavalent chromium by cytochromes MtrC and OmcA of *Shewanella oneidensis* MR-1. *Applied and Environmental Microbiology* 77: 4035–4041.

REFERENCES

Bennett, J.A., Attard, G.A., Deplanche, K., Casadesus, M., Huxter, S.E., Macaskie, L.E., Wood, J. 2012. Improving selectivity in 2-butyne-1,4-diol hydrogenation using biogenic Pt catalysts. *ACS Catalysis* 2: 504–511.

Bennett, J.A., Mikheenko, I.P., Deplanche, K., Shannon, I.J., Wood, J., Macaskie, L.E. 2013. Nanoparticles of palladium supported on bacterial biomass: New re-usable heterogeneous catalyst with comparable activity to homogeneous colloidal Pd in the Heck reaction. *Applied Catalysis B: Environmental* 140–141: 700–707.

Bigall, N.C., Reitzig, M., Naumann, W., Simon, P., van Pee, K.H., Eychmuller, A. 2008. Fungal templates for noble-metal nanoparticles and their application in catalysis. *Angewandte Chemie-International Edition* 47: 7876–7879.

Billings, N., Ramirez Millan, M., Caldara, M., Rusconi, R., Tarasova, Y., Stocker, R., Ribbeck, K. 2013. The extracellular matrix component Psl provides fast-acting antibiotic defense in *Pseudomonas aeruginosa* biofilms. *Plos Pathogens* 9: e1003526.

Bishop, P.L. 2007. The role of biofilms in water reclamation and reuse. *Water Science and Technology* 55: 19–26.

Borole, A.P., Reguera, G., Ringeisen, B., Wang, Z.W., Feng, Y.J., Kim, B.H. 2011. Electroactive biofilms: current status and future research needs. *Energy & Environmental Science* 4: 4813–4834.

Bunge, M., Søbjerg, L.S., Rotaru, A.-E., Gauthier, D., Lindhardt, A.T., Hause, G. et al. 2010. Formation of palladium(0) nanoparticles at microbial surfaces. *Biotechnology and Bioengineering* 107: 206–215.

Cao, B., Ahmed, B., Beyenal, H. 2010. Immobilization of uranium in groundwater using biofilms. In *Emerging Environmental Technologies* (ed. Shah, V.) Springer Sciences+Business Media B.V., pp. 1–34.

Cao, B., Ahmed, B., Kennedy, D.W., Wang, Z., Shi, L., Marshall, M.J. et al. 2011. Contribution of extracellular polymeric substances from *Shewanella* sp. HRCR-1 biofilms to U(VI) immobilization. *Environmental Science & Technology* 45: 5483–5490.

Cegelski, L., Marshall, G.R., Eldridge, G.R., Hultgren, S.J. 2008. The biology and future prospects of antivirulence therapies. *Nature Reviews Microbiology* 6: 17–27.

Chen, M.S., Kumar, D., Yi, C.W., Goodman, D.W. 2005. The promotional effect of gold in catalysis by palladium-gold. *Science* 310: 291–293.

Costerton, J.W., Cheng, K.J., Geesey, G.G., Ladd, T.I., Nickel, J.C., Dasgupta, M., Marrie, T.J. 1987. Bacterial biofilms in nature and disease. *Annual Review of Microbiology* 41: 435–464.

Cousins, B.G., Allison, H.E., Doherty, P.J., Edwards, C., Garvey, M.J., Martin, D.S., Williams, R.L. 2007. Effects of a nanoparticulate silica substrate on cell attachment of *Candida albicans*. *Journal of Applied Microbiology* 102(3): 757–765.

Davies, D. 2003. Understanding biofilm resistance to antibacterial agents. *Nature Reviews Drug Discovery* 2: 114–122.

De Corte, S., Sabbe, T., Hennebel, T., Vanhaecke, L., De Gusseme, B., Verstraete, W., Boon, N. 2012. Doping of biogenic Pd catalysts with Au enables dechlorination of diclofenac at environmental conditions. *Water Research* 46: 2718–2726.

Deplanche, K., Woods, R.D., Mikheenko, I.P., Sockett, R.E., Macaskie, L.E. 2008. Manufacture of stable palladium and gold nanoparticles on native and genetically engineered flagella scaffolds. *Biotechnology and Bioengineering* 101: 873–880.

Deplanche, K., Caldelari, I., Mikheenko, I.P., Sargent, F., Macaskie, L.E. 2010. Involvement of hydrogenases in the formation of highly catalytic Pd(0) nanoparticles by bioreduction of Pd(II) using Escherichia coli mutant strains. *Microbiology* 156: 2630–2640.

Deplanche, K., Merroun, M.L., Casadesus, M., Tran, D.T., Mikheenko, I.P., Bennett, J.A. et al. 2012. Microbial synthesis of core/shell gold/palladium nanoparticles for applications in green chemistry. *Journal of the Royal Society Interface* 9: 1705–1712.

Dhanjal, S., Cameotra, S. 2010. Aerobic biogenesis of selenium nanospheres by Bacillus cereus isolated from coalmine soil. *Microbial Cell Factories* 9: 52.

Dobias, J., Suvorova, E.I., Bernier-Latmani, R. 2011. Role of proteins in controlling selenium nanoparticle size. *Nanotechnology* 22: 195605.

Donlan, R.M. 2001. Biofilms and device-associated infections. *Emerging Infectious Diseases* 7: 277–281.

Flemming, H. 2011. The perfect slime. *Colloids and Surfaces B: Biointerfaces* 86: 251–259.

Flemming, H., Wingender, J. 2010a. The biofilm matrix. *Nature Reviews in Microbiology* 8: 623–633.

Flemming, H.C., Wingender, J. 2010b. The biofilm matrix. *Nature Reviews Microbiology* 8: 623–633.

Fredrickson, J.K., Romine, M.F., Beliaev, A.S., Auchtung, J.M., Driscoll, M.E., Gardner, T.S. et al. 2008. Towards environmental systems biology of Shewanella. *Nature Reviews Microbiology* 6: 592–603.

Gadd, G.M. 2010. Metals, minerals and microbes: geomicrobiology and bioremediation. *Microbiology* 156: 609–643.

Galy, O., Latour-Lambert, P., Zrelli, K., Ghigo, J.M., Beloin, C., Henry, N. 2012. Mapping of bacterial biofilm local mechanics by magnetic microparticle actuation. *Biophysical Journal* 103: 1400–1408.

Giwercman, B., Jensen, E.T., Høiby, N., Kharazmi, A., Costerton, J.W. 1991. Induction of beta-lactamase production in Pseudomonas aeruginosa biofilm. *Antimicrobial Agents and Chemotherapy* 35: 1008–1010.

Golmohamadi, M., Clark, R.J., Veinot, J.G.C., Wilkinson, K.J. 2013. The role of charge on the diffusion of solutes and nanoparticles (silicon nanocrystals, nTiO(2), nAu) in a biofilm. *Environmental Chemistry* 10: 34–41.

Gristina, A.G., Hobgood, C.D., Webb, L.X., Myrvik, Q.N. 1987. Adhesive colonization of biomaterials and antibiotic resistance. *Biomaterials* 8: 423–426.

Gurunathan, S., Kalishwaralal, K., Vaidyanathan, R., Venkataraman, D., Pandian, S.R.K., Muniyandi, J. et al. 2009. Biosynthesis, purification and characterization of silver nanoparticles using Escherichia coli. *Colloids and Surfaces B: Biointerfaces* 74: 328–335.

Habimana, O., Steenkeste, K., Fontaine-Aupart, M.P., Bellon-Fontaine, M.N., Kulakauskas, S., Briandet, R. 2011. Diffusion of nanoparticles in biofilms is altered by bacterial cell wall hydrophobicity. *Applied and Environmental Microbiology* 77: 367–368.

He, S., Guo, Z., Zhang, Y., Zhang, S., Wang, J., Gu, N. 2007. Biosynthesis of gold nanoparticles using the bacteria *Rhodopseudomonas capsulata*. *Materials Letters* 61: 3984–3987.

Hennebel, T., Van Nevel, S., Verschuere, S., De Corte, S., De Gusseme, B., Cuvelier, C. et al. 2011. Palladium nanoparticles produced by fermentatively cultivated bacteria as catalyst for diatrizoate removal with biogenic hydrogen. *Applied Microbiology & Biotechnology* 91: 1435–1445.

Hetrick, E.M., Shin, J.H., Paul, H.S., Schoenfisch, M.H. 2009. Anti-biofilm efficacy of nitric oxide-releasing silica nanoparticles. *Biomaterials* 30: 2782–2789.

Heugebaert, T.S.A., De Corte, S., Sabbe, T., Hennebel, T., Verstraete, W., Boon, N., Stevens, C.V. 2012. Biodeposited Pd/Au bimetallic nanoparticles as novel Suzuki catalysts. *Tetrahedron Letters* 53: 1410–1412.

Huh, A.J., Kwon, Y.J. 2011. 'Nanoantibiotics': A new paradigm for treating infectious diseases using nanomaterials in the antibiotics resistant era. *Journal of Controlled Release* 156: 128–145.

Jain, N., Bhargava, A., Majumdar, S., Tarafdar, J., Panwar, J. 2011. Extracellular biosynthesis and characterization of silver nanoparticles using Aspergillus flavus NJP08: A mechanism perspective. *Nanoscale* 3: 635–641.

Jiang, S., Lee, J.-H., Kim, M.-G., Myung, N.V., Fredrickson, J.K., Sadowsky, M.J., Hur, H.-G. 2009. Biogenic formation of As-S nanotubes by diverse shewanella strains. *Applied and Environmental Microbiology* 75: 6896–6899.

Jung, J.H., Park, T.J., Lee, S.Y., Seo, T.S. 2012. Homogeneous biogenic paramagnetic nanoparticle synthesis based on a microfluidic droplet generator. *Angewandte Chemie* 51: 5634–5637.

Kalathil, S., Lee, J., Cho, M. 2011a. Electrochemically active biofilm-mediated synthesis of silver nanoparticles in water. *Green Chemistry* 13: 1482–1485.

Kalathil, S., Lee, J., Cho, M.H. 2011b. Electrochemically active biofilm-mediated synthesis of silver nanoparticles in water. *Green Chemistry* 13: 1482–1485.

Kalathil, S., Khan, M.M., Banerjee, A.N., Lee, J., Cho, M.H. 2012. A simple biogenic route to rapid synthesis of Au@TiO2 nanocomposites by electrochemically active biofilms. *Journal of Nanoparticle Research* 14: 1051.

Kalathil, S., Lee, J., Cho, M.H. 2013. Gold nanoparticles produced *in situ* mediate bioelectricity and hydrogen production in a microbial fuel cell by quantized capacitance charging. *Chemsuschem* 6: 246–250.

Kalimuthu, K., Suresh Babu, R., Venkataraman, D., Bilal, M., Gurunathan, S. 2008. Biosynthesis of silver nanocrystals by Bacillus licheniformis. *Colloids and Surfaces B: Biointerfaces* 65: 150–153.

Khan, M.M., Ansari, S.A., Amal, M.I., Lee, J., Cho, M.H. 2013. Highly visible light active Ag@TiO2 nanocomposites synthesized using an electrochemically active biofilm: a novel biogenic approach. *Nanoscale* 5: 4427–4435.

Kim, E.-S., Hwang, G., Gamal El-Din, M., Liu, Y. 2012. Development of nanosilver and multi-walled carbon nanotubes thin-film nanocomposite membrane for enhanced water treatment. *Journal of Membrane Science* 394–395: 37–48.

Konishi, Y., Tsukiyama, T., Tachimi, T., Saitoh, N., Nomura, T., Nagamine, S. 2007a. Microbial deposition of gold nanoparticles by the metal-reducing bacterium Shewanella algae. *Electrochimica Acta* 53: 186–192.

Konishi, Y., Ohno, K., Saitoh, N., Nomura, T., Nagamine, S., Hishida, H. et al. 2007b. Bioreductive deposition of platinum nanoparticles on the bacterium Shewanella algae. *Journal of Biotechnology* 128: 648–653.

Kryukov, A.I., Stroyuk, A.L., Zin'chuk, N.N., Korzhak, A.V., Kuchmii, S.Y. 2004. Optical and catalytic properties of Ag2S nanoparticles. *Journal of Molecular Catalysis a-Chemical* 221: 209–221.

Kumar, C., Mamidyala, S. 2011. Extracellular synthesis of silver nanoparticles using culture supernatant of Pseudomonas aeruginosa. *Colloids and Surfaces B: Biointerfaces* 84: 462–466.

Kwak, S.Y., Kim, S.H., Kim, S.S. 2001. Hybrid organic/inorganic reverse osmosis (RO) membrane for bactericidal anti-fouling. 1. Preparation and characterization of TiO2 nanoparticle self-assembled aromatic polyamide thin-film-composite (TFC) membrane. *Environmental Science and Technology* 35: 2388–2394.

Lewis, K. (2001) Riddle of biofilm resistance. *Antimicrobial Agents and Chemotherapy* 45: 999–1007.

Li, Q., Mahendra, S., Lyon, D.Y., Brunet, L., Liga, M.V., Li, D., Alvarez, P.J. 2008. Antimicrobial nanomaterials for water disinfection and microbial control: potential applications and implications. *Water Research* 42: 4591–4602.

Li, W.-R., Xie, X.-B., Shi, Q.-S., Zeng, H.-Y., Ou-Yang, Y.-S., Chen, Y.-B. 2010. Antibacterial activity and mechanism of silver nanoparticles on Escherichia coli. *Applied Microbiology and Biotechnology* 85: 1115–1122.

Li, Z., Chung, S.W., Nam, J.M., Ginger, D.S., Mirkin, C.A. 2003. Living templates for the hierarchical assembly of gold nanoparticles. *Angewandte Chemie* 42: 2306–2309.

Liu, X.W., Wang, D.S., Li, Y.D. 2012. Synthesis and catalytic properties of bimetallic nanomaterials with various architectures. *Nano Today* 7: 448–466.

Liu, X., Qi, S., Li, Y., Yang, L., Cao, B., Tang, C.Y. 2013. Synthesis and characterization of novel antibacterial silver nanocomposite nanofiltration and forward osmosis membranes based on layer-by-layer assembly. *Water Research* 47: 3081–3092.

Lloyd, J., Byrne, J., Coker, V. 2011. Biotechnological synthesis of functional nanomaterials. *Current Opinion in Biotechnology* 22: 509–515.

Mah, T.F., O'Toole, G.A. 2001a. Mechanisms of biofilm resistance to antimicrobial agents. *Trends in Microbiology* 9: 34–39.

Mah, T.F.C., O'Toole, G.A. 2001b. Mechanisms of biofilm resistance to antimicrobial agents. *Trends in Microbiology* 9: 34–39.

Marambio-Jones, C., Hoek, E.V. 2010. A review of the antibacterial effects of silver nanomaterials and potential implications for human health and the environment. *Journal of Nanoparticle Research* 12: 1531–1551.

Marra, A. 2004. Can virulence factors be viable antibacterial targets? *Expert Review of Anti-Infective Therapy* 2: 61–72.

Marshall, M.J., Beliaev, A.S., Dohnalkova, A.C., Kennedy, D.W., Shi, L., Wang, Z. et al. 2006. c-Type cytochrome-dependent formation of U(IV) nanoparticles by Shewanella oneidensis. *PLoS Biology* 4: e268.

Martin, K.J., Nerenberg, R. 2012. The membrane biofilm reactor (MBfR) for water and wastewater treatment: Principles, applications, and recent developments. *Bioresource Technology* 122: 83–94.

McDougald, D., Rice, S.A., Barraud, N., Steinberg, P.D., Kjelleberg, S. 2012. Should we stay or should we go: mechanisms and ecological consequences for biofilm dispersal. *Nature Reviews Microbiology* 10: 39–50.

Meerburg, F., Hennebel, T., Vanhaecke, L., Verstraete, W., Boon, N. 2012. Diclofenac and 2-anilinophenylacetate degradation by combined activity of biogenic manganese oxides and silver. *Microbial Biotechnology* 5: 388–395.

Mohammed Fayaz, A., Girilal, M., Rahman, M., Venkatesan, R., Kalaichelvan, P.T. 2011. Biosynthesis of silver and gold nanoparticles using thermophilic bacterium Geobacillus stearothermophilus. *Process Biochemistry* 46: 1958–1962.

Morones, J.R., Elechiguerra, J.L., Camacho, A., Holt, K., Kouri, J.B., Ramirez, J.T., Yacaman, M.J. 2005. The bactericidal effect of silver nanoparticles. *Nanotechnology* 16: 2346–2353.

Narayanan, K.B., Sakthivel, N. 2010. Biological synthesis of metal nanoparticles by microbes. *Advances in Colloid and Interface Science* 156: 1–13.

Ng, C., Sivakumar, K., Liu, X., Madhaiyan, M., Ji, L., Yang, L. et al. 2013*a*. Influence of outer membrane c-type cytochromes on particle size and activity of extracellular polymeric substances produced by *Shewanella oneidensis*. *Biotechnology and Bioengineering* 110: 1831–1837.

Ng, C.K., Tan, T.K.C., Song, H., Cao, B. 2013*b*. Reductive formation of palladium nanoparticles by *Shewanella oneidensis*: Role of outer membrane cytochromes and hydrogenases. *RSC Advances* 3: 22498–22503.

Ng, C.K., Sivakumar, K., Liu, X., Madhaiyan, M., Ji, L., Yang, L. et al. 2013*c*. Influence of outer membrane c-type cytochromes on particle size and activity of extracellular nanoparticles produced by Shewanella oneidensis. *Biotechnology and Bioengineering* 110: 1831–1837.

Peulen, T.O., Wilkinson, K.J. 2011. Diffusion of nanoparticles in a biofilm. *Environmental Science & Technology* 45: 3367–3373.

Prosser, B.L., Taylor, D., Dix, B.A., Cleeland, R. 1987. Method of evaluating effects of antibiotics on bacterial biofilm. *Antimicrobial Agents and Chemotherapy* 31: 1502–1506.

Rasko, D.A., Sperandio, V. 2010. Anti-virulence strategies to combat bacteria-mediated disease. *Nature Reviews Drug Discovery* 9: 117–128.

Redwood, M.D., Deplanche, K., Baxter-Plant, V.S., Macaskie, L.E. 2008. Biomass-supported palladium catalysts on *Desulfovibrio desulfuricans* and *Rhodobacter sphaeroides*. *Biotechnology and Bioengineering* 99: 1045–1054.

Roe, D., Karandikar, B., Bonn-Savage, N., Gibbins, B., Roullet, J.B. 2008. Antimicrobial surface functionalization of plastic catheters by silver nanoparticles. *Journal of Antimicrobial Chemotherapy* 61: 869–876.

Sathish, K.S., Amutha, R., Arumugam, P., Berchmans, S. 2011. Synthesis of gold nanoparticles: an ecofriendly approach using Hansenula anomala. *ACS Applied Materials & Interfaces* 3: 1418–1425.

Secinti, K.D., Ozalp, H., Attar, A., Sargon, M.F. 2011. Nanoparticle silver ion coatings inhibit biofilm formation on titanium implants. *Journal of Clinical Neuroscience* 18: 391–395.

Shi, L., Chen, B., Wang, Z., Elias, D.A., Mayer, M.U., Gorby, Y.A. et al. 2006. Isolation of a high-affinity functional protein complex between OmcA and MtrC: Two outer membrane decaheme c-type cytochromes of Shewanella oneidensis MR-1. *Journal of Bacteriology* 188: 4705–4714.

Shigeta, M., Tanaka, G., Komatsuzawa, H., Sugai, M., Suginaka, H., Usui, T. 1997. Permeation of antimicrobial agents through *Pseudomonas aeruginosa* biofilms: a simple method. *Chemotherapy* 43: 340–345.

Singh, R., Paul, D., Jain, R.K. 2006. Biofilms: implications in bioremediation. *Trends in Microbiology* 14: 389–397.

Sintubin, L., Verstraete, W., Boon, N. 2012. Biologically produced nanosilver: current state and future perspectives. *Biotechnology and Bioengineering* 109(10): 2422–2436.

Smith, K., Hunter, I.S. 2008. Efficacy of common hospital biocides with biofilms of multidrug resistant clinical isolates. *Journal of Medical Microbiology* 57: 966–973.

Stewart, P.S. 1996. Theoretical aspects of antibiotic diffusion into microbial biofilms. *Antimicrobial Agents and Chemotherapy* 40: 2517–2522.

Stewart, P.S., Costerton, J.W. 2001. Antibiotic resistance of bacteria in biofilms. *Lancet* 358: 135–138.

Suresh, A.K., Pelletier, D.A., Wang, W., Moon, J.W., Gu, B., Mortensen, N.P. et al. 2010. Silver nanocrystallites: biofabrication using Shewanella oneidensis, and an evaluation of their comparative toxicity on gram-negative and gram-positive bacteria. *Environmental Science & Technology* 44: 5210–5215.

Suresh, A.K., Pelletier, D.A., Wang, W., Broich, M.L., Moon, J.W., Gu, B. et al. 2011. Biofabrication of discrete spherical gold nanoparticles using the metal-reducing bacterium Shewanella oneidensis. *Acta Biomaterialia* 7: 2148–2152.

Sweeney, R.Y., Mao, C., Gao, X., Burt, J.L., Belcher, A.M., Georgiou, G., Iverson, B.L. 2004. Bacterial biosynthesis of cadmium sulfide nanocrystals. *Chemistry & Biology* 11: 1553–1559.

Tseng, B.S., Zhang, W., Harrison, J.J., Quach, T.P., Song, J.L., Penterman, J. et al. 2013. The extracellular matrix protects *Pseudomonas aeruginosa* biofilms by limiting the penetration of tobramycin. *Environmental Microbiology* 15(10): 2865–2878.

Yadav, V., Sharma, N., Prakash, R., Raina, K., Bharadwaj, L., Tejo Prakash, N. 2008. Generation of Selenium containing Nano-structures by soil bacterium, Pseudomonas aeruginosa. *Biotechnology* 7: 299–304.

Yates, M.D., Cusick, R.D., Logan, B.E. 2013. Extracellular palladium nanoparticle production using *Geobacter sulfurreducens*. *ACS Sustainable Chemistry & Engineering* 1(9): 1165–1171.

Zhang, L., Jiang, Y., Ding, Y., Povey, M., York, D. 2007. Investigation into the antibacterial behaviour of suspensions of ZnO nanoparticles (ZnO nanofluids). *Journal of Nanoparticle Research* 9: 479–489.

Zodrow, K., Brunet, L., Mahendra, S., Li, D., Zhang, A., Li, Q., Alvarez, P.J. 2009. Polysulfone ultrafiltration membranes impregnated with silver nanoparticles show improved biofouling resistance and virus removal. *Water Research* 43: 715–723.

5

EXTREMOPHILES AND BIOSYNTHESIS OF NANOPARTICLES: CURRENT AND FUTURE PERSPECTIVES

Jingyi Zhang, Jetka Wanner, and Om V. Singh

Division of Biological and Health Sciences, University of Pittsburgh, Bradford, PA, USA

5.1. INTRODUCTION

Nanotechnology is a fast-developing cutting-edge technology with wide-ranging applications in different areas of science and technology. It is defined as the creation, manipulation, and implication of material on a nanoscale (0.1–1000 nm). The prefix "nano" is derived from Greek meaning "dwarf" and refers to the extremely small measurement of one-billionth of a meter (10^{-9} m) in size. For over a decade, nanotechnology has progressed in diverse fields such as nanomaterials, nanomedicine, and nanorobotics.

Among nanomaterials, nanoparticles (NPs) are considered the building blocks of nanotechnology (Rotello and Shenhar, 2003; Davis et al., 2008; Rai and Duran, 2011). Most metallic NPs range in size from 0.1 to 1000 nm and comprise different shapes, including triangular, spherical, rod, and other irregular shapes (Narayanan and Sakthivel, 2010; Thakkar et al., 2010). Table 5.1 summarizes the major types of NPs. Because of their extremely small size and large surface area, NPs have unusual characteristics that make them valuable in many fields including medicine and industry.

Bio-Nanoparticles: Biosynthesis and Sustainable Biotechnological Implications, First Edition.
Edited by Om V. Singh.
© 2015 John Wiley & Sons, Inc. Published 2015 by John Wiley & Sons, Inc.

TABLE 5.1. Major Types of Nanoparticles in Nanotechnology

Types of NPs	Applicability	Mode of synthesis	References
Metal nanoparticles	Magnetic separation, preconcentration of target analytes, targeted drug delivery, vehicles for gene and drug delivery and more importantly diagnostic imaging, etc.	Biological and chemical synthesis methods	Mukherjee et al. (2008); Thakkar et al. (2010)
Semiconductor nanoparticles	Single electron transistors, tunnel junctions, magnetic spin valves, integrated circuits, optoelectronic devices, memory devices, etc.	Physical synthesis methods	Ding et al. (2006)
Superparamagnetic nanoparticles	Imaging, target drug delivery, gene therapy, biomolecular separation, early cancer diagnosis, etc.	Chemical synthesis methods	Arruebo et al. (2007); Mahmoudi et al. (2011)
Organic nanoparticles	Cross-membrane permeation, drug delivery, liposomal-nanovesicles, antibody conjugation, reporter particles in immunoassays, nano-micelles, etc.	Chemical synthesis methods	Chen et al. (2005); Kim et al. (2009); Ramasamy et al. (2009)

In the field of medicine, researchers have demonstrated the applications of NPs in disease detection, drug delivery, and even tissue regeneration (Han et al., 2007; Sapir et al., 2014). Silver (Ag) NPs can be used as biomarkers to detect infectious states before patients show symptoms (Burke et al., 2013). In another study, highly fluorescent europium (III)-chelate doped NPs (c. 107 nm) coated with anti-hexon antibodies helped in the direct detection of adenoviruses (Saini and Everts, 2011). Fluorescent silica NPs have also been used as ultrasensitive detection agents for viruses and bacteria. In gene therapy, siRNA molecules coated with gold (Au) NPs have been used as nucleic acid delivery vehicles (Oishi et al., 2006). Magnetic NPs can serve as magnetic labels in cell sheets and aid blood vessel reconstruction in tissue engineering (Ito et al., 2005). Epithelial ovarian cancer, a common malignancy of the female genital tract, can be treated with the use of AuNPs which inhibit the progression of ovarian growth and metastasis (Bamberger and Perrett, 2002; Bhattacharya and Mukherjee, 2008).

INTRODUCTION

The NPs currently available for industrial use encompass a wide diversity of applications. A synthetic self-repairing and pressure-sensitive skin composed of nickel NPs and polymer has been developed for use in emerging fields such as biomimetic prosthetics and soft robotics. Based on their antibacterial properties, silver-NP-coated polyurethane foams have been used in an antibacterial water filtration system that was significantly effective: no bacterium was detected in the output water when the input water had a bacterial load of $1-10 \times 10^5$ CFU mL^{-1} (Jain and Pradeep, 2005; Phong et al., 2009). AgNPs are utilized in electronics, nanowires, and electric circuits (Alshehri et al., 2012). Such enormous use of NPs relies upon fabrication and synthesis of NPs with well-defined morphology and controlled monodispersity. The traditional methods for NP synthesis usually involve harsh chemicals and a complex downstream process, which result in high processing costs and the generation of toxic by-products (Thakkar et al., 2010). Regulatory agencies have raised concerns about the safety of chemically synthesized NPs. One study sponsored by the United States Environmental Protection Agency (EPA) explored the effects of titanium oxide (TiO_2) nanoparticles on two aquatic species: *Daphnia magna* and *Japanese medaka*. The results showed that under sunlight, NPs causes acute phototoxicity in the aquatic organisms (Ma et al., 2012). Biological synthesis of NPs therefore appears to be the best alternative to secure the future of nanotechnology.

The biological synthesis of NPs is carried out by naturally occurring microorganisms (bacteria and fungi) and plant extracts, and has many advantages over chemical and physical methods (Klaus et al., 1999; Ahmad et al., 2002, 2007; Das et al., 2012). Although the biosynthesis of NPs with well-defined sizes and shapes is a challenging task, biosynthesized NPs allow eco-friendly downstream processing and safest passage of usage. The search is underway for an ideal category of microorganisms to biosynthesize morphologically defined NPs in a fast, easy, and inexpensive way. Microorganisms that thrive in extreme environments (i.e., extremophiles) have received a great deal of attention (Tian and Hua, 2010; Oren, 2008, 2013; Singh, 2013; Gabani and Singh, 2013). The extreme conditions in which extremophiles thrive can be physical (e.g., temperature, pressure, or radiation) or chemical (e.g., salinity, pH, or occurrence of wide variety of chemicals). The survival of extremophiles under non-standard conditions has led to the assumption that their metabolic products and enzymes have been optimized for these conditions.

Novel routes by which extremophiles can be employed to generate NPs have yet to be discovered. However, a fundamental understanding of the bioreduction of metallic ions in standard microorganisms would assist in determining the physical and chemical parameters required for continuous production of well-defined NPs from extremophiles in a fermentative solution. Therefore, this chapter discusses the extremophiles explored for biosynthesis of NPs and the mechanisms by which standard microorganisms biosynthesize NPs in order to correlate unexplored routes for producing NPs from extremophiles. In view of the tremendous industrial potential of NPs created using extremophiles, the chapter also sheds light on the specifics of fermentation media and recovery of NPs from the microbiological process using standard microorganisms. A brief discussion of limitations and challenges helps to further outline the future of extremophile-mediated nanotechnology.

5.2. SYNTHESIS OF NANOPARTICLES

The synthesis of NPs can be classified into two categories: top-down and bottom-up. In the top-down approach, the source materials are "milled" by mechanical devices (e.g., an ordinary or a planetary ball mill) from bulk-sized down to nanoscale (Skaff and Emrick, 2004). Bottom-up synthesis processes are subcategorized into vapor (e.g., pyrolysis, inert gas condensation) and liquid phases (e.g., sol-gel, solvothermal) fabrication (Skaff and Emrick, 2004). Both types of methods provide monodisperse NPs that are compatible for binding with certain ligands. However, these methods are fraught with problems including the use of toxic solvents, the generation of hazardous by-products, low production yield, and high processing cost. Therefore, an environmentally benign nanoparticle production technology is still needed. The biosynthesis of NPs using fungi, bacteria, viruses, algae, and plants has attracted interest (Ahmad et al., 2007; Narayanan and Sakthivel, 2010; Thakkar et al., 2010; Ma et al., 2012). Biological synthesis of NPs may be the most suitable, cost-effective, and eco-friendly way to produce them.

5.2.1. Microorganisms: An Asset in Nanoparticle Biosynthesis

Microorganisms, bacteria, algae, fungi, yeasts, viruses, and protozoans play a crucial role in the ecosystem: they act as decomposers in nutrient recycling. Among other natural resources, bacteria have been intensively researched for biological NP synthesis due to the relative ease of manipulating them. Many bacteria can produce NPs, with various morphologies sized from 0.1 nm to 1000 nm, at room temperature and under mild culture conditions (Bai and Zhang, 2009; Balaji et al., 2009; Narayanan and Sakthivel, 2010; Thakkar et al., 2010). The microbial synthesis of NPs occurs either intra- or extracellularly. The several known mechanisms include biosorption and bioreduction with aid from proteins (reductase, etc.) inside or outside the bacterial cells (Fig. 5.1).

Microbial biosynthesis of NPs is not uncommon (Table 5.2); however, the particle size and morphology depends on several physical and chemical parameters such as the pH of the solution and incubation time. Lin and colleagues (2005) described the silver reduction ability of *Lactobacillus* sp. strain A09. In the study, *Lactobacillus* sp. strain A09 was able to reduce Ag^+ to Ag^0 by biosorption and bioreduction and achieve an optimum production rate at pH 5 and 30 °C within 24 h. He et al. (2007) have reported that *Rhodopseudomonas capsulata* bacteria can produce AuNPs of different shapes and sizes in different pH conditions. Interestingly, the AuNPs are mainly spherical and in the size range 10–20 nm at pH 7; however, at pH 4, triangular nanoplates in the size range of 40–50 nm became predominant.

5.2.2. Extremophiles in Nanoparticle Biosynthesis

Microbes live in every part of the biosphere, including the atmosphere, soil, hot springs, the ocean floor, caves, etc. Microorganisms living in environments that are lethal to normal life forms are called extremophiles. Extremophiles have the ability

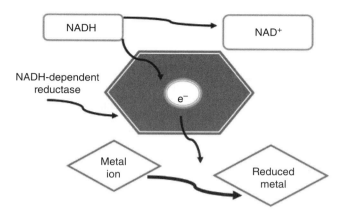

Figure 5.1. Bioreduction of metal ions on the surface of the cell wall due to the electrostatic interaction between the metal ions and positively charged groups in the enzymes of the cell wall. The enzymatic reduction of the metal ions soon follows, which are reduced by NADH-dependent reductase such as enzymes. The reductase gains electrons from NADH and oxidized NAD⁺. When the enzymes oxidize the metal ions simultaneously reduce, causing the formation of nanoparticles. *See insert for color representation of the figure.*

to thrive in extreme conditions such as: desert; deep sediment; high and low temperature, pH, and pressure; and various types of radiation (Singh, 2013). There have been attempts to exploit extremophiles for their ability to synthesize NPs, as summarized in Table 5.3.

Metallic nanoparticles (MNPs) have been biosynthesized by various microorganisms (Ahmad et al., 2003; Kowshik et al., 2002; Lengke et al., 2006). Metallotolerant microbes exist in diverse bacterial groups, but are mostly aerobic and facultative aerobic chemo-heterotrophic microorganisms in the *Bacillus*, *Pseudomonas*, *Escherichia*, and *Staphylococcus* genera (Silver and Phung, 1996; Bruins et al., 2000; Clausen, 2000). Among metallotolerant microbes, *Ferroplasma* sp. and *Cupriavidus metallidurans* were found capable of tolerating high levels of dissolved heavy metals in solution, such as gold, silver, copper, cadmium, and zinc.

Inspired by the amazing flexibility of these microorganisms to survive under unusual conditions, a number of studies have explored their metal-resistance and self-detoxification mechanisms (Altimira et al., 2012; Lima de Silva et al., 2012; Maynaud et al., 2013; Muehe et al., 2013; Wang et al., 2013). These remarkable functions are significantly valuable for the biological synthesis of NPs. Metal-resistant bacteria isolated from gold mines are able to precipitate metals. In the earliest studies, bacteria from gold mines such as *Pedomicrobium*-like budding bacteria (Mann, 1992) and *Acidithiobacillus thiooxidans* spp. (Lengke and Southam, 2005, 2006; Lengke et al., 2006, 2007), were reported to accumulate gold. Heterotrophic sulfate-reducing bacterial enrichment from a gold mine in South Africa was used to destabilize gold (I)-thiosulfate complex ($Au(S_2O_3)_2^{3-}$) to elemental gold particles with dimensions <10 nm.

TABLE 5.2. Standard Microorganisms Involved in the Biosynthesis of Nanoparticles

Microorganisms	Types of NPs	References
Bacteria		
Plectonema boryanum	Au	Lengke et al. (2006, 2007)
Acidithiobacillus thiooxidans	Au	Lengke and Southam (2005)
Marinobacter Pelagius sp.	Au	Sharma et al. (2012)
Rhodopseudomonas capsulata	Au	He et al. (2007)
Pseudomonas stutzeri	Ag and Au	Joerger et al. (2000), Gericke and Pinches (2006)
Corynebacterium	Ag	Zhang et al. (2005)
Lactobacillus	Ag and Au	Lin et al. (2005), Nair and Pradeep (2002)
Morganella sp.		Parikh et al. (2008)
Sulfurospirillum barnesii, Selenihalanaerobacter shriftii, Bacillus selenitireducens	Se	Oremland et al. (2004)
Escherichia coli	CdS	Sweeney et al. (2004)
Bacillus licheniformis	Ag	Kalimuthu et al. (2008)
Shewanella algae	Au	Konishi et al. (2007)
Desulfovibrio desulfuricans	Pd	Ping et al. (2002)
Shewanella oneidensis	Pd	de Windt et al. (2005, 2006)
Fungi		
Penicillium brevicompactum	Au	Mishra et al. (2011)
Penicillium fellutanum	Ag	Kathiresan et al. (2009)
Fusarium oxysporum	Au, Ag, Au–Ag, Zr, Pt	Ahmad et al. (2002), Riddin et al. (2006), Bansal et al. (2004), Karbasian et al. (2008)
Fusarium semitectum	Ag	Basavaraja et al. (2008)
Humicola sp.	$CuAlO_2$	Ahmad et al. (2007)
Yarrowia lipolytica	Au	Pimprikar et al. (2009)
Phanerochaete chrysosporium	Ag	Vigneshwaran et al. (2006)
Schizosaccharomyces pombe	CdS	Kowshik et al. (2002)
Aspergillus flavus	Ag	Vigneshwaran et al. (2007)
Trichoderma asperellum	Ag	Mukherjee et al. (2008)

Parikh et al. (2008) investigated AgNP biosynthesis by *Morganella* sp. isolated from the insect gut. This strain was found to be highly resistant to silver cations. It was observed that when *Morganella* sp. was exposed to silver nitrate ($AgNO_3$) at 37 °C, crystalline AgNPs of size 20 ± 5 nm were produced extracellularly. The authors postulated that the molecular mechanisms of silver resistance and its gene products might play a key role in the process of synthesis of AgNPs. In the study, three homologous genes (namely *silE*, *silP*, and *silS*) were identified. The homologous *silE* was identified as a periplasmic silver-binding protein; the homologous *silP* was identified as a cation-transporting P-type ATPase; and the homologous *silS* was identified as a

TABLE 5.3. Extremophiles in Biosynthesis of Nanoparticles

Extremophiles	Extremophilic conditions	NPs	References
Halococcus salifodinae BK3	Halophilic	Ag	Srivastava et al. (2013)
Hormoconis resinae	Psychrophilic	Ag	Varshney et al. (2009)
Ureibacillus thermosphaericus	Thermophilic, thrive under temperatures of 60–80 °C, optimal growth achieved at 65 °C.	Au	Abbasalizadeh et al. (2012)
Ureibacillus thermosphaericus	Thermophilic, thrive under temperatures of 60–80 °C, optimal growth achieved at 65 °C.	Ag	Juibari et al. (2011)
Thermomonospora sp.	Thermophilic, thrive under temperatures of 32–65 °C, optimal growth achieved at 55 °C.	Au	Ahmad et al. (2003)

two-component membrane sensor kinase of the silver-resistance mechanism (Parikh et al., 2008). A few bacteria have been reported to produce different types of NPs, and even bimetallic alloys. Nair and Pradeep (2002) found that *Lactobacillus* sp. isolated from buttermilk could produce gold, silver, and gold-silver alloy NPs from chloroauric acid ($HAuCl_4$) and $AgNO_3$.

Bacterial production of NPs based upon other metallic elements has been intensively studied (Sweeney et al., 2004; Konishi et al., 2007; Sharma et al., 2012). Sweeney et al. (2004) demonstrated that cadmium sulfide semiconductor nano-crystals were spontaneously formed when *Escherichia coli* bacteria were incubated with cadmium chloride ($CdCl_2$) and sodium sulfide (Na_2S) (Sweeney et al., 2004). A few selenide- and selenate-respiring bacteria, such as *Sulfurospirillum barnesii*, *Bacillus selenitireducens*, and *Selenihalanaerobacter shriftii*, were reported to produce stable uniform selenium nanospheres of diameter c. 300 nm (Oremland et al., 2004). Konishi et al. (2007) found platinum NPs of c. 5 nm deposited in *Shewanella* algae cells. *Desulfovibrio desulfuricans* NCIMB 8307 was found able to produce palladium NPs by Ping et al. (2002). Sharma et al. (2012) reported AuNP synthesis using *Marinobacter pelagius* sp. isolated from marine water. The optimum production of AuNPs was achieved at temperatures of 30–37 °C and pH conditions of 7.0–8.0, and their size varied within the range 2–10 nm. The NPs formed were mostly spherical with occasionally triangular morphologies, indicating a satisfactory level of monodispersity (Sharma et al., 2012).

One of the salt-loving Haloarchaea, *Halococcus salifodinae* BK3, has been exploited for biosynthesis of AgNPs (Srivastava et al., 2013). These microorganisms are found in most of the thalassohaline and athalassohaline lakes where salinity reaches up to 300 g L^{-1} (Zafrilla et al., 2010). They maintain their osmotic balance with their surroundings by building up the potassium ion concentration within their cells (Oren, 2008). Crystalline-structured AgNPs with spherical morphology were produced intracellularly. Silver nitrate was found to be toxic to *Halococcus salifodinae* BK3 above 0.5 mM. At the beginning, the growth of *H. salifodinae* was found to be low. However, the silver-adapted *H. salifodinae* developed silver tolerance and the growth rate increased over time (Srivastava et al., 2013). Silver ions from an $AgNO_3$ solution were reduced into Ag^0 by nitrate reductase with NADH as the electron donor. In an earlier study, the metal-resistant genes were noted in model organism *Halobacterium* sp. strain NRC-1 (Ng et al., 2000) and it is believed that they genetically control the Ag+ reduction ability and the presence of related enzymes (Ng et al., 2000).

Varshney et al. (2009) demonstrated a novel route to synthesize AgNPs using the fungus *Hormoconis resinae*. *Hormoconis resinae* is known as an extremophile because it can live in tanks of diesel or jet fuel, consuming alkanes and traces of water. In the study, *Hormoconis resinae* was incubated with $AgNO_3$ solution, and after 6 hours the biomass turned a dark red color which indicated the formation of AgNPs. The NPs synthesized were polydisperse. Triangle-shaped and spherical particles and clusters were observed under transmission electron microscopy (TEM). The sizes ranged from 10 nm to 80 nm.

The extremophilic bacteria species *Ureibacillus thermosphaericus*, isolated from the Ramsar geothermal hot springs located in Mazandaran province, Iran was also found to produce AgNPs (Juibari et al., 2011). The organism was able to synthesis highly stable AgNPs at temperatures of 60–80 °C. Another thermophilic species of actinomycete, *Thermomonospora* sp., which thrives at 32–65 °C, was found to synthesize AuNPs extracellularly (Ahmad et al., 2003). The AuNPs were characterized by UV-vis spectroscopy and the gold surface plasmon band occurred at 520 nm. The synthesized nanogold was polydisperse and ranged in size from 8 to 40 nm.

5.3. MECHANISM OF NANOPARTICLE BIOSYNTHESIS

Bacteria have developed numerous detoxification mechanisms to survive in toxic metallic environments. These mechanisms include dissimilatory oxidation or reduction, complexation, precipitation, and more (Das and Marsili, 2011). The microbial ability to produce NPs is based on their resistance to most toxic heavy metals, mineral detoxification, and membrane ion efflux (Lengke et al., 2006). The mechanism of NP biosynthesis has yet to be understood; however, one of the most fundamental processes involved in mineral detoxification is bioreduction through an organic matrix (i.e., proteins or other macromolecules) that controls the nucleation and growth of inorganic structures (Nair and Pradeep, 2002; Lin et al., 2005; He et al., 2007; Parikh et al., 2008; Sharma et al., 2012). Most of the microorganisms that produce NPs do so via biosequestration and reduction of minerals mediated by enzymes. The bioreduction involves proteins, carbohydrates, and biomembranes.

In bioreduction, the toxicity of metal ions is reduced or eliminated by changing the redox state of the metal ion and/or precipitation of the metals intracellularly. The first step of this process is the trapping of metal ions on the surface of the cell wall; it is assumed that this occurs due to the electrostatic interaction between the metal ions and positively charged groups in the enzymes of the cell wall. The enzymatic reduction of the metal ions soon follows through the use of specific reducing enzymes such as NADH-dependent reductase, which is a main factor in biosynthetic processes. As summarized in Figure 5.1, the NADH-dependent enzyme reductase gains electrons from NADH and oxidizes to NAD$^+$. While the enzyme oxidizes the metal ions simultaneously reduce, causing the formation of NPs.

In one example, exposing the fungus *Fusarium oxysporum* to aqueous Cd^{2+} and SO$_4^{2-}$ ions led to the extracellular formation of extremely stable CdSNPs in solution (Ahmad et al., 2002). The presence of proteins (detected by UV-vis spectra at 280 nm) in the solution suggested a possible sulfate-reducing enzyme-based process for the synthesis of CdSNPs from Cd^{2+} and SO$_4^{2-}$ ions. The total protein from the microbial aqueous solution revealed four unique protein bands after the exposure to CdSO$_4$. To confirm the suggestion further, the protein was extracted through dialysis, ATP and NADH were added, and it was exposed to CdSO$_4$; the formation of CdSNPs was then observed. The studies were extended to the formation of other MNPs, i.e., PbS, ZnS, and MoS$_2$. The same organism was also reported to biosynthesize platinum NPs (PtNPs). Riddin et al. (2006) reported that an enzyme secreted in fungal mycelia was the key to PtNP synthesis. It was suggested that a hydrogenase enzyme, one of the biofactors freely secreted into the medium, carried out the platinum salt reduction. Also, the pH level of the solution was related to changes in the pKa values of crucial amino acids within the active sites of the enzyme. Platinum ions can exchange with protons, either with –COOH or –NH^{+3} or both, and a subsequent reduction would form Pt$^{(0)}$. Furthermore, the mechanism of the hydrogenase enzyme regenerates the carboxylate/amino protons, thereby allowing the cycle to be repeated and the size of the particle to grow.

Reductases in microorganisms, such as sulfate reductase and nitrate, are crucial in many bioreduction processes of NPs (Riddin et al., 2006; Han et al., 2007; Karbasian et al., 2008). A series of reductions leads to the production of ZnSNPs in bacterium *Rhodobacter sphaeroides*. Mechanistically, soluble sulfate enters the microbial cell via diffusion and sulfate permease. Sulfate is reduced to sulfite by ATP sulfurylase and phosphoadenosine phosphosulfate reductase. This sulfite is reduced to sulfide by sulfite reductase. The sulfide further reacts with O-acetyl serine to synthesize cysteine via O-acetylserine thiolyase. Cysteine produces S^{2-} by a cysteine desulfhydrase in the presence of zinc. S^{2-} reacts with the soluble zinc salt to form ZnSNPs, which then are secreted out into the solution (Fig. 5.2; Bai et al., 2009).

Metal ions can also be reduced by nitrate-dependent reductase. Using *Fusarium oxysporum* again as an example, nitrate-dependent reductase is conjugated with an electron donor, reduces the metal ion, and changes it to its elemental form. Because rapid extracellular synthesis occurs over a short amount of time, complex electron shuttle materials may be involved in the biosynthetic process. In the case of *Enterobacteriaceae* species, the culture supernatant synthesizes NPs by reducing Ag$^+$ to elemental silver. The addition of piperitone partially inhibits the reduction, which shows the involvement of nitroreductase enzymes in the reduction process (Ramezani et al., 2010). Nitrate reductase plays a major role in ferric iron reduction as well as Ag$^+$ reduction (Durán et al., 2005).

Proteins play an integral role in NP formation. Fungi, due to their secretion of greater amounts of protein, might produce significantly greater quantities of NPs than

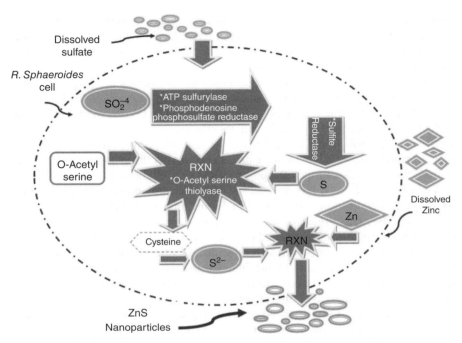

Figure 5.2. Mechanism of formation of ZnS nanoparticles in *Rhodobacter sphaeroides*. A series of reductions leads to the production of ZnS nanoparticles. Soluble sulfate enters the cell wall via diffusion and permease, and is then reduced to sulfite by ATP sulfurylase and phosphoadenosine phosphosulfate reductase. This sulfite is reduced to sulfide by sulfite reductase. The sulfide reacts with O-acetyl serine to synthesize cysteine via O-acetylserine thiolyase. Cysteine produces S^{2-} by a cysteine desulfhydrase in the presence of zinc. S^{2-} reacts with the soluble zinc salt and the ZnS nanoparticles are being synthesized and released from the cells into the solution.

bacteria (Mohanpuria et al., 2008). In order to produce zirconium NPs, some fungi secrete proteins that are capable of hydrolyzing compounds with zirconium ions. One study indicated that the proteins involved in NP reduction were cationic, with high molecular weights. The extracellularly produced NPs were stabilized by the proteins and reducing agents secreted by most fungi. In fact, a minimum of four high-molecular-weight proteins released by the fungal biomass were found to be associated with NPs (Mohanpuria et al., 2008). In addition, the reduction of metal ions and surface binding of the proteins to the NPs did not compromise the tertiary structure of the proteins (Ramezani et al., 2010).

In another study, protein depletion experiments were performed to confirm the role of cellular proteins in the control of synthesis rate and morphology. Nearly 18 AgNP-bound proteins were identified, most coming from the aforementioned oxidoreductive machinery of microorganism cells (Barwal et al., 2011). As these experiments were conducted with *Chlamydomonas reinhardtii*, some of these proteins were part of the metallic stress including ATP synthase, RuBP carboxylase, and an oxygen-evolving enhancer protein. Other proteins involved with NP synthesis include sedoheptulose-1, 7-bisphosphatase, carbonic anhydrase, and ferredoxin NADP$^+$ reductase. Nuclear histone is a protein of interest, as it is known to reduce silver ammonia to AgNPs. Another interesting protein is superoxide dismutase, which is able to reduce Ag$^+$ to elemental silver (Barwal et al., 2011).

Another factor in the biological mechanism of NP formation is glutathione. Glutathione is a tripeptide that contains a –SH group. Because of its antioxidant nature, it plays an important role in cell detoxification by removing organic peroxides and free radicals (Lim et al., 2008). Yeast cells under heavy metal stress produce more glutathione and glutathione-like compounds called phytochelatins. The resulting metal thiolate complex formation neutralizes toxicity and traps metal ions within the cells. Sulfide anions are incorporated into the cadmium–glutathione complexes, and NPs are formed (Ramezani et al., 2010).

5.4. FERMENTATIVE PRODUCTION OF NANOPARTICLES

In order to successfully implement nanotechnology via microbiology, microorganisms require a suitable fermentative medium contained with essential carbon sources and inorganic nutrients. The other growth factors such as aeration, pH modulation, and salt compositions vary from one organism to another. Variations in chemical and physical growth factors can have different effects on the resulting microorganisms and their primary and secondary metabolites.

In recent years, the extremophiles have been considered for their potential to create value-added products of commercial interest. Different fermentation media are being used to grow a variety of microorganisms, including extremophiles, to support biosynthesis of NPs. In a few cases, the fermentation media have been specially formulated to support biomass growth along with efficient liberation of primary and secondary metabolites (i.e., reductive enzymes and proteins for NP stabilization; Lengke et al., 2005; Lengke and Southman, 2006; Balaji et al., 2009). Table 5.4 summarizes a few of the fermentation media used to grow a variety of microorganisms for NP biosynthesis.

An extremophilic yeast strain was found to produce AbNPs, 20 nm in diameter and AuNPs, 20–100 nm in diameter when grown in yeast nitrogen base with casamino acid (YNBC) liquid medium. Glucose was added as a carbon source at pH 2.5 under shaking at 160 rpm for the growth of the microorganisms (Mourato et al., 2011).

The modulation of aeration has a potential effect on biomass in aerobic microorganisms. *Aspergillus flavus* grew at 200 rpm at 37 °C for 5 days, producing AgNPs of approximately 8 nm (Vigneshwaran et al., 2007). *Rhizopus stolonifer* was grown in malt glucose yeast peptone at 180 rpm and incubated at 40 °C for 72 hours. The resulting NPs were monodisperse, spherical, and 3–20 nm in size (Banu et al., 2011).

The pH of the medium where the organisms are grown has an important effect on the size and shape of NPs. The AuNPs produced by *Rhodopseudomonas capsulata*

TABLE 5.4. Optimized Fermentation Media for Microbial Synthesis of Nanoparticles

Type of medium	Organism	NP types	Reference
Lennox L (LB broth)	*Pseudomonas stutzeri*	Ag	Klaus et al. (1999)
2% tryptone, 1% yeast extract, and 2% glucose (pH 5.6)	*Schizosaccharomyces pombe*	CdS	Kowshik et al. (2002)
Luria–Bertani (LB) medium	*Shewanella oneidensis*	Pd	de Windt et al. (2005, 2006)
Thiosulfate growth medium contained (g L^{-1}): (NH$_4$)$_2$SO$_4$: 0.3; KH$_2$PO$_4$: 0.1; MgSO$_4$·7H$_2$O: 0.4; CaCl$_2$: 0.25; FeSO$_4$·7H$_2$O: 0.018; Na$_2$S$_2$O$_3$·5H$_2$O: 4.93	*Acidithiobacillus thiooxidans*	Au	Lengke et al. (2005)
1% (w/v) soya peptone and 0.5% (w/v) beef extract	*Corynebacterium* sp.	Ag	Zhang et al. (2005)
MGYP medium, 1023 M K2SiF6 (pH 3.1) and K2TiF6 (pH 3.5)	*Fusarium oxysporum*	Si, Ti	Bansal et al. (2004)
Universal yeast medium (YM) (0.3% yeast extract, 0.3% malt extract, 0.5% peptone, 1% glucose, 0.15% agar) or nutrient broth	*Pseudomonas stutzeri* NCIMB 13420, *Bacillus subtilis* DSM 10, *Pseudomonas putida* DSM 291, *Schizosaccharomyces pombe* DSM 2791, *Schizosaccharomyces pombe* DSM 70576, *Pichia jadinii* UOFS Y-0156, *Pichia jadinii* UOFS Y-0520, *Verticillium dahlia* DSM 63083, *Verticillium luteoalbum* DSM 63545, and *Fusarium oxysporum*	Au	Gericke and Pinches (2006)
SRB medium: tryptone 10.0, yeast extract 1.0, iron sulfate (FeSO$_4$ 7H$_2$O) 0.5, magnesium sulfate (MgSO$_4$ 7H$_2$O) 2.0, L-ascorbic acid 1.5, sodium lactate (60%) 6 mL L^{-1}, sodium thiosulfate (Na$_2$S$_2$O$_3$ 5H$_2$O) 8.5, iron acetate (Fe(C$_2$H$_3$O$_2$)$_2$) 8.5, sodium chloride (NaCl) 8.5, agar 8.5, sodium sulfide (Na$_2$S 9H$_2$O) 8.5	Sulfate-reducing bacteria (SRB)	Au	Lengke et al. (2006)
50 mL nutrient broth media, hydrogen tetrachloroaurate	*Pseudomonas aeruginosa*	Au	Husseiny et al. (2007)
Malt extract agar medium	*Phaenerochaete chrysosporium*	Ag	Vigneshwaran et al. (2006)
Liquid media containing (g L^{-1}) KH$_2$PO$_4$: 7.0; K$_2$HPO$_4$: 2.0; MgSO$_4$ 7H$_2$O: 0.1; (NH$_2$)SO$_4$: 1.0; yeast extract: 0.6; glucose: 10.0	*Fusarium semitectum*	Ag	Basavaraja et al. (2008)

Medium	Organism	Nanoparticle	Reference
Medium containing puryate, yeast extract, NaCl, NH_4Cl and K_2HPO_4	Rhodopseudomonas capsulate	Au	He et al. (2007)
Yeast malt broth containing dextrose ($10\,g\,L^{-1}$); peptic digest of animal tissue ($5\,g\,L^{-1}$); yeast extract ($3\,g\,L^{-1}$); malt extract ($3\,g\,L^{-1}$, pH 6.2)	Aspergillus flavus	Ag	Vigneshwaran et al. (2007)
Malt extract 3 g, glucose 10 g, yeast extract 3 g and peptone 5 g per 1 L of distilled water	Fusarium oxysporum	Ag	Karbasian et al. (2008)
MM1 media mineral salts medium ($g\,L^{-1}$) K_2HPO_4: 2.5; KNO_3: 5.0; $MgSO_4 \cdot 7H_2O$: 1.00; $MnSO_4 \cdot H_2O$: 0.001; $CuSO_4 \cdot 5H_2O$: 0.003; $ZnSO_4 \cdot 7H_2O$: 0.01; $Na_2MoO_3 \cdot 2H_2O$: 0.0015; $FeCl_3$: 0.02; Na_2SO_4: 1.00; glucose: 40.00	Cladosporium cladosporioides	Ag	Balaji et al. (2009)
50 mL of the growth medium, consisting of $5\,g\,L^{-1}$ malt extract powder and $10\,g\,L^{-1}$ glucose	Coriolus versicolor	Ag	Sanghi and Verma (2009)
Liquid broth containing ($g\,L^{-1}$): KH_2PO_4: 7; K_2HPO_4: 2; $MgSO_4 \cdot 7H_2O$: 0.1; $(NH_2)SO_4$: 1; yeast extract: 0.6; glucose: 10	Trichoderma viride	Ag	Fayaz et al. (2009)
Sabouraud's glucose agar medium containing 20 g glucose, 10 g peptone, 20 g agar, 500 mL aged seawater, 500 mL distilled water	Penicillium fellutanum	Ag	Kathiresan et al. (2009)
Czapek's-dox (CD) agar ($g\,L^{-1}$): Na_2NO_3: 2; K_2HPO_4: 1; $MgSO_4$: 0.5; KCl: 0.5; $FeSO_4$: 0.01; sucrose: 30; agar: 15; pH 6.5	Hormoconis resinae (Extremophile)	Ag	Varshney et al. (2009)
YNB-glucose medium containing 7.0 yeast nitrogen base (YNB); glucose, 10.0 g per litre distilled water	Yarrowia lipolytica	Au	Pimprikar et al. (2009)
Ultra-pure Milli-Q water used to dissolve salt, minerals, and nutrients; nutrient agar; nutrient Broth	Bacillus licheniformis	Ag, Au	Sriram et al. (2012)
Luria–Bertani agar medium (LBA), LB broth, 1 mM $AgNO_3$ and 1 mM $Na_2S_2O_3$	Shewanella oneidensis	Ag_2S	Debabov et al. (2013)
25% NaCl, 0.5% tryptone, 0.3% yeast extract, 2% $MgSO_4 7H_2O$, 0.5% KCl, 2% agar	Halococcus salifodinae (Extremophile)	Ag	Srivastava et al. (2013)

at pH 7 are mainly spherical and about 10–20 nm in size. However, when *R. capsulata* was grown at pH 4, 60% of the AuNPs were triangular in the size range 50–400 nm and 40% were spherical (He et al., 2007). Because the pH regulates the proton concentration in medium, adjusting the pH adjusts the morphology of the NPs extracellularly. This poses a challenge for mass production of NPs, as the shapes of the NPs determine their potential uses due to their surface area and ability to bind with the external moieties (He et al., 2007). At pH 8, AgNPs produced by *Penicillium* are fairly stable due to electrostatic repulsion. In contrast, when AgNPs were produced with *Penicillium* in an acidic environment, the low negative zeta potentials indicated that the resulting NPs were very unstable (Sadowski et al., 2008).

Other factors which may affect the synthesis of NPs are metal sources and time of fermentation/culture. In a study by Bai and Zhang (2009) to synthesize lead sulfide NPs, *Rhodobacter sphaeroides* was grown in malic sodium supplemented with $MgSO_4$, yeast extract, and three different metals: $Pb(Ac)_2$, $PbCl_2$, and $Pb(NO_3)_2$. The average size of the NPs synthesized from each of these metals was 6.8, 8.5, and 11.2 nm, respectively. Varying culture times also resulted in varying sizes of NPs: times of 24, 32, and 40 hrs produced NPs with average sizes of 3.8, 10.8, and 43.2 nm, respectively. Each growth factor in the fermentation media therefore affected the morphological appearance of NPs (Bai and Zhang, 2009). A controlled set of experiments, as for other fermentation types, may eventually make it possible to produce NPs of a required shape and length from microorganisms.

5.5. NANOPARTICLE RECOVERY

As mentioned in Section 5.2.1, NPs can be synthesized either intracellularly or extracellularly. Whether the microorganisms are extremophiles or mesophiles, NP recovery requires a standardized procedure. In most cases, the procedure for recovering extracellularly synthesized NPs is simple and straightforward: they can be recovered from the solution by a series of centrifugation or filtration processes at room temperature without a specific need for low temperature (Sadowski et al., 2008; Banu et al., 2011; Mishra et al., 2011). After recovery it is recommended to store biologically synthesized NPs in the dark at a low temperature (4 °C) or at room temperature (22–25 °C).

For intracellularly synthesized NPs, the microbial cells have to be lysed to harvest the NPs. The synthesis of AuNPs with the tropical marine yeast *Yarrowia lipolytica* is intracellular. These NPs were recovered by incubating cells at 20 °C for 48 hours in a lysis buffer. The lysate was centrifuged ($6000 \times g$), and supernatant was filtered through a 0.22 μm filter membrane to remove any remaining cells, securing NPs on the membrane (Pimprikar et al., 2009). Another method is to disrupt the microbial cells by sonication. Au- and AgNPs from *Bacillus licheniformis* were recovered by sonication using an ultrasonic processor for 20 seconds, pausing for 30 seconds, and repeating in phosphate buffer saline (PBS). The sonicated biomass was filtered with a 0.22 μm filter membrane, and NPs were recovered on the membrane (Sriram et al., 2012).

5.6. CHALLENGES AND FUTURE PERSPECTIVES

During the past three decades, studies have shown that many types of microorganisms, including fungi and bacteria, can synthesize various types of NPs in different shapes and sizes. Biosynthesis of NPs is a clean, non-toxic, and environmentally friendly method compared to the chemical and physical methods. The biosynthesis of NPs involves both intracellular and extracellular pathways. Extracellular synthesis involves a number of reducing and capping agents, the effects of which on the shape and size of NPs must be clarified since the size of NPs plays a key role in their application. For example, in the case of HIV-1 infections, silver NPs interact in a size-dependent manner. The dispersity of the NPs should therefore also be considered as a determining factor during the synthesis of NPs (Rai and Duran, 2011).

It is mostly assumed that microbes produce NPs as a consequence of the detoxification or bioreduction pathway. However, current and future perspectives on extremophile metabolism, genetics, and physiology will likely yield new breakthroughs in biotechnological applications. With regard to employing extremophiles in nanotechnology, future research in the refinement of fermentation media for NP size specificity, cell–cell communication to enhance extracellular enzymatic activity, and high-throughput innovation toward NP stability by exploring microbial genomics and proteomics can lead to novel insights into environments that were once thought to harbor no forms of life. High-throughput screening methods are available to detect the microbial metabolome (primary and secondary metabolites) using advanced proteomics- and metabolomics-based approaches (Singh, 2006).

Future development to trace the pathways of NP biosynthesis and their interaction with macromolecules will depend on joint ventures between chemistry, physics, and life science. Standard methodologies and protocols must also be developed for the biosynthesis of NPs, since current experiments are carried out under varying conditions (temperature, airflow rate, relative humidity, measurement techniques, etc.). The lack of a standard synthesis protocol limits the validation of the procedure and makes further studies more difficult.

5.7. CONCLUSION

NPs may help us develop technologies to address many unresolved concerns of humankind, such as diseases and energy dependence. Even though they are a fraction of a hair width in size, they have become popular in science, gaining their own recognition in the area of nanotechnology. Commercial applications of nanotechnology have created demand for a method of synthesizing well-defined NPs that is both economically feasible for industry and environmentally friendly. In the past few years, a movement has started to synthesize NPs with microorganisms, as opposed to chemical synthesis processes. Bacteria, extremophiles, fungi, and even plants can produce NPs. Extremophiles have a strong potential for future advancements in bio-nanotechnology and extermination of toxic compounds generated during chemical synthesis of NPs. Because of the versatility of extremophiles and their metabolic products (i.e., enzymes and proteins) that may be involved in the stabilization of NPs under varying environmental conditions, they could potentially be used in therapeutic applications for different disease types. In the years to come, the exploitation of a variety of extremophiles will indubitably advance the nanotechnology field toward non-toxic and environmentally friendly biosynthesis of NPs.

REFERENCES

Abbasalizadeh, S., Salehi Jouzani, G., Motamedi Juibari, M., Azarbaijani, R., Parsa Yeganeh, L., Ahmad Raji, M., Mardi, M., Salekdeh, G.H. 2012. Draft genome sequence of Ureibacillus thermosphaericus strain thermo-BF, isolated from Ramsar hot springs in Iran. *Journal of Bacteriology* 194(16): 4431.

Ahmad, A., Mukherjee, P., Mandal, D., Senapati, S., Khan, M.I., Kumar, R., Sastry, M. 2002. Enzyme mediated extracellular synthesis of CdS nanoparticles by the fungus, *Fusarium oxysporum*. *Journal of American Chemical Society* 124(41): 12108–12109.

Ahmad, A., Senapati, S., Khan, M.I., Kumar, R., Sastry, M. 2003. Extracellular biosynthesis of monodisperse gold nanoparticles by a novel extremophilic actinomycete, *Thermomonospora* sp. *Langmuir* 19: 3550–3553.

Ahmad, A., Jagadale, T., Dhas, V., Khan, S., Patil, S., Paricha, R., Ravi, V., Ogale, S. 2007. Fungus-based synthesis of chemically difficult-to-synthesize multifunctional nanoparticles of CuAlO2. *Advanced Materials* 19: 3295–3299.

Alshehri, A.H., Jakubowska, M., Młożniak, A., Horaczek, M., Rudka, D., Free, C., Carey, J.D. 2012. Enhanced electrical conductivity of silver nanoparticles for high frequency electronic applications. *American Chemical Society* 4(12): 7007–7010.

Altimira, F., Yáñez, C., Bravo, G., González, M., Rojas, L.A., Seeger, M. 2012. Characterization of copper-resistant bacteria and bacterial communities from copper-polluted agricultural soils of central Chile. *BMC Microbiology* 12: 193.

Arruebo, M., Fernandes-Pacheco, R., Velasco, B., Marquina, C., Arbiol, J., Irusta, S., Ricardo, I., Santamaria, J. 2007. Antibody-functionalized hybrid superparamegnetic nanoparticles. *Advanced Functional Materials* 17: 1473–1479.

Bai, H., Zhang, Z. 2009. Microbial synthesis of semiconductor lead sulfide nanoparticles using immobilized *Rhodobacter sphaeroides*. *Materials Letters* 63: 764–766.

Bai, H.J., Zhang, Z.M., Guo, Y, Yang, G.E. 2009. Biosynthesis of cadmium sulfide nanoparticles by photosynthetic bacteria Rhodopseudomonas palustris. *Colloids and Surfaces B: Biointerfaces* 70:142–146.

Balaji, D.S., Basavaraja, S., Deshpande, R., Mahesh, D.B., Prabhakar, B.K., Venkataraman, A. 2009. Extracellular biosynthesis of functionalized silver nanoparticles by strains of *Cladosporium cladosporioides*. *Colloids and Surfaces B: Biointerfaces* 68: 88–92.

Bamberger, E.S., Perrett, C.W. 2002. Angiogenesis in epithelium ovarian cancer. *Molecular Pathology* 55(6): 348–359.

Bansal, V., Rautaray, D., Ahmad, A., Sastry, M. 2004. Biosynthesis of zirconia nanoparticles using the fungus *Fusarium oxysporum*. *Journal of Materials Chemistry* 14: 3303–3305.

Banu, A., Rathod, V., Ranganath, E. 2011. Silver nanoparticle production by *Rhizopus stolonifer* and its antibacterial activity against extended spectrum β-lactamase producing (ESBL) strains of *Enterobacteriaceae*. *Materials Research Bulletin* 46: 1417–1423.

Barwal, I., Ranjan, P., Kateriya, S., Yadav, S.C. 2011. Cellular oxidoreductive proteins of *Chlamydomonas reinhardtii* control the biosynthesis of silver nanoparticles. *Journal of Nanobiotechnology* 9: 56.

Basavaraja, S., Balaji, S.D., Lagashetty, A., Rajasab, A.H., Venkataraman, A. 2008. Extracellular biosynthesis of silver nanoparticles using the fungus *Fusarium semitectum*. *Materials Research Bulletin* 43: 1164–1170.

Bhattacharya, R. Mukherjee, P. 2008. Biological properties of "naked" metal nanoparticles. *Advanced Drug Delivery Review* 60(11): 1289–1306.

Bruins, M.R., Kapil, S., Oehme, F.W. 2000. Microbial resistance to metals in the environment. *Ecotoxicology and Environmental Safety* 45: 198–207.

Burke, T., Fales, A.M., Ginsburg, G.S., Tuan, V., Woods, C.W., Wang, H., Zaas, A.K. 2013. Surface-enhanced raman scattering molecular sentinel nanoprobes for viral infection diagnostics. *Analytica Chimica Acta* 786: 153–158.

Chen, C., Baeumner, A.J., Durst, R. 2005. Protein G-liposomal nanovesicles as universal reagents for immunoassays. *Talanta* 67(1): 205–211.

Clausen, C.A. 2000. Isolating metal-tolerant bacteria capable of removing copper, chromium, and arsenic from treated wood. *Waste Management and Research* 18: 264–268.

Das, K.R., Gogoi, N., Babu, P.J., Sharma, P., Mahanta, C., Bora, U. 2012. The synthesis of gold nanoparticles using *Amaranthus spinosus* leaf extract and study of their optical properties. *Advances in Materials Physics and Chemistry* 2: 275–281.

Das, S.K., Marsili, E. 2011. Bioinspired metal nanoparticle: synthesis, properties and application. In: *Nanomaterials* (Rahman, M.M., ed.). InTech, Europe, pp. 253–278.

Davis, P.H., Morrisey C.P., Tuley, S.M.V., Bingham, C.I. 2008. Synthesis and stabilization of colloidal gold nanoparticle suspensions for SERS. In *Nanoparticles: Synthesis, Stabilization, Passivation, and Functionalization* (Nagarajan, R., Alan, H.T. eds) (pp 2-14). American Chemical Society, Washington DC, pp. 2–14.

de Windt, W., Aelterman, P., Verstraete, W. 2005. Bioreductive deposition of palladium nanoparticles on Shewanella oneidensis with catalytic activity towards reductive dechlorination of polychlorinated biphenyls. *Environmental Microbiology* 7: 314–325.

de Windt, W., Boon, N., van den Bulcke, J., Rubberecht, L., Prata, F., Mast, J., Hennebel, T., Verstraete, W. 2006. Biological control of the size and reactivity of catalytic Pd(0) produced by *Shewanella oneidensis*. *Antonie van Leeuwenhoek* 90: 377–389.

Debabov, V.G., Voeikova, T.A., Shebanova, A.S., Shaitan, K.V., Emel'yanova, L.K., Fayaz, A.M., Balaji, K., Kalaichelvan, P.T., Venkatesan, R. 2009. Fungal based synthesis of silver nanoparticles – an effect of temperature on the size of particles. *Colloids Surfaces B: Biointerfaces* 74: 123–126.

Debabov, V.G., Voeikova, T.A., Shebanova, A.S., Shaitan, K.V., Emel'yanova, L.K., Novikova, L.M., Kirpichnikov, M.P. 2013. Bacterial synthesis of silver sulfide nanoparticles. *Nanotechnologies in Russia* 8: 269–276.

Ding, Y., Dong, Y., Bapat, A., Nowak, J.D., Carter, C.B., Kortshagen, U.R., Campbell, S.A. 2006. Single nanoparticle semiconductor devices. *IEEE Transactions on Electron Devices* 53(10): 2525–2531.

Durán, N., Marcato, P., Alves, O.L., De Souza, G.I.H., Esposito, E. 2005. Mechanistic aspects of biosynthesis of silver nanoparticles by several *Fusarium oxysporum* strains. *Journal of Nanobiotechnology* 3: 8.

Fayaz, A.M., Balaji, K., Girilal, M., Kalaichelvan, P.T., Venkatesan, R. 2009. Mycobased synthesis of silver nanoparticles and their incorporation into sodium alginate films for vegetable and fruit preservation. *Journal of Agriculture and Food Chemistry* 57: 6246–6252.

Gabani, P., Singh, O.V. 2013. Radiation-resistant extremophiles and their potential in biotechnology and therapeutics. *Applied Microbiology and Biotechnology* 97(3): 993–1004.

Gericke, M., Pinches, A. 2006. Microbial production of gold nanoparticles. *Gold Bulletin* 39: 22–28.

Han, G., Ghosh, P., Rotello, V.M. 2007. Multi-functional gold nanoparticles for drug delivery. In: *Bio-Applications of Nanoparticles* (Chan, W.C.W., ed.). Landes Bioscience and Springer Science, New York, pp. 45–56.

He, S., Guo, Z., Zhang, Y., Zhang, S., Wang, J., Gu, N. 2007. Biosynthesis of gold nanoparticles using the bacteria *Rhodopseudomonas capsulate*. *Material Matters* 61(18): 3984–3987.

Husseiny, M.I., Abd El-Aziz, M., Badr, Y., Mahmoud, M.A. 2007. Biosynthesis of gold nanoparticles using *Pseudomonas aeruginosa*. *Spectrochimica Acta* 67: 1003–1006.

Ito, A., Ino, A., Hayashida, M., Kobayashi, T., Matsunuma, H., Kagami, H., Ueda, M., Honda, H. 2005. Novel methodology for fabrication of tissue-engineered tubular constructs using magnetite nanoparticles and magnetic force. *Tissue Engineering* 11: 1553–1561.

Jain, P., Pradeep, T. 2005. Potentiao of silver nanoparticles-coated polyurethane foam as an antibacterial water filter. *Biotechnology and Bioengineering* 90: 59–63.

Joerger, R., Klaus, T., Granqvist, C.G. 2000. Biologically produced silver carbon composite materials for optically functional thin-film coatings. *Advanced Materials* 12: 407–409.

Juibari, M.M., Abbasalizadeh, S., Jouzani, Gh.S., Noruzi, M. 2011. Intensified biosynthesis of silver nanoparticles using a native extremophilic *Ureibacillus thermosphaericus* strain. *Materials Letters* 65: 1014–1017.

Kalimuthu, K., Babu, R.S., Venkataraman, D., Bilal, M., Gurunathan, S. 2008. Biosynthesis of silver nanoparticles by *Bacillus licheniformis*. *Colloids Surfaces B: Biointerfaces* 65: 150–153.

Karbasian, M., Atyabi, M., Siadat, S.D., Momen, S.B., Norouzian, D. 2008. Optimizing nano-silver formation by *Fusarium oxyporum* PTCC 5115 employing response surface methology. *American Journal of Agricultural and Biological Science* 3: 433–437.

Kathiresan, K., Manivannan, S., Nabeel, M.A., Dhivya, B. 2009. Studies on silver nanoparticles synthesized by a marine fungus, *Penicillium fellutanum* isolated from coastal mangrove sediment. *Colloids Surfaces B: Biointerfaces* 71: 133–137.

Kim, M.S., Hoon, H., Khang, G., Lee, H.B. 2009. Polymeric nano micelles as a drug carrier. In *Nanoscience in Biomedicine* (Shi, D., ed.), Springer, New York

Klaus, T., Joerger, R., Olsson, E., Granqvist, C.G. 1999. Silver-based crystalline nanoparticles, microbially fabricated. *Proceedings of the National Academy of Science, USA* 96: 13611–13614.

Konishi, Y., Tsukiyama, T., Saitoh, N., Nomura, T., Nagamine, S., Takahashi, Y., Uruga, T. 2007. Direct determination of oxidation state of gold deposits in metal-reducing bacterium *Shewanella* algae using X-ray absorption near-edge structure spectroscopy (XANES). *Journal of Bioscience and Bioengineering* 103: 568–571.

Kowshik, M., Dashmukh, N., Vogel, W., Urban, J., Kulkarni, S.K., Paknikar, K.M. 2002. Microbial synthesis of semiconductor CdS nanoparticles, their characterization, and their use in the fabrication of an ideal diode. *Biotechnology and Bioengineering* 78: 583–588.

Lengke, M.F., Southam, G. 2005. The effect of thiosulfate-oxidizing bacteria on the stability of the gold-thiosulfate complex. *Geochimica et Cosmochimica Acta* 69(15): 3759–3772.

Lengke, M.F., Southam, G. 2006. Bioaccumulation of gold by sulfate-reducing bacteria cultured in the presence of gold(I)-thiosulfate complex. *Geochimica et Cosmochimica Acta* 70: 3646–3661.

Lengke, M.F., Ravel, B., Fleet, M.E., Wanger, G., Gordon, R.A., Southam, G. 2006. Mechanisms of gold bioaccumulation by filamentous cyanobacteria from gold(III)-chloride complex. *Environmental Science and Technology* 4920: 6304–6309.

Lengke, M.F., Ravel, B., Fleet, M.E., Wanger, G., Gordon, R.A., Southam, G. 2007. Precipitation of gold by the reaction of aqueous gold(III) chloride with cyanobacteria at 25–80 °C: studies by x-ray absorption spectroscopy. *Canadian Journal of Chemistry* 8590: 651–659.

Lim, I.S., Mott, D., Ip, W., Njoki, P., Pan, Y., Zhou, S., Zhong, C. 2008. Interparticle interactions in glutathione mediated assembly of gold nanoparticles. *Langmuir* 24: 8857–8863.

Lima de Silva, A.A., de Carvalho, M.A., de Souza, S.A., Dias, P.M., da Silva Filho, R.G., de Meirelles Saramago, C.S., de Melo Bento, C.A., Hofer, E. 2012. Heavy metal tolerance (Cr, Ag AND Hg) in bacteria isolated from sewage. *Brazilian Journal of Microbiology* 43(4): 1620–1631.

Lin, Z., Zhou, C., Wu, J., Wang, L. 2005. A further insight into the mechanism of Ag^+ biosorption by *Lactobacillus* sp. strain A09. *Spectrochimica Acta* 61: 1195–1200.

Ma, H., Brennan, A., Diamond, S.A. 2012. Phototoxicity of TiO_2 nanoparticles under solar radiation to two aquatic species: *daphnia magna* and *Japaneses medaka*. *Environmental Toxicology and Chemistry* 31(7): 1621–1629.

Mahmoudi, M., Arbab, A.S., Milani, A.S. 2011. Synthesis of SIPONs, nanotechnology science and technology: superparamagnetic iron oxide nanoparticles: synthesis, surface engineering. In: *Cytotoxicity and Biomedical Applications*. Nova Science Publishers, Inc., New York.

Mann, S. 1992. Bacteria and the Midas touch. *Nature* 357: 358–360.

Maynaud, G., Brunel, B., Mornico, D., Durot, M., Severac, D., Dubois, E., Navarro, E., Cleyet-Marel, J.C., Le Quéré, A. 2013. Genome-wide transcriptional responses of two metal-tolerant symbiotic *Mesorhizobium* isolates to zinc and cadmium exposure. *BMC Genomics* 14: 292.

Mishra, A., Tripathy, S., Wahab, R., Jeong, S., Hwang, I., Yang, Y., Yun, S. 2011. Microbial synthesis of gold nanoparticles using the fungus *Penicillium brevicompactum* and their cytotoxic effects against mouse mayo blast cancer C2C12 cells. *Applied Microbiology and Biotechnology* 92(3): 617–630.

Mohanpuria, P., Rana, N.K., Yadow, S.K. 2008. Biosynthesis of nanoparticles: Technological concepts and future applications. *Journal of Nanoparticle Research* 10: 507–517.

Mourato, A., Gadanho, M., Lino, A., Tenreiro, R. 2011. Biosynthesis of crystalline silver and gold nanoparticles by extremophilic yeasts. *Bioinorganic Chemistry and Applications* 2011: 1–8.

Muehe, E.M., Obst, M., Hitchcock, A., Tyliszczak, T., Behrens, S., Schröder, C., Byrne, J.M., Michel, F.M., Krämer, U., Kappler, A. 2013. Fate of Cd during Microbial Fe(III) Mineral Reduction by a Novel and Cd-Tolerant *Geobacter* Species. *Environmental Science and Technology* 47(24): 14099–14109.

Mukherjee, P., Roy, M., Dey, G.K., Mukherjee, P.K., Ghatak, J., Tyagi, A.K., Kale, S.P. 2008. Green synthesis of highly stabilized nanocrystalline silver particles by a non-pathogenic and agriculturally important fungus *T. asperellum*. *Nanotechnology* 19: 1–7.

Nair, B., Pradeep, T. 2002. Coalescence of nanoclusters and formation of submicron crystallites assisted by *Lactobacillus* strain. *Crystal Growth & Design* 2(4): 293–298.

Narayanan, K.B., Sakthivel, N. 2010. Biological synthesis of metal nanoparticles by microbes. *Advances in Colloid and Interface Science* 156: 1–13.

Ng, W.V., Kennedy, S.P., Mahairas, G.G., et al. 2000. Genome sequence of *Halobacterium* species NRC-1. *Proceedings of the National Academy of Science, USA* 97: 12176–12181.

Oishi, M., Nakaogami, J., Ishii, T., Nagasaki, Y. 2006. Smart PEGylated gold nanoparticles for the cytoplasmic delivery of siRNA to induce enhanced gene silencing. *Chemistry Letters* 35: 1046–1047.

Oremland, R.S., Herbel, M.J., Blum, J.S., Langley, S., Beveridge, T.J., Ajayan, P.M., Sutto, T., Ellis, A.M., Curran, S. 2004. Structural and spectral features of selenium nanospheres produced by Se-respiring bacteria. *Applied Environmental Microbiology* 70: 52–60.

Oren, A. 2008. Microbial life at high salt concentrations: phylogenetic and metabolic diversity. *Saline Systems* 4: 2.

Oren, A. 2013. Life at high salt concentrations, intracellular KCl concentrations, and acidic proteomes. *Frontiers in Microbiology* 4: 315.

Parikh, R.Y., Singh, S., Prasad, B.L.V., Sastry, M., Shouche, Y.S. 2008. Extracellular synthesis of crystalline silver nanoparticles and molecular evidence of silver resistance from *Morganella* sp.: towards understanding biochemical synthesis mechanism. *BioChemBio* 9: 1415–1422.

Pimprikar, P.S., Joshi, S.S., Kumar, A.R., Zinjarde, S.S., Kulkarni, S.K. 2009 Influence of biomass and gold salt concentration on nanoparticle synthesis by the tropical marine yeast Yarrowia lipolytica NCIM 3589. *Colloids Surfaces B: Biointerfaces* 74: 309–316.

Ping, Y., Rowson, N.A., Farr, P.J., Harris, I.R., Macaskie, L.E. 2002. Bioreduction and biocrystallization of palladium by *Desulfovibrio desulfuricans* NCIMB 8307. *Biotechnology and Bioengineering* 80(4): 369–380.

Phong, N.T.P., Thanh, N.V.K., Phuong, P.H. 2009. Fabrication of antibacterial water filter by coating silver nanoparticles on flexible polyurethane foams. *Journal of Physics: Conference Series* 187: 012079.

Rai, M., Duran, N. 2011. *Metal Nanoparticles in Microbiology*. Springer, New York.

Ramasamy, T., Khankasamy, U., Ruttala, H., Kona, K. 2009. Nanocochleate – a new drug delivery system. *European Journal of Pharmaceutical Sciences* 34: 991–1009.

Ramezani, F., Ramezani, M., Talebi, S. 2010. Mechanistic aspects of biosynthesis of nanoparticles by several microbes. In *Proceedings of 2nd International Conference on Nanomaterials - Research & Application (Nanocon)*, Czech Republic, pp. 12–14.

Riddin, T.L., Gericke, M., Whiteley, C.G. 2006. Analysis of the intra- and extracellular formation of plantinum nanoparticle by *Fusarium oxysporum* f. sp. *lycopersici* using response surface methodology. *Nanotechnology* 7: 3482–3489.

Rotello, R., Shenhar, V.M. 2003. Nanoparticles: scaffolds and building blocks. *Accounts of Chemical Research* 36(7): 549–561.

Sadowski, Z., Maliszewska, I.H., Grochowalska, B., Polowczyk, I., Kozlecki, T. 2008. Synthesis of silver nanoparticles using microorganisms. *Material Science Poland* 26: 2.

Saini, V., Everts, M. 2011. Viral biology and nanotechnology. In *Handbook of Nanophysics: Nanomedicine and Nanorobotics* (Sattler, K.D., ed.) Boca Raton, CRC Press, 2-1–2-9.

Sanghi, R., Verma, P. 2009. Biomimetic synthesis and characterization of protein capped nanoparticles. *Bioresource Technology* 100: 501–504.

Sapir, Y., Polyak, B., Cohen, S. 2014. Cardiac tissue engineering in magnetically actuated scaffolds. *Nanotechnology* 25(1): 014009.

Sharma, N., Pinnaka, A.K., Raje, M., Fnu, A., Bhattacharyya, M.S., Choudhury, A.R. 2012. Exploitation of marine bacteria for production of gold nanoparticles. *Microbial Cell Factories* 11(86): 1–6.

Silver, S., Phung, L.T. 1996. Bacterial heavy metal resistance: new surprise. *Annual Review of Microbiology* 50: 753–789.

Singh, O.V. 2006. Proteomics and metabolomics, the molecular make-up of toxic aromatic pollutant bioremediation. *Proteomics* 6: 5481–5492.

Singh, O.V. 2013. *Extremophiles: Sustainable Resources and Biotechnological Implications.* Wiley-Blackwell, New Jersey, USA.

Skaff, H., Emrick, T. 2004 Semiconductor nanoparticles. In: *Nanoparticles: Building Blocks for Nanotechnology* (Rotello, V., ed.) Springer Science+Business Media, New York.

Sriram, M.I., Kalishwaralal, K., Gurunathan, S. 2012. Biosynthesis of silver and gold nanoparticles using bacillus licheniformis. *Nanoparticles in Biology and Medicine: Methods and Protocols, Methods in Molecular Biology* 906: 33–43.

Srivastava, P., Bragança, J., Ramanan, S.R., Kowshik, M. 2013. Synthesis of silver nanoparticles using haloarchaeal isolate Halococcus salifodinae BK3. *Extremophiles* 17(5): 821–831.

Sweeney, R.Y., Mao, C., Gao, X., Burt, J.L., Belcher, A.M., Georgiou, G. et al. 2004. Bacterial biosynthesis of cadmium sulfide nanocrystals. *Chemical Biology* 11: 1553–1559.

Thakkar, K.N., Mhatre, S.S., Parikh, M.S. 2010. Biological synthesis of metallic nanoparticles. *Nanomedicine: Nanotechnology, Biology and Medicine* 6(2): 257–262.

Tian, B., Hua, Y. 2010. Carotenoid biosynthesis in extremophilic *Deinococcus-Thermus* bacteria. *Trends in Microbiology* 18(11): 512–520.

Varshney, R., Mishra, A.N., Bhadauria, S., Gaur, M.S. 2009. A novel microbial route to synthesize silver nanoparticles using fungus *Hormoconis resinae*. *Digest Journal of Nanomaterials and Biostructures* 4(2): 349–355.

Vigneshwaran, N., Kathe, A.A., Varadajan, P.V., Nachane, R.P., Balasubramanya, R.H. 2006. Biomimetics of silver nanoparticles by white rot fungus, *Phaenerochaete chrysosporium*. *Colloids Surfaces B: Biointerfaces* 53: 55–59.

Vigneshwaran, N., Ashtaputre, N.M., Varadarajan, P.V., Nachane, R.P., Paralikar, K.M., Balasubramanya, R.H. 2007. Biological synthesis of silver nanoparticles using the fungus *Aspergillus flavus*. *Materials Letters* 61: 1413–1418.

Wang, C., Wang, C.Y., Zhao, X.Q., Chen, R.F., Lan, P., Shen, R.F. 2013. Proteomic analysis of a high aluminum tolerant yeast Rhodotorula taiwanensis RS1 in response to aluminum stress. *Biochimica et Biophysica Acta* 1834(10): 1969–1975.

Zafrilla, B., Martínez-Espinosa, R.M., Alonso, M.A., Bonete, M.J. 2010. Biodiversity of Archaea and floral of two inland saltern ecosystems in the Alto Vinalopó Valley, Spain. *Saline Systems* 6(10): 1–12.

Zhang, H., Li, Q., Lu, Y., Sun, D., Lin, X., Deng, X., He, N., Zheng, S. 2005. Biosorption and bioreduction of diamine silver complex by *Corynebacterium*. *Journal of Chemical Technology and Biotechnology* 80: 285–290.

6

BIOSYNTHESIS OF SIZE-CONTROLLED METAL AND METAL OXIDE NANOPARTICLES BY BACTERIA

Chung-Hao Kuo, David A. Kriz, Anton Gudz, and Steven L. Suib

Department of Chemistry, University of Connecticut, Storrs, CT, USA

6.1. INTRODUCTION

Nanotechnology has a significant influence in science and the two connect on several levels, such as bulk materials and atomic or molecular structures. Unique properties of nanomaterials have been considered in various applications such as optoelectronics (Wang and Herron, 1991; Colvin et al., 1994), catalysis (Hoffman et al., 1992; Schmid, 1992), single-electron transistors (Klein et al., 1997; Wller, 1998), and magnetic (Shi et al., 1996) and biomedical devices (Dagani, 2002). The key area of nanotechnology research is concerned with the development of controllable and reliable methods to produce nanomaterials. Besides the physical and chemical approaches to synthesizing nanomaterials, nature also helps us to find a way to produce nano-sized particles via microorganisms. Less-expensive and more environmentally friendly methods (Gade et al., 2010) using bacteria (Nair and Pradeep, 2002; Lengke et al., 2006a,b; Husseiny et al., 2007; Shahverdi et al., 2007), fungi (Ahmad et al., 2003; Durán et al., 2005; Kumar et al., 2007; Parikh et al., 2008; Gajbhiye et al., 2009; Govender et al., 2010), and plants (Gardea-Torresday et al., 2002; Shankar et al., 2003; Huang et al., 2007; Narayan and Sakthivel, 2008; Rai et al., 2008;

Bio-Nanoparticles: Biosynthesis and Sustainable Biotechnological Implications, First Edition.
Edited by Om V. Singh.
© 2015 John Wiley & Sons, Inc. Published 2015 by John Wiley & Sons, Inc.

Parashar et al., 2009; Song and Kim, 2009; Tavera-Davila et al., 2009) for nanomaterial synthesis have been studied. In order to survive under detrimental conditions rich in metal ions, microorganisms have developed a unique way to convert metal ions to elemental nanoparticles (NPs) as a detoxification process.

Extensive studies using bacteria to produce a variety of NPs (both intracellularly and extracellularly) have been performed. Most commonly, secreted enzymes play an important role in the formation mechanism of NPs by bacteria. Extracellular synthesis of NPs usually occurs when bacteria secrete reductive enzymes on the surface of cell wall or outside the cell. Since no additional processing such as ultra-sonication or the addition of detergents in order to remove NPs from bacteria is needed, extracellular synthesis has more practical applications than the intracellular method. This article will summarize the main production methods of different elemental NPs from a variety of bacteria. The uniqueness of the synthesis mechanisms as well as a discussion of the morphologies of NPs is also included.

6.2. INTRACELLULAR SYNTHESIS OF METAL NANOPARTICLES BY BACTERIA

"Biomineralization" is the term used to refer to the utilization of microorganisms for the extraction of precious metals from mineral ores (Mann, 1992). Biological methods such as this one are regarded as relatively safe, cost-effective, sustainable, and environmentally friendly. In 1992, J. R. Watterson discovered that gold nuggets distributed in Alaska form due to microbially induced chemical deposition (Watterson, 1992). Further studies provided evidence that specific, Pedomicrobium-like budding bacteria accumulate gold by reductive processes and iron or manganese by oxidation processes (Beveridge and Murray, 1980). An overview of intracellularly biosynthesized metal NPs is presented in Table 6.1. *Bacillus subtilis* 168 is a gram-positive bacterium, also known as the hay bacillus or grass bacillus. *Bacillus subtilis* can reduce Au^{3+} to Au^0 in acidic environments. The Au^{3+} can penetrate the bacterial cell wall which consists of amine and carboxyl groups and be converted to hexagonal–octahedral gold crystals (Southam and Beveridge, 1994). As another example, sulfate-reducing bacteria (SRB) are able to destabilize the $Au(S_2O_3)_2^{3-}$ ions and precipitate gold. The observed H_2S released at the end point of metabolism in active SRB environments contributes to elemental gold formation. The size of the individual precipitated gold NPs was less than 10 nm, and each of these particles exhibit crystalline octahedral and sub-octahedral morphology (Lengke and Southam, 2006).

A large diversity of Fe(III)-reducing bacteria and archaea have the ability to precipitate gold by reducing Au(III) to Au(0) using hydrogen as the electron donor. The dissimilatory metal reducer, Au(III) reductase, is able to transfer electrons to Au(III). Most of the Au(III) reductase appears to be located near the outer edge of the cell surface because Au(0) precipitates extracellularly (Kashefi et al., 2001). However, in other cases such as *G. ferrireducens*, the gold NPs are found intracellularly, suggesting that the Au(III) reductase is present within the periplasmic space. Mesophilic bacterium *Shewanella algae* also have a similar ability to reduce $AuCl_4^-$ to gold

TABLE 6.1. Use of Bacteria in the Production of Metal Nanoparticles Intracellularly

Bacteria	NPs	Localization/morphology	Size (nm)	Reference
Bacillus subtilis 168	Au	Intracellularly/hexagonal–octahedral crystals	—	Beveridge and Murray (1980)
Sulfate-reducing bacteria	Au	Intracellularly/octahedral and suboctahedral crystals	<10	Lengke and Southam (2006)
Fe-reducing bacteria	Au	Near outer edge of the cell surface	—	Kashefi et al. (2001)
G. ferrireducens	Au	In the periplasmic space/irregular shape particles	—	Kashefi et al. (2001)
Shewanella algae	Au	In the cell wall/spherical particles of pH <7; On the surface of the cell wall, pH <2.8; Both intracellular and extracellular, pH <2	10–20; 15–200; 20	Konishi et al. (2007a)
Plectonema borynum UTEX 485	Au	Using $Au(S_2O_3)_2^{3-}$ as reactant, the formed AuNPs are inside the cell wall with cubic crystal morphology; using $AuCl_4^-$ as reactant, the formed AuNPs are octahedral	<10–25; <10	Lengke et al. (2006a,b)
Escherichia coli (*E. coli*) DH5α	Au	On the surface of cell wall/triangles and quasi-hexagon particles	c. 100	Du et al. (2007)
Pseudomonas stutzeri AG259	Ag	In the periplasmic space/triangular, hexagonal, and spheroidal crystals	35–40	Joerger et al. (2000)
Lactobacillus sp. A09	Ag	On the surface of the cell wall/irregular shape particles	—	Fu et al. (2000)
Corynebacterium strain SH09	Ag	On the surface of the cell wall	10–15	Zhang et al. (2005)
Silver-binding peptides AG3 and AG4	Ag	Flat plate-like crystals	15–18	Naik et al. (2002)
Cillus sp.	Ag	Inside of the cell membrane	10–15	Pugazhenthiran et al. (2009)
Stenotrophomonas maltophilia (SeITE02)	Se	In the cell cytoplasm or extracellular space	few hundred	Di Gregorio et al. (2005)
Enterobacter cloacae SLD1a-1	Se	Outside of the cell membrane/irregular particles	<0.1	Rosi and Frankenberger (1997)
Shewanella algae	Pt	In the periplasm space/spherical particles	<5	Konishi et al. (2007b)
Desulfovibrio desulfuricans NCLMB 8307	Pt	In the cell wall/spherical particles	50	Yong et al. (2002)
Shewanella oneidensis MR-1	Pd	On the surface of cell wall/spherical particles	—	Windt et al. (2005)

(Konishi et al., 2007a). The pH of the solution is an important factor in controlling the particle morphology and the location of gold deposition. At pH 7, the size of gold NPs is around 10–20 nm and located in the cell wall of *Shewanella algae*. At pH 2.8, larger-sized gold NPs (15–200 nm) were found on the surface of the bacteria cells. At a lower pH of 2.0, c. 20 nm spherical NPs were deposited intracellularly and c. 350 nm particles were deposited extracellularly (Table 6.1). A possible formation mechanism involves the release of enzymes into solution that reduces $AuCl_4^-$ to gold. Unlike the gold NPs inside the cell, nucleation and agglomeration of gold NPs occurs in the aqueous solution outside the cell, leading to the formation of larger gold particles. Under cyanobacterial and abiotic conditions, the Cyanobacteria *Plectonema boryanum* UTEX 485 reacted with $Au(S_2O_3)_2^{3-}$, and $AuCl_4^-$ initially promoted the precipitation of NPs of amorphous gold(I)-sulfide at the cell walls and ultimately deposited metallic gold (Lengke et al., 2006a). From the $Au(S_2O_3)_2^{3-}$ solution, the precipitated gold NPs were cubic (100) in morphology and <10–25 nm in size (Fig. 6.1). On the other hand, using $AuCl_4^-$ solution led to octahedral (111) gold platelets (c. 1–10 µm) in solution and NPs of gold (<10 nm) within bacterial cells (Lengke et al., 2006b). Additionally, there have been recent reports that the typical gram-negative bacterium, *Escherichia coli* (*E. coli*) DH5a, can assist the electrochemistry of hemoglobin to bioreduce the chloroauric acid and modulate the formation of gold nanoparticles. The gold NPs range from 20 to 40 nm in size and aggregated larger gold particles (c. 100 nm) can also be found (Du et al., 2007; Table 6.1).

Silver NPs show effective antimicrobial activity against bacteria, even against highly multi-resistant strains such as *Staphylococcus aureus*. The documented record of using silver NPs can be tracked back to 1000 BC (Li et al., 1997; Gupta and Silver, 1998; Panáček et al., 2006). One of the earliest modern studies of biomimetic approaches to silver nanoparticle synthesis was made by Slawson et al. (1992). The authors used silver-resistant bacteria strain, *Pseudomonas stutzeri* AG259, to accumulate silver NPs within periplasmic space. The silver NPs are 35–40 nm in size and exhibit triangular, hexagonal, and spheroidal morphologies (Joerger et al., 2000). *Lactobacillus* sp. A09, a genus of microaerophilic rod-shaped bacteria, is another silver nanoparticle producer. Under optimum pH values, *Lactobacillus* sp. A09 is capable of biosorption and bioreduction of Ag^+ to Ag^0 polycrystalline NPs on the surface of cells (Fu et al., 2000). *Corynebacterium* strain SH09 also has strong biosorption ability for $[Ag(NH_3)_2]^+$ and can reduce $[Ag(NH_3)_2]^+$ to Ag^0 (Zhang et al., 2005). The AgNPs have been found on the cell wall surface, with the average size being 10–15 nm. A possible AgNP formation mechanism involves $Ag(NH_3)_2^+$ binding to a specific functional group on the cell surface before being reduced to Ag^0. A specific organic matrix (proteins and/or other biological macromolecules) that controls metal NP nucleation and the growth of the inorganic structure was used to produce silver NP patterning arrays (Naik et al., 2002). The *in vitro* biosynthesis of silver NPs using silver-binding peptides was identified from a combinatorial phage display peptide library. The AG3 and AG4 peptides contain amino acid moieties that assist in the reduction of the silver ions. The crystals exhibited a flat plate-like morphology, with a thickness of approximately 15–18 nm. Additionally, the silver-resistant airborne

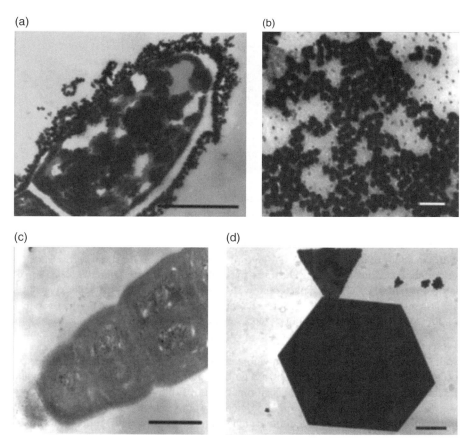

Figure 6.1. TEM micrographs of whole mounts of cyanobacteria *Plectonema boryanum* UTEX 485 cultured in the presence of Au(S$_2$O$_3$)$_2^{3-}$ or AuCl$_4^-$ solutions. (a) TEM micrographs of a thin section of cyanobacteria cells reacted with Au(S$_2$O$_3$)$_2^{3-}$ ions to form nanoparticles of gold and gold sulfide deposited on the sheaths and inside the cell. (b) Cubic nanoparticles of gold precipitated with smaller particles of gold sulfide also formed extracellularly in the Au(S$_2$O$_3$)$_2^{3-}$ containing solution. (c) Image of a thin-section of cyanobacteria cells reacted with AuCl$_4^-$ ions showing nanoparticles of gold deposited inside cells. (d) Octahedral gold platelets from cyanobacteria –AuCl$_4^-$ experiments at incubation. Source: Adapted with permission from Langmuir 2006, 22, 2780–2787. Copyright © 2006, American Chemical Society.

bacteria *Bacillus* sp. was found to produce 10–15 nm silver NPs in the periplasmic space and inner cell membranes (Pugazhenthiran et al., 2009). *Lactobacillus* strains, extracted from common buttermilk, are able to generate gold, silver, and gold-silver alloy NPs upon exposure to the precursor ions (Nair and Pradeep, 2002). The crystal growth occurs via intracellular reduction of the metal ions. The metal NPs found in the cell were formed due to diffusion occurring after their formation outside the cell

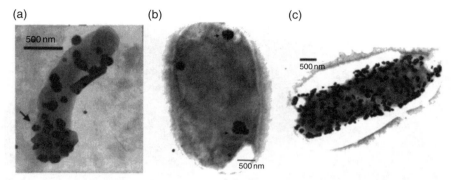

Figure 6.2. TEM micrographs of *Lactobacillus* strains cultured in the presence of Au or Ag precursor ions to form Au, Ag, or Au–Ag alloy nanoparticles. (a) TEM micrographs of *Lactobacillus* strains cell with Au nanoparticles deposited on the cell surface. Several crystal morphologies are manifested. Arrows point to regions where the cell wall is being pushed with the edges of the crystals. (b) TEM micrographs of *Lactobacillus* strain with silver clusters or nanoparticles. The image illustrates the coalescence of clusters and the formation of large crystallites. (c) TEM of a bacterium with Au–Ag alloy crystallites. Smaller crystallites are also seen outside the bacteria. Source: Adapted with permission from Crystal Growth & Design 2002, 2, 293–298. Copyright © 2002, American Chemical Society.

wall (Fig. 6.2). The bacteria remain viable even after crystal growth. The particle size of gold NPs is in the range of 20–50 nm with hexagonal and triangular morphologies. Bacterial precipitation of AgNPs is summarized in Table 6.1.

Selenium exists naturally in biological systems and has beneficial effects in trace amounts. However, selenium can be highly toxic in ppm concentrations in living organisms. The toxicity of selenium is relative to the intrinsic bioavailability of its different chemical species (Vinceti et al., 2001). For example, the selenium (Se) oxyanions selenite (SeO_3^{2-}) and selenate (SeO_4^{2-}) are water soluble and highly toxic. On the other hand, elemental selenium (Se^0) is insoluble and harmless. Nature therefore finds its own way to detoxify the selenium in bacteria by transforming selenium salts to elemental selenium. *Stenotrophomonas maltophilia* (SeITE02) showed the ability to resist selenite (SeO_3^{2-}) concentrations up to 50 mM in the growth medium (Gergorio et al., 2005; Table 6.1). The strain SeITE02 can reduce selenite to micronsize Se^0 particles in the cell cytoplasm or extracellular space. Another example is facultative bacterium *Enterobacter cloacae* SLD1a-1 which reduces selenium oxyanions to elemental selenium in microaerophilic environments (Losi and Frankenberger, 1997). The reduction of SeO_4^{2-} proceeds through intermediate SeO_3^{2-} then precipitates as elemental selenium particles of less than 0.1 mm in diameter. *Desulfovibrio desulfuricans* (DSM 1924) (Tomei et al., 1995), *Pseudomonas stutzeri* Isolate (Lortie et al., 1992), and *Escherichia coli* (*E. coli*) (Gerrard et al., 54) all showed the ability to reduce selenium oxyanions to elemental selenium in the cell wall or expel the NPs outside the cell after reduction. For platinum group metal (PGM) production, resting cells of *Shewanella algae* were proposed to deposit PtNPs by $PtCl_6^{2-}$ reduction under neutral pH in the presence of lactate as the electron donor (Konishi et al., 2007b; Table 6.1). The black precipitates were found to be less than 5 nm and located in the periplasm. Sulfate-reducing bacterium *Desulfovibrio desulfuricans* NCLMB 8307 also showed the ability to recover platinum group metals such as palladium, platinum, and rhodium (Yong et al., 2002). The reduction of Pd^{2+} to Pd^0 occurred using formate or H_2 as an electron donor at pH 2–7. The deposited PdNPs were near 50 nm in size within the cell wall. *Shewanella oneidensis* MR-1 is another example of bioreductively precipitated elemental Pd^0 (Windt et al., 2005). The recovery of Pd associated with biomass was greater than 90% when the organic electron donors lactate or formate were used.

6.3. EXTRACELLULAR SYNTHESIS OF METAL NANOPARTICLES BY BACTERIA

The location of biosynthesized NPs depends on the reductive species of the bacteria. Reductive enzymes or functional groups that are located inside the cell or on the cell wall lead to intracellular nanoparticles. Formed NPs are therefore restricted to certain areas and are sometimes not easily extracted from the cell. The NPs seen extracellularly usually form when reductive enzymes dissolve in the solvent media or when bacteria secrete the enzymes out of the cell. Extracellular synthesis of NPs by bacteria provides a better route to synthesizing NPs for application purposes. Recent studies on extracellular bacterial NP synthesis are summarized in Table 6.2. Prokaryote bacteria *Rhodopseudomonas capsulata* is an example capable of generating extracellular biosynthesized AuNPs (He et al., 2007). The synthesized AuNPs are quite stable, and their shape is controllable by adjusting pH values. The proposed mechanism of AuNP formation is bacteria *R. capsulata* secreting reductive $NADH^-$ and NADH-dependent enzymes which then reduce Au^{3+} to Au^0 and subsequently form the gold nanoparticles. The spherical AuNPs with 10–20 nm sizes were found under pH 7 and Au nano-plates formed under pH 4 (Table 6.2). The cell-free extract (CFE) of *R. capsulata* was also utilized for the production of Au nanostructures (He et al,. 2008). Different morphologies of Au nanostructures can be obtained by changing Au precursor concentration in the reaction. At lower Au concentrations, spherical 10–20 nm AuNPs were produced, whereas Au nanowires formed at higher concentrations. Another analysis of extracellular biosynthesis of AuNPs, using cell supernatant *P. aeruginosa* ATCC 90271, *P. aeruginosa* (2) and *P. aeruginosa* (1), was performed (Husseiny et al., 2007). Various sizes of AuNPs were obtained within the range 5–40 nm.

Similar to the traditional chemical reductive process of forming AuNPs, Wen et al. (2009) used *Bacillus megatherium* D01 as the reducing agent and alkyl thiol as the capping ligand to form AuNPs. The Au^{3+} first adsorbed on the surface of *B. megatherium* D01, and was then reduced. Without the addition of alkyl thiol during the reaction, the formed AuNPs aggregated to form irregular shapes with a broad size distribution (0.5–8 nm) under longer reaction time. Better controlled AuNPs size (1–3 nm) and morphology could be obtained by adding alkyl thiol, even after 6 hours of reaction time. Dried cells of the bacterium *Aeromonas* sp. SH10 showed the ability

TABLE 6.2. Use of Bacteria in the Production of Metal Nanoparticles Extracellularly

Bacteria	NPs	Localization/morphology	Size (nm)	Reference
Rhodopseudomonas capsulata	Au	Extracellularly/spherical particles	10–20	He et al. (2007)
Extracts (CFE) of *R. capsulata*	Au	Extracellularly/spherical particles	10–20	He et al. (2008)
P. aeruginosa ATCC 90271	Au	Extracellularly/irregular spherical particles	5–40	Husseiny et al. (2007)
Bacillus megatherium D01	Au	Extracellularly/spherical particles	1–3	Wen et al. (2009)
Aeromonas sp. SH10	Ag	Extracellularly/spherical particles	6.4	Fu et al. (2006)
Enterobacteria strains	Ag	Extracellularly/spherical particles	c. 50	Shahverdi et al. (2007)
Bacillus licheniformis	Au	Extracellularly/spherical particles	c. 50	Kalishwaralal et al. (2008)
Cellulose (BC) membranes from *Acetobacter xylinum*	Ag	Coated on the fiber surface/ irregular shapes	8	Barud et al. (2008)
Sulfurospirillum barnesii	Se	On the exterior of the cell envelope and intracellular/ nanoshpere	200–400	Oremland et al. (2004)
Bacillus selenitireducens	Se	On the exterior of the cell envelope and intracellular/ nanoshpere	200–400	Oremland et al. (2004)
Selenihalanaerobacter shriftii	Se	On the exterior of the cell envelope and intracellular/ nanoshpere	200–400	Oremland et al. (2004)
Sulfurospirillum barnesii	Te	Extracellularly/irregularly shaped nanospheres	<50	Baesman et al. (2007)
Bacillus selenitireducens	Te	Extracellularly/nanorods to rosettes shapes	10–100	Baesman et al. (2007)
Plectonema boryanum UTEX 485	Pt	Extracellularly/branched dendritic morphologies	300	Lengke et al. (2006c)

to reduce Ag^{3+} to Ag^0 and also accelerated reaction times in the presence of OH^- (Fu et al., 2006). The obtained AgNPs were 6.4 nm in size and were well-stabilized in the solution. The culture supernatants of different *Enterobacteria* strains were also found to reduce Ag^+ to AgNPs in a very short time (Shahverdi et al., 2007). Preliminary investigations showed that piperitone slightly inhibited the reductive activity of the bacteria, and nitroreductase enzymes might be involved in the Ag^+ ion reduction process. In a similar study of non-pathogenic bacterium *Bacillus licheniformis*, the prepared culture supernatants were shown to produce AgNPs in the range of 50 nm (Kalishwaralal et al., 2008). Interestingly, the use of bacterial cellulose (BC) membranes obtained from *Acetobacter xylinum* cultures accompanied with triethanolamine (as a stabilizer and reducing agent) can produce well-dispersed spherical AgNPs, which are attached to the BC fiber (Barud et al., 2008; Table 6.2).

For the extracellular bioproduction of selenium, Oremland et al. (2004) examined three selenate- and selenite-respiring bacteria: *Sulfurospirillum barnesii*, *Bacillus selenitireducens*, and *Selenihalanaerobacter shriftii*. Uniform and stable SeNPs with sizes of c. 300 nm formed extracellularly from selenium oxyanions. The enormous differences in the structural orientations and optical properties of synthesized SeNPs via the three different bacterial species are believed to be due a corresponding diversity of enzymatic reactions involved. Different constitutive enzymes in different concentrations within the bacteria cells may therefore determine the property of elemental Se^0 after being reduced from Se^{4+}. Notably, the biosynthesized SeNPs have unique properties which cannot be reproduced by chemical reductive processes (Table 6.2).

Tellurium is a metalloid element that is chemically similar to selenium. Oremland and his group studied the use of selenium-reducing *Sulfurospirillum barnesii* and *Bacillus selenitireducens* applied to the production of elemental tellurium (Baesman et al., 2007). Similarly to the reductive process that produces SeNPs, tellurium oxyanions were successfully precipitated as TeNPs. *Bacillus selenitireducens* bacteria produced TeNPs in the initial shape of nanorods of c. 10 nm diameter and 200 nm length. Larger rosettes with widths of c. 100 nm and lengths of 1000 nm also formed after longer reaction times. *Sulfurospirillum barnesii* formed small, irregularly shaped nanospheres (with diameter less than 50 nm) that coalesced into larger composite aggregates. For the production of precious platinum NPs, the interaction of cyanobacteria *Plectonema boryanum* UTEX 485 with platinum(IV) chloride produced Pt^{2+}-bearing NPs (Lengke et al., 2006c). The size of these NPs was less than 0.3 µm at 25 °C and less than 0.2 nm at 60–80 °C. Pure Pt^0 NPs were also obtained and recrystallized to form branched dendritic morphologies at elevated temperatures.

6.4. SYNTHESIS OF METAL OXIDE AND SULFIDE NANOPARTICLES BY BACTERIA

Many studies have been performed recently utilizing different bacteria for intracellular and extracellular bioproduction of magnetic iron oxide and iron sulfide NPs. Table 6.3 summarizes some of the reports pertaining to the synthesis of metal oxide or sulfide NPs mediated by bacteria. *Aquaspirillum magnetotacticum* extracted from sediment, for example, has shown the ability to produce Fe_3O_4 NPs mostly intracellularly (Mann et al., 1984). The formed Fe_3O_4 NPs showed well-ordered single domain crystals with octahedral prism morphology. Compared to the *A. magnetotacticum* synthesis procedure, Bazylinski et al. (1988) discovered the MV-1 strain to synthesize Fe_3O_4 NPs using magnetite bacterium under anaerobic conditions. The Fe_3O_4 NPs, of size of $40 \times 40 \times 60$ nm, were converted by using hydrous ferric oxide as a precursor. Bacteria with magnetic behavior, which are widely existent in nature,

TABLE 6.3. Use of Bacteria in the Production of Metal Oxide and Sulfide Nanoparticles

Bacteria	NPs	Localization/morphology	Size (nm)	Reference
Aquaspirillum magnetotacticum	Fe_3O_4	Intracellularly dominate/octahedral prism crystals	40–50	Mann et al. (1984)
Netotactic bacterium MV-1 strain	Fe_3O_4	Intracellularly/single domain magnetite crystals	40× 40×60	Bazylinski et al. (1988)
Magnetospirillum magnetotacticum	Fe_3O_4	Intracellularly/chain-like particles	47.1	Philipse and Maas (2002)
Geobacter metallireducens (GS-15)	Fe_3O_4	Extracellularly/ultrafine-grained particles	10–50	Lovley et al. (1987)
Thermophilic bacterium TOR-39	Fe_3O_4	Extracellularly/irregular particles	23–55	Zhang et al. (1998)
Sulfate-reducing bacteria	Fe_3O_4	On the cell surface	a few	Watson et al. (1999)
Clostridium thernoaceticum	CdS	Inside the cell wall and extracellularly/ irregular particles	–	Cunningham and Lundie (1993)
Klebsiella pneumoniae	CdS	On the cell surface/spherical particles	5–200	Smith et al. (1998)
Rhodopseudomonas palustris	CdS	Extracellularly/spherical particles	c. 8	Bai et al. (2009)
Bacterial cellulose nanofibers from *Gluconoacetobacter xylinus*	CdS	On the fiber surface	30	Li et al. (2009)
Desulfobacteriaceae	ZnS	Inside the cell/aggregated spherical particles	2–5	Bai et al. (2006)
Rhodobacter sphaeroides	ZnS	Extracellularly/spherical particles	8	Bai et al. (2006)
Rhodobacter sphaeroides	PbS	Extracellularly/spherical particles	c. 10.5	Bai and Zhang (2009)

are called magnetotactic bacteria. Various observations suggest that magnetotactic bacteria (those capable of magnetotaxis) move downward along the geomagnetic field lines to oxygen-deficient areas that are more favorable for growth. Strain MS-1 magnetic cells contain a chain of electron-dense crystals inside the cell. The microscopic evidence indicated that dried cells of magnetic bacteria contain major elements such as phosphorus, sulfur, potassium, and iron (Blakemore et al, 1979). Non-magnetic cells with similar DNA composition to Stain MS-1 showed much lower iron adsorption in controlled experiments. In order to conduct a detailed characterization of Fe_3O_4 NPs in magnetotactic bacteria, Philipse and Maas (2002) used *Magnetospirillum magnetotacticum* to grow chain-like Fe_3O_4 NPs. The chain-like morphology of Fe_3O_4 NPs was aligned along the longitudinal axis of the cell (Fig. 6.3). The amount of accumulated Fe_3O_4 varied from 0 to 45 particles, with mean equivalent sphere diameter of 47.1 nm in a single cell. Another study showed that the biogenically formed chain-like magnetic particles can be controlled by cultivating *M. magnetotacticum* under applied microelectromagnets (Lee et al., 2004; Table 6.3).

The manipulation of placing magnetic NPs in a desired position provides the possibility of incorporating these particles into customized structures. Few examples are provided for extracellular biosynthesis of magnetic NPs. Iron-reducing bacteria, such as *Geobacter metallireducens* (GS-15), are able to produce magnetic particles under anaerobic conditions (Lovley et al., 1987). The growth of non-magnetotactic bacteria *G. metallireducens* utilizes ferric ions as reductants for organic matter oxidation. Ultrafine-grained Fe_3O_4 NPs with sizes in the range 10–50 nm were produced by reducing ferric ions to amorphous ferric oxide, followed by

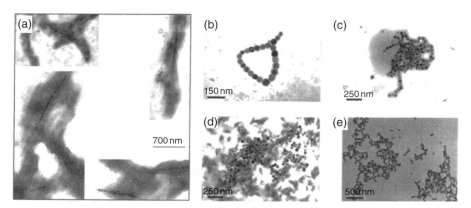

Figure 6.3. TEM micrographs of *Magnetospirillum magnetotacticum* cultured in the presence of $FeCl_3$ solution to form Fe_3O_4 magnetic nanoparticles. (a) TEM micrographs of *Magnetospirillum magnetotacticum* cell with Fe_3O_4 nanoparticles intracellularly. The formed Fe_3O_4 nanoparticles aligned in a chain parallel to the long axis of the bacteria. (b–e) TEM micrographs of Fe_3O_4 magnetite colloids extracted from cells. Note the magnetic flux closure rings in (b). The tendency to form string-like aggregates can be clearly seen in (c–e). Source: Adapted with permission from Langmuir 2002, 18, 9977–9984. Copyright © 2002, American Chemical Society.

subsequent formation of the magnetic NPs. Thermophilic bacterium (a strain called TOR-39 isolated from the deep subsurface) also showed the ability to produce magnetic Fe_3O_4 NPs extracellularly (Zhang et al., 1998; Table 6.3). Under controlled pH and Eh (oxidation potential) conditions, strain TOR-39 can reduce Fe^{3+} oxyhydroxide to Fe_3O_4 NPs with sizes in the range 26.8–55.4 nm. Compared to magnetic particles produced by *G. metallireducens*, the Fe_3O_4 NPs produced by the strain TOR-39 exhibited morphologies and sizes similar to particles produced intracellularly by magnetotactic bacteria. Furthermore, another study showed that *Thermoanaerobacter ethanolicus* (TOR-39) can even produce metal-doped magnetic Fe_3O_4 NPs by incubation (Roh et al., 2001). In this study, metals ions are mixed with iron oxyhydroxide solutions, using such metals as Co, Cr, and Ni as dopants.

In some cases, similar iron sulfide NPs have also been found in magnetotactic bacteria. Metal sulfide nanoparticles have wide applications in many areas. In terms of the iron sulfide NPs, magnetotactic bacteria under sulfide-enriched environments can produce magnetic Fe_3S_4 or non-magnetic FeS_2 nanocrystals intracellularly. Under highly reducing environments, magnetotactic bacteria slowly lost their magnetotactic behavior by preferential formation of FeS_2 NPs (Mann et al., 1990). Watson et al. (1999) also used sulfate-reducing bacteria to produce magnetic Fe_3S_4 particles of various nanosizes (Table 6.3).

Cadmium is one of the major industrial pollutants. Several possible mechanisms have been proposed to ensure bacteria survival when they are exposed to an environment with toxic concentration of cadmium ions (Silver, 1992). *Clostridium thernoaceticum* has shown the ability to precipitate CdS NPs in $CdCl_2$ solution (Cunningham and Lundie, 1993). The difference between starved and non-starved cells is that the non-starved cells tend to precipitate cadmium extracellularly, and only minor CdSNPs were found in the cell wall. Starved cells, on the other hand, exhibited localized amorphous precipitates with small electron-dense granules. In terms of bio-semiconductor application, Smith et al. (1998) used *Klebsiella pneumonia* bacterium to synthesize CdSNPs and spectroscopically characterized its behavior (Table 6.3). Spherical CdSNPs in the size range 5–200 nm were biosynthesized, and showed similar photoactive behavior as chemically synthesized nanoparticles. Bacterium *Klebsiella aerogenes* formed CdSNPs on the surface of cell walls in the size range 20–200 nm (Holmes et al., 1995). Cadmium phosphate can also be precipitated when the bacteria is grown in the presence of cadmium ions using phosphate-buffered media.

Photosynthetic bacterium *Rhodopseudomonas palustris* was shown to form CdSNPs extracellularly using $CdSO_4$ as a precursor (Bai et al., 2009). The CdSNPs were uniformly distributed, with an average size of c. 8 nm. In addition, the study showed that cysteine desulfhydrase (C-S-lyase), an intracellular enzyme located in the cytoplasm, was responsible for the formation of nanocrystals. Bacterial cellulose (BC) nanofibers derived from strain *Gluconoacetobacter xylinus* also showed the ability to form CdSNPs (Li et al., 2009). The ether and hydroxyl functional groups on the surface not only enhanced CdSNPs formation ability, but also stabilized CdSNPs, preventing aggregation. The well-distributed 30 nm CdSNPs were attached to the fiber wall. Other metal sulfides, such as ZnS nanomaterials, can be biosynthesized using natural biofilms from sulfate-reducing bacteria *Desulfobacteriaceae*. Spherical aggregates of 2–5 nm ZnSNPs were found within the biofilm. *Rhodobacter sphaeroides* also showed the ability to form ZnS and PbS NPs under different conditions (Bai et al., 2006; Bai and Zhang, 2009; Table 6.3). Using $ZnSO_4$ as a precursor, the obtained ZnSNPs exhibit unique optical properties. The size of these NPs is within the range 8–100 nm and can be controlled by adjusting bacteria culture time. PbSNPs, with a size of c. 10.5 nm, can also be formed outside the bacterium from various lead salts.

6.5. CONCLUSION

The capability of bacteria to synthesize versatile nanoparticles represents an effective and environmentally benign alternative to chemical and physical procedures, especially for elemental metal nanoparticle productions. So far, most of the studies have focused on the discovery of new bacteria for the production of nanomaterials, as well as limited discussions about precise size control of nanoparticles and applications. The elucidation of exact mechanisms still requires significant effort. Regardless, the potential for using bacteria to produce inorganic nanomaterials in the near future is promising.

REFERENCES

Ahmad, A., Mukherjee, P., Senapati. P., Mandal, D., Khan, M.I., Kumar, R., Santry, M. 2003. Extracellular biosynthesis of silver nanoparticles using the fungus *Fusarium oxysporum*. *Colloids and Surfaces B: Biointerfaces* 28: 313–318.

Baesman, S.M., Bullen, T.D., Dewald, J., Zhang, D., Curran, S., Islam, F.S., Beveridge, T.J., Oremland, R.S. 2007. Formation of tellurium nanocrystals during anaerobic growth of bacteria that use Te oxyanions as respiratory electron acceptors. *Applied and Environmental Microbiology* 73: 2135–2143.

Bai, H.-J., Zhang, Z.-M. 2009. Microbial synthesis of semiconductor lead sulfide nanoparticles using immobilized Rhodobacter sphaeroides. *Materials Letters* 63: 764–766.

Bai, H.-J., Zhang, Z.-M., Gong, J. 2006. Biological synthesis of semiconductor zinc sulfide nanoparticles by immobilized *Rhodobacter sphaeroides*. *Biotechnology Letters* 28: 1135–1139.

Bai, H.J., Zhang, Z.M., Guo, Y., Yang, G.E. 2009. Biosynthesis of cadmium sulfide nanoparticles by photosynthetic bacteria *Rhodopseudomonas palustris*. *Colloids and Surfaces B: Biointerfaces* 70: 142–146.

Barud, H.S., Barrios, C., Regiani, T., Marques, R.F.C., Verelst, M., Dexpert-Ghys, J., Messaddeq, Y., Ribeiro, S. J. L. 2008. Self-supported silver nanoparticles containing bacterial cellulose membranes. *Materials Science and Engineering C* 28: 515–518.

Bazylinski, D.A., Frankel, R.B., Jannasch, H.W. 1988. Anaerobic magnetite production by a marine, magnetotactic bacterium. *Nature* 334: 518–519.

Beveridge, T.J., Murray, R.G.E. 1980. Sites of metal deposition in the cell wall of *Bacillus subtilis*. *Journal of Bacteriology* 141: 876–887.

Blakemore, R.P., Maratea, D., Wolfe, R.S. 1979. Isolation and pure culture of a freshwater magnetic spirillum in chemically defined medium. *Journal of Bacteriology* 140: 720–729.

Colvin, V.L., Schlamp, M.C., Alivisatos, A.P. 1994. Light-emitting diodes made from cadmium selenide nanocrystals and a semiconducting polymer. *Nature* 370: 354–357.

Cunningham, D.P., Lundie, L.L. 1993. Precipitation of cadmium by *Clostridium thermoaceticum*. *Applied and Environmental Microbiology* 59: 7–14.

Dagani, R. 2002. Therein for coverage of the new applications envisaged for nanomaterials. *Chemical and Engineering News* 28 February.

Di Gregorio, S.D., Lampis, S., Vallini, G. 2005. Selenite precipitation by a rhizospheric strain of *Stenotrophomonas* sp. isolated from the root system of *Astragalus bisulcatus*: a biotechnological perspective. *Environment International* 31: 233–241.

Du, L., Jiang, H., Liu, X., Wang, E. 2007. Biosynthesis of gold nanoparticles assisted by *Escherichia coli* DH5α and its application on direct electrochemistry of hemoglobin. *Electrochemistry Communications* 9: 1165–1170.

Durán, N., Marcato, P.D., Alves, O.L., De Souza, G.I.H., Esposito, E. 2005. Mechanistic aspects of biosynthesis of silver nanoparticles by several *Fusarium oxysporum* strains. *Journal of Nanobiotechnology* 3: 8.

Fu, J., Liu, Y., Gu, P., Tang, D., Lin, Z., Yao, B., Weng, S. 2000. Spectroscopic characterization on the biosorption and bioreduction of Ag (I) by *Lactobacillus* sp. A09. *Acta Physico-Chimica Sinica* 16: 779–782.

Fu, M., Li, Q., Sun, D., Lu, Y., He, N., Xu, D., Wang, H., Huang, J. 2006. Rapid preparation process of silver nanoparticles by bioreduction and their characterizations. *Chinese Journal of Chemical Engineering* 14: 114–117.

Gade, A., Ingle, A., Whiteley, C.G., Rai, M. 2010. Mycogenic metal nanoparticles: progress and applications. *Biotechnology Letters* 32: 593–600.

Gajbhiye, M., Kesharwani, J., Ingle, A., Gade, A., Rai, M. 2009. Fungus-mediated synthesis of silver nanoparticles and their activity against pathogenic fungi in combination with fluconazole. *Nanomedicine NBM* 5: 382–386.

Gardea-Torresdey, J.L., Parsons, J.G., Gomez, E., Peralta-Videa, J., Troiani, H.E., Santiago, P., Yacaman, M.J. 2002. Formation and growth of Au nanoparticles inside live alfalfa plants. *Nano Letters* 3: 397–401.

Gerrard, T.L., Telford, J.N., Williams, H.H. 1974. Detection of selenium deposits in Escherichia coli by electron microscopy. *Journal of Bacteriology* 119: 1057–1060.

Govender, Y., Riddin, T.L., Gericke, M., Whiteley, C.G. 2010. On the enzymatic formation of platinum nanoparticles. *Journal of Nanoparticle Research* 12: 261–271.

Gupta, A., Silver, S. 1998. Silver as a biocide: will resistance become a problem? *Nature Biotechnology* 16: 888.

He, S., Guo, Z., Zhang, Y., Zhang, S., Wang, J., Gu, N. 2007. Biosynthesis of gold nanoparticles using the bacteria *Rhodopseudomonas capsulata*. *Materials Letters* 61: 3984–3987.

He, S., Zhang, Y., Guo, Z., Gu, N. 2008. Biological synthesis of gold nanowires using extract of *Rhodopseudomonas capsulata*. *Biotechnology Progress* 24: 476–480.

Hoffman, A.J., Mills, G., Yee, H., Hoffman, M.R. 1992. Q-sized cadmium sulfide: synthesis, characterization, and efficiency of photoinitiation of polymerization of several vinylic monomers. *Journal of Physical Chemistry* 96: 5546–5552.

Holmes, J.D., Smith, P.R., Evans-Gowing, R., Richardson, D.J., Russell, D.A., Sodeau, J.R. 1995. Energy-dispersive X-ray analysis of the extracellular cadmium sulfide crystallites of *Klebsiella aerogenes*. *Archves of Microbiology* 163: 143–147.

Huang, J., Chen, C., He, N., Hong, J., Lu, Y., Qing, B.L., Shao, W., Sun, D., Wang, X.H., Wang, Y., Yiang, X. 2007. Biosynthesis of silver and gold nanoparticles by novel sundried Cinnamomum camphora leaf. *Nanotechnology* 18: 105–106.

Husseiny, M.I., El-Azz, M.A., Badr, Y., Mahmoud, M.A. 2007. Biosynthesis of gold nanoparticles using *Pseudomonas aeruginosa*. *Spectrochimica Acta Part A: Molecular and Biomolecular Spectroscopy* 67: 1003–1006.

Joerger, R., Klaus, T., Granqvist, C.G. 2000. Biologically produced silver–carbon composite materials for optically functional thin-film coatings. *Advanced Materials* 12: 407–409.

Kalishwaralal, K., Deepak, V., Ramakumarpandian, S., Nellaiah, H., Sangiliyandi, G. 2008. Extracellular biosynthesis of silver nanoparticles by the culture supernatant of *Bacillus licheniformis*. *Materials Letters* 62: 4411–4413.

Kashefi, K., Tor, J.M., Nevin, K.P., Lovley, D.R. 2001. Reductive precipitation of gold by dissimilatory Fe (III)-reducing bacteria and archaea. *Applied Environmental Microbiology* 67: 3275–3279.

Klein, D.L., Roth, R., Lim, A.K.L., Alivisatos, A.P., McEuen, P.L. 1997. A single-electron transistor made from a cadmium selenide nanocrystal. *Nature* 389: 699–701.

Konishi, Y., Tsukiyama, T., Tachimi, T., Saitoh, N., Nomura, T., Nagamine, S. 2007a. Bioreductive deposition of platinum nanoparticles on the bacterium Shewanella algae. *Electrochimica Acta* 53: 186–192.

Konishi, Y., Ohno, K., Saitoh, N., Nomura, T., Nagamine, S., Hishida, H., Takahashi, Y., Uruga, T. 2007b. Microbial deposition of gold nanoparticles by the metal-reducing bacterium Shewanella algae. *Journal of Biotechnology* 128: 648–653.

Kumar, S.A., Abyaneh, M.K., Gosavi, S., Kulkarni, S.K., Pasricha, R., Ahmad, A. & Khan, M. 2007. Nitrate reductase-mediated synthesis of silver nanoparticles from $AgNO_3$. *Biotechnology Letters* 29: 439–445.

Lee, H., Purdon, A.M., Chu, V., Westervelt, R.M. 2004. Controlled assembly of magnetic nanoparticles from magnetotactic bacteria using microelectromagnets arrays. *Nano Letters* 4: 995–998.

Lengke, M., Southam, G. 2006. Bioaccumulation of gold by sulfate-reducing bacteria cultured in the presence of gold (I)-thiosulfate complex. *Geochimica et Cosmochimica Acta* 70: 3646–3661.

Lengke, M., Fleet, M.E., Southam, G. 2006a. Morphology of gold nanoparticles synthesized by filamentous cyanobacteria from gold (I)-thiosulfate and gold (III)-chloride complexes. *Langmuir* 22: 2780–2787.

Lengke, M., Ravel, B., Fleet, M.E., Wanger, G., Gordon, R.A., Southam, G. 2006b. Synthesis of platinum nanoparticles by reaction of filamentous cyanobacteria with platinum (IV)-chloride complex. *Environmental Science and Technology* 40: 6304–6309.

Lengke, M.F., Fleet, M.E., Southam, G. 2006c. Mechanisms of gold bioaccumulation by filamentous cyanobacteria from gold (III)-chloride complex. *Langmuir* 22: 7318–7323.

Li, X.Z., Nikaido, H., Williams, K.E. 1997. Silver-resistant mutants of Escherichia coli display active efflux of Ag+ and are deficient in porins. *Journal of Bacteriology* 179: 6127–6132.

Li, X., Chen, S., Hu, W., Shi, S., Shen, W., Zhang, X., Wang, H. 2009. In situ synthesis of CdS nanoparticles on bacterial cellulose nanofibers. *Carbohydrate Polymers* 76: 509–512.

Lortie, L., Gould, W.D., Rajan, S., McCready, R.G.L., Cheng, K.-J. 1992. Reduction of selenate and selenite to elemental selenium by a Pseudomonas stutzeri isolate. *Applied and Environmental Microbiology* 58: 4042–4044.

Losi, M.E., Frankenberger, W.T. 1997. Reduction of selenium oxyanions by Enterobacter cloacae SLD1a-1: isolation and growth of the bacterium and its expulsion of selenium particles. *Applied and Environmental Microbiology* 63: 3079–3084.

Lovley, D.R., Stolz, J.F., Nord, G.L., Phillips, E.J.P. 1987. Anaerobic production of magnetite by a dissimilatory iron-reducing microorganism. *Nature* 330: 252–254.

Mann, S. 1992. Bacteria and the Midas touch. *Nature* 357: 358–360.

Mann, S., Frankel, R.B., Blakemore, R.P. 1984. Structure, morphology and crystal growth of bacterial magnetite. *Nature* 310: 405–407.

Mann, S., Sparks, N H C., Frankel, R.B., Bazylinski, D.A., Jannasch, H.W. 1990. Biomineralization of ferrimagnetic greigite (Fe3S4) and iron pyrite (FeS2) in a magnetotactic bacterium. *Nature* 343: 258–261.

Naik, R.R., Stringer, S.J., Agarwal, G., Jones, S.E., Stone, M.O. 2002. Biomimetic synthesis and patterning of silver nanoparticles. *Nature Materials* 1: 169–172.

Nair, B., Pradeep, T. 2002. Coalescence of nanoclusters and formation of submicron crystallites assisted by Lactobacillus strains. *Crystal Growth and Design* 2: 293–298.

Narayan, K.B., Sakthivel, N. 2008. Coriander leaf mediated biosynthesis of gold nanoparticles. *Materials Letters* 62: 4588–4590.

Oremland, R.S., Herbel, M.J., Blum, J.S., Langley, S., Beveridge, T.J., Ajayan, P.M., Sutto, T., Ellis, A.V., Curran, S. 2004. Structural and spectral features of selenium nanospheres produced by Se-respiring bacteria. *Applied and Environmental Microbiology* 70: 52–60.

Panáček, A., Kvítek, L., Prucek, R., Kolář, M., Večeřová, R., Pizúrová, N., Sharma, V.K., Nevěčná, T., Zbořil, R. 2006. Silver colloid nanoparticles: synthesis, characterization, and their antibacterial activity. *Journal of Physical Chemistry B* 110: 16248–16253.

Parashar, V., Parashar, R., Sharma, B., Pandey, A.C. 2009. Parthenium leaf extract mediated synthesis of silver nanoparticles: a novel approach towards weed utilization. *Digest Journal of Nanomaterials and Biostructures* 4: 45–50.

Parikh, R.Y., Singh, S., Prasad, B.L.V., Patole, M.S., Sastry, M., Shouche, Y.S. 2008. Extracellular synthesis of crystalline silver nanoparticles and molecular evidence of silver resistance from *Morganella* sp.: towards understanding biochemical synthesis mechanism. *Chembiochemistry* 9: 1415–1422.

Philipse, A.P., Maas, D. 2002. Magnetic colloids from magnetotactic bacteria: chain formation and colloidal stability. *Langmuir* 18: 9977–9984.

Pugazhenthiran, N., Anandan, S., Kathiravan, G., Prakash, N.K.U., Crawford, S., Ashokkumar, M. 2009. Microbial synthesis of silver nanoparticles by *Bacillus* sp. *Journal of Nanoparticle Research* 11: 1811–1815.

Rai, M.K., Yadav, A.P., Gade, A.K. 2008. CRC 675: Current trends in phytosynthesis of metal nanoparticles. *Critical Reviews in Biotechnology* 28: 277–284.

Roh, Y., Lauf, R.J., McMillan, A.D., Zhang, C., Rawn, C.J., Bai, J., Phelps, T.J. 2001. Microbial synthesis and the characterization of metal-substituted magnetites. *Solid State Communications* 118: 529–534.

Schmid, G. 1992. Large clusters and colloids. Metals in the embryonic state. *Chemical Reviews* 92: 1709–1727.

Shahverdi, A.R., Minaeian, S., Shahverdi, H.R., Jamalifar, H., Nohi, A.A. 2007. Rapid synthesis of silver nanoparticles using culture supernatants of Enterobacteria: A novel biological approach. *Process Biochemistry* 42: 919–923.

Shankar, S.S., Ahmad, A., Sastry, M. 2003. Geranium leaf assisted biosynthesis of silver nanoparticles. *Biotechnology Progress* 19: 1627–1631.

Shi, J., Gider, S., Babcock, K., Awschalom, D.D. 1996. Magnetic clusters in molecular beams, metals, and semiconductors. *Science* 271: 937–941.

Silver, S. 1992. Plasmid-determined metal resistance mechanisms: range and overview. *Plasmid* 27: 1–3.

Slawson, R.M., Trevors, J.T., Lee, H. 1992. Silver accumulation and resistance in *Pseudomonas stutzeri*. *Archves of Microbiology* 158: 398.

Smith, P.R., Holmes, J.D., Richardson, D.J., Russell, D.A., Sodeau, J.R. 1998. Photophysical and photochemical characterisation of bacterial semiconductor cadmium sulfide particles. *Journal of the Chemical Society, Faraday Transactions* 94: 1235–1241.

Song, J.Y., Kim, B.S. 2009. Rapid biological synthesis of silver nanoparticles using plant leaf extracts. *Bioprocess and Biosystems Engineering* 32: 79–84.

Southam, G., Beveridge, T.J. 1994. The in vitro formation of placer gold by bacteria. *Geochimica et Cosmochimica Acta* 58: 4527–4530.

Tavera-Davila, L., Liu, H. B., Herrera-Becerra, R., Canizal, G., Balcazar, M., Ascencio, J.A. 2009. Analysis of Ag nanoparticles synthesized by bioreduction. *Journal of Nanoscience and Nanotechnology* 9: 1785–1791.

Tomei, F.A., Barton, L.L., Lemanski, C.L., Zocco, T.G., Fink, N.H., Sillerud, L.O. 1995. Transformation of selenate and selenite to elemental selenium by *Desulfovibrio desulfuricans*. *Journal of Industrial Microbiology* 14: 329–336.

Vinceti, M., Wei, E.T., Malagoli, C., Bergomi, M., Vivoli, G. 2001. Adverse health effects of selenium in humans. *Reviews on Environmental Health* 16: 233–251.

Wang, Y., Herron, N. 1991. Nanometer-sized semiconductor clusters: materials synthesis, quantum size effects, and photophysical properties. *Journal of Physical Chemistry* 95: 525–532.

Watson, J.H.P., Ellwood, D.C., Soper, A.K., Charnock, J. 1999. Nanosized strongly-magnetic bacterially produced iron sulfide materials. *Journal of Magnetism and Magnetic Materials* 203: 69–72.

Watterson, J.R. 1992. Preliminary evidence for the involvement of budding bacteria in the origin of Alaskan placer gold. *Geology* 20: 315–318.

Weller, H. 1998. Transistors and light emitters from single nanoclusters. *Angewandte Chemie International Edition* 37: 1658–1659.

Wen, L., Lin, Z., Gu, P., Zhou, J., Yao, B., Chen, G., Fu, J. 2009. Extracellular biosynthesis of monodispersed gold nanoparticles by a SAM capping route. *Journal of Nanoparticle Research* 11: 279–288.

Windt, D.D., Aelterman, P., Verstraete, W. 2005. Bioreductive deposition of palladium (0) nanoparticles on *Shewanella oneidensis* with catalytic activity towards reductive dechlorination of polychlorinated biphenyls. *Environmental Microbiology* 7: 314–325.

Yong, P., Rowsen, N.A., Farr, J.P.G., Harris, I.R., Macaskie, L.E. 2002. Bioreduction and biocrystallization of palladium by *Desulfovibrio desulfuricans* NCIMB 8307. *Biotechnology and Bioengineering* 80: 369–379.

Zhang, C., Vali, H., Romanek, C.S., Phelps, T.J., Liu, S. V. 1998. Formation of single-domain magnetite by a thermophilic bacterium. *American Mineralogist* 83: 1409–1418.

Zhang, H., Li, Q., Lu, Y., Sun, D., Lin, X., Deng, X., He, N., Zheng, S. 2005. Biosorption and bioreduction of diamine silver complex by *Corynebacterium*. *Journal of Chemical Technology and Biotechnology* 80: 285–290.

7

METHODS OF NANOPARTICLE BIOSYNTHESIS FOR MEDICAL AND COMMERCIAL APPLICATIONS

Shilpi Mishra

Department of Biological Sciences, Alabama State University, Montgomery, AL, USA

Saurabh Dixit

Center for Nanobiotechnology Research, Alabama State University, Montgomery, AL, USA

Shivani Soni

Department of Biological Sciences, Alabama State University, Montgomery, AL, USA

7.1. INTRODUCTION

Nanotechnology is a technology that deals with nanometer-sized objects (Feynman, 1991). Nanoparticles have one dimension that measures 100 nm or less (Narayanan and Sakthivel, 2010). It is a most promising technology that can be applied to almost all spheres of life, ranging from electronics, medical, pharmaceutical, defense, transportations, and heat transfer, to sports and aesthetics (Glomm, 2005; Chan, 2006; Boisselier

---

Bio-Nanoparticles: Biosynthesis and Sustainable Biotechnological Implications, First Edition.
Edited by Om V. Singh.
© 2015 John Wiley & Sons, Inc. Published 2015 by John Wiley & Sons, Inc.

and Astruc, 2009). A decade ago, nanoparticles were studied because of their size-dependent physical and chemical properties (Murray et al., 2000). Nanoparticles exhibit enhanced properties of specific characteristics such as size, distribution, and morphology. They have a greater surface area per weight than larger particles, which causes them to be more reactive to some other molecules.

Due to their significant role in chemical, biological, and electronic fields, the synthesis of nanoparticles has generated much interest. The development of many synthetic procedures for uniform nanometer-sized nanoparticles is essential for many advanced applications, because the monodispersity of metal nanoparticles influences their precise size and size-dependent properties. There are numerous methods available including chemical, physical, and biological protocols for the synthesis of nanoparticles. These methods are not only expensive, but also hazardous to ecosystems and human life, and often result in the synthesis of a mixture of nanoparticles with poor morphology. Chemical synthesis is advantageous as it takes a short period of time to synthesize a large quantity of nanoparticles. However, in this method capping agents such as thioglycerol (TG) mercaptoethanol (ME) and sodium hexaetaphosphate (SHMP) are required for size stabilization of the nanoparticles. Moreover, chemical generally reagents used for nanoparticle synthesis and stabilization are toxic and lead to the formation of by-products that are harmful to the environment. Therefore, the development of reliable experimental protocols for the synthesis of nanomaterials over a range of chemical compositions, sizes, and high monodispersity is one of the challenging issues in nanotechnology.

There is an ever-growing need to develop ecofriendly, non-toxic, and cheaper pathways for nanoparticle synthesis. The synthesis of metal nanoparticles of various sizes and shapes and their colloidal stabilization through bimolecular immobilization is essential due to their wide-ranging applications in biotechnology. Utilization of non-toxic chemicals, biodegradable solvents, and renewable materials are some of the key issues that push the idea of green synthesis of these useful nanoparticles. Microorganisms (Sastry et al., 2003; Konishi et al., 2004; Lengke and Southam, 2006; Birla et al., 2009; Gajbhiye et al., 2009), plants (Shankar et al., 2004; Chandran et al., 2006; Huang et al., 2007; Bar et al., 2009a, b; Jha et al., 2009; Song and Kim, 2009), and plant extracts have been extensively used by researchers in the biosynthesis of nanoparticles.

Biosynthesis of nanoparticles is also considered to be a bottom-up technique, where the oxidation/reduction is the main reaction that occurs during the production. Nanoparticles are biosynthesized when the microorganisms grab target ions from their environment and then turn the metal ions into the element metal through enzymes generated by the cell activities. It can be classified as either intracellular and extracellular (Ahmad et al., 2003a, b; Mukherjee et al., 2008; Rai et al., 2009; Shaligram et al., 2009) synthesis according to the location where nanoparticles are formed (Simkiss and Wilbur, 1989).

The intracellular method usually involves the use of bacteria and actinomycetes. In this method the metal ions are transported into the microbial cell at ambient temperature and pressure to form nanoparticles in the presence of enzymes such as nitrate reductase (Nair and Pradeep, 2002; Ahmad et al., 2003a, b). In case of fungi, fungal mycelium is treated with metal salt solution followed by incubation for 24 hour, leading to the intracellular synthesis of nanoparticles (Mukherjee et al., 2001). In this method the fungal mycelium is harvested by centrifuging and

INTRODUCTION

subsequently freeze-drying, and this freeze-dried mycelium is immersed in metal salt solution and kept on a shaker (Chen et al., 2003).

The extracellular synthesis of nanoparticles involves trapping the metal ions on the surface of the cells and reducing ions in the presence of enzymes such as hydrogenase. This extracellular enzyme shows excellent redox properties and can act as an electron shuttle in the metal reduction. It was evident that electron shuttles or other reducing agents (e.g., hydroquinones) released by microorganisms are capable of reducing ions to nanoparticles (Zhang et al., 2011). For the extracellular synthesis of nanoparticles, the bacterial culture is centrifuged at 8000×g and the supernatant is challenged with metal salt solution (Kalishwaralal et al., 2008a, b; Das et al., 2009; Ogi et al., 2010). The filtrate of the fungal mycelium is treated with metal salt solution and incubated for 24 hours in extracellular synthesis (Fayaz et al., 2009; Shaligram et al., 2009).

In this chapter we describe some of the organisms used in the biosynthesis of nanomaterials and describe the required inherent properties that for the production of nanoparticles of desired characteristics. Some organisms which have been used for the production of nanoparticles are listed in Table 7.1.

TABLE 7.1. List of Various Biological Entities used in the Biosynthesis of Metal Nanoparticles

Biological entities	Size (nm)	Reference
Bacteria		
Pseudomonas stutzeri AG259	200	Klaus et al. (1999)
Bacillus subtilis	5–60	Saifuddin et al. (2009)
Klebsiella pneumoniae (culture supernatant)	50	Klaus et al. (1999)
Escherichia coli	1–100	Gurunathan et al. (2009)
Rhodopseudomonas capsulata	10–20	He et al. (2007)
Pseudomonas aeruginosa	15–30	Husseiny et al. (2007)
Bacillus licheniformis (culture supernatant)	50	Kalishwaralal et al. (2008a, b; 2009)
Brevibacterium casei	50	Kalishwaralal et al. (2010)
Fungi		
Fusarium oxysporum	5–50	Ahmad et al. (2003b)
Aspergillus fumigatus	5–25	Bhainsa and D'Souza (2006)
Verticillium	20	Mukherjee et al. (2001)
Trichoderma viride	5–40	Fayaz et al. (2009)
Aspergillus flavus	8.92±1.61	Vigneshwaran et al. (2007)
Verticillium sp.	12–25	Mukherjee at al. (2001)
Plants		
Azadirachta indica	50–100	Shankar et al. (2004)
Cinnamomum camphora leaf	55–80	Huang et al. (2007)
Carica papaya	60–80	Mude et al. (2009)
Aloe vera (Plant)	15.2±4.2	Chandran et al. (2006)
Emblica officinalis	10–20, 15–25	Ankamwar et al. (2005a)

7.2. BIOSYNTHESIS OF NANOPARTICLES USING BACTERIA

The use of microbial cells for the synthesis of nano-sized materials has emerged as a novel approach for the synthesis of metal nanoparticles. Although efforts directed towards the biosynthesis of nanomaterials are recent, the interactions between microorganisms and metals have been well documented and the ability of microorganisms to extract and/or accumulate metals is exploited in commercial biotechnological processes such as bioleaching and bioremediation (Gericke and Pinches, 2006). Bacteria have been most extensively researched for synthesis of nanoparticles because of their fast growth rates and relative ease of genetic manipulation.

7.2.1. Synthesis of Silver Nanoparticles by Bacteria

Klaus and his co-workers used a bacteria named *Pseudomonas stutzeri* AG259 isolated from silver mines to synthesize silver nanoparticles. These bacteria reduced a concentrated aqueous solution of silver nitrate to Ag^+ ions (Klaus et al., 1999). This results in the formation of silver nanoparticles (AgNPs) of well-defined size and distinct topography within the periplasmic space of the bacteria (Klaus et al., 1999; Joerger et al., 2000).

Similar observations were made when culture supernatant of non-pathogenic bacterium *Bacillus licheniformis* was added to aqueous silver ions, resulting in the biosynthesis of highly stable and well-dispersed silver nanoparticles in the size range 40–50 nm by bioreduction (Kalishwaralal et al., 2008*b*). *Bacillus licheniformis* is a gram-positive theromophilic bacterium commonly found in soil. A change in its color from yellow to brown after incubating the culture supernatant with silver nitrate indicates the formation of silver nanoparticles. Kalishwaral and co-workers (Kalishwaralal et al., 2008*a, b*) have investigated the process to synthesize silver nanoparticles by sonification of bacteria *Bacillus lichemiformis*, isolated from sewage. The average particle size was found to be around 50 nm. The enzyme involved in the fabrication of nanoparticles can be nitrate reductase, present in *B. licheniformis* (Sadowski et al., 2008). A better and faster method to biosynthesize silver nanoparticles has been described by Shahverdi et al. (2007) and Mokhtari et al. (2009) by using *Enterobacteria* (*Klebsiella pneumoniae*), a gram-negative bacteria associated with intestinal infections. The culture supernatant obtained from *Klebsiella pneumoniae* was treated with aqueous silver nitrate solution, resulting in the formation of silver nanoparticles ranging in size 28.2–122 nm with an average size of 52.5 nm (Shahverdi et al., 2007). Investigations by Mokhtari et al. (2009) showed that piperitone (3 methyl-6-1 methylethyl)-2 cyclohexan-1-one) can be responsible for silver ion reduction to metal. This conclusion supports the hypothesis that nitroreductase enzymes may be involved in the reduction process of silver ion.

Saifuddin et al. (2009) used an entirely different approach to biosynthesize silver nanoparticles. They reported the extracellular biosynthesis of monodispersed Ag nanoparticles (5–50 nm) using supernatants of *Bacillus subtilis* and microwave irradiation in water. The samples (supernatant and $AgNO_3$ solution) were subjected to several short burst of microwave irradiation at the frequency of 2.45 GHz, power output of c. 100 W in a following cyclic mode of on 15 sec, off 15 sec to prevent overheating. The formation of nanoparticles by this method was not only extremely rapid, but also provided uniform heating around the nanoparticles and assisted the digestive ripening of particles with no aggregation. Nair and Pradeep (2002) reported that common *Lactobacillus* strains found in buttermilk assisted in the growth of microscopic gold, silver, and gold-silver alloy crystals of well-defined morphology.

7.2.2. Synthesis of Gold Nanoparticles by Bacteria

In demand for curing various diseases, gold nanoparticles (AuNPs) have a rich history in chemistry dating back to ancient Roman times where they were used to stain glasses for decorative purposes. Almost 150 years ago, Michael Faraday observed for the first time the unique properties of colloidal gold solutions, which were different from that of bulk gold (Hayat, 1989).

Gold nanoparticles were first produced by *B. subtilis* by reduction of Au^{3+} ions when it was incubated with gold chloride (Beveridge and Murray, 1980). Because of the ability to reduce Au(III) ions in anaerobic conditions and in the presence of hydrogen gas, the bacterium *Shewanella algae* (Konishi et al., 2004) has also been used to biosynthesize gold nanoparticles (Douglas and Beveridge, 1998). The different nanoparticle sizes may be attributed to the cell growth and the incubation conditions of the metal. In an analogous study, Shiying et al. (2007) showed that the bacteria *Rhodopseudomonas capsulata* produces gold nanoparticles of different sizes and shapes. They found that *R. capsulata* biomass incubated with aqueous chlorauric acid ($HAuCl_4$) solution with pH values of 4–7 formed spherical gold nanoparticles in the range of 10–20 nm at pH 7. Husseiny et al. (2007) used the bacterial cell supernatant of *Pseudomonas aeruginosa* for the reduction of gold ions, resulting in extracellular biosynthesis of gold nanoparticles. The cell filtrate provides better control over size and polydispersity of nanoparticles. Further, it was documented that extracellular synthesis of nanoparticles using cell filtrate could be beneficial to intracellular synthesis (Mohanpuria et al., 2008). Lengke et al. (2006) demonstrated the synthesis of gold nanostructures by filamentous cyanobacteria. They succeeded in biosynthesizing the nanoparticles in different shapes (spherical, cubic, and octahedral) from Au(I)-thiosulfate and Au(III)-chloride complexes and analyzed their formation mechanisms (Lengke and Southam, 2006; Lengke et al., 2006). Monodisperse gold nanoparticles have been synthesized by alkalotolerant *Rhodococcus* sp. under extreme biological conditions such as alkaline and slightly elevated temperature conditions (Ahmad et al., 2003*b*).

7.2.3. Synthesis of other Metallic Nanoparticles by Bacteria

Heavy metals are known to be toxic to microorganism life. Konishi et al. (2007) reported the synthesis of platinum nanoparticles by using metal ion-reducing bacterium *Shewanella algae*. Resting cells of *S. algae* were able to reduce aqueous $PtCl_6^{2-}$ ions into elemental platinum at room temperature and neutral pH within

60 min when lactate was provided as the electron donor. Platinum nanoparticles of about 5 nm were located in the periplasm. Sinha and Khare (2011) synthesized mercury nanoparticles using *Enterobacter* sp. cells. *Pyrobaculum islandicum*, an anaerobic hyperthermophilic microorganism, was reported to reduce many heavy metals including U(VI), Tc(VII), Cr(VI), Co(III), and Mn(IV) with hydrogen as the electron donor (Kashefi and Lovley, 2000). Similarly, in the presence of an exogenous electron donor, sulfate-reducing bacterium *Desulfovibrio desulfuricans* NCIMB 8307 has been shown to synthesize palladium nanoparticles (Yong et al., 2002).

Magnetic nanoparticles are an important type of compound nanoparticles synthesized by microbes. They are recently gaining interest due to their unique microconfiguration and properties such as superparamagnetic and high coercive force, as well as their many applications in biological separation and biomedicine. Magnetic nanoparticles such as Fe_3O_4 (magnetite) and Fe_2O_3 (maghemite) are known to be biocompatible. They have been actively investigated for targeted cancer treatment (magnetic hyperthermia), stem cell sorting and manipulation, guided drug delivery, gene therapy, DNA analysis, and magnetic resonance imaging (MRI) (Fan et al, 2009). Zhou et al. (2009a, b) synthesized magnetic Fe_3O_4 materials with mesoporous structure as a template by coprecipitation, using yeast cells. Roh et al. (2001) used *Magnetospirillum magneticum* to produce magnetic nanoparticles. Magnetotactic bacteria such as *M. magneticum* produce two types of particles; some produce magnetic (Fe_3O_4) nanoparticles in chains and some produce greigite (Fe_3S_4) nanoparticles, while others produce both types (Bazylinski and Frankel, 2004).

7.3. BIOSYNTHESIS OF NANOPARTICLES USING ACTINOMYCETE

Sastry et al. (2003) observed that when the extremophilic actinomycete *Thermomonospora* sp. is exposed to gold ions the metal ions are reduced extracellularly, yielding gold nanoparticles with a much improved polydispersity. However, Ahmed et al. (2003a) discovered an efficient method of biosynthesizing the monodisperse gold nanoparticles. Scientists carried out the reduction of $AuCl_4^-$ ions with extremophilic *Thermomonospora* sp. biomass in an alkaline medium and at a slightly elevated temperature (Ahmad et al., 2003b). The reduction of metal ions and the stabilization of the gold nanoparticles were believed to occur by an enzymatic process (Ahmad et al., 2003a, b). Based on the hypothesis extreme, biological conditions such as alkaline and elevated temperature conditions are favorable for the synthesis of nanoparticles. Alkalotolerant *Rhodococcus* sp. has been used for intracellular synthesis of good quality monodisperse gold nanoparticles. They observed that the concentration of nanoparticles were more on the cytoplasmic membrane than on the cell wall. This could be due to reduction of the metal ions by enzymes present in the cell wall and on the cytoplasmic membrane but not in the cytosol. These metal ions were not toxic to the cells which are producing them, and continued to multiply even after the biosynthesis of gold nanoparticles (Ahmad et al., 2003a).

7.4. BIOSYNTHESIS OF NANOPARTICLES USING FUNGI

Compared to bacteria, fungi could be used as a source for the production of large quantities of nanoparticles. Because of their tolerance and metal bioaccumulation ability, fungi take center stage in studies on biological generation of metallic nanoparticles (Sastry et al., 2003). This is due to the fact that fungi secrete greater amounts of proteins which directly translate to higher productivity of nanoparticle formation (Mohanpuria et al., 2008). A number of different genera of fungi have been investigated in this effort, and it has been shown that fungi are extremely good candidates as "nanofactories" for the synthesis of metal nanoparticles. Biosynthesis of nanoparticles by fungi takes place extracellularly, intracellularly, and on the surface of the cell wall.

Extracellular synthesis has been observed in *Fusarium oxysporum* with silver and gold-silver nanoparticles (Ahmad et al., 2003*a*). The reduction of silver ions by *F. oxysporum* strains has been attributed to a nitrate-dependent reductase and a shuttle quinine extracellular process. Mukherjee et al. (2001) studied the synthesis of intracellular AgNPs using the fungus *Verticillium*. The authors observed that exposure of the fungal biomass to aqueous Ag^+ ions results in the intracellular reduction of the metal ions and formation of 25 ± 12 nm AgNPs. However, a novel biological method for the intra- and extracellular synthesis of silver nanoparticles using the fungi *Verticillium* and *Fusarium oxysporum*, respectively, has been documented. This has opened up an exciting possibility where the nanoparticles may be entrapped in biomass in the form of a film or produced in solution, both having important commercial potential (Senapati et al., 2004). The fungus *Aspergillus flavus* also resulted in the accumulation of silver nanoparticles on the surface of its cell wall when incubated with silver nitrate solution (Vigneshwaran et al., 2007). Endophytic fungus *Collitotrichum* sp., which grows on the leaves of geranium, was used for the synthesis of stable and various-shaped gold nanoparticles. Extracellularly produced nanoparticles were stabilized by the proteins and reducing agents secreted by the fungus and leaves of geranium. Reducing and capping agents obtained from geranium leaves are terpenoids; in the case of fungus they were polypeptides/enzymes (Shankar et al., 2003). Instead of fungi culture, isolated proteins (glutathione and cystein residue amines) from them have been used successfully in the production of nanoparticles. Nanocrystalline zirconia was produced at room temperature by cationic proteins. These proteins were similar in nature to silicatein, secreted by *F. oxysporum*, and were capable of hydrolyzing aqueous ZrF_6^{2-} ions extracellularly (Bansal et al., 2004).

Jha and co-workers found a green low-cost and reproducible method using *Saccharomyces cerevisiae* to biosynthesize Sb_2O_3 nanoparticles (Jha and Prasad, 2009). Bansal et al. (2005) used *F. oxysporum* (fungus) to produce SiO_2 and TiO_2 nanoparticles from aqueous anionic complexes SiF_6^{2-} and TiF_6^{2-}, respectively. They also prepared tetragonal $BaTiO_3$ and quasi-spherical ZrO_2 nanoparticles from *F. oxysporum* with a size range of 4–5 nm and 3–11 nm, respectively (Bansal et al., 2004, 2006).

Bharde et al. (2006) reported the synthesis of magnetic nanoparticles by *F. oxysporum* and *Verticillium* sp. at room temperature. Both fungi secreted proteins which were capable of hydrolyzing iron precursors extracellularly to form iron oxides predominantly in the magnetite (Fe_3O_4) phase. A nitrate-dependent reductase and a shuttle quinone from several *F. oxysporum* strains were found to be involved in extracellular production of silver nanoparticles or silver hydrosol. With the aim of determining the mechanism of synthesis of nanoparticles, a NADPH-dependent nitrate reductase and phytochelatin isolated from *F. oxysporum* has been used for *in vitro* silver nanoparticle production (Kumar et al., 2007).

Bhainsa and D'Souza (2006) demonstrated the biosynthesis of monodispersed silver nanoparticles in a few minutes using *Aspergillus fumigatus*. This was the first report of such rapid synthesis of nanoparticles using fungus. The production was even faster compared to the physical and chemical processes of nanoparticles synthesis. This process could therefore be suitable for developing a biological process for the large-scale production of nanoparticles.

7.5. BIOSYNTHESIS OF NANOPARTICLES USING PLANTS

The biosynthesis of nanoparticles using plants or plant extracts is known to be simple, eco-friendly, and economically viable compared to microbial systems such as bacteria and fungi. This is not only due to their pathogenicity, but also the chemical and physical methods used for the synthesis of metal nanoparticles (Shankar et al., 2004; Ankamwar et al., 2005*a*; Chandran et al., 2006; Parashar et al., 2009).

The use of plants for the preparation of nanoparticles could be advantageous because it does not require elaborate processes such as intracellular synthesis and multiple purification steps, or the maintenance of microbial cell cultures (Bhattacharya and Gupta, 2005). Several plants and their parts have been successfully used for the extracellular synthesis of metal nanoparticles (Mohanpuria et al., 2008).

Shankar et al. (2003) reported the extracellular biosynthesis of silver nanoparticles by reducing silver ions with geranium (*Pelargonium graueolens*) leaf extract. These nanoparticles were highly stable and crystalline in solution. It is not only individual pure metallic Ag and Au which have been biosynthesized, but Shankar et al. (2004) also reported the formation of bimetallic Ag/Au nanopaticles (50–100 nm) using *Azadirachta indica* leaf broth. With the use of *Emblica officinalis* fruit extract as reducing agent, the extracellular synthesis of highly stable Ag and Au nanoparticles was achieved (Ankamwar et al., 2005*b*). Along with stability, control over the shape of nanoparticles production using plants and plant extracts is also possible. This has been shown by the rapid synthesis of stable gold nanotriangles at high concentrations using tamarind leaf extract as reducing agent (Ankamwar et al., 2005*b*).

Other plants showing great potential in the synthesis of nanoparticles include *Aloe vera* (Chandran et al., 2006) and *Cinnamomum camphora* (Huang et al., 2007); leaf extracts from these have been used as a reducing agent to synthesize gold nanotriangles as well as silver nanoparticles. The marked difference in shape control between gold and silver nanoparticles was attributed to the comparative advantage of protective biomolecules and reductive biomolecules. The slow reduction of aqueous gold ions ($HAuCl_4$), along with the shape-directing effects of constituents (carbonyl compounds) by varying the amount of extract in the reaction medium, played a key role in the formation of gold nanotriangles.

Nagaraj et al. (2013) used the fruit extract of *Ananas comosus* (L.) to synthesize gold nanoparticles from $AuCl_4$. In this study, pineapple extract was used to obtain phytochemically derived reducing agents for the production and stabilization of gold nanoparticles. When stirred with an aqueous solution (1 mM of $AgNO_3$ to 90 mL of distilled water at room temperature) of silver nitrate at 70 °C, leaf extracts of *Rosmarinus officinalis* L. (rosemary) lead to the synthesis of silver nanoparticles (Sulaiman et al., 2013). Having enhanced antimicrobial activity against human urinary tract infections, many other plants have also been used to biosynthesize nanoparticles such as *Acalypha indica*, *Pelargonium graveolens*, *Parthenium hysterophorus*, *Gliricidia sepium*, hypocotyls, collar, and bark of *Rhizophora mucronata*, and other species of mangrove plants (Shankar et al., 2003; Raut et al., 2009; Safaepour et al., 2009; Krishnaraj et al., 2010).

7.6. CONCLUSIONS

It has been concluded that various kinds of physical and chemical methods that are employed for the synthesis of nanoparticles require both strong and weak chemical reducing agents and protective agents (sodium borohydride, sodium citrate, and alcohols). These agents are are mostly toxic, flammable, and difficult to dispose of without harming the environment, and also demonstrate a low production rate (Mohanpuria et al., 2008; Rai et al., 2009; Sharma et al., 2009).

On the other hand, biological methods provide a wide range of resources for the synthesis of nanoparticles. The rate of reduction of metal ions using biological agents is found to be much faster, and can also take place under ambient temperature and pressure conditions. By using the organisms from simple bacteria to highly complex eukaryotes in the reaction mixture, the production of nanoparticles of desired shape and size can be obtained. The biomass used for the synthesis of nanoparticles is simpler to handle and is easily disposed of without harming the environment. Considering the above points, the biosynthesis of nanoparticles would appear to be superior to physical and chemical methods of synthesis due to its environmently friendly approach and low cost.

REFERENCES

Ahmad, A., Mukherjee, P., Senapati, S., Mandal, D., Khan, M.I., Kumar, R., Sastry, M. 2003a. Extracellular biosynthesis of silver nanoparticles using the fungus *Fusarium oxysporum*. *Colloids and Surfaces B* 28: 313–318.

Ahmad, A., Senapati, S., Khan, M.I., Kumar, R., Ramani, R., Shrinivas, V., Sastry, M. 2003b. Intracellular synthesis of gold nanoparticles by a novel alkalotolerant actinomycete, *Rhodococcus species*. *Nanotechnology* 14: 824–828.

Ankamwar, B., Damle, C., Absar, A., Mural, S. 2005a. Biosynthesis of gold and silver nanoparticles using *Emblica Officinalis* fruit extract, their phase transfer and transmetallation in an organic solution. *Journal of Nanoscience and Nanotechnology* 10: 1665–1671.

Ankamwar, B., Chaudhary, M., Mural, S. 2005b. Gold nanotriangles biologically synthesized using tamarind leaf extract and potential application in vapor sensing. *Synthesis and Reactivity in Inorganic, Metal-Organic, and Nano-Metal Chemistry* 35: 19–26.

Bansal, V., Rautaray, D., Ahmad, A., Sastry, M. 2004. Biosynthesis of zirconia nanoparticles using the fungus *Fusarium oxysporum*. *Journal of Materials Chemistry* 14: 3303–3305.

Bansal, V., Rautaray, D., Bharde, A. et al. 2005. Fungus-mediated biosynthesis of silica and titania particles. *Journal of Materials Chemistry* 15: 2583–2589.

Bansal, V., Poddar, P., Ahmad, A., Sastry, M. 2006. Roomtemperature biosynthesis of ferroelectric barium titanate nanoparticles. *Journal of the American Chemical Society* 128: 11958–11963.

Bar, H., Bhui, D.K., Sahoo, G.P., Sarkar, P., De, S.P., Misra, A. 2009a. Green synthesis of silver nanoparticles using latex of *Jatropha curcas*. *Colloids and Surfaces A: Physicochemical and Engineering Aspects* 339: 134–139.

Bar, H., Bhui, D.K., Sahoo, G.P., Sarkar, P., Pyne, S., Misra, A. 2009b. Green synthesis of silver nanoparticles using seed extract of Jatropha curcas. *Colloids and Surfaces A: Physicochemical and Engineering Aspects* 348: 212–216.

Bazylinski, A.B., Frankel, B.R. 2004. Magnetosome formation in procayotes. *Nature Reviews Microbiology* 2: 213–230.

Bharde, A., Rautaray, D., Bansal, V., Ahmad, A., Sarkar, I., Yusuf, S.M., Sanyal, M., Sastry, M. 2006. Extracellular biosynthesis of magnetite using fungi. *Small* 2: 135–141.

Bhainsa, K. C., D'Souza, S. F. 2006. Extracellular biosynthesis of silver nanoparticles using the fungus *Aspergillus fumigatus*. *Colloids and Surfaces B* 47:160–164.

Bhattacharya, D., Gupta, R.K. 2005. Nanotechnology and potential of microorganisms. *Critical Reviews in Biotechnology* 25: 199–204.

Beveridge, T.J., Murray, R.G.E. 1980. Site of metal deposition in the cell wall of *Bacillus subtilis*. *Journal of Bacteriology* 141: 876–887.

Birla, S.S., Tiwari, V.V., Gade, A.K., Ingle, A.P., Yadav, A.P., Rai, M.K. 2009. Fabrication of silver nanoparticles by Phoma glomerata and its combined effect against *Escherichia coli Pseudomonas aeruginosa* and *Staphylococcus aureus*. *Letters in Applied Microbiology* 48: 173–179.

Boisselier, E., Astruc, D. 2009. Gold nanoparticles in nanomedicine: preparation, imaging, diagnostics, therapies and toxicity. *Chemical Society Reviews* 38: 1759–1782.

Chan, W.C.W. 2006. Bionanotechnology progress and advances. *Biology Blood Marrow Transplantation* 12: 87–91.

Chandran, S.P., Chaudhary, M., Pasricha, R., Ahmad, A., Sastry, M. 2006. Synthesis of gold nanotriangles and silver nanoparticles using *Aloe vera* plant extract. *Biotechnology Progress* 22: 577–583.

Chen, J.C., Lin, Z.H., Ma, X.X. 2003. Evidence of the production of silver nanoparticles via pretreatment of Phoma sp 32883 with silver nitrate. *Letters in Applied Microbiology* 37: 105–108.

Das, S.K., Das, A.R., Guha, A.K. 2009. Gold nanoparticles: microbial synthesis and applications in water hygiene management. *Langmuir* 25: 8192–8199.

Douglas, S., Beveridge, T.J. 1998. Mineral formation by bacteria in natural microbial communities. *FEMS Microbiology Ecology* 26: 79–88.

Fan, T.X., Chow, S.K., Zhang, D. 2009. Biomorphic mineralization: from biology to materials. *Progress in Materials Science* 54: 542–659.

Fayaz, A.M., Balaji, K., Girilal, M., Kalaichelvan, P.T., Venkatesan, R. 2009. Mycobased synthesis of silver nanoparticles and their incorporation into sodium alginate films for vegetable and fruit preservation. *Journal of Agricultural and Food Chemistry* 57: 6246–6252.

Feynman, R. 1991. There's plenty of room at the bottom. *Science* 254: 1300–1301.

Gajbhiye, M., Kesharwani, J., Ingle, A., Gade, A., Rai, M. 2009. Fungus-mediated synthesis of silver nanoparticles and their activity against pathogenic fungi in combination with fluconazole. *Nanomedicine: Nanotechnology, Biology and Medicine* 5: 382–386.

Gericke, M., Pinches, A. 2006. Biological synthesis of metal nanoparticles. *Hydrometallurgy* 83: 132–140.

Glomm, R.W. 2005. Functionalized nanoparticles for application in biotechnology. *Journal of Dispersion Science and Technology* 26: 389–314.

Gurunathan, S., Kalishwaralal, K., Vaidyanathan, R. et al. 2009. Biosynthesis, purification and characterization of silver of silver nanoparticles using Escherichia coli. *Colloids and Surfaces B* 74: 328–335.

Hayat, M.A. 1989. *Colloidal Gold: Principles, Methods, and Applications*. Academic Press, San Diego, USA.

He, S., Guo, Z., Zhang, Y., Zhang, S., Wang, J., Gu, N. 2007. Biosynthesis of gold nanoparticles using the bacteria *Rhodopseudomonas capsulate*. *Materials Letters* 61: 3984–3987.

Huang, J., Li, Q., Sun, D., Lu, Y., Su, Y., Yang, X., Wang, H., Wang, Y., Shao, W., He, N., Hong, J., Chen, C. 2007. Biosynthesis of silver and gold nanoparticles by novel sundried *Cinnamomum camphora* leaf. *Nanotechnology* 18: 105104–105114.

Husseiny, M.I., El-Aziz, M.A., Badr, Y., Mahmoud, M.A. 2007. Biosynthesis of gold nanoparticles using *Pseudomonas aeruginosa*. *Spectrochimica Acta A: Molecular and Biomolecular Spectroscopy* 67: 1003–1006.

Jha, A.K., Prasad, K. 2009. A green low-cost biosynthesis of Sb_2O_3 nanoparticles. *Biochemical Engineering Journal* 43: 303–306.

Jha, A.K., Prasad, K., Kulkarni, A.R. 2009. Plant system: nature's nanofactory. *Colloids and Surfaces B: Biointerfaces* 73: 219–223.

Joerger, R., Klaus, T., Pettersson, J., Granqvist, C.G. 2000. Digestion method for silver accumulated in micro-organisms. *Fresenius' Journal of Analytical Chemistry* 366: 311–312.

Kalishwaralal, K., Deepak, V., Ramkumarpndian, S., Nellaiah, H., Sangiliyandi, G. 2008a. Extracellular biosynthesis of silver nanoparticles by the culture supernatant of *Bacillus licheniformis*. *Materials Letters* 62: 4411–4413.

Kalishwaralal, K., Suresh, B.R., Venkataraman, D., Bilal, M., Gurunathan, S. 2008b. Biosynthesis of silver nanocrystals by *Bacillus licheniformis*. *Colloids and Surfaces B: Biointerfaces* 65: 150–153.

Kalishwaralal, K., Deepak, V., Ram Kumar Pandian, S., Gurunathan, S. 2009. Biosynthesis of gold nanocubes from *Bacillus lichemiformis*. *Bioresource Technology* 100: 5356–5358.

Kalishwaralal, K., Deepak, V., Pandian, S.R.K., Kottaisamy, M., BarathManiKanth, S., Kartikeyan, B., Gurunathan, S. 2010. Biosynthesis of silver and gold nanoparticles using *Brevibacterium casei*. *Colloids and Surfaces B* 77(2): 257–262.

Kashefi, K., Lovley, D.R. 2000. Reduction of Fe(III), Mn(IV), and toxic metals at 100 °C by *Pyrobaculum islandicum*. *Applied and Environmental Microbiology* 66: 1050–1056.

Klaus, T., Joerger, R., Olsson, E., Granqvist, C.G. 1999. Silver-based crystalline nanoparticles, microbially fabricated. *Proceedings of the National Academy of Sciences, USA* 96: 13611–13614.

Konishi, Y., Nomura, T., Tsukiyama, T., Saioth, N. 2004. Microbial preparation of gold nanoparticles by anaerobic bacterium. *Transactions of the Materials Research Society of Japan* 29: 2341–2343.

Konishi, Y., Ohno, K., Saitoh, N. et al. 2007. Bioreductive deposition of platinum nanoparticles on the bacterium *Shewanella algae*. *Journal of Biotechnology* 128: 648–653.

Kumar, S.A., Abyaneh, M.K., Gosavi, S.W., Kulkarni, S.K., Pasricha, R., Ahmad, A., Khan, M.I. 2007. Nitrate reductase-mediated synthesis of silver nanoparticles from AgNO3. *Biotechnology Letters* 29: 439–445.

Krishnaraj, C., Jagan, E.G., Rajasekar, S., Selvakumar, P., Kalaichelvan, P.T., Mohan, N. 2010. Methods of nanoparticle biosynthesis for medical and commercial applications. *Colloids and Surfaces B: Biointerfaces* 78(50): 1–21.

Lengke, M., Southam, G. 2006. Bioaccumulation of gold by sulfate-reducing bacteria cultured in the presence of gold(I)-thiosulfate complex. *Geochimica et Cosmochimica Acta* 70: 3646–3661.

Lengke, M., Ravel, B., Fleet, M.E., Wanger, G., Gordon, R.A., Southam, G. 2006. Mechanisms of gold bioaccumulation by filamentous cyanobacteria from gold(III)-chloride complex. *Environmental Science and Technology* 40: 6304–6309.

Mohanpuria, P., Rana, N.K., Yadav, S.K. 2008. Biosynthesis of nanoparticles: technological concepts and future applications. *Journal of Nanoparticle Research* 10: 9275–9280.

Mokhtari, N., Deneshpojouh, S., Seyedbagheri, S., Atashdehghan, R., Abdi, K., Sarkar, S., Minaian, S., Shahverdi, R.H., Shahverdi, R.A. 2009. Biological synthesis of very small silver nanoparticles by culture supernatant of *Klebsiella pneumonia*. The effects of visible-light irradiation and the liquid mixing process. *Materials Research Bulletin* 44: 1415–1421.

Mude, N., Avinash, I., Aniket G., Mahendra, R. 2009. Synthesis of silver nanoparticles using callus extract of Carica papaya. *Journal of Plant Biochemistry and Biotechnology* 18: 0971.

Mukherjee, P., Ahmad, A., Mandal, D., Senapati, S., Sainkar, S.R., Khan, M.I., Ramani, R., Parischa, R., Ajayakumar, P.V., Alam, M., Sastry, M., Kumar, R. 2001. Bioreduction of AuCl4-ions by the fungus *Verticillium sp* and surface trapping of the gold nanoparticles formed. *Angewandte Chemie International Edition* 40: 3585–3588.

Mukherjee, P., Roy, M., Dey, G.K., Mukherjee, P.K., Ghatak, J., Tyagi, A.K., Kale, S.P. 2008. Green synthesis of highly stabilized nanocrystalline silver particles by a non-pathogenic and agriculturally important fungus *T. asperellum*. *Nanotechnology* 19: 103–110.

Murray, C.B., Kagan, C.R., Bawendi, M.G. 2000. Synthesis and characterization of monodisperse nanocrystals and close-packed nanocrystal assemblies. *Annual Review of Materials Science* 30: 545–610.

Nagaraj, B., Sobczak-Kupiec, A., Malina, D., Yathirajan, H.S., Keerthi, V.R., Chandrashekar, N., Dinkar, S., Liny, P. 2013. Plant mediated synthesis of gold nanoparticles using fruit extracts of *ananas comosus* (L.) (pineapple) and evaluation of biological activities. *Advanced Materials Letters* 4: 332–337.

Nair, B., Pradeep, T. 2002. Coalescence of nanoclusters and formation of submicron crystallites assisted by *Lactobacillus strains*. *Crystal Growth and Design* 4: 295–298.

Narayanan, K. B., Sakthivel, N. 2010. Biological synthesis of metal nanoparticles by microbes. *Advances in Colloid and Interface Science* 156: 1–13.

Ogi, T., Saitoh, N., Nomura, T., Konishi, Y. 2010. Room-temperature synthesis of gold nanoparticles and nanoplates using *Shewanella* algae cell extract. *Journal of Nanoparticles Research* 9: 9822–9828.

Parashar, V., Parashar, R., Sharma, B., Pandey, A.C. 2009. Parthenium leaf extract mediated synthesis of silver nanoparticles: a novel approach towards weed utilization. *Digest Journal of Nanomaterials and Biostructures* 4: 45–50.

Rai, M., Yadav, A., Gade, A. 2009. Silver nanoparticles as a new generation of antimicrobials. *Biotechnology Advances* 27: 76–83.

Raut R.W., Jaya, R.L., Niranjan, S.K., Sahebrao, B.M.D.K. 2009. Phytosynthesis of silver nanoparticles using *Glirricidia sepium* (Jacq.). *Current Nanoscience* 5: 117–122.

Roh, Y., Lauf, R.J., McMillan, A.D., Zhang, C., Rawn, C.J., Bai, J., Phelps, T.J. 2001. Microbial synthesis and characterization of metal-substituted magnetites. *Solid State Communications* 118:529–534.

Sadowski, Z., Maliszewska, I., Polowczyk, I., Kozlecki, T., Grochowalska, B. 2008. Biosynthesis of colloidal-silver particles using microorganisms. *Polish Journal of Chemistry* 82: 377–382.

Safaepour, M., Shahverdi, A.R., Shahverdi, H.R., Khorramizadeh, M.R., Gohari, A.R. 2009. Green synthesis of small silver nanoparticles using geraniol and its cytotoxicity against Fibrosarcoma-Wehi 164 Avicenna. *Journal of Medical Biotechnology* 1: 111.

Saifuddin, N., Wang, W.C., Nur Yasumira, A.A. 2009. Rapid biosynthesis of silver nanoparticles using culture supernatant of bacteria with microwave irradiation. *E-Journal of Chemistry* 6: 61–70.

Sastry, M., Ahmad, A., Khan, I.M., Kumar, R. 2003. Biosynthesis of metal nanoparticles using fungi and actinomycete. *Current Science* 85: 162–170.

Senapati, S., Mandal, D., Ahmad, A., Khan, M.I., Sastry, M., Kumar, R. 2004. Fungus mediated synthesis of silver nanoparticles: a novel biological approach. *Indian Journal of Physics* 78: 101–105.

Shahverdi, R.A., Minaeian, S., Shahverdi, R.H., Jamalifar, H., Nohi, A.A. 2007. Rapid synthesis of silver nanoparticles using culture supernatants of Enterobacteria: A novel biological approach. *Process Biochemistry* 42: 919–923.

Shaligram, N.S., Bule, M., Bhambure, R., Singhal, R.S., Singh, S.K., Szakacs, G., Pandey, A. 2009. Biosynthesis of silver nanoparticles using aqueous extract from the compactin producing fungal strain. *Process Biochemstry* 44: 939–943.

Shankar, S.S., Ahmad, A., Sastry, M. 2003. Geranium leaf assisted biosynthesis of silver nanoparticles. *Biotechnology Progress* 19: 1627–1631.

Shankar, S.S., Rai, A., Ankamwar, B., Singh, A., Ahmad, A. et al. 2004. Biological synthesis of triangular gold nanoprisms. *Nature Materials* 3: 482–488.

Sharma, V.K., Yngard, R.A., Lin, Y. 2009. Silver nanoparticles: green synthesis and their antimicrobial activities. *Advances in Colloid Interface Science* 145: 83–96.

Shiying, H., Zhirui, G., Zhanga, Y., Zhanga, S., Wanga, J., Ning, G. 2007. Biosynthesis of gold nanoparticles using the bacteria *Rhodopseudomonas capsulata*. *Materials Letters* 61: 3984–3987.

Simkiss, K., Wilbur, K.M. 1989. *Biomineralization: Cell Biology and Mineral Deposition*. Academic Press, New York.

Sinha, A., Khare, S.K. 2011. Mercury bioaccumulation and simultaneous nanoparticle synthesis by *Enterobacter sp.* cells. *Bioresource Technology* 102: 4281–4284.

Song, J.Y., Kim, B.S. 2009. Rapid biological synthesis of silver nanoparticles using plant leaf extracts. *Bioprocess and Biosystems Engineering* 44: 1133–1138.

Sulaiman, G.M., Mohammad, A.A.W., Abdul-wahed, H., Ismail, M.M. 2013. Biosynthesis, antimicrobial and cytotoxic effects of silver nanoparticles using rosmarinus officinalis extract. *Digest Journal of Nanomaterials and Biostructures* 8: 273–280.

Vigneshwaran, N., Ashtaputre, N.M., Varadarajan, P.V., Nachane, R.P., Paralikar, K.M., Balasubramanya, R.H. 2007. Biological synthesis of silver nanoparticles using the fungus *Aspergillus flavus*. *Materials Letters* 61: 1413–1418.

Yong, P., Rowson, N.A., Farr, J.P.G., Harris, I.R., Macaskie, L.E. 2002. Bioreduction and biocrystallization of palladium by *Desulfovibrio desulfuricans NCIMB 8307*. *Biotechnology and Bioengineering* 80: 369–379.

Zhang, X., Yan, S., Tyagi, R.D., Surampalli, R.Y. 2011. Synthesis of nanoparticles by microorganisms and their application in enhancing microbiological reaction rates. *Chemosphere* 82: 489–494.

Zhou, W., He, W., Zhong, S. et al. 2009a. Biosynthesis and magnetic properties of mesoporous Fe3O4 composites. *Journal of Magnetism and Magnetic Materials* 321: 1025–1028.

Zhou, W., He, W., Zhang, X. et al. 2009b. Biosynthesis of iron phosphate nanopowders. *Powder Technology* 194: 106–108.

8

MICROBIAL SYNTHESIS OF NANOPARTICLES: AN OVERVIEW

Sneha Singh

Department of Bio-Engineering, Birla Institute of Technology, Mesra, Ranchi, Jharkand, India

Ambarish Sharan Vidyarthi

Institute of Engineering and Technology, Lucknow, Uttar Pradesh, India

Abhimanyu Dev

Department of Pharmaceutical Sciences and Technology, Birla Institute of Technology, Mesra, Ranchi, Jharkand, India

The quintessence of nanotechnology finds its full expression in the words of R.P. Feynman:

> The principles of physics, as far as I can see, do not speak against the possibility of moving things atom by atom. It is not an attempt to violate any laws; it is something, in principle, that can be done; but in practice, it hasn't been done because we are too big.

> These words marked the dawn of an technologically advanced era which changed the outlook of the world from big to small (Feynman, 1960).

8.1. INTRODUCTION

Much research has been focused on nanoparticles (dimension <100 nm) over the last decade due to their exclusive physico-chemical properties that differ from their bulk counterparts. These unique properties have been harnessed for application in diverse fields such as materials science, information technology, nanotechnology, electronics, physics, chemistry, biology, medicine, and medical diagnostics (Schmid et al., 1999; Sun et al., 2000; Kamat, 2002; Daniel and Astruc, 2004; Salata, 2004; Okazaki and Moers, 2005; Koplin and Simon, 2007).

Although the recent surge in nanoparticles (NPs) is new, their history dates back to ancient civilization where beautiful paintings on glass emitting a ruby-red color were due to gold nanoparticles (AuNPs) trapped in the glass matrix. The decorative glaze used for pottery and artifacts also relied upon the interaction of spherical NPs with light. One of the first scientific reports explaining the properties of this glaze was Michael Faraday's pioneering work *Experimental relations of gold (and other metals) to light* (Faraday, 1857). Later in 1959 Richard P. Feynman, a physicist at Cal Tech, delivered a speech on the topic "There is plenty of room at the bottom", forecasting the advent of nanomaterials and suggesting that scaling down to nano-levels and starting from the bottom was the key to future technology and advancement (Appenzeller, 1991). The term nanotechnology was introduced much later by Tokyo Science University Professor Norio Taniguchi (Taniguchi, 1974). With advances in science and technology, the physico-chemical properties exhibited by these NPs are understood and novel nanomaterials have emerged with different catalytic, electronic, optical, mechanical, and magnetic properties (Jenekhe et al., 1997; Schmid et al., 1999; Sun et al., 2000; Daniel and Astruc, 2004; Koplin and Simon, 2007) as compared to their larger counterparts. This difference in the physico-chemical properties of nanomaterials can be attributed to their extremely small size, high surface-to-volume ratio, aspect ratio, and quantum confinement.

The advancement of research on the synthesis of NPs from natural living sources has attracted considerable attention in the field of nanobiotechnology and its application. The different synthetic techniques for NP synthesis are fraught with many drawbacks such as the hazardous by-products produced from the toxic compounds used in the process, which are damaging to the environment, as well as imperfections in surface structure and formation of mixed-shape NPs that require low-yield and expensive purification procedures such as differential centrifugation (Murphy, 2002). It is also well known that the shape, morphology, and size of NPs play an important role in controlling their special properties (Kamat, 2002; El-Sayed, 2001). Unique optical responses are also observed and known to be a function of the various sizes and shapes of NPs. For example, symmetric spherical NPs exhibit a single scattering peak while anisotropic shapes exhibit multiple scattering peaks in the visible wavelengths as a result of highly localized charge polarizations at corners and edges (Mie, 1908). Accordingly, there is an essential requirement to develop high-yield, low-cost, non-toxic, simple, and environmentally benign procedures for the green synthesis of NPs. Consequently, the biological approach for synthesis of NPs is important, especially in employing microorganisms as novel green "nanofactories". The adoption of biological methods in the synthesis of NPs is expected to yield novel structural entities of the desired morphology.

8.2. NANOPARTICLES SYNTHESIS INSPIRED BY MICROORGANISMS

Microorganisms have the potential to synthesize NPs intracellularly or extracellularly under ambient conditions without toxic chemicals and stringent conditions; even the properties of such NPs are similar to chemically synthesized materials (Baüerlein, 2003). In the recent years, a variety of NPs have been synthesized in an eco-friendly manner by using microorganisms. Such particles inherently show a variation in location, chemical composition, shape and size, depending upon the mechanism involved in bioreduction.

By taking inspiration from nature, where living organisms naturally produce inorganic materials through a biologically guided process known as biomineralization, a superior approach for the assembly of nanomaterials can be developed (Mann, 1993). The biomineralization processes exploit biomolecular templates that interact with the inorganic material at the nanoscale, resulting in extremely efficient and highly controlled synthesis. Microorganisms are endowed with many natural occurrences of inorganic nanomaterials, either intra- or extracellularly (Simkiss and Wilbur, 1989; Mann, 1996). This has encouraged their use as possible green alternative for nanoparticle synthesis routes compared to synthetic methods. The best-known example is perhaps the production of magnetite NPs from magnetotactic bacteria (*Magnetospirillum magnetotacticum*, *Magnetobacterium bavaricum*, *Magnetospirillum gryphiswaldense*) (Lovley et al., 1987; Spring and Schleifer, 1995; Dickson, 1999; Lang et al., 2007) in sediments of a variety of aqueous environments. These bacteria have specialized organelles called magnetosomes that store magnetic nanocrystals (NCs) composed of greigite (Fe_3S_4) or magnetite (Fe_3O_4). The magnetosome membrane (MM) helps in anchoring the NCs at a particular location in the cell, and also acts as the locus of biological control for the nucleation and growth of the magnetosome crystals (Fig. 8.1). These particles are arranged in chains and function as biological compass needles that enable them to steer along an oxygen gradient in aquatic environments under the influence of the Earth's geomagnetic field (Blakemore, 1975; Simmons et al., 2006; Frankel et al., 2007; Faivre and Schüler, 2008). Grünberg et al. (2001) suggested that the "mam" (MM) genes, which are conserved in a large gene cluster within several magnetotactic bacteria (*Magnetospirillium* species and strain MC-1), may be involved in magnetic biomineralization. The surface layer (S-layer) bacteria are known to produce gypsum and calcium carbonate layers (Pum and Sleytr, 1999; Sleytr et al., 1999).

Further examples of biological nanostructures include diatoms, unicellular algae that have the amazing capability to produce an enormous variety of biosilicate structures (Mann, 1993; Oliver et al., 1995; Kröger et al., 1999). These are commonly found in fresh and marine water reservoirs. Each diatom species is characterized by an ornamented cell walls, which are mostly made of amorphous polysilic acid. The genesis of nano-patterned biosilica (10 and 1000 nm) occurs in a specialized membrane-bound compartment, termed the silica deposition vesicle, within the cell catalyzed by silaffins and polyamines (Poulsen et al., 2003). Similar to diatoms, sponges belonging to phylum Porifera are the oldest-known metazoans that produce a variety

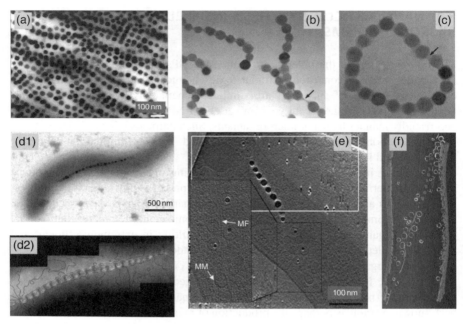

Figure 8.1. Isolated magnetosome particles form (a) straight chains in weak ambient magnetic fields but (b) bent chains or (c) flux-closure rings in zero fields adhered to each other by junctions of organic material (arrows). (d) Magnetic microstructure of a magnetosome chain. (d1) TEM bright-field image of a single bacterial cell of *M. magnetotacticum*; (d2) magnetic induction map recorded using off-axis electron holography from the magnetosome chain of the same cell. (e) Cryoelectron tomography image shows organization of the magnetosome chain and the cytoskeletal magnetosome filament (MF) in *M. gryphiswaldense*. (f) Tomographic reconstruction of a magnetic cell showing the cytoplasmic membrane, empty vesicles, growing and mature magnetite crystals, and the magnetosome filament. Source: Faivre, D. and Schüler, D. Magnetotactic bacteria and magnetosomes. *Chemical Reviews*, 108(11), 4875–4898. Copyright © 2008, American Chemical Society.

of intricate skeletal elements. They have the special ability to build robust structures of amorphous silica, calcium phosphates and carbonates that form structural entities in vertebrates and invertebrates in conjunction with organic polymers (proteins, lipids, or polysaccharides). Apart from these, simpler organisms such as bacteria, yeast, and fungi have also developed highly specialized strategies for biomineral synthesis over the course of evolution.

The role of microorginsms in bioremediation, biotransformation, biosorption, and biomineralization are well known; however, the microbial-based biogenesis of nanoparticles is a relatively new and largely unexplored area of research. The biogenic potential of different prokaryotes (bacteria and actinomycetes), eukaryotes (such as yeast and fungi), and viruses have been explored for the synthesis of NPs (Table 8.1). The status of research on biosynthesis of NPs using a diverse range of microorganisms and their various aspects is discussed in the following sections.

TABLE 8.1. Use of Microorganisms for the Synthesis of Different Nanoparticles

Microrganism	NPs	Location/morphology	Size (nm)	Reference
Bacteria				
Bacillus subtilis	Au	Cell inside/octahedral	5–25	Beveridge and Murray (1980)
Lactobacillus sp.	Au/Ag Au-Ag	Cell inside/hexagonal triangular	20–50	Nair and Pradeep (2002)
S. algae	Au	Periplasmic space cell envelope	10–20 15–200	Konishi et al. (2004)
Rhodopseudomonas capsulata	Au	Cell surface/spherical, triangular	10–20 50–400	He et al. (2007)
B. megatherium	Au	Spherical	1.9 ± 0.8	Wen et al. (2009)
B. licheniformis	Au	Extracellular/spherical	38	Singh et al. (2014)
Rhodobacter capsulatus	Au	Plasmamembrane extracellular	–	Feng et al. (2007)
P. stutzeri AG259	Ag, Ag$_2$S	Periplasmic space/triangular, hexagon	3–200	Klaus et al. (1999), Joerger et al. (2000)
Corynebacterium sp. SH09	Ag	Cell wall	10–15	Zhang et al. (2005)
S. Typhimurium	Ag	Extracellular	87 ± 30	Ghorbani (2013)
Enterobacteria (*K. Pneumoniae*, *E. cloacae*)	Ag	Extracellular	28–122	Shahverdi et al. (2007)
Morganella sp.	Ag	Spherical	20 ± 5	Parikh et al. (2008)
B. licheniformis	Ag	Intracellular/extracellular	40–50	Kalishwaralal et al. (2008)
Lactobacillus casei	Ag	Cell surface/cytoplasm/cell outside	25–100	Korbekandi et al. (2012)
Clostridium thermoaceticum	CdS	Cell surface	–	Cunningham and Lundie (1993)
E. coli	CdS	–	–	Sweeney et al. (2004)
Rhodopseudomonas palustris	CdS	Extracellular/spherical	8.01 ± 0.25	Bai et al. (2009a)
Rhodobacter sphaeroides	ZnS	Extracellular/spherical	8	Bai et al. (2006)
	PbS		10.5 ± 0.15	Bai et al. (2009b)
Lactobacillus sp.	Sb$_2$O$_3$	ND	3–12	Jha et al. (2009a)
Serratia marcescens	Sb$_2$S$_3$	Cell inside	<35	Bahrami et al. (2012)
Shewanella	Pt	Periplasm	5–10	Konishi et al. (2007)
S. algae				
Desulfovibrio desulfuricans	Pd	Cell surface	–	Yong et al. (2002)
S. oneidensis MR-1	Pd	Periplasm, cell wall	–	Windt et al. (2005)

(*continued*)

TABLE 8.1. (Continued)

Microrganism	NPs	Location/morphology	Size (nm)	Reference
Lactobacillus sp.	TiO$_2$	—	8–35	Jha et al. (2009b)
Brevibacterium casei	Co$_3$O$_4$	Extracellular	5–7	Kumar et al. (2008)
Actinobacter sp.	Magnetite	Extracellular/quasi-spherical	10–40	Bharde et al. (2005)
Shewanella putrefaciens CN 32	UO$_2$	Extracellular	—	Fredrickson et al. (2002)
Shewanella oneidensis MR-1	UO$_2$	Extracellular	1–5	Marshall et al. (2006)
Bacillus subtilis	Se	Spherical	5–400	Wang et al. (2010)
Bacillus cereus	Se	Cell surface/cytoplasm	150–200	Dhanjal and Cameotra (2010)
Tetrathiobacter kashmirensis	Se	Extracellular	ND	Hunter and Manter (2008)
Bacillus thurigiensis	Co	Spherical/oval	85	Marimuthu et al. (2013)
Fungi				
Verticillium sp.	Au	Cell wall/spherical, triangular, hexagonal	20 ± 8	Mukherjee et al (2001a)
	Ag	Cytoplasm/ quasi-hexagonal/ cell inside	25 ± 12	Mukherjee et al. (2001b)
Colletotrichum sp.	Au	Spherical	8–40	Shankar et al. (2003)
Trichothecium	Au	Extracellular and intracellular cell wall/ cytoplasmic membrane	5–200 10–25	Ahmad et al. (2005)
Fusarium oxysporum	Ag	Extracellular	20–50	Duran et al. (2005)
Fusarium semitectum	Ag	Spherical	10–60	Basavaraja et al. (2008)
Trichoderma asperellum	Ag	Extracellular	13–18	Mukherjee et al. (2008)
Phoma glomerata	Ag	Spherical	60–80	Birla et al. (2009)
Penicillium brevicompactum WA 2315	Ag	Extracellular	58.35 ± 17.88	Shaligram et al. (2009)
Aspergillus flavus NJP08	Ag	Extracellular		Jain et al. (2011)
Neurospora crassa	Ag, Au	Extracellular and intracellular spherical/ ellipsoidal	11	Castro-Longoria et al. (2011)
	Ag–Au		32	
Rhizopus oryzae	Au	Extracellular	c. 20	Das et al. (2012)
Fusarium oxysporum	CdS	Extracellular	5–20	Ahmad et al. (2002)
Fusarium oxysporum	Pt	Spherical/extracellular	5–30	Syed and Ahmad (2012)
Aspergillus aeneus NJP 12	ZnO	Extracellular	100–140	Jain et al. (2013)
Fusarium oxysporum	Magnetite	Quasi-spherical	20–50	Bharde et al. (2006)
Verticillium sp.		Cubo octahedral	10–40	
Fusarium oxysporum	Zr	Extracellular/quasi-spherical	3–11	Bansal et al. (2004)

Microorganism	NPs	Location/morphology	Size (nm)	Reference
Fusarium oxysporum	SrCo₃	Quasi-linear	10–50	Rautaray et al. (2004)
Fusarium oxysporum	Si	Quasi-spherical	5–15	Bansal et al. (2005)
	Ti	Spherical	6–13	
Humicola sp.	CeO	Spherical	12–20	Khan and Ahmad (2013)
Actinomycetes				
Thermonospora sp.	Au	Extracellular	8–40	Ahmad et al. (2003a)
Rhodococcus sp.	Au	Intracellular (cytoplasmic membrane, cell wall)	5–15	Ahmad et al. (2003b)
Nocardiopsis sp. MBRC-1	Ag	Spherical	45±0.15	Manivasagan et al. (2013)
Yeast				
Candida glabrata	CdS	Intracellular	–	Dameron (1989)
Schizosaccharomyces pombe	CdS	Intracellular	1–1.5	Kowshik et al. (2002)
MKY3	Ag	Extracellular	2–5	Kowshik et al. (2003)
Yarrowia lipolytica	Au	Cell wall	15	Agnihotri et al. (2009)
Extremophilic yeast	Ag	Cell wall	<20	Mourato et al. (2011)
	Au	Extracellular	30–100	
Hansenulla anomala	Au	Extracellular	14–40	Kumar et al. (2011)
Candida guilliermondii	Au	Extracellular	50–70	Mishra et al. (2011)
	Ag	Spherical	10–20	
Virus				
Tobacco mosaic virus	FeO, CdS, PbS, SiO₂	Cell surface	ND	Shenton et al. (1999)
M13 bacteriophage	ZnS	–	10–20	Lee et al. (2002)
M13 viral capsid	ZnS	–	3–5	Mao et al. (2003)
	CdS		c. 20	
M13 bacteriophage	CoPt	1D structure	10±5	Mao et al. (2004)
	FePt			
Influenza virus matrix protein M1	ZnS	Peptide template surface	4	Banerjee et al. (2005)
Cowpea chlorotic mottle virus (SubE(yeast), (HRE- SubE), wild type)	Au	Cell surface	9.2±3.9 8.3±2.6 23.8±14.5	Slocik et al. (2005)

8.2.1. Bacteria in NPs Synthesis

Among the microorganisms, bacteria have gained most attention in the area of NPs synthesis. The biosynthesis of AuNPs was reported by Beveridge and Murray in 1980 using *Bacillus subtilis* 168, using auric chloride as a precursor under ambient temperature and pressure (Beveridge and Murray, 1980; Southam and Beveridge, 1994; Fortin and Beveridge, 2000). It was observed that the organic phosphate compounds mediated the bioreduction of the Au^{3+} ions *in vitro* to octahedral AuNPs of 5–25 nm size within the bacterial cell, possibly as bacteria–Au-complexing agents (Southam and Beveridge, 1996). The exposure of lactic acid bacteria *Lactobacillus* strains present in the whey of buttermilk to gold and silver ions resulted to the large-scale production of Au, Ag and Au-Ag alloy NPs intracellularly without affecting the viability of the bacterial cells (Nair and Pradeep, 2002). The NCs formed were triangular, hexagonal in shape and 20–50 nm in size. Fe(III)-reducing bacteria *Shewanella algae* in the presence of hydrogen gas and anaerobic environments at 25 °C completely reduced Au^{3+} ions to AuNPs in the periplasmic space at pH 7.0 of size 15–200 nm, and on the cell envelope at pH 2.8 of size 10–20 nm (Konishi et al., 2004). Similarly, room temperature and pH-dependent synthesis of AuNPs by *Rhodopseudomans capsulate* has also been reported (He et al., 2007). Aqueous chloroaurate ions were reduced to AuNPs after 48 h of incubation on the cell surface of the bacteria. It was found that the shapes of AuNPs was controlled by the pH of the solution, which were mainly spherical and of size 10–20 nm at pH 7.0. However, with a change in pH of the solution, various shapes and sizes were formed. At pH 4.0, triangular NPs of size 50–400 nm and spherical particles of size 10–50 nm were also formed. The reduction of Au^{3+} to Au^0 was mediated by the transfer of electrons from NADH by NADH-dependent reductase. The binding of aqueous chloroaurate ions on the bacterial cell surface was arbitrated by different positively charged cell surface functional groups such as amino, sulfhydryl, and carboxylic (He et al., 2007).

The bioreduction of chloroauric acid to AuNPs was also reported in photosynthetic bacterium, *Rhodobacter capsulatus*, with biosorption capacity of 92.43 mg $HAuCl_4$ per gram dry weight in the logarithmic phase of its growth. The bioreduction of Au^{3+} to Au^0 was on the plasma membrane and also extracellularly with the help of carotenoids and NADPH-dependent enzymes embedded in plasma membrane and/or secreted extracellularly (Feng et al., 2007). Synthesis of monodispersed AuNPs by self-assembled monolayer capping route using *B. megatherium* D01 has been reported. The authors investigated the effect of thiol as an external stabilizer upon shape, size, and dispersity of AuNPs. The bioreduction occurred at 26 °C and the stable AuNPs obtained were small (1.9±0.8 nm) and spherical in shape (Wen et al., 2009). Recently, cell lysate supernatant (CLS) of *Bacillus licheniformis* was used for the reduction of gold ions at 37 °C resulting in extracellular biosynthesis of AuNPs (Singh et al., 2014). The obtained NPs of average size 38 nm were also stable for two months at room temperature without the addition of any external capping agent. Further, the AuNPs exhibited antimicrobial activity against both Gram-negative and Gram-positive bacteria. The authors also suggested that use of viable cells for biosynthesis was not mandatory as the bacteria retained its bioreduction potential even in lysed state.

Further, extracellular synthesis of NPs using CLS could be advantageous over intracellular synthesis, as the downstream processing for product recovery and the cost involved is minimized, favoring scale-up synthesis.

Further, a number of bacterial strains have also shown great affinity towards synthesis of silver nanoparticles (AgNPs) in a similar manner to that of AuNPs biosynthesis. *Pseudomonas stutzeri* AG259, isolated from a silver mine, was capable of reducing aqueous silver nitrate solutions to AgNPs with a diameter of 3–200 nm with various morphologies as equilateral, triangles, and hexagons in the periplasmic space. The bacterium also produced a small percentage of monoclinic crystalline α-form silver sulfide acanthite (Ag_2S) crystallite particles with the composition of silver and sulfur in the ratio 2:1 (Klaus et al., 1999; Joerger et al., 2000). Physicochemical parameters (pH, incubation time, growth in the light or dark, and composition of the culture medium) have an impact on the size and morphology of the NPs (Joerger et al., 2000). Zhang et al. (2005) reported AgNP synthesis from dried cells of *Corynebacterium* sp. SH09 using diamine silver complex $[Ag(NH_3)_2]^+$ as a precursor. The NPs were formed on the cell wall of the bacterium at 60 °C after 72 h, with a size range of 10–15 nm. Their study demonstrated that the ionized carboxyl of amino acid residues and an amide of the peptide chains were most likely the main groups trapping $[Ag(NH_3)_2]^+$ onto the bacterium cell wall and later, with the help of the reducing groups, such as aldehyde or ketone, bioreduced $[Ag(NH_3)_2]^+$ and formed Ag^0 nuclei which further grew into Ag^0 nanocrystals. Further, extracellular biosynthesis of AgNPs by the culture supernatants of different Enterobacteria (*Klebsiella pneumonia*, *Enterobacter cloacae*, and *E. coli*) strains have also been reported by Shahverdi et al. (2007). The NPs were of size 28–122 nm with an average size of 52 nm. The authors demonstrated that upon addition of piperitone (3-methyl-6-(1-methylethyl)-2-cyclohexen-1-one) to culture supernatant, the reduction of Ag^+ ions to Ag^0 was partially inhibited . Piperitone was found to have inhibitory effect on the expression of nitroreductase (an oxygen-insensitive flavoprotein) in *Enterobacteriaceae* under aerobic condition (Rafii et al., 2005), suggesting the possible involvement of the nitroreductase enzymes in the reduction process of silver ions (Shahverdi et al., 2007).

Silver is normally toxic to microbial cells; however, in low concentrations, it is tolerated by the microbes due to the different detoxifying mechanisms employed by the cell. The involvement of small periplasmic proteins in detoxification of silver has been documented. Parikh et al. (2008) demonstrated the synthesis of spherical AgNPs of size 20 ± 5 nm using *Morganella* sp., a silver-resistant bacterium isolated from insect guts. Three silver-resistant gene homologs – silE, silP, and silS – functioning as periplasmic silver-binding protein, cation-transporting P-type ATPase, and two-component membrane sensor kinase, respectively, were identified in *Morganella* sp. These genes were anticipated to be involved in the silver-resistance mechanism in bacteria, thereby leading to AgNP formation. The silver binding proteins become adsorbed to the silver ions at the cell surface and via efflux pumps propels the incoming metal, thereby protecting the cytoplasm from toxicity (Li et al., 1997; Gupta and Silver, 1998). Silver nanoparticles synthesis from biomass and culture supernatant of *Bacillus licheniformis* isolated from sewage samples have been reported (Kalishwaralal et al., 2008). The NPs obtained via the reduction of Ag^+ ions to AgNPs were 40–50 nm in size after incubation for 24 h at 37 °C. The authors suggested the bioreduction to be enzymatically catalyzed by NADH and NADH-dependent enzymes, especially nitrate reductase which exist in *Bacillus licheniformis*. Recently AgNPs synthesis from *Lactobacillus casei subsp. casei* has been reported, where different parameters were optimized to enhance the yield of NPs and the productivity of the biosynthetic approach (Korbekandi et al., 2012). The AgNPs obtained were of 25–100 nm size and located on the cell surface, inside the cytoplasm, and outside the cells. The process was catalyzed by nitrate reductase, glucose as electron donor, and silver nitrate as inducer (Korbekandi et al., 2012). AgNPs have also been synthesized using culture supernatant of bacterium *Salmonella typhimurium*. The authors reported that, by the use of visible light irradiation, AgNPs were synthesized rapidly in just 4 min using silver sulfate as a precursor and the light source as inducer (Ghorbani, 2013).

Apart from gold and silver nanoparticles, much attention has been paid towards developing a biosynthesis approach for the synthesis of semiconductor (the so-called quantum dots) NCs such as CdS, ZnS, and PbS along with other metals and radionuclides. Cunningham and Lundie (1993) demonstrated the precipitation of CdS NCs from *Clostridium thermoaceticum* at the cell surface as well as in the medium using $CdCl_2$ as precursor in the presence of cysteine hydrochloride in the growth medium, which acted as a sulfide source. A further effect of different growth phases upon CdS NCs formation from *E. coli* was also demonstrated by Sweeney et al. (2004). An increase of 20-fold was noticed in NCs formation with stationary phase cultures as compared to that of late logarithmic phase. In addition, extracellular synthesis of CdS NCs at room temperature using purple non-sulfur photosynthetic bacterium *Rhodopseudomonas palustris* has been reported. The process was catalyzed by cysteine desulfhydrase (C-S lyase) and the NCs obtained were spherical in shape of a face-centered cubic (FCC) lattice and 8.01 ± 0.25 nm in size (Bai et al., 2009a). In a different approach for biosynthesis, the use of immobilized microbes and enzymes has gained importance recently as they offer simpler scale-up and downstream processing. Further immobilization of the biomass in solid structures, polymeric matrix, on covalent bonds in vector compounds, or on cell cross-linking creates material of the proper size, mechanical strength, rigidity, and porosity necessary for metal accumulation. The immobilized biomass or enzymes can be reactivated and reused in a manner similar to ion exchange resins and activated carbon. In this context, immobilized *Rhodobacter sphaeroides* have been used to extracellularly synthesize spherical zinc sulfide NCs of 8 nm size (Bai et al., 2006) and lead sulfide NCs of 10.5 ± 0.15 nm size (Bai et al., 2009b).

Jha et al. (2009a) reported the synthesis of antimony trioxide (Sb_2O_3) NPs of 3–12 nm in size using *Lactobacillus* sp. at room temperature. The authors advocated involvement of oxidoreductase, a pH-sensitive enzyme, and the level of rH_2 (the negative logarithm of the partial pressure of gaseous hydrogen) in the medium to influence the synthesis of Sb_2O_3 NPs. Similarly, antimony sulfide NPs were synthesized intracellularly using marine bacterium *Serratia marcescens*. The NPs of less than 35 nm size were recovered using liquid nitrogen and two solvent extraction systems (Bahrami et al., 2012). These kinds of semiconductor NCs have very important applications in optical devices, electronics, and biotechnologies (Murray et al., 2000).

Due to high oxidation temperature and strong resistance to corrosion, NPs of platinum-group are widely used in catalytic applications, fuel cell, and data storage (Sun and Murray, 2000). Platinum nanoparticles (PtNPs) of 5–10 nm size were synthesized in the periplasm of *Shewanella algae* at room temperature and neutral pH. The $PtCl_6^{2-}$ ions were reduced to elemental platinum within 60 min in the presence of lactate as electron donor, thereby changing the color of reaction mixture from pale yellow to black (Konishi et al., 2007). Further, a mixed and uncharacterized consortium of sulfate-reducing bacteria was able to synthesize platinum nanoparticles with the help of two hydrogenase enzymes. First, the Pt^{4+} ions were reduced to Pt^{2+} by an oxygen-sensitive cytoplasmic hydrogenase and later the formed ions were reduced to Pt^0 NPs by an oxygen-tolerant periplasmic hydrogenase that was inhibited by Cu^{2+} (Riddin et al., 2009).

Another platinum group metal, palladium, was biosynthesized by sulfate-reducing bacterium *Desulfovibrio desulfuricans* NCIMB 8307. The bacterium anaerobically bioreduced palladium (2+) ions to palladium NPs on the cell surface in the presence of formate as electron donor within minutes at neutral pH (Yong et al., 2002). These findings were in line with the work of Windt et al. (2005) where iron-reducing bacterium *Shewanella oneidensis* MR-1 happened to reduced Pd^{2+} to Pd^0 nanoparticles in the periplasm and on the cell wall, also in the presence of exogenous formate as the electron donor. This cell-associated nanoPd is used as a catalyst in the dechlorination of polychlorinated biphenyl 21 (2, 3, 4-chloro biphenyl).

Next in the series are TiO_2 nanoparticles of 8–35 nm size, synthesized at room temperature from *Lactobacillus* sp. It was suggested that pH and partial pressure of gaseous hydrogen (rH_2) or redox potential of the culture solution had a strong influence on the bioreduction process (Jha et al., 2009b). *Brevibacterium casei*, a metal-tolerant marine bacterium, has shown an immense capability to extracellularly bioreduce cobalt acetate to cubic spinel-structured single-crystalline bio-functionalized ferromagnetic Co_3O_4 in the size range 5–7 nm at room temperature (Kumar et al., 2008). Magnetic nanocrystals were also reported to be synthesized by *Actinobacter* sp., a non-magnetotactic aerobic bacterium. The bioreduction was extracellular at room temperature in fully aerobic conditions and the NCs obtained had quasi-spherical morphology of size 10–40 nm. The involvement of two proteins of molecular weight 120 and 70 kDa in the bioreduction of potassium ferricyanide and potassium ferrocyanide as iron precursors to magnetite has been also suggested (Bharde et al., 2005).

The extracellular formation of uranium oxide (UO_2) nanoparticles involving the outer membrane cytochromes by *Shewanella putrefaciens* CN32 was also reported (Fredrickson et al., 2002); however, the mechanism was explained later by Marshall et al. (2006). The authors demonstrated that c-type cytochromes of *Shewanella oneidensis* MR-1, a dissimilatory metal-reducing bacterium are necessary for the biotransformation of U^{6+} and their extracellular deposition as UO_2 nanoparticles of size 1–5 nm. Marshall and co-workers also ascertained that MtrC, a decaheme c-type cytochrome, acts as electron donor to U^{6+}. Unlike U^{6+} which is soluble, U^{4+} is insoluble and well documented to be biosynthesized by different bacteria. It is well known that radionuclides in soluble form pose a serious threat to living systems by contaminating the environment. In order to overcome this problem, the potential of microbes as mini-nanofactories for transforming the soluble form to insoluble can be harnessed.

Spherical selenium nanoparticles (SeNPs) of 50–400 nm size were synthesized using *Bacillus subtilis* (Wang et al., 2010). After synthesis, spherical monoclinic SeNPs can be transformed into highly anisotropic, one-dimensional (1D) trigonal structure by incubating at room temperature for 24 hours. Similarly, aerobic synthesis of SeNPs by biotransformation of toxic selenite (SeO_3^{2-}) ions into red elemental selenium (Se^0) using *Bacillus cereus* isolated from coalmine soil was also reported (Dhanjal and Cameotra, 2010). The bioreduction process was rapid; within 2–3 h SeNPs of size 150–200 nm were synthesized in the cytoplasm of the bacterial cell and subsequently exudated to the cell surface and outside via NADH/NADPH-dependent reductase (Fig. 8.2). In addition, *Tetrathiobacter kashmirensis* was used to bioreduce selenite to elemental red selenium under aerobic conditions (Hunter and Manter, 2008). The results suggested that a 90 kDa protein present in the cell-free extract of the bacterium was responsible for this biotransformation.

Recently, *Bacillus thuringiensis*-mediated synthesis of cobalt nanoparticles (CoNPs) was reported (Marimuthu et al., 2013). The CoNPs were synthesized using cobalt acetate as precursor and cetyltrimethylammoniumbromide (CTAB) as external stabilizing agent. The particles were spherical, oval in shape, of average size 85 nm, and polydisperse in nature. The authors also investigated the larvicidal potential of CoNPs against malaria vector *Anopheles subpictus* and dengue vector *Aedes aegypti* (Diptera: Culicidae).

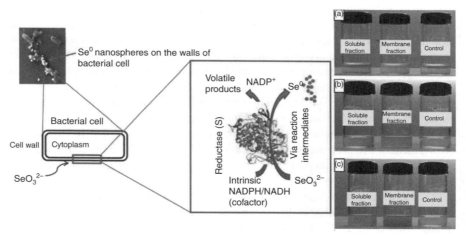

Figure 8.2. Illustration of proposed mechanism for biogenesis of selenium nanospheres at different time intervals: (a) selenite reduction at 0 h; (b) formation of red elemental selenium in membrane fraction after 3–4 h of incubation; and (c) in soluble fraction after 12 h of incubation. Source: Dhanjal, S. and Cameotra, S. S. (2010). Aerobic biogenesis of selenium nanospheres by *Bacillus cereus* isolated from coalmine soil. *Microbial Cell Factories*, 9, 52–52. See insert for color representation of the figure.

8.2.2. Fungi in NPs Synthesis

Due to their strong tolerance and biosorption ability for metals, fungi are taking the center stage in bio-nanoparticle research. Various authors have observed that fungi are easy to handle during synthesis of NPs in term of flow pressure as well as agitation in the bioreactor. The concept of formation of these NPs is mainly based on the enzyme capable of reduction. Further, the secretions of these extracellular reductive and capping proteins are quite common in fungi, making downstream processing quite easy (Narayanan and Sakthivel, 2010). Various studies have been reported for the synthesis of NPs through fungi, in both the intracellular and extracellular environment.

Intracellular synthesis of AuNPs using the biomass of *Verticillium* sp. upon exposure to aqueous $HAuCl_4$ solution was reported (Mukherjee et al., 2001a). The gold-loaded biomass exhibited a surface plasmon resonance (SPR) at around 550 nm, and transmission electron microscopy (TEM) analysis revealed the average size of AuNPs to be 20 ± 8 nm. The NPs were located on the cell wall with spherical, triangular, and hexagonal morphology and in the cytoplasmic membrane with quasi-hexagonal morphology of the fungal mycelia. Similarly, the authors also demonstrated intracellular synthesis of AgNPs using the same species, and suggested that the gold and silver ions initially bind to the cell surface via electrostatic interaction. Upon successful adsorption, the enzymes present on the cell wall reduce the metal ion to zero valency metal nuclei and subsequently to NPs (Mukherjee et al., 2001b).

Shankar et al. (2003) reported AuNPs synthesis from an endophytic fungus *Colletotrichum* sp. isolated from the leaves of geranium plant (*Pelargonium graveolens*). The NPs were polydispersed, spherical, and within the size range of 8–40 nm. Fourier transform infrared (FTIR) spectroscopy analysis revealed strong bands at 1658, 1543, and 1240 cm^{-1} which correspond to amide I, II, and III bands of proteins. The stability of AuNPs was speculated to be conferred by glutathiones binding either through free amine group or cysteine residues (Gole et al., 2001).

Ahmad et al. (2005) reported reaction-condition-related synthesis of AuNPs using the alkalo-tolerant fungus *Trichothecium* sp. biomass. It was observed that under stationary conditions the biomass resulted in extracellular AuNPs of 5–200 nm in size, while under shaking condition intracellular AuNPs of 10–25 nm in size were synthesized on the cell wall as well as on the cytoplasmic membrane of the fungus. The NPs were of FCC structure and demonstrated a varied morphology of polydispersed, spherical, rod-like, and triangular. The process was enzymatically catalyzed by the enzymes released by the fungus into the medium.

Extracellular biogenesis of AgNPs was demonstrated by Durán et al. (2005) using *Fusarium oxysporum*. The NPs were of 20–50 nm in size at 28 °C and the reduction of Ag^+ ions was influenced by a nitrate-dependent reductase and a quinone electron shuttle process. When silver ions were incubated with the culture filtrate of *Fusarium semitectum*, AgNPs of 10–60 nm size of spherical shape and polydispersity were synthesized (Basavaraja et al., 2008). *Trichoderma asperellum*, a nonpathogenic fungus and a known biocontrol agent, has also shown the potential to bioreduce silver ions to AgNPs in the size range of 13–18 nm with well-defined morphology and stability, with pseudo-zero-order kinetic mechanism (Mukherjee et al., 2008). Further extracellular synthesis of spherical AgNPs of size range 60–80 nm using silver nitrate as precursor by the fungal filtrate of *Phoma glomerata* has been reported (Birla et al., 2009). The authors also investigated the antibacterial activity of the AgNPs against *E. coli*, *P. aeruginosa* and *Staphylococcus aureus*.

Shaligram et al. (2009) reported extracellular synthesis of stable AgNPs using the fungus *Penicillium brevicompactum* WA 2315. The fungal supernatant was used as a source for bioreduction of silver ions to AgNP with an average size of 58.35 ± 17.88 nm. The fungal proteins released in the supernatant played a major role in the synthesis of NPs. FTIR spectrum of the freeze-dried powder AgNPs sample showed bands at 3356cm^{-1} and 2922cm^{-1}, assigned to the stretching vibration, while bands at 1622cm^{-1} and 1527cm^{-1} corresponded to the bending vibration of primary and secondary amines. Two bands observed at 1412cm^{-1} and 1029cm^{-1} were assigned to the C-N stretching vibrations of aromatic and aliphatic amines. The results obtained confirm the involvement of proteins in the biosynthesis of AgNPs.

The green biosynthesis of AgNPs using cell-free filtrate of *Aspergillus flavus* NJP 08 has been attempted recently (Jain et al., 2011). The authors stressed that the two major proteins were responsible for the synthesis and stability of AgNPs formed in extracellular environment (Fig. 8.3). Through ammonium sulfate precipitation and dialysis, two major proteins of molecular weight 32 and 35 kDa were purified from the extracellular filtrate. SDS-PAGE results suggested that the enzyme reductase (probably a 32 kDa protein) was responsible for the synthesis of AgNPs from the aqueous silver ions. The role of fungal secreted proteins as capping ligands for imparting stability to AgNPs after their synthesis has also been advocated.

In another study, biosynthesis of mono- and bimetallic silver and gold nanoparticles was observed with non-pathogenic filamentous fungus *Neurospora crassa* using different ratios of silver and gold ions by Castro-Longoria et al. (2011). It was observed that AgNPs were formed both extracellularly and intracellularly while AuNPs were only formed intracellularly (Fig. 8.3). The metallic NPs synthesized were observed throughout the cell area and were mainly spherical/ellipsoidal in shape, with an average size of 11 nm for AgNPs and 32 nm for AuNPs. Recently Das et al. (2012) reported biosynthesis of AuNPs using the cell-free protein extract of *Rhizopus oryzae* that served as both a reducing and a stabilizing agent. The NPs exhibited a characteristic absorption band at 538 nm, as observed by UV-Vis spectroscopy. The bio-nanoparticles were well dispersed without agglomeration, of size 20 nm, and were stable for three months at room temperature. The findings of the study also suggest that the carboxyl and amine groups of phosphoproteins were responsible for reduction and subsequent stabilization of AuNPs.

In addition to gold and silver, the biogenic potential of fungus has also been explored for other metals and oxide NPs. Attempts have also been made to synthesize semiconductor NPs from fungal species. When exposed to $CdSO_4$ solution, the biomass free filtrate of *Fusarium oxysporum* lead to the production of stable CdSNPs extracellularly, which was confirmed by the bright yellow color of the solution after the reaction. UV-Visible spectroscopic analysis showed a characteristic absorption

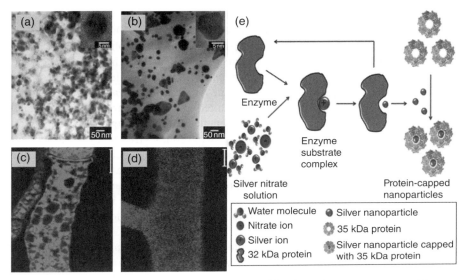

Figure 8.3. Synthesis of silver and gold nanoparticles from *Neurospora crassa* biomass after 24 h of incubation in aqueous solutions of $AgNO_3$ and $HAuCl_4$: (a) silver nanoparticles; (b) gold nanoparticles; (c) fungal hypha with silver nanoparticles scanned under confocal microscopy (Abs/Em 420/515–530 nm); (d) hypha with gold nanoparticles scanned under confocal microscopy (Abs/Em 543/574–691 nm; scale bar: 10 μm); and (e) the role of extracellular proteins in the synthesis of silver nanoparticles from *Aspergillus flavus*. Source: (a) Castro-Longoria, E., Vilchis-Nestor, A. R., and Avalos-Borja, M. (2011). Biosynthesis of silver, gold and bimetallic nanoparticles using the filamentous fungus *Neurospora crassa*. *Colloids and Surfaces B: Biointerfaces*, 83(1), 42–48. Copyright © 2011, Elsevier. (b) Jain, N., Bhargava, A., Majumdar, S., Tarafdar, J., and Panwar, J. (2011). Extracellular biosynthesis and characterization of silver nanoparticles using *Aspergillus flavus* NJP08: a mechanism perspective. *Nanoscale*, 3(2), 635–641. Copyright © 2011, Royal Society of Chemistry. *See insert for color representation of the figure.*

peak for CdSNPs at 450 nm. TEM analysis reported the well-dispersed CdS nanoparticles of size 5–20 nm. However, CdSNPs were not formed when the fungal biomass was exposed to aqueous $CdNO_3$ solution, even for an extended period of time, suggesting the possible induction of sulfate reductase enzyme by the presence of $CdSO_4$ into the solution. Further, four protein bands were found in the aqueous extract of the fungal biomass through polyacrylamide gel electrophoresis, suggesting the possible involvement of the proteins in the biogenesis of CdSNCs and that the stability was due to the attachment of proteins on the surface of NPs, thereby preventing their coalescence (Ahmad et al., 2002).

Extracellular biosynthesis of spherical PtNPs in the size range 5–30 nm by *Fusarium oxysporum* has been attempted recently, using hexachloroplatinic acid (H_2PtCl_6) as precursor at 25–27 °C. The synthesis of NPs was due to the proteins present in the medium and PtNPs were characterized by the appearance of an absorption band at 270–280 nm. The findings suggest that the synthesis and stability of NPs was governed by enzyme-mediated process (Syed and Ahmad, 2012).

Similarly, different fungal species isolated from rhizospheric soil of plants thriving around a zinc mine in India were screened for their metal tolerance and ability to synthesize zinc oxide (ZnO) nanoparticles under ambient conditions (Jain et al., 2013). Results indicated that the isolate *Aspergillus aeneus* NJP 12 exhibited the maximum tolerance to zinc ions and resulted in extracellular biogenesis of ZnO nanoparticles. The ZnO nanoparticles, as obtained from the fungal cell-free filtrate, were spherical in shape and 100–140 nm in size. From FTIR studies it was evident that the proteins present in the filtrate acted as a stabilizing agent to the NPs as the freeze-dried sample of zinc oxide NPs exhibited characteristic bands at $1625\,cm^{-1}$ and $1550\,cm^{-1}$ of amide I and amide II proteins. Further, the authors also demonstrated the process to be non-enzymatic as it was found that the interaction between amino acid obtained from the denatured fungal proteins and Zn ions was responsible for the synthesis of NPs.

Extracellular synthesis of magnetite NPs from *Fusarium oxysporum* and *Verticillium* sp. using ferricyanide/ferrocyanide as precursor at room temperature has been reported (Bharde et al., 2006). The magnetite obtained from *F. oxysporum* was of quasi-spherical morphology, of 20–50 nm in size, and attached with fungal proteins; magnetite obtained from *Verticillum sp.* demonstrated cubo-octahedral morphology and a size range of 10–40 nm. The hydrolysis of anionic precursor and capping of the magnetite NPs was mediated by the secretion of cationic proteins of molecular weight 55 kDa and 13 kDa.

The biogenic potential of *F. oxysporum* was further extended towards synthesizing zirconia nanoparticles, which have important technological applications. The fungus at room temperature was challenged by the aqueous solution of K_2ZrF_6, resulting in the extracellular reduction of zirconium hexafluoride (ZrF_6^{2-}) anions to crystalline zirconia nanoparticles with the help of cationic protein of molecular weight 24–28 kDa (Bansal et al., 2004). This fungus was also reported to synthesize strontianite ($SrCO_3$) NCs of needle-like, quasi-linear morphology using aqueous Sr^{3+} ions and carbonate ions supplied by the fungus itself (Rautaray et al., 2004). This procedure was therefore referred to as "total biological synthesis". Similarly, this fungus also synthesized silica and crystalline titania nanoparticles using SiF_6^{2-} and TiF_6^{2-} anionic complexes as precursors at room temperature (Bansal et al., 2005).

Recently, cerium oxide nanoparticles were synthesized by the mycelia of thermophilic fungus *Humicola* sp. using aqueous solution of cerium (III) nitrate hydrate ($CeN_3O_9.6H_2O$) as precursor. The NPs exhibited strong absorption bands at 300 and 400 nm, were spherical, had a size range of 12–20 nm, and were naturally stabilized by the proteins secreted by the fungus (Khan and Ahmad, 2013).

8.2.3. Actinomycetes in NPs Synthesis

A few actinomycetes have also been explored for their NP fabrication ability. The extremophilic actinomycete *Thermomonospora* sp. reduced Au^{3+} ions to monodispersed AuNPs of 8–40 nm size extracellularly. It was suggested that the reduction of metal ions and the stability of NPs was due to an enzymatic process and that the monodispersity of AuNPs was attributed mainly due to the extreme alkalinity (pH 9.0) and high temperature (50 °C) conditions used for the synthesis of NPs (Ahmad et al., 2003a). In contrast, an alkalotolerant actinomycete *Rhodococcus* sp. resulted in the intracellular synthesis of monodispersed AuNPs of 5–15 nm size, where the NPs were more concentrated on the cytoplasmic membrane than on the cell wall. The authors suggested that this could be due to the presence of enzymes involved in the bioreduction of the metal ions in the cell wall and on the cytoplasmic membrane, but not in the cytosol. Gold ions did not have a toxic effect on the biomass either, as they continued to multiply even after gold nanoparticles synthesis (Ahmad et al., 2003b).

AgNP synthesis from a novel *Nocardiopsis* sp. MBRC-1 culture supernatant at room temperature was recently attempted (Manivasagan et al., 2013). The nanocrystallites exhibited an absorption peak at around 420 nm, were spherical in shape, and had an average particle size of 45 ± 0.15 nm. Through FTIR analysis it was revealed that the reduction and stabilization of AgNPs was mainly due to the enzyme nitrate reductase present in the culture supernatant of the isolate. The authors also investigated the antimicrobial and cytotoxic effects of the biosynthesized AgNPs.

8.2.4. Yeast in NPs Synthesis

Among the eukaryotes, yeasts have been explored mostly for fabrication of semiconductor NPs. Intracellular synthesis of CdSNCs from *Candida glabrata* was reported by Dameron et al. (1989). The reduction of Cd^+ ion to CdSNCs was mediated via the degradation of the Cd-PC complex (PC refers for phytochelatins having a repeat sequence of (γ-Glu-Cys)n Gly where $n = 2-6$; Reese and Winge, 1988; Dameron et al., 1989). Similarly, when *Schizosaccharomyces pombe* was challenged by Cd^+ ions, the result was the intracellular synthesis of biogenic CdSNCs in the size range of 1–1.5 nm exhibiting an absorbance maximum at 305 nm (Kowshik et al., 2002). The findings from this study also suggest that the biogenesis of CdSNCs was growth-phase-dependent and that the yeast cells in the mid-log phase of growth resulted in maximum NCs synthesis. The mechanism involve an enzyme phytochelatin synthase which synthesizes phytochelatins (PC), helping to chelate the cytoplasmic Cd^+ ions to phytochelatins, forming a low-molecular-weight Cd-PC complex. With the help of HMT1 (an ATP-binding cassette-type vacuolar membrane protein), these complexes are transported across the vacuolar membrane. With the addition of sulfide, high-molecular-weight $PCCdS^{2-}$ forms, allowing them to sequester into vacuole (Reese and Winge, 1988; Ortiz et al., 1995).

The involvement of phytochelatins in bioreduction was also confirmed through size-exclusion chromatography. The Cd was attached to a protein fraction of molecular weight 25–67 kDa which corresponds to the theoretical molecular weight of CdSNPs of 35 kDa coated with phytochelatins (Krumov et al., 2007). The crystallites exhibited size-dependent tenability of fluorescence spectrum, enabling easy recognition and ideal diode characteristics (Kowshik et al., 2002).

Further extracellular synthesis of AgNPs by silver-tolerant yeast strain MKY3 has been reported. The NPs obtained were of size 2–5 nm and formed when the yeast cells

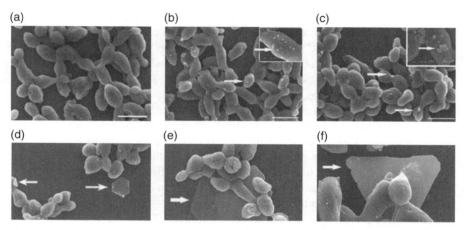

Figure 8.4. Representative SEM images of *Yarrowia lipolytica* cells incubated with 1 mM HAuCl$_4$: (a) control; (b) after 15 min; (c) after 3 h; (d) after 12 h; (e) after 24 h at magnification 5000× and (f) after 30 h at 10 000×. (b, c) Inset images at 30,000×. Scale bar: 5 μm (a–e) and 1 μm (f). Source: Agnihotri, M., Joshi, S., Kumar, A. R., Zinjarde, S., and Kulkarni, S. (2009). Biosynthesis of gold nanoparticles by the tropical marine yeast Yarrowia lipolytica NCIM 3589. *Materials Letters*, 63(15), 1231–1234. Copyright © 2009, Elsevier.

were challenged with Ag$^+$ ions in the exponential growth phase, separated through differential thawing of the samples (Kowshik et al., 2003). Yeast cells have also demonstrated size- and shape-controlled synthesis of AuNPs by optimizing different parameters for growth and cellular activities of the cell (Gericke and Pinches, 2006).

The bioreduction potential of non-conventional tropical marine yeast *Yarrowia lipolytica* NCIM 3589 was also demonstrated (Agnihotri et al., 2009). AuNPs were synthesized at a diverse range of pH values (2, 7 and 9), where acidic pH favored NCs while basic pH resulted in NPs of 15 nm size, associated with the cell wall (Fig. 8.4).

In a recent report, the biosynthesis of AgNPs and AuNPs were achieved using an extremophilic yeast strain isolated from acid mine drainage in Portugal. The authors investigated the growth potential of the isolate in the presence of metal ions, the ability of biomass, and the culture supernatant for NPs synthesis. The findings suggest that isolate responded well in the presence of Ag$^+$ ions rather than high levels of Au$^+$ ions. The reaction was carried out at 22 °C resulting in AgNPs of less than 20 nm with characteristic absorbance at 420 nm, whereas synthesized AuNPs were of size 30–100 nm with an absorbance of 550 nm. The proteins present in the supernatant were found to be responsible for the formation and stabilization of the AgNPs, while the involvement of the cell wall was essential for AuNPs synthesis (Mourato et al., 2011). Similarly, biogenic AuNPs were synthesized using yeast *Hansenula anomala* in the presence of gold salt as precursor and amine-terminated polyamidoamine dendrimer (G4 and G5) as stabilizer, with the average size of the synthesized particles 14 nm and 40 nm respectively. The potential of AuNPs to function as an antimicrobial agent and as biological ink to be used in fingerprint analysis has also been investigated (Kumar et al., 2011).

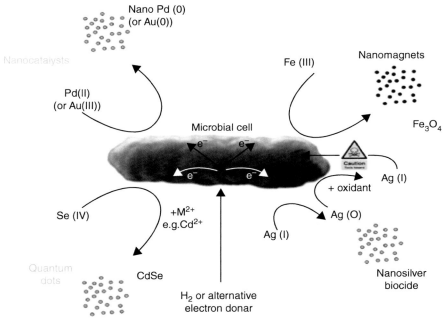

Figure 1.1. Mechanisms of microbial fabrication of nanobiominerals, catalyzed by enzymatic reductive biotransformations of redox active metals, driven by a suitable electron donor such as hydrogen. In some cases, for example transformations of Fe(III) minerals and Se(IV), redox mediators such as AQDS (anthraquinone-2,6 disulfonate) are utilized to increase the kinetics of metal reduction and hence nanobiomineral formation. Source: Lloyd, J.R., Byrne, J.M., Coker, V.S. 2011. Biotechnological synthesis of functional nanomaterials. *Current Opinion in Biotechnology* 22: 509–515. Copyright © 2011, Elsevier.

Bio-Nanoparticles: Biosynthesis and Sustainable Biotechnological Implications, First Edition.
Edited by Om V. Singh.
© 2015 John Wiley & Sons, Inc. Published 2015 by John Wiley & Sons, Inc.

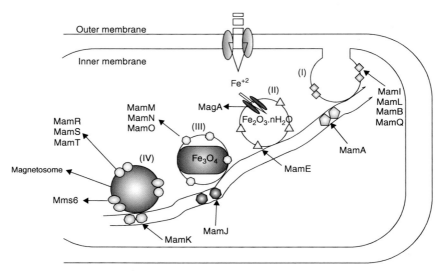

Figure 1.2. Magnetosome bio-mineralization in magnetotactic bacteria (MTB). (I) MamI, MamL, MamB, and MamQ proteins initiate the membrane invagination and form a vesicular membrane around the magnetosome structure. (II) The protease-independent function of MamE recruits other proteins such as MamK, MamJ, and MamA to align magnetosomes in a chain. (III) Iron uptake occurs via MagA, a transmembrane protein, and initiation of magnetic crystal bio-mineralization occurs through MamM, MamN, and MamO proteins. (IV) Finally, MamR, MamS, MamT, MamP, MamC, MamD, MamF, MamG, the protease-dependent function of MamE, and Mms6, a membrane tightly bounded by GTPase, regulate crystal growth and determine morphology of the produced magnetic nanoparticles. Source: Faramarzi, M.A., Sadighi, A. 2013. Insights into biogenic and chemical production of inorganic nanomaterials and nanostructures. *Advances in Colloid and Interface Science* 189–190: 1–20. Copyright © 2013, Elsevier.

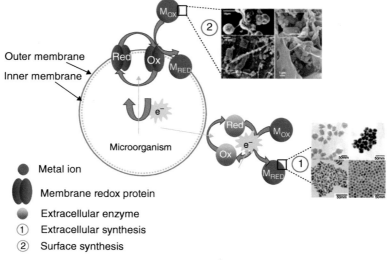

Figure 1.4. Biomineralization process for nanoparticle synthesis. Source: Das, S.K., Marsili, E. A green chemical approach for the synthesis of gold nanoparticles: characterization and mechanistic aspect. *Reviews in Environmental Science and Biotechnology* 9: 199–204. Copyright © 2010, Springer.

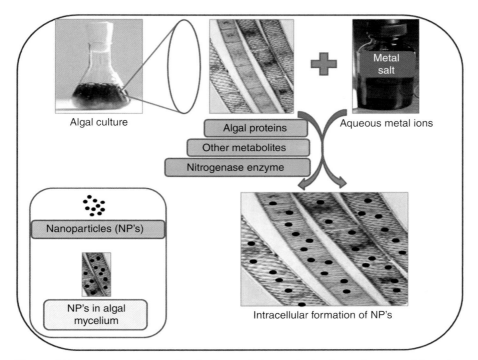

Figure 1.5 General mechanism for the intracellular synthesis of metal nanoparticles using algae.

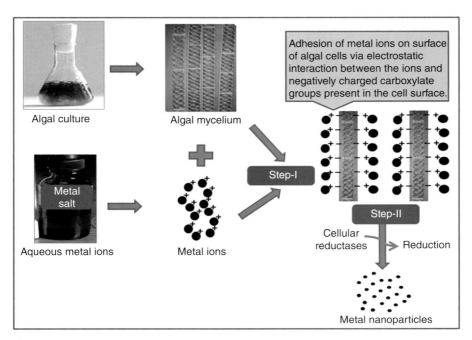

Figure 1.6. General mechanism for the extracellular synthesis of metal nanoparticles using algae.

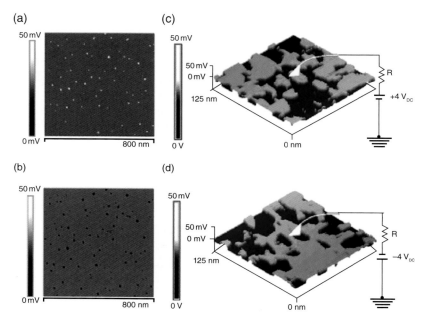

Figure 2.4. (a, b) Low- and (c, d) higher-magnification surface potential microscopy (SPM) images of BaTiO$_3$ nanoparticles synthesized using *Fusarium oxysporum*. SPM images from ferroelectric BaTiO$_3$ nanoparticles obtained in potential mode after application of +4 V (a, c) and −4 V (b, d) external DC bias voltages, where the reversal in image contrast on reversal of bias voltage is observed. Source: Bansal, V.; Poddar, P.; Ahmad, A.; Sastry, M., Room-Temperature Biosynthesis of Ferroelectric Barium Titanate Nanoparticles. *Journal of American Chemical Society*, 128(36), 11958–11963. Copyright © 2006, American Chemical Society.

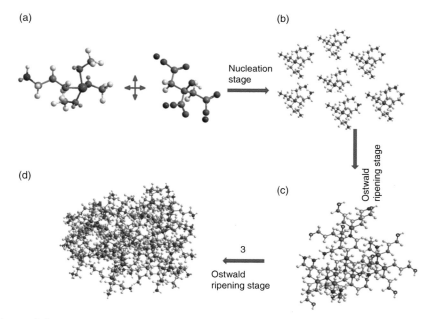

Figure 3.1. Bottom-up -one-pot synthesis of SiNPs. Source: Redrawn from Yiling Zhong, Fei Peng, Feng Bao, Siyi Wang, Xiaoyuan Ji, Liu Yang, Yuanyuan Su, Shuit-Tong Lee, Yao He. Large-Scale Aqueous Synthesis of Fluorescent and Biocompatible Silicon Nanoparticles and Their Use as Highly Photostable Biological Probes. *Journal of the American Chemical Society* 135(22): 8350–8356. Copyright © 2013, American Chemical Society.

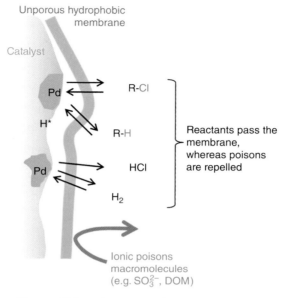

Figure 10.3. Polymer-coated catalysts particle.

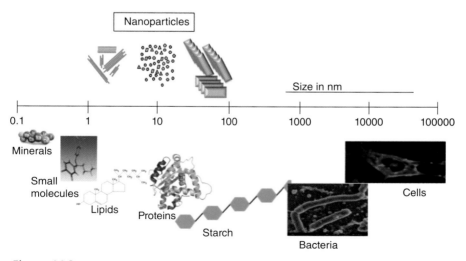

Figure 14.2. Nanomaterials and nanoparticles in context to other biological molecules.

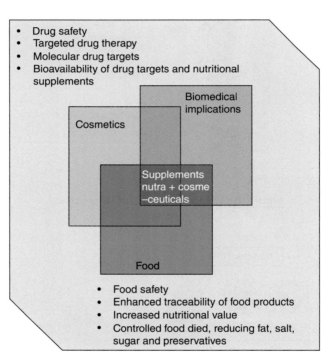

Figure 14.3. Projected benefits of nanotechnology in food- and biomedical-related sectors.

Candida guilliermondii were reported to synthesize gold and AgNPs extracellularly (Mishra et al., 2011). The NPs exhibited distinct surface plasmon peaks at 530 nm for AuNPs and 425 nm for AgNPs. The particles were spherical and well dispersed with face-centered cubic structures in the size range of 50–70 nm and 10–20 nm, respectively. The authors also investigated the antimicrobial potential of the NPs against five pathogenic bacteria; they found that biogenic NPs exhibited potent antimicrobial activity, especially against *Staphylococcus aureus*, while the synthetic NPs showed no inhibitory effect against any of the pathogenic strains.

8.2.5. Virus in NPs Synthesis

Intriguingly, a few reports on the virus-mediated assembly of majorly semiconducting NCs have also been published recently (Shenton et al., 1999; Lee et al., 2002; Mao et al., 2003, 2004; Banerjee et al., 2005; Slocik et al., 2005). In this context, the hollow protein tube of *Tobacco mosaic virus* (TMV) was used as a template for the synthesis of a range of nanotubes through different processes, namely: iron oxides by oxidative hydrolysis; CdS and PbS by co-crystallization; and SiO_2 by sol-gel condensation. Glutamate and aspartate present on the external surface of the virus assisted the assembly of particles over the protein template (Shenton et al., 1999). Genetically engineered *M13 bacteriophage*-based liquid crystal system was used for the assembly of zinc sulfide (ZnS) NCs of 10–20 nm diameter (Lee et al., 2002). The peptides were selected through pIII phage display library for their ability to nucleate ZnS and later expressed in *M13 bacteriophage* to form the basis of the self-ordering system. When challenged with ZnS solution precursors, the A7 phage resulted in A7-ZnS biofilm formation of 15 mm thickness, which aligns to form the liquid crystal system. Additionally, the fabrication of viral film was a reversible process, and the biofilm was found to be stable at room temperature for about 7 months without losing their ability to infect a bacterial host with minimal titre loss. The authors also suggested that the potential of these genetically engineered viruses with specific recognition, as well as a liquid crystalline self-ordering system, can be harnessed to create newer pathways to organize electronic, optical, and magnetic materials and store high-density engineered DNA (Lee et al., 2002).

Similarly, genetically controlled synthesis of quantum dot nanowires (including heterostructures and superlattices) was reported using self-assembled viral capsids of genetically engineered viruses as biological templates following peptide-templated growth mechanism. The peptides A7 (Cys-Asn-Asn-Pro-Met-His-Gln-Asn-Cys) and J140 (Ser-Leu-Thr-Pro-Leu-Thr-Thr-Ser-His-Leu-Arg-Ser) were selected by using a pIII phage display library for their ability to nucleate ZnS and CdSNCs, respectively, and were expressed as pVIII fusion proteins into the crystalline capsid of the *M13 bacteriophages*. In the presence of semiconductor precursor solutions, these organized template peptides (A7/J140-pVIIIM13) synthesized ZnSNCs on the viral capsid with a hexagonal wurtzite structure of size 3–5 nm or wurtzite CdS assembled as nanowires of 20 nm. Further, heterogeneous nanowires (ZnS–CdS) were also obtained with a dual peptide virus engineered to express A7 and J140 on the same viral capsid (Mao et al., 2003).

Apart from ZnS and CdS NCs, viral assembly of ferromagnetic alloys (CoPt and FePt) has also been reported using genetically engineered *M13 bacteriophages*. The specific peptides for each nanoparticle nucleation were selected through an evolutionary screening process and expressed on the highly ordered filamentous capsid of the *M13 bacteriophage*. The peptides were identified as FP12 (HNKHLPSTQPLA) for FePt and CP7 (CNAGDHANC) for CoPt systems. The obtained nanowires were crystalline in nature and had a one-dimnesional (1D) structure of $10 nm \pm 5\%$ diameter (Mao et al., 2004).

Banerjee et al. (2005) demonstrated the phase-controlled synthesis of ZnS nanocrystals and nanowires of average size 4 nm at room temperature using the Zn finger-like peptides as template consisting of VAL-CYS-ALA-THR-CYS-GLU-GLN-ILE-ALA-ASP-SER-GLN-HIS-ARG-SER-HIS-ARG-GLN-MET-VAL, M1 peptide sequences, synthesized based on the peptide motif of the *Influenza Virus* Matrix Protein M1. The change in pH was essential to obtain the desired phase and size control and for the number of nucleation sites in M1 peptides to grow ZnS nanocrystals, thereby tuning the band gap of the resulting nanotube. In a comprehensive mechanistic study, Slocik et al. (2005) demonstrated AuNP synthesis using wild-type and engineered viral template of *Cowpea chlorotic mottle* virus as unmodified SubE (yeast) and engineered HRE peptide epitopes (AHHAHHAAD) as (HRE)-SubE and wild type. The viral capsid enabled the bioreduction of $AuCl_{4-}$, with the help of surface tyrosine residues, to AuNPs with average sizes of 9.2 ± 3.9, 8.3 ± 2.6, and 23.8 ± 14.5 nm which decorated the viral surface (Fig. 8.5). The authors also observed that the viral template was not able to reduce other metallic ions (Ag^+, Pt^{4+}, Pd^{4+}, and an insoluble Au^I complex) to the zero-valent state.

8.3. MECHANISMS OF NANOPARTICLES SYNTHESIS

The exact mechanism for the synthesis of NPs using biological agents has not yet been determined, as different biological agents react differently with metal ions and there are also different biomolecules responsible for the synthesis of NPs. In addition, the mechanism for intra- and extracellular synthesis of NPs is different in various biological agents.

Due to their complex structure organization, microorganisms have evolved various methods to counter the heavy metal stress through metabolism-dependent and metabolism-independent processes. Metabolism-dependent uptake of metal ions is often the result of the active defense system of the microbe, which becomes triggered in the presence of the toxic metals. The formation of NPs is mediated by the viable cells and is always intracellular. In metabolism-independent processes, the uptake of metal ions is however mediated by the physico-chemical interaction between the metal and the functional groups present on the microbial cell surface. This interaction is based on ion exchange, oxidation-reduction, physical adsorption, and chemical sorption, which are not dependent on the cell metabolism. The process is rapid and can be reversible (Kuyucak and Volesky, 1988).

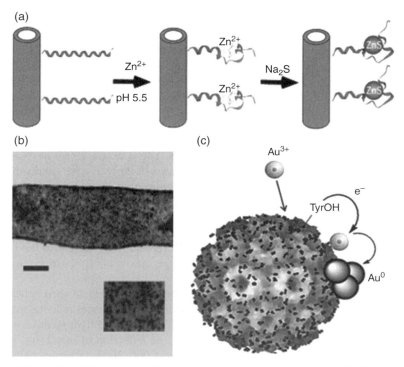

Figure 8.5. (a) Unfolding M1 peptides on the nanotube template showing the ZnS nanocrystal growth as a function of pH. (b) TEM image of ZnS nanocrystals on the M1 peptide nanotube template grown at pH 5.5. Inset image at high magnification; scale bar 70 nm. (c) Viral capsid-mediated reduction of gold ions to gold nanoparticles. Source: (a) Banerjee, I. A., Yu, L., and Matsui, H. (2005). Room-temperature wurtzite ZnS nanocrystal growth on Zn finger-like peptide nanotubes by controlling their unfolding peptide structures. *Journal of the American Chemical Society*, 127(46), 16002–16003. Copyright © The American Chemical Society, 2006. (b) Slocik, J. M., Naik, R. R., Stone, M. O., and Wright, D. W. (2005). Viral templates for gold nanoparticle synthesis. *Journal of Materials Chemistry*, 15(7), 749–753. Copyright © Royal Society of Chemistry, 2005. *See insert for color representation of the figure.*

The microbial cell wall is mainly composed of polysaccharides, glycoproteins, and glycolipids that interact with the metals (Geesey and Jang, 1990). Apart from these, the cell wall of microbes also harbors various metal-binding components such as carboxyl, sulfate, phosphate, and amino groups (McLean and Beveridge, 1990). Various metal-binding proteins and peptides are known that have strong affinity towards metal binding and also become induced by their presence in near proximity. Among these, the most extensively studied are metallothioneins and metal γ-glutamyl peptides (phytochelatins). These are short peptides that are involved in heavy-metal detoxification in algae, plants, and some fungi and yeasts (Mehra and Winge, 1991; Gadd, 1993).

Essential aspects of bio-nanoparticle formation are the macromolecules DNA, protein, or peptide template that serves as the locus of control for morphology and the congregation procedure of NPs. It was suggested by Niemeyer (2001) that it is the electrostatic and topographic properties of biological macromolecules, and their derived supramolecular complexes, that influences the synthesis and assembly of organic and inorganic components. A wide range of biological entities such as DNA (Mirkin et al., 1996; Braun et al., 1998), protein cages (Wong et al., 1998), viroid capsules (Douglas and Young, 1998), bacterial rhapidosomes (Pazirandeh et al., 1992), biolipid tubules (Archibald and Mann, 1993), bacterial S-layers (Shenton et al., 1997), and multicellular superstructures (Davis et al., 1997) have already been exploited for the template-mediated synthesis of inorganic NPs and supramolecular structures. The use of protein cages as the mini-bioreactor for NPs synthesis serves as an ideal template for confining particle growth and assembly with homogenous distribution; it also acts as a stabilizer to avert particle aggregation. Due to the extensive use of proteins as specific biomolecular recognition elements in the synthesis of bio-nanoparticles, these will serve as the "factory of the future" for the nanofabrication of spatially defined aggregates via bottom-up assembly of NPs (Mann et al., 2000).

The enzymatic route to biosynthesis has also been well established, where oxidoreductase, NADH/NADPH-dependent reductase, nitrate/nitrite reductase, sulfate and sulfite reductase, hydrolase, cysteine desulfhydrase, and hydrogenases are the major classes of enzyme involved in mediating the reduction of metal ions to NPs (Ahmad et al., 2002; Durán et al., 2005; Bai et al., 2006, 2009a; Bharde et al., 2006; He et al., 2007; Kalishwaralal et al., 2008; Jha et al., 2009*a*; Nangia et al., 2009; Riddin et al., 2009).

8.4. PURIFICATION AND CHARACTERIZATION OF NANOPARTICLES

As discussed in Sections 8.2.1–8.2.5, bio-nanoparticles can be obtained intra-/extracellularly from the microorganisms depending upon their site of bioreduction and subsequent precipitation of metal ions precursor. The biggest challenge *en route* to biosynthesis is the efficient recovery of the synthesized NPs following different downstream procedures. As is evident, intracellular biogenesis makes the job of downstream processing and recovery of NPs difficult and also defeats the purpose of developing a simple, rapid and economical bioprocess. Surface-trapped NPs or those which are formed inside the cell require an additional processing step, such as ultrasonication (Ganesh Babu and Gunasekaran, 2009), cell disruption techniques using detergents (Nangia et al., 2009), or lysis buffer (Samadi et al., 2009), for their release into the surrounding medium. The released particles in suspension need further clarification followed by concentration steps to obtain cell-free NPs. A final purification step is also required in order to free the surface of NPs from cellular contaminants in order to obtain purified NPs. The recovery procedure also results in substantial reductions in the quantity of NPs obtained. However, in the case of extracellular biosynthesis, the downstream processing steps are minimized considerably, thereby reducing the time and cost involved in recovering the NPs (Singh et al., 2014). The recovery of purified NPs is easier with the advanced concentration and purification techniques available. The focus of research on NPs biosynthesis has recently shifted towards development of an efficient extracellular process for easy recovery and scale-up synthesis for NPs.

The obtained NPs also need to be characterized in order to validate their physico-chemical characteristics using a range of diverse techniques including: UV-vis spectroscopy; dynamic light scattering (DLS); zeta potential measurement; x-ray diffractometry (XRD); fourier transform infrared spectroscopy (FTIR); x-ray photoelectron spectroscopy (XPS); atomic force microscopy (AFM); and scanning and transmission electron microscopy (SEM, TEM). The advent of these advanced techniques has helped to resolve different parameters such as particle shape, size, surface area, size dispersion, crystallinity, composition and scattering properties (Fig. 8.6). The initial confirmation of the formed NPs is ascertained by visual observation for color change, and the extinction spectra of metallic NPs is recorded by UV-vis spectroscopy allowing the concentration and aggregation level to be estimated. Secondly, through dynamic light scattering techniques, the particle size distribution of the NPs can be determined. In order to identify the functional groups present on the NPs surface and to ascertain the chemical composition, the FTIR technique is used. Finally, the crystallinity of NPs is determined by x-ray diffraction and knowledge of the particles size, shape, morphology, height, volume and 3D image is determined by TEM, SEM, and AFM, respectively.

8.5. CONCLUSION

The various disadvantages associated with conventional nanoparticle synthesis and an increasing environmental awareness have lead to the emergence of a new, reliable, eco-friendly area of nanomaterial research in the form of microbial nanoparticle synthesis. Considering the enormous research effort in this area, it is likely that in the future scale-up synthesis of NPs from microbial minifactories will be realized. This chapter has provided an overview of microbial-inspired methods for NPs fabrication as well as a discussion on mechanistic aspects, purification, and characterization of NPs. The recent advances in the biogenesis of NPs, precursors involved, use of reducing and stabilizing agents with various examples and case studies have been described.

As outlined in this review, bacterial-mediated NPs synthesis has received most attention compared to other microbes. It is also known that genetic manipulation and control are easier in bacteria and viruse than in eukaryotes with rapid NPs synthesis. However, fungi are easy to work with during downstream processing and scale-up production. Regardless of the selection of microbe, for efficient and size-controlled synthesis it is essential to establish the biochemical and molecular mechanism for bio-nanoparticle formation. The advent of extracellular synthesis using cell filtrate, supernatant, lysate, and biological fractions suggests that bioreduction potential

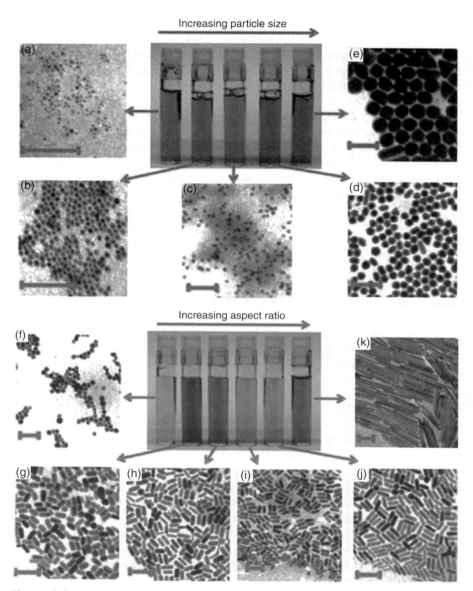

Figure 8.6. Photographs of aqueous solutions of gold nanospheres (upper panels) and gold nanorods (lower panels) as a function of increasing dimensions with respective transmission electron microscopy images of the nanoparticles (all scale bars 100 nm). (a–e) For nanospheres, the size varies in the range 4–40 nm. For nanorods, the aspect ratio (f–j) is 1.3–5 for short rods and (k) 20 for long rods. Source: Prashant K, Jain, Kyeong Seok Lee, Ivan H El-Sayed, and Mostafa A. El Sayed. Calculated Absorption and Scattering Properties of Gold Nanoparticles of Different Size, Shape, and Composition: Applications in Biological Imaging and Biomedicine. *The Journal of Physical Chemistry B*. Copyright © the American Chemical Society, 2006. *See insert for color representation of the figure*.

persists even in lysed state for microbes, indicating the potential involvement of cellular components for NPs synthesis rather than the whole cell.

Further research into the use of isolated purified enzymes, biological templates, and genetically engineered microbes may lead to the establishment of protocols for enhanced biosynthesis, easy downstream processing, scale-up, and commercialization of the process. Future research is also warranted to investigate the efficacy and functionality of the bio-nanoparticles as compared to conventionally synthesized synthetic nanoparticles. More research needs to be focused on the reaction kinetics, physico-chemical parameters optimization, and mechanistics of nanoparticle synthesis, which may lead to the fine-tuning of the process for optimum yield with strict control over size and shape of the synthesized bio-nanoparticles.

REFERENCES

Agnihotri, M., Joshi, S., Kumar, A.R., Zinjarde, S., Kulkarni, S. 2009. Biosynthesis of gold nanoparticles by the tropical marine yeast *Yarrowia lipolytica* NCIM 3589. *Materials Letters* 63(15): 1231–1234.

Ahmad, A., Mukherjee, P., Mandal, D., Senapati, S., Khan, M.I., Kumar, R., Sastry, M. 2002. Enzyme mediated extracellular synthesis of CdS nanoparticles by the fungus, *Fusarium oxysporum*. *Journal of the American Chemical Society* 124(41): 12108–12109.

Ahmad, A., Senapati, S., Khan, M.I., Kumar, R., Sastry, M. 2003*a*. Extracellular biosynthesis of monodisperse gold nanoparticles by a novel extremophilic actinomycete, *Thermomonospora* sp. *Langmuir* 19(8): 3550–3553.

Ahmad, A., Senapati, S., Khan, M.I., Kumar, R., Ramani, R., Srinivas, V., Sastry, M. 2003*b*. Intracellular synthesis of gold nanoparticles by a novel alkalotolerant actinomycete, *Rhodococcus* species. *Nanotechnology* 14(7): 824.

Ahmad, A., Senapati, S., Khan, M.I., Kumar, R., Sastry, M. 2005. Extra-/intracellular biosynthesis of gold nanoparticles by an alkalotolerant fungus, *Trichothecium* sp. *Journal of Biomedical Nanotechnology* 1(1): 47–53.

Appenzeller, T. 1991. The man who dared to think small. *Science* 254(5036): 1300–1300.

Archibald, D.D., Mann, S. 1993. Template mineralization of self-assembled anisotropic lipid microstructures. *Nature* 364(6436): 430–433.

Bahrami, K., Nazari, P., Sepehrizadeh, Z., Zarea, B., Shahverdi, A.R. 2012. Microbial synthesis of antimony sulfide nanoparticles and their characterization. *Annals of Microbiology* 62(4): 1419–1425.

Bai, H.-J., Zhang, Z.-M., Gong, J. 2006. Biological synthesis of semiconductor zinc sulfide nanoparticles by immobilized *Rhodobacter sphaeroides*. *Biotechnology Letters* 28(14): 1135–1139.

Bai, H.J., Zhang, Z.M., Guo, Y., Yang, G.E. 2009*a*. Biosynthesis of cadmium sulfide nanoparticles by photosynthetic bacteria *Rhodopseudomonas palustris*. *Colloids and Surfaces B: Biointerfaces* 70(1): 142–146.

Bai, H.-J., Zhang, Z.-M. 2009*b*. Microbial synthesis of semiconductor lead sulfide nanoparticles using immobilized *Rhodobacter sphaeroides*. *Materials Letters* 63(9): 764–766.

Banerjee, I.A., Yu, L., Matsui, H. 2005. Room-temperature wurtzite ZnS nanocrystal growth on Zn finger-like peptide nanotubes by controlling their unfolding peptide structures. *Journal of the American Chemical Society* 127(46): 16002–16003.

Bansal, V., Rautaray, D., Ahmad, A., Sastry, M. 2004. Biosynthesis of zirconia nanoparticles using the fungus *Fusarium oxysporum*. *Journal of Materials Chemistry* 14(22): 3303–3305.

Bansal, V., Rautaray, D., Bharde, A., Ahire, K., Sanyal, A., Ahmad, A., Sastry, M. 2005. Fungus-mediated biosynthesis of silica and titania particles. *Journal of Materials Chemistry* 15(26): 2583–2589.

Basavaraja, S., Balaji, S., Lagashetty, A., Rajasab, A., Venkataraman, A. 2008. Extracellular biosynthesis of silver nanoparticles using the fungus *Fusarium semitectum*. *Materials Research Bulletin* 43(5): 1164–1170.

Bäuerlein, E. 2003. Biomineralization of unicellular organisms: an unusual membrane biochemistry for the production of inorganic nano-and microstructures. *Angewandte Chemie International Edition* 42(6): 614–641.

Beveridge, T., Murray, R. (1980). Sites of metal deposition in the cell wall of *Bacillus subtilis*. *Journal of Bacteriology* 141(2): 876–887.

Bharde, A., Wani, A., Shouche, Y., Joy, P.A., Prasad, B.L., Sastry, M. 2005. Bacterial aerobic synthesis of nanocrystalline magnetite. *Journal of the American Chemical Society* 127(26): 9326–9327.

Bharde, A., Rautaray, D., Bansal, V., Ahmad, A., Sarkar, I., Yusuf, S.M., Sanyal, M., Sastry, M. 2006. Extracellular biosynthesis of magnetite using fungi. *Small* 2(1): 135–141.

Birla, S., Tiwari, V., Gade, A., Ingle, A., Yadav, A., Rai, M. 2009. Fabrication of silver nanoparticles by Phoma glomerata and its combined effect against *Escherichia coli, Pseudomonas aeruginosa* and *Staphylococcus aureus*. *Letters in Applied Microbiology* 48(2): 173–179.

Blakemore, R. 1975. Magnetotactic bacteria. *Science* 190(4212): 377–379.

Braun, E., Eichen, Y., Sivan, U., Ben-Yoseph, G. 1998. DNA-templated assembly and electrode attachment of a conducting silver wire. *Nature* 391(6669): 775–778.

Castro-Longoria, E., Vilchis-Nestor, A.R., Avalos-Borja, M. 2011. Biosynthesis of silver, gold and bimetallic nanoparticles using the filamentous fungus *Neurospora crassa*. *Colloids and Surfaces B: Biointerfaces* 83(1): 42–48.

Cunningham, D.P., Lundie, L. 1993. Precipitation of cadmium by *Clostridium thermoaceticum*. *Applied and Environmental Microbiology* 59(1): 7–14.

Dameron, C., Reese, R., Mehra, R., Kortan, A., Carroll, P., Steigerwald, M., Brus, L., Winge, D. 1989. Biosynthesis of cadmium sulphide quantum semiconductor crystallites. *Nature* 338: 596–597.

Daniel, M., Astruc, D. 2004. Gold nanoparticles: assembly, supramolecular chemistry, quantum-size-related properties, and applications toward biology, catalysis, and nanotechnology. *Chemical Reviews* 104(1): 293–346.

Das, S.K., Dickinson, C., Lafir, F., Brougham, D.F., Marsili, E. 2012. Synthesis, characterization and catalytic activity of gold nanoparticles biosynthesized with *Rhizopus oryzae* protein extract. *Green Chemistry* 14(5): 1322–1334.

Davis, S.A., Burkett, S.L., Mendelson, N.H., Mann, S. 1997. Bacterial templating of ordered macrostructures in silica and silica-surfactant mesophases. *Nature* 385: 420–423.

Dhanjal, S., Cameotra, S.S. 2010. Aerobic biogenesis of selenium nanospheres by *Bacillus cereus* isolated from coalmine soil. *Microbial Cell Factories* 9: 52–52.

Dickson, D.P. 1999. Nanostructured magnetism in living systems. *Journal of Magnetism and Magnetic Materials* 203(1): 46–49.

Douglas, T., Young, M. 1998. Host-guest encapsulation of materials by assembled virus protein cages. *Nature* 393(6681): 152–155.

Durán, N., Marcato, P.D., Alves, O.L., De Souza, G.I., Esposito, E. 2005. Mechanistic aspects of biosynthesis of silver nanoparticles by several *Fusarium oxysporum* strains. *Journal of Nanobiotechnology* 3(8): 1–7.

El-Sayed, M.A. 2001. Some interesting properties of metals confined in time and nanometer space of different shapes. *Accounts of Chemical Research* 34(4): 257–264.

Faivre, D., Schüler, D. 2008. Magnetotactic bacteria and magnetosomes. *Chemical Reviews* 108(11): 4875–4898.

Faraday, M. 1857. The Bakerian lecture: experimental relations of gold (and other metals) to light. *Philosophical Transactions of the Royal Society of London* 147: 145–181.

Feng, Y., Yu, Y., Wang, Y., Lin, X. 2007. Biosorption and bioreduction of trivalent aurum by photosynthetic bacteria Rhodobacter capsulatus. *Current Microbiology* 55(5): 402–408.

Feynman, R.P. 1960. There's plenty of room at the bottom. *Engineering and Science* 23(5): 22–36.

Fortin, D., Beveridge, T.J. 2000. From biology to biotechnology and medical applications. In *Biomineralization* (ed. E. Baeuerlein), pp. 7–22. Weinheim: Wiley-VCH.

Frankel, R.P., Williams, T.J., Bazylinski, D.A. 2007. Magneto-aerotaxis. In *Magnetoreception and Magnetosomes in Bacteria* (ed. D. Schüler), p. 1. Heidelberg: Springer.

Fredrickson, J.K., Zachara, J.M., Kennedy, D.W., Liu, C., Duff, M.C., Hunter, D.B., Dohnalkova, A. 2002. Influence of Mn oxides on the reduction of uranium (VI) by the metal–reducing bacterium *Shewanella putrefaciens*. *Geochimica et Cosmochimica Acta* 66(18): 3247–3262.

Gadd, G.M. 1993. Interactions of fungi with toxic metals. *New Phytologist* 124(1): 25–60.

Ganesh Babu, M., Gunasekaran, P. 2009. Production and structural characterization of crystalline silver nanoparticles from *Bacillus cereus* isolate. *Colloids and Surfaces B: Biointerfaces* 74(1): 191–195.

Geesey, G., Jang, L. 1990. Extracellular polymers for metal binding. In *Microbial Mineral Recovery* (eds H.L. Ehrlich and C.L. Brierley), pp. 223–247. New York: McGraw–Hill.

Gericke, M., Pinches, A. 2006. Biological synthesis of metal nanoparticles. *Hydrometallurgy* 83(1): 132–140.

Ghorbani, H.R. 2013. Biosynthesis of silver nanoparticles using *Salmonella typhimurium*. *Journal of Nanostructure in Chemistry* 3(1): 29.

Gole, A., Dash, C., Soman, C., Sainkar, S., Rao, M., Sastry, M. 2001. On the preparation, characterization, and enzymatic activity of fungal protease-gold colloid bioconjugates. *Bioconjugate Chemistry* 12(5): 684–690.

Grünberg, K., Wawer, C., Tebo, B.M., Schüler, D. 2001. A large gene cluster encoding several magnetosome proteins is conserved in different species of magnetotactic bacteria. *Applied and Environmental Microbiology* 67(10): 4573–4582.

Gupta, A., Silver, S. 1998. Silver-resistant mutants of *Escherichia coli* display active efflux of Ag+ and are deficient in porins. *Nature Biotechnology* 16(10): 888.

He, S., Guo, Z., Zhang, Y., Zhang, S., Wang, J., Gu, N. 2007. Biosynthesis of gold nanoparticles using the bacteria *Rhodopseudomonas capsulata*. *Materials Letters* 61(18): 3984–3987.

Hunter, W.J., Manter, D.K. 2008. Bio-reduction of selenite to elemental red selenium by *Tetrathiobacter kashmirensis*. *Current Microbiology* 57(1): 83–88.

Jain, N., Bhargava, A., Majumdar, S., Tarafdar, J., Panwar, J. 2011. Extracellular biosynthesis and characterization of silver nanoparticles using *Aspergillus flavus* NJP08: a mechanism perspective. *Nanoscale* 3(2): 635–641.

Jain, N., Bhargava, A., Tarafdar, J.C., Singh, S.K., Panwar, J. 2013. A biomimetic approach towards synthesis of zinc oxide nanoparticles. *Applied Microbiology and Biotechnology* 97(2): 859–869.

Jain, P.K., Lee, K.S., El-Sayed, I.H., El-Sayed, M.A. 2006. Calculated absorption and scattering properties of gold nanoparticles of different size, shape, and composition: applications in biological imaging and biomedicine. *The Journal of Physical Chemistry B* 110(14): 7238–7248.

Jenekhe, S.A., Zhang, X., Chen, X.L., Choong, V.-E., Gao, Y., Hsieh, B.R. 1997. Finite size effects on electroluminescence of nanoscale semiconducting polymer heterojunctions. *Chemistry of Materials* 9(2): 409–412.

Jha, A.K., Prasad, K., Prasad, K. 2009a. Biosynthesis of Sb_2O_3 nanoparticles: A low-cost green approach. *Biotechnology Journal* 4(11): 1582–1585.

Jha, A.K., Prasad, K., Kulkarni, A.R. 2009b. Synthesis of TiO_2 nanoparticles using microorganisms. *Colloids and Surfaces B: Biointerfaces* 71(2): 226–229.

Joerger, R., Klaus, T., Granqvist, C. 2000. Biologically produced silver-carbon composite materials for optically functional thin-film coatings. *Advanced Materials* 12(6): 407–409.

Kalishwaralal, K., Deepak, V., Ramkumarpandian, S., Nellaiah, H., Sangiliyandi, G. 2008. Extracellular biosynthesis of silver nanoparticles by the culture supernatant of *Bacillus licheniformis*. *Materials Letters* 62(29): 4411–4413.

Kamat, P.V. 2002. Photophysical, photochemical and photocatalytic aspects of metal nanoparticles. *The Journal of Physical Chemistry B* 106(32): 7729–7744.

Khan, S.A., Ahmad, A. 2013. Fungus mediated synthesis of biomedically important cerium oxide nanoparticles. *Materials Research Bulletin* 48: 4134–4138.

Klaus, T., Joerger, R., Olsson, E., Granqvist, C.-G. 1999. Silver-based crystalline nanoparticles, microbially fabricated. *Proceedings of the National Academy of Sciences* 96(24): 13611–13614.

Konishi, Y., Nomura, T., Tsukiyama, T., Saitoh, N. 2004. Microbial preparation of gold nanoparticles by anaerobic bacterium. *Transactions of the Materials Research Society of Japan* 29: 2341–2344.

Konishi, Y., Ohno, K., Saitoh, N., Nomura, T., Nagamine, S., Hishida, H., Takahashi, Y., Uruga, T. 2007. Bioreductive deposition of platinum nanoparticles on the bacterium *Shewanella algae*. *Journal of Biotechnology* 128(3): 648–653.

Koplin, E., Simon, U. 2007. Application of metal nanoclusters in nanoelectronics. In *Metal Nanoclusters in Catalysis and Materials Science: The Issue of Size Control* (eds B. Corain, G. Schmid, and N. Toshima), pp. 107–128. Weinheim, Germany: Elsevier.

Korbekandi, H., Iravani, S., Abbasi, S. 2012. Optimization of biological synthesis of silver nanoparticles using *Lactobacillus casei subsp. casei*. *Journal of Chemical Technology and Biotechnology* 87(7): 932–937.

Kowshik, M., Deshmukh, N., Vogel, W., Urban, J., Kulkarni, S., Paknikar, K. 2002. Microbial synthesis of semiconductor CdS nanoparticles, their characterization, and their use in the fabrication of an ideal diode. *Biotechnology and Bioengineering* 78(5): 583–588.

Kowshik, M., Ashtaputre, S., Kharrazi, S., Vogel, W., Urban, J., Kulkarni, S., Paknikar, K. 2003. Extracellular synthesis of silver nanoparticles by a silver-tolerant yeast strain MKY3. *Nanotechnology* 14(1): 95.

Kröger, N., Deutzmann, R., Sumper, M. 1999. Polycationic peptides from diatom biosilica that direct silica nanosphere formation. *Science* 286(5442): 1129–11.

Krumov, N., Oder, S., Perner-Nochta, I., Angelov, A., Posten, C. 2007. Accumulation of CdS nanoparticles by yeasts in a fed-batch bioprocess. *Journal of Biotechnology* 132(4): 481–486.

Kumar, S., Amutha, R., Arumugam, P., Berchmans, S. 2011. Synthesis of gold nanoparticles: an ecofriendly approach using *Hansenula anomala*. *ACS Applied Materials and Interfaces* 3(5): 1418–1425.

Kumar, U., Shete, A., Harle, A.S., Kasyutich, O., Schwarzacher, W., Pundle, A., Poddar, P. 2008. Extracellular bacterial synthesis of protein-functionalized ferromagnetic Co_3O_4 nanocrystals and imaging of self-organization of bacterial cells under stress after exposure to metal ions. *Chemistry of Materials* 20(4): 1484–1491.

Kuyucak, N., Volesky, B. 1988. Biosorbents for recovery of metals from industrial solutions. *Biotechnology Letters* 10(2): 137–142.

Lang, C., Schüler, D., Faivre, D. 2007. Synthesis of magnetite nanoparticles for bio-and nanotechnology: genetic engineering and biomimetics of bacterial magnetosomes. *Macromolecular Bioscience* 7(2): 144–151.

Lee, S.-W., Mao, C., Flynn, C.E., Belcher, A.M. 2002. Ordering of quantum dots using genetically engineered viruses. *Science* 296(5569): 892–895.

Li, X.-Z., Nikaido, H., Williams, K.E. 1997. Silver-resistant mutants of *Escherichia coli* display active efflux of Ag^+ and are deficient in porins. *Journal of Bacteriology* 179(19): 6127–6132.

Lovley, D.R., Stolz, J.F., Nord, G.L., Jr., Phillips, E.J.P. 1987. Anaerobic production of magnetite by a dissimilatory iron-reducing microorganism. *Nature* 330: 252–254.

Manivasagan, P., Venkatesan, J., Senthilkumar, K., Sivakumar, K., Kim, S.-K. 2013. Biosynthesis, antimicrobial and cytotoxic effect of silver nanoparticles using a novel *Nocardiopsis* sp. MBRC-1. *BioMed Research International* 2013: 1–9.

Mann, S. 1993. Molecular tectonics in biomineralization and biomimetic materials chemistry. *Nature* 365(6446): 499–505.

Mann, S. 1996. Biominerilization and biomimetic materials chemistry. In *Biomimetic Materials Chemistry* (ed. S. Mann), pp. 1–40. New York: Wiley.

Mann, S., Shenton, W., Li, M., Connolly, S., Fitzmaurice, D. 2000. Biologically programmed nanoparticle assembly. *Advanced Materials* 12(2): 147–150.

Mao, C., Flynn, C.E., Hayhurst, A., Sweeney, R., Qi, J., Georgiou, G., Iverson, B., Belcher, A.M. 2003. Viral assembly of oriented quantum dot nanowires. *Proceedings of the National Academy of Sciences* 100(12): 6946–6951.

Mao, C., Solis, D.J., Reiss, B.D., Kottmann, S.T., Sweeney, R.Y., Hayhurst, A., Georgiou, G., Iverson, B., Belcher, A.M. 2004. Virus-based toolkit for the directed synthesis of magnetic and semiconducting nanowires. *Science* 303(5655): 213–217.

Marimuthu, S., Rahuman, A.A., Kirthi, A.V., Santhoshkumar, T., Jayaseelan, C., Rajakumar, G. 2013. Eco-friendly microbial route to synthesize cobalt nanoparticles using *Bacillus thuringiensis* against malaria and dengue vectors. *Parasitology Research*, 112(12): 4105–4112.

Marshall, M.J., Beliaev, A.S., Dohnalkova, A.C., Kennedy, D.W., Shi, L., Wang, Z., Boyanov, M.I., Lai, B., Kemner, K.M., McLean, J.S. 2006. C-Type cytochrome–dependent formation of U (IV) nanoparticles by *Shewanella oneidensis*. *PLoS Biology* 4(8): 1324–1333.

McLean, R.J.C., Beveridge, T.J. 1990. Metal-binding capacity of bacterial surfaces and their ability to form mineralized aggregates. In *Microbial Mineral Recovery* (eds H.L. Ehrlich and C.L. Brierley), pp. 185–222. New York: McGraw–Hill.

Mehra, R.K., Winge, D.R. 1991. Metal ion resistance in fungi: molecular mechanisms and their regulated expression. *Journal of Cellular Biochemistry* 45(1): 30–40.

Mie, G. 1908. Contributions to the optics of turbid media, particularly of colloidal metal solutions. *Annalen der Physik* 25(3): 377–445.

Mirkin, C.A., Letsinger, R.L., Mucic, R.C., Storhoff, J.J. 1996. A DNA-based method for rationally assembling nanoparticles into macroscopic materials. *Nature* 382: 607–609.

Mishra, A., Tripathy, S.K., Yun, S.I. 2011. Bio-synthesis of gold and silver nanoparticles from *Candida guilliermondii* and their antimicrobial effect against pathogenic bacteria. *Journal of Nanoscience and Nanotechnology* 11(1): 243–248.

Mourato, A., Gadanho, M., Lino, A.R., Tenreiro, R. 2011. Biosynthesis of crystalline silver and gold nanoparticles by extremophilic yeasts. *Bioinorganic Chemistry and Applications* 2011: 1–8.

Mukherjee, P., Ahmad, A., Mandal, D., Senapati, S., Sainkar, S.R., Khan, M.I., Ramani, R., Parischa, R., Ajayakumar, P.V., Alam, M., Sastry, M., Kumar, R. 2001a. Bioreduction of $AuCl_4^-$ ions by the fungus, *Verticillium* sp. and surface trapping of the gold nanoparticles formed. *Angewandte Chemie International Edition* 40(19): 3585–3588.

Mukherjee, P., Ahmad, A., Mandal, D., Senapati, S., Sainkar, S.R., Khan, M.I., Ramani, R., Parischa, R., Ajaykumar, P.V., Alam, M., Kumar, R., Sastry, M. 2001b. Fungus-mediated synthesis of silver nanoparticles and their immobilization in the mycelial matrix: a novel biological approach to nanoparticle synthesis. *Nano Letters* 1(10): 515–519.

Mukherjee, P., Roy, M., Mandal, B., Dey, G., Mukherjee, P., Ghatak, J., Tyagi, A., Kale, S. 2008. Green synthesis of highly stabilized nanocrystalline silver particles by a non-pathogenic and agriculturally important fungus *T. asperellum*. *Nanotechnology* 19(7): 075103.

Murphy, C.J. 2002. Nanocubes and nanoboxes. *Science* 298(5601): 2139–2141.

Murray, C.B., Kagan, C.R., Bawendi, M.G. 2000. Synthesis and characterization of monodisperse nanocrystals and close-packed nanocrystal assemblies. *Annual Review of Materials Science* 30(1): 545–610.

Nair, B., Pradeep, T. 2002. Coalescence of nanoclusters and formation of submicron crystallites assisted by *Lactobacillus* strains. *Crystal Growth and Design* 2(4): 293–298.

Nangia, Y., Wangoo, N., Sharma, S., Wu, J.-S., Dravid, V., Shekhawat, G., Raman Suri, C. 2009. Facile biosynthesis of phosphate capped gold nanoparticles by a bacterial isolate *Stenotrophomonas maltophilia*. *Applied Physics Letters* 94(23): 233901–233903.

Narayanan, K.B., Sakthivel, N. 2010. Biological synthesis of metal nanoparticles by microbes. *Advances in colloid and Interface Science* 156(1): 1–13.

Niemeyer, C.M. 2001. Nanoparticles, proteins, and nucleic acids: biotechnology meets materials science. *Angewandte Chemie International Edition* 40(22): 4128–4158.

Okazaki, S., Moers, J. 2005. Lithography. In *Nanoelectronics and Information Technology* (ed. R. Waser), pp. 221–247. Weinheim, Germany: Wiley VCH.

Oliver, S., Kuperman, A., Coombs, N., Lough, A., Ozin, G.A. 1995. Lamellar aluminophosphates with surface patterns that mimic diatom and radiolarian microskeletons. *Nature* 378(6552): 47–50.

Ortiz, D.F., Ruscitti, T., McCue, K.F., Ow, D.W. 1995. Transport of metal-binding peptides by HMT1, a fission yeast ABC-type vacuolar membrane protein. *Journal of Biological Chemistry* 270(9): 4721–4728.

Parikh, R.Y., Singh, S., Prasad, B., Patole, M.S., Sastry, M., Shouche, Y.S. 2008. Extracellular synthesis of crystalline silver nanoparticles and molecular evidence of silver resistance from *Morganella* sp.: towards understanding biochemical synthesis mechanism. *ChemBioChem* 9(9): 1415–1422.

Pazirandeh, M., Baral, S., Campbell, J. 1992. Metallized nanotubules derived from bacteria. *Biomimetics* 1(1): 41–50.

Poulsen, N., Sumper, M., Kröger, N. 2003. Biosilica formation in diatoms: characterization of native silaffin-2 and its role in silica morphogenesis. *Proceedings of the National Academy of Sciences* 100(21): 12075–12080.

Pum, D., Sleytr, U.B. 1999. The application of bacterial S-layers in molecular nanotechnology. *Trends in Biotechnology* 17(1): 8–12.

Rafii, F., Hehman, G., Shahverdi, A. 2005. Factors affecting nitroreductase activity in the biological reduction of nitro compounds. *Current Enzyme Inhibition* 1(3): 223–230.

Rautaray, D., Sanyal, A., Adyanthaya, S.D., Ahmad, A., Sastry, M. 2004. Biological synthesis of strontium carbonate crystals using the fungus *Fusarium oxysporum*. *Langmuir* 20(16): 6827–6833.

Reese, R., Winge, D. 1988. Sulfide stabilization of the cadmium-gamma-glutamyl peptide complex of *Schizosaccharomyces pombe*. *Journal of Biological Chemistry* 263(26): 12832–12835.

Riddin, T., Govender, Y., Gericke, M., Whiteley, C. 2009. Two different hydrogenase enzymes from sulphate-reducing bacteria are responsible for the bioreductive mechanism of platinum into nanoparticles. *Enzyme and Microbial Technology* 45(4): 267–273.

Salata, O.V. 2004. Applications of nanoparticles in biology and medicine. *Journal of Nanobiotechnology* 2(1): 3.

Samadi, N., Golkaran, D., Eslamifar, A., Jamalifar, H., Fazeli, M.R., Mohseni, F.A. 2009. Intra/extracellular biosynthesis of silver nanoparticles by an autochthonous strain of *Proteus mirabilis* isolated from photographic waste. *Journal of Biomedical Nanotechnology* 5(3): 247–253.

Schmid, G., Bäumle, M., Geerkens, M., Heim, I., Osemann, C., Sawitowski, T. 1999. Current and future applications of nanoclusters. *Chemical Society Reviews* 28(3): 179–185.

Shahverdi, A.R., Minaeian, S., Shahverdi, H.R., Jamalifar, H., Nohi, A.A. 2007. Rapid synthesis of silver nanoparticles using culture supernatants of Enterobacteria: A novel biological approach. *Process Biochemistry* 42(5): 919–923.

Shaligram, N.S., Bule, M., Bhambure, R., Singhal, R.S., Singh, S.K., Szakacs, G., Pandey, A. 2009. Biosynthesis of silver nanoparticles using aqueous extract from the compactin producing fungal strain. *Process Biochemistry* 44(8): 939–943.

Shankar, S.S., Ahmad, A., Pasricha, R., Sastry, M. 2003. Bioreduction of chloroaurate ions by geranium leaves and its endophytic fungus yields gold nanoparticles of different shapes. *Journal of Materials Chemistry* 13(7): 1822–1826.

Shenton, W., Pum, D., Sleytr, U.B., Mann, S. 1997. Synthesis of cadmium sulphide superlattices using self-assembled bacterial S-layers. *Nature* 389(6651): 585–587.

Shenton, W., Douglas, T., Young, M., Stubbs, G., Mann, S. 1999. Inorganic-organic nanotube composites from template mineralization of tobacco mosaic virus. *Advanced Materials* 11(3): 253–256.

Simkiss, K., Wilbur, K.M. (1989). *Biomineralization*. San Diego: Elsevier.

Simmons, S.L., Bazylinski, D.A., Edwards, K.J. 2006. South-seeking magnetotactic bacteria in the Northern Hemisphere. *Science* 311(5759): 371–374.

Singh, S., Vidyarthi, A.S., Nigam, V.K., Dev, A. 2014. Extracellular facile biosynthesis, characterization and stability of gold nanoparticles by *Bacillus licheniformis*. *Artificial Cells Nanomedicine and Biotechnology* 42(1): 6–12.

Sleytr, U.B., Messner, P., Pum, D., Sara, M. 1999. Crystalline bacterial cell surface layers (S layers): from supramolecular cell structure to biomimetics and nanotechnology. *Angewandte Chemie International Edition* 38(8): 1034–1054.

Slocik, J.M., Naik, R.R., Stone, M.O., Wright, D.W. 2005. Viral templates for gold nanoparticle synthesis. *Journal of Materials Chemistry* 15(7): 749–753.

Southam, G., Beveridge, T.J. 1994. The in vitro formation of placer gold by bacteria. *Geochimica et Cosmochimica Acta* 58(20): 4527–4530.

Southam, G., Beveridge, T.J. 1996. The occurrence of sulfur and phosphorus within bacterially derived crystalline and pseudocrystalline octahedral gold formed in vitro. *Geochimica et Cosmochimica Acta* 60(22): 4369–4376.

Spring, S., Schleifer, K.-H. 1995. Diversity of magnetotactic bacteria. *Systematic and Applied Microbiology* 18(2): 147–153.

Sun, S., Murray, C., Weller, D., Folks, L., Moser, A. 2000. Monodisperse FePt nanoparticles and ferromagnetic FePt nanocrystal superlattices. *Science* 287(5460): 1989–1992.

Sweeney, R.Y., Mao, C., Gao, X., Burt, J.L., Belcher, A.M., Georgiou, G., Iverson, B.L. 2004. Bacterial biosynthesis of cadmium sulfide nanocrystals. *Chemistry and Biology* 11(11): 1553–1559.

Syed, A., Ahmad, A. 2012. Extracellular biosynthesis of platinum nanoparticles using the fungus *Fusarium oxysporum*. *Colloids and Surfaces B: Biointerfaces* 97: 27–31.

Taniguchi, N. 1974. On the basic concept of nano-technology. *International Conference on Production Engineering*. Tokyo, Japan: Society of Precision Engineering.

Wang, T., Yang, L., Zhang, B., Liu, J. 2010. Extracellular biosynthesis and transformation of selenium nanoparticles and application in H_2O_2 biosensor. *Colloids and Surfaces B: Biointerfaces* 80(1): 94–102.

Wen, L., Lin, Z., Gu, P., Zhou, J., Yao, B., Chen, G., Fu, J. 2009. Extracellular biosynthesis of monodispersed gold nanoparticles by a SAM capping route. *Journal of Nanoparticle Research* 11(2): 279–288.

Windt, W.D., Aelterman, P., Verstraete, W. 2005. Bioreductive deposition of palladium (0) nanoparticles on *Shewanella oneidensis* with catalytic activity towards reductive dechlorination of polychlorinated biphenyls. *Environmental Microbiology* 7(3): 314–325.

Wong, K.K., Douglas, T., Gider, S., Awschalom, D.D., Mann, S. 1998. Biomimetic synthesis and characterization of magnetic proteins (magnetoferritin). *Chemistry of Materials* 10(1): 279–285.

Yong, P., Rowson, N.A., Farr, J.P.G., Harris, I.R., Macaskie, L.E. 2002. Bioreduction and biocrystallization of palladium by *Desulfovibrio desulfuricans* NCIMB 8307. *Biotechnology and Bioengineering* 80(4): 369–379.

Zhang, H., Li, Q., Lu, Y., Sun, D., Lin, X., Deng, X., He, N., Zheng, S. 2005. Biosorption and bioreduction of diamine silver complex by *Corynebacterium*. *Journal of Chemical Technology and Biotechnology* 80(3): 285–290.

9

MICROBIAL DIVERSITY OF NANOPARTICLE BIOSYNTHESIS

Raveendran Sindhu, Ashok Pandey, and Parameswaran Binod

CSIR-National Institute for Interdisciplinary Science and Technology, Pappanamcode, Trivandrum, Kerala, India

9.1. INTRODUCTION

Nanotechnology involves designing and producing objects and structures smaller than 100 nm in size. Nanomaterials are an increasingly important product of nanotechnology, having wider applications in health care, electronics, cosmetics, and other areas. These materials can be of different shapes such as rods, tubes, or fibers, and the physical and chemical properties of these materials often differ from those of bulk materials.

The process for the synthesis of nanomaterials with defined dimensions, structures, and composition is one of the most important areas in nanotechnology. Chemical and physical synthesis are the most widely used methods for the synthesis of metal nanoparticles, which involves a number of steps taking place in liquid and gas phase under controlled reaction conditions. Moreover, the use of toxic chemicals limits their biomedical applications. The development of a non-toxic and eco-friendly method for the synthesis of nanoparticles is therefore of utmost importance. One method of tackling this issue is the use of biological synthesis of these particles using microorganisms.

9.2. MICROBIAL-MEDIATED NANOPARTICLES

Several microbes are known to produce enzymes that convert specific metal ions into their nano-size. The nanoparticles produced by a biogenic enzymatic process are far superior to those produced by chemical methods. Developments in the biological

---

Bio-Nanoparticles: Biosynthesis and Sustainable Biotechnological Implications, First Edition.
Edited by Om V. Singh.
© 2015 John Wiley & Sons, Inc. Published 2015 by John Wiley & Sons, Inc.

TABLE 9.1. List of Microorganisms Capable of Producing Various Nanoparticles

Type of nanoparticle	Microorganism	Reference
Gold	*Turbinaria conoides*	Rajeshkumar et al. (2013)
	Trichothecium sp.	Absar et al. (2005)
	Arthrobacter globiformis	Kalabegishvili et al. (2012)
	Klebsiella pneumoniae	Malarkodi et al. (2013)
Silver	*Pseudomonas stutzeri*	Kumar et al. (2007)
	Bacillus subtilis	Saifuddin et al. (2009)
	Lactobacillus acidophilus	Mohseniazar et al. (2011)
	Idiomarina sp.	Seshadri et al. (2012)
	Cochiobolus lunatus	Salunkhe et al. (2011)
Selenium	*Bacillus cereus*	Dhanjal and Cameotra (2010)
	Klebsiella pneumoniae	Fesharaki et al. (2010)
Silica	*Acinetobacter* sp.	Singh et al. (2008)
Cadmium	*E. coli*	Kang et al. (2008)
	Schizosaccharomyces pombe	Kowshik et al. (2002a)
	Klebsiella pneumoniae	Mousavi et al. (2012)
Palladium	*Geobacter sulfurreducens*	Coker et al. (2010)
	Citrobacter braakii	Hennebel et al. (2011)
	Staphylococcus aureus	Vani et al. (2011)
	Aspergillus aeneus	Jain et al. (2012)
Lead	*Aspergillus* sp.	Pavani et al. (2012)
	Torulopsis sp.	Kowshik et al. (2002b)
	Klebsiella aerogenes	Holmes et al. (1995)
Iron	*Aspergillus oryzae*	Tarafdar and Raliya (2013)
Copper	*Pseudomonas stutzeri*	Varshney et al. (2010)
	Streptomyces sp.	Usha et al. (2010)
	Penicillium sp.	Honary et al. (2012)
Cerium	*Staphylococcus aureus*	Negahdary et al. (2012)

synthesis of nanoparticles on a commercial scale are still in their infancy but attracting the attention of material scientists throughout the world. The microorganism-assisted synthesis of nanoparticles is a safe and economical process. The following sections review the developments in microbial-mediated metal nanoparticles. Table 9.1 lists the various microorganisms which produce nanoparticles.

9.2.1. Gold

Gold nanomaterials (AuNMs) have attracted attention in biomedicine due to their unique optical and electrical properties, high chemical and thermal stability, and good biocompatibility and potential applications in various life-sciences-related applications including biosensing, bioimaging, drug delivery for cancer diagnosis, and therapy (Jiang et al., 2012). The drawback of chemically synthesized AuNMs (i.e., toxic by-products) necessitates the development of greener methods for nanoparticle synthesis (Tikariha et al., 2012). Covalently modified gold nanoparticles have attracted a great deal of interest as drug delivery vehicles. Their predictable and reliable surface modification chemistry, usually through gold–thiol binding, makes the desired functionalization of nanoparticles quite possible and accurate. A variety of therapeutic molecules have been attached in this manner, including various oligonucleotides for gene therapy, bactericidal compounds, and anticancer drugs.

Rajeshkumar et al. (2013) reported synthesis of gold nanoparticles by marine brown algae *Turbinaria conoides*. The nanoparticles showed antibacterial activity against *Streptococcus* sp., *B. subtilis*, and *Klebsiella pneumoniae*. The extracellular and intracellular biosynthesis of gold nanoparticles by fungus *Trichothecium* sp. was reported by Absar et al. (2005). The gold ions react with the *Trichothecium* sp. fungal biomass under stationary conditions, resulting in the rapid extracellular formation of gold nanoparticles; reaction of the biomass under shaking conditions resulted in intracellular growth of the gold nanoparticles. The biosynthesis of gold nanoparticles using marine alga *Sargassum wightii* was reported by Singaravelu et al. (2007). The stable gold nanoparticles were obtained by reduction of aqueous $AuCl_4^-$ ions by extract of marine algae.

The synthesis of gold nanoparticles by two novel strains of *Arthrobacter* sp. 61B and *Arthrobacter globiformis* 151B isolated from basalt rocks in Georgia was studied by Kalabegishvili et al. (2012). It was shown tha, the extracellular formation of nanoparticles took place after 1.5–2 days. The results indicate that the concentration of gold accumulated by bacterial biomass grows rapidly at the beginning, followed by an insignificant increase during the following few days. He et al. (2007) demonstrated that the bacteria *Rhodopseudomonas capsulata* are capable of producing gold nanoparticles extracellularly, and that these gold nanoparticles are quite stable in solution. Gericke and Pinches (2006) reported intracellular gold nanoparticle production by *Pseudomonas stutzeri* NCIMB 13420, *Bacillus subtilis* DSM 10, and *Pseudomonas putida* DSM 291.

Chauhan et al. (2011) reported the biogenic synthesis of gold nanoparticles employing a cytosolic extract of *C. albicans*, which was simple, economically viable, and environmentally friendly. The study revealed that the shape and size of the nanoparticles formed govern the characteristic features of their spectra. The technique can be extended for rapid, specific, and cost-effective detection of various cancers, hormones, pathogenic microbes, and their toxins if a specific antibody is available.

In a study conducted by Malarkodi et al. (2013) the extracellular biosynthesis of gold nanoparticles was achieved by an easy biological procedure using *Klebsiella pneumoniae* as the reducing agent. After exposing the gold ions to *K. pneumoniae*, rapid reduction of gold ions is observed and leads to the formation of gold nanoparticles in colloidal solution. The method exploits a cheap and easily available biomaterial for the synthesis of metallic nanoparticles (Malarkodi et al., 2013). *Pseudomonas fluorescens* was capable of producing gold nanoparticles extracellularly and were quite stable in the solution.

Applications of such eco-friendly nanoparticles in bactericidal, wound-healing, and other medical and electronic applications makes this method potentially exciting for the large-scale synthesis of other nanomaterials (Rajasree and Suman, 2012). An efficient, simple, and environmentally friendly biosynthesis of gold nanoparticles (AuNPs) mediated by fungal proteins of *Coriolus versicolor* was reported by Sanghi and Verma (2010). The size of the gold nanoparticles using atomic force microscopy (AFM) studies was found to be in the range 5–30 nm. These nanoparticles were found to be highly stable as, even after prolonged storage for over 6 months, they did not show aggregation. This study represents an important advancement in the use of fungal protein for the extracellular synthesis of functional gold nanoparticles by an eco-friendly technique.

Marine microorganisms are unique in their ability to tolerate high salt concentrations and can evade toxicity of different metal ions. However, these marine microbes have not been sufficiently explored for their metal nanoparticle synthesis ability. Sharma et al. (2012) reported the possibility of using the marine bacterial strain of *Marinobacter pelagius* to achieve a fast rate of nanoparticles synthesis, which may be of interest for future process development of AuNPs. This is the first report of AuNP synthesis by marine bacteria. The "green" route of biosynthesis of gold nanoparticles involves a simple, economically viable, and eco-friendly process, which offers a great advantage over an intracellular process of synthesis from the point of view of applications in medicine, catalysis, and electronics.

9.2.2. Silver

Silver nanoparticles synthesis by microbes is as a result of their defense mechanism. The resistance caused by the bacterial cell for silver ions in the environment is responsible for its nanoparticle synthesis. The silver ions in nature are highly toxic to the bacterial cells. The cellular machinery helps in the conversion of reactive silver ions to stable silver atoms. Parameters such as temperature and pH play an important role in their production. More nanoparticles are synthesized by microbes under alkaline conditions than under acidic conditions. After pH 10, cell deaths occur however (Saklani and Jain, 2012). The first evidence of the synthesis is from *Pseudomonas stutzeri* AG259, a bacterial strain that was originally isolated from silver mine (Kumar et al., 2007).

Gupta et al. (2008) revealed that the *in situ* formation of Ag nanoparticles in a grafted polymer network of cotton fabric was an effective method for the preparation of antibacterial fabrics. The uniform distribution of narrow dispersed Ag nanoparticles is a major advantage of this method. These fabrics show biocidal action against the bacteria *E. coli*, thus showing great potential for its use as an antiseptic dressing or bandage, products which are in high demand for biomedical applications. Saifuddin et al. (2009) describe a novel combinatorial synthesis approach which is rapid, simple, and "green" for the synthesis of metallic nanostructures of noble metals such as silver (Ag). They uses a combination of culture supernatant of *Bacillus subtilis* and microwave irradiation in water in the absence of a surfactant or soft template. It was found that exposure of culture supernatant of *Bacillus subtilis* and microwave irradiation to silver ions lead to the formation of silver nanoparticles. The formation of nanoparticles by this method is a rapid process and requires no toxic chemicals, and the nanoparticles are stable for several months.

Silver nanoparticle production from the three algal species *Nannochloropsis oculata*, *Dunaliella salina*, and *Chlorella vulgaris* and three Lactobacilli including *L. acidophilus*, *L. casei*, and *L. reuteri* were evaluated by Mohseniazar et al. (2011). The study revealed that all three Lactobacilli were found to be promising in the production of silver nanoparticles.

Extracellular biosynthesis of highly stabilized AgNPs using *R. stolonifer* and the antibacterial effect of biosynthesized AgNPs against *P. aeruginosa* (MDR strain) were reported (Rathod and Ranganath, 2011). Extracellular synthesis of silver nanoparticles by *Klebsiella pneumoniae*, *Escherichia coli*, and *Enterobacter cloacae* was reported by Minaeian et al. (2008).

Among fungi, *Aspergillus* sp. is the most cost-effective source of biomaterial for biosynthesis of silver nanoparticles. In the study carried out by Sundaramoorthi et al. (2009), silver nanoparticles were synthesized extracellularly using *Aspergillus niger* and evaluated for their wound-healing properties. This study shows that the silver nanoparticles synthesized from *Aspergillus niger* possess effective wound-healing activity when compared with silver nitrate.

The synthesis of intracellular AgNPs by a marine bacterium *Idiomarina* sp. PR58-8 was reported by Seshadri et al. (2012). This is the first report on heavy metal resistance and synthesis of metal nanoparticles in the *Idiomarina* genus. The use of *Cochliobolus lunatus* cell mass for accumulation of silver and subsequent formation of silver nanoparticles is a promising new approach. This approach of biological synthesis has advantages over other methods as it ensures specificity in the formation of silver nanoparticles, their size, shape, uniform crystallographic orientation, monodispersity, and maximum stability (Salunkhe et al., 2011).

9.2.3. Selenium

Selenium possesses several applications in medicine (e.g., anti-cancer), chemistry, and electronics. The effects of selenium nanoparticles on microorganisms remain largely unknown to date.

Dobias et al. (2011) evaluated the potential for harnessing the association of bacterial proteins to biogenic selenium nanoparticles (SeNPs) to control the size distribution and the morphology of the resultant SeNPs. Proteomic studies were carried out, comparing proteins associated with biogenic SeNPs produced by *E. coli* to chemically synthesized SeNPs as well as magnetite nanoparticles. The study revealed that four proteins (AdhP, Idh, OmpC, AceA) bound specifically to SeNPs, and observed a narrower size distribution as well as more spherical morphology when the particles were synthesized chemically in the presence of proteins. A more detailed study of AdhP (alcohol dehydrogenase propanol-preferring) confirmed the strong affinity of this protein for the SeNP surface. It was also revealed that this protein controlled the size distribution of the SeNPs and yielded a narrow size distribution with a three-fold decrease in the median size. These results confirm that protein is an important tool in the industrial-scale synthesis of SeNPs of uniform size and properties.

Biosynthesis of amorphous SeO nanospheres under aerobic conditions offers advantages over chemical processes, in which amorphous SeO is produced under environmentally harmful conditions. Dhanjal and Cameotra (2010) produced nanoparticles by *Bacillus cereus*, and produce high levels of selenium oxyions and extracellular nanospheres of selenium which can be easily separated from the biomass by a simple centrifugation step. This green route of biosynthesis of selenium nanospheres is a simple, economically viable, and an eco-friendly process resulting in nearly monodispersed highly stable selenium nanospheres. Fesharaki et al. (2010) first reported the biogenesis of selenium nanoparticles using *Klebsiella pneumoniae*. A wet sterilization process was used to disrupt the bacterial cells containing selenium particles, and these nanoparticles were chemically stable during the sterilization process.

9.2.4. Silica

Silica nano-crystals play an important role in materials such as resins, catalysts, and molecular sieves. The chemical synthesis of silica and silica-based materials are relatively expensive and eco-hazardous, often requiring extremes of temperature, pressure, and pH. There has therefore been increasing demand to develop more environmentally friendly processes to make these materials. In this context, a biotechnological process such as toxic metal remediation that uses microorganisms such as bacteria and yeast seems promising (Stephen and Maenaughton, 1999).

Singh et al. (2008) demonstrated the synthesis of silicon/silica nanoparticle composites by the bacterium *Acinetobacter* sp. The formation of silicon/silica nanocomposite is shown to occur when the bacterium is exposed to K2SiF6 precursor under ambient conditions. This bacterium has been shown to synthesize iron oxide and iron sulfide nanoparticles, and it is hypothesized that this bacterium secretes reductases and oxidizing enzymes which lead to Si/SiO_2 nanocomposite synthesis. The synthesis of silica nanoparticles by bacteria demonstrates the versatility of the organism, and the formation of elemental silicon by this environmentally friendly process expands further the scope of microorganism-based nanomaterial synthesis. The proteins responsible for such formations have been identified.

9.2.5. Cadmium

Semiconductor nano-crystals which have unique optical, electronic, and optoelectronic properties have potential application in the emerging field of nano-electronics. Kang et al. (2008) reported phytochelatin-mediated intracellular synthesis of CdS nano-crystals in engineered *E. coli*. By controlling the population of the capping PCs, *E. coli* cells were engineered as an eco-friendly biofactory to produce uniformly sized PC-coated CdS nano-crystals. This is the first systematic approach toward the tunable synthesis of semiconductor nano-crystals by genetically engineering bacteria. *Schizosaccharomyces pombe* produced CdS nano-particles intracellularly (Kowshik et al., 2002a). The first report on the production of semiconductor nano-crystal synthesis in bacteria was published by Sweeney et al. (2004). The study revealed that *E. coli* has the endogenous ability to direct the growth of nano-crystals. Parameters such as growth phase and strain type are essential for initiating nano-crystal growth.

Cell-associated biosynthesis of cadmium sulfide (CdS) nano-particles has been reported to be rather slow and costly. El-Raheen et al. (2012) reported a rapid and low-cost biosynthesis of CdS using culture supernatants of *Escherichia coli* ATCC 8739, *Bacillus subtilis* ATCC 6633, and *Lactobacillus acidophilus* DSMZ 20079T. The CdS nano-particles synthesis were performed at room temperature and were formed within 24h. The process of extracellular and fast biosynthesis may help in the development of an easy and eco-friendly route for the synthesis of CdS nanoparticles.

The synthesis of CdS nanoparticles using *Escherichia coli* PTCC 1533 and *Klebsiella pneumoniae* PTCC 1053 was reported by Mousavi et al. (2012). The synthesis occurred after 96h of incubation at room temperature (30°C) and pH 9. The nanoparticles are found to be polydisperse in the size range 5–200 nm.

9.2.6. Palladium

Precious metals supported on ferrimagnetic particles have a diverse range of uses in catalysis. Coker et al. (2010) demonstrated a novel biotechnological route for the synthesis of a heterogeneous catalyst, consisting of reactive palladium nanoparticles arrayed on a nanoscale biomagnetite support. The magnetic support was synthesized at ambient temperature by the Fe (III)-reducing bacterium, *Geobacter sulfurreducens*. The palladium nanoparticles were deposited on the nanomagnetite using a simple one-step method due to an organic coating priming the surface for Pd adsorption, which was produced by the bacterial culture during the formation of the nanoparticles.

A new biological method to produce nanopalladium is the precipitation of Pd on a bacterium. This bio-Pd can be applied as catalyst in dehalogenation reactions. Large amounts of hydrogen are required as electron donors in these reactions, resulting in considerable cost. The study carried out by Hennebel et al. (2011) demonstrates how bacteria is cultivated under fermentative conditions and can be used to reductively precipitate bio-Pd catalysts and generate the electron donor hydrogen. This could avoid the costs coupled to hydrogen supply. Batch reactors with nanoparticles formed by *Citrobacter braakii* showed the highest diatrizoate dehalogenation activity.

9.2.7. Zinc

In prokaryotic systems, cell death due to interactions between reactive oxygen species (ROS) and proteins, DNA, or membrane structures can be induced by oxidative stress. There are concerns of toxicity and safety issues regarding the expanding growth of nanotechnology and nano-biotechnology, and related industrial products. Due to their wide range of practical applications including their use in sunscreens and cosmetics, these recent indications of the toxic nature of nanoscale metal oxides such as zinc oxide (ZnO) is a current focus of the Nanotechnology Safety Initiative under the National Institute of Environmental Health and Safety. Experimental observations have explained the antibacterial behavior of ZnO nanoparticles. Zinc oxide nanoparticles can be used as an effective biocide for Gram-positive bacteria *Staphylococcus aureus*. From the results obtained in the study conducted by Vani et al. (2011), it is understood that the proteins are the important biological molecules which are fundamental to the proper functioning of cells in the microorganisms.

Aspergillus aeneus isolate NJP12 has been shown to have a high zinc metal tolerance ability and a potential for extracellular synthesis of ZnO nanoparticles under ambient conditions (Jain et al., 2012). The study demonstrates that interactions between amino acids and metal ions are responsible for the synthesis of metal nanoparticles. This is the first study which proposes the involvement of simply amino acids in the fungal-mediated synthesis of metal nanoparticles.

Xie et al. (2011) demonstrated that ZnO nanoparticles exhibited remarkable antibacterial activity and lethal effect against *C. jejuni*, even at low concentrations. ZnO nanoparticles induced significant morphological changes, measurable membrane leakage, and substantial increases in oxidative stress gene expression in *C. jejuni*. The mechanism of ZnO inactivation of bacteria involves the direct interaction between ZnO nanoparticles and cell surfaces, which affects the permeability of membranes where nanoparticles enter and induce oxidative stress in bacterial cells, subsequently resulting in the inhibition of cell growth and eventually in cell death.

9.2.8. Lead

Responding to the current demand to develop green technologies in material synthesis, Pavani et al. (2012) reported the natural synthesis of lead particles by *Aspergillus* species. The fungal strain was grown in a medium containing different concentrations of lead (0.2–1.5 mM) to determine its resistance to heavy metals. The organism was found to utilize some mechanism and accumulate lead particles outside and inside the cell. The possible mechanism may be the trapping of the lead ions on the surface of the fungal cells via electrostatic attractions between lead ions and negatively charged carboxylate groups present in the cell wall of mycelia. This was followed by the entry of lead ions into the cell and then a reduction by the enzymes present in the cell wall and inside the cell. The results indicate that reductases or cytochromes that are present inside the cell and cell wall may be responsible for the synthesis of lead nanoparticles inside the cell and cell wall. Intracellular synthesis of nanoscale PbS crystallites when exposed to aqueous Pb^{2+} ions was reported by Kowshik et al. (2002*b*) for *Torulopsis* species and by Holmes et al. (1995) for *Klebsiella aerogenes*. Exposure to Cd ions resulted in the intracellular formation of CdS nanoparticles in the 20–200 nm size range. The biogenic process in *Aspergillus* species provide opportunities for: better management of bioremediation of contamination; better understanding and control over size and polydispersity of the nanoparticles; and better understanding of the biochemical and molecular mechanisms in the synthesis of the nanoparticles. This is the first report on the biogenesis of lead nanoparticles using *Aspergillus* species (Pavani et al., 2012).

9.2.9. Iron

Biologically induced and controlled mineralization mechanisms are the two modes through which the microorganisms synthesize iron oxide nanoparticles. In biologically induced mineralization (BIM) mode, environmental factors such as pH, pO2, pCO2, redox potential, and temperature govern the synthesis of iron oxide nanoparticles. In contrast, biologically controlled mineralization (BCM) processes initiate the microorganism itself to control the synthesis. BIM can be observed in the Fe(III)-reducing bacterial species of *Shewanella*, *Geobacter*, and *Thermoanaerobacter* and sulfate-reducing bacterial species of *Archaeoglobus fulgidus* and *Desulfuromonas acetoxidans*, whereas BCM mode can be observed in the magnetotactic bacteria (MTB) such as *Magnetospirillum magnetotacticum* and *M. gryphiswaldense* and sulfate-reducing magnetic bacteria (*Desulfovibrio magneticus*). Tarafdar and Raliya (2013) reported an efficient, eco-friendly green approach to synthesize iron nanoparticles using the fungi *Aspergillus oryzae* TFR9. These useful features of the biosynthesized iron nanoparticles may benefit in the agriculture, biomedical, and engineering sectors.

9.2.10. Copper

The study conducted by Varshney et al. (2010) reports the synthesis of copper nanoparticles which involves non-pathogenic bacteria *Pseudomonas stutzeri*, isolated from soil. These copper nanoparticles are further characterized for size and shape distributions by ultraviolet-visible spectroscopy, x-ray diffraction, and high-resolution transmission electron microscopy techniques. It is an easy, fast, and cost-effective technique and does not involve any harmful and environmentally toxic chemicals used in conventional chemical reduction methods. Aqueous solutions of Cu nanoparticles with very good stability have been synthesized. The biomolecules present in the biomass not only reduce the metal ions, but also stabilize the metal nanoparticles by preventing them from being oxidized after the preparation. Usha et al. (2010) demonstrated the synthesis of metal oxide nanoparticles by a *Streptomyces* sp. isolated from a Pichavaram mangrove site in India. It was found that when exposed to *Streptomyces* sp. copper sulfate and zinc nitrate are reduced in solution, thereby leading to the formation of metal oxide nanoparticles. The possible mechanism for the reduction of metal ions is via the reductase enzyme.

Honary et al. (2012) proposed a green process for the extracellular production of copper oxide nanoparticles. Copper NPs were synthesized and stabilized using *Penicillium aurantiogriseum*, *Penicillium citrinum*, and *Penicillium waksmanii* isolated from soil. The study revealed that some secreted proteins from fungi are capable of hydrolyzing metal precursors to form metal oxides extracellularly.

9.2.11. Cerium

Cerium oxide nanoparticles have found numerous applications in the biomedical industry due to their strong antioxidant properties. Industrial applications include its use as a polishing agent, ultraviolet absorbing compound in sunscreen, solid electrolytes in solid oxide fuel cells, as a fuel additive to promote combustion, and in automotive exhaust catalysts. In the biomedical industry CeO_2 NPs are gaining much attention because of their antioxidant properties. Their applications range from fighting inflammation and cancer to radiation protection of cells (Shah et al., 2012).

The antibacterial activities of different concentrations of CeO_2 nanoparticles were investigated by Negahdary et al. (2012). *Staphylococcus aureus* was used as the test organism. Growth inhibition results were observed when the bacterial cells were incubated with nanoparticles during the liquid and solid cultures. A hydrothermal synthesis approach was used to prepare cerium oxide nanoparticles of defined sizes in order to eliminate complications originating from the use of organic solvents and surfactants.

9.2.12. Microbial Quantum Dots

One of the fastest-growing and most exciting areas of nanotechnology is the use of quantum dots (QDs) in biology. QDs are nanoparticles that range from 2 to 100 nm in diameter and act both as semiconductors and fluorophores (Bruchez et al. 1998). Structurally, QDs are composed of a metal core that determines their color, an inorganic shell that helps to enhance stability and brightness, and a polymer layer or functional group that enhances water solubility and conjugation capacity. The unique optical properties of QDs make them appealing as *in vivo* and *in vitro* fluorophores in a variety of biological investigations, in which traditional fluorescent labels based on organic molecules fall short of providing long-term stability and simultaneous detection of multiple signals. The ability to make QDs water soluble and target them to specific biomolecules has led to promising applications in cellular labeling, deep-tissue imaging, assay labeling, and as efficient fluorescence resonance energy transfer donors (Medintz et al., 2005).

Mahendra et al. (2008) reported QD weathering and release of toxic core components following the degradation of surface coatings after exposure to moderate acidic and alkaline conditions. The results suggest that QDs may be safely used in a variety of applications at circum-neutral pH. The release of toxic inorganic constituents during their weathering under acidic or alkaline conditions in the human body or the environment may cause unintended harm that might be difficult to predict with only short-term toxicity tests. Long-term toxicity and biocompatibility tests that include QD transformations under various environmental conditions are therefore essential to guarantee safety in their intended applications without compromising risks to public or environmental health.

Zinc oxide quantum dots (ZnOQDs) are nanoparticles of purified powdered ZnO. Jin et al. (2009) evaluated for antimicrobial activity against *Listeria monocytogenes*, *Salmonella enteritidis*, and *Escherichia coli* O157:H7. The ZnOQDs were utilized as a powder, bound in a polystyrene film (ZnO-PS), or suspended in a polyvinylpyrrolidone gel (ZnO-PVP). Bacteria cultures were inoculated into culture media or liquid egg white (LEW) and incubated at 22 °C. The inhibitory capacity of ZnOQDs against three pathogens were concentration dependent and also related to type of application. The higher the concentration of ZnO used, the higher the antibacterial effect was achieved. This study compared the effectiveness of ZnO in powder, PVP-capped, film, and coating forms. The availability of ZnOQDs in media was important for antibacterial efficiency. The results of this study demonstrated several approaches (powder, PVP capped, film, and coating) for incorporation of ZnO into food systems.

A panel of QDs conjugated to molecules that label bacteria specifically according to strain, metabolism, surrounding conditions, or other factors would be extremely useful for a wide range of applications. One potential use is to study complex microbial populations, such as biofilms. Associations of microbes into biofilms result in properties that are very different from those of the individual cells, with resulting environmental, medical, and technological implications. QDs containing cadmium telluride (CdTe-QDs) offer great potential in therapeutic targeting and in medical and molecular imaging due to their spectral properties such as broad absorption.

Nguyen et al. (2013) examined the dose effects of cadmium telluride quantum dots (CdTe-QDs) from two commercial sources on model macrophages (J774A.1) and colonic epithelial cells (HT29). The effects on cellular immune signaling responses were measured following sequential exposure to QDs and *Pseudomonas aeruginosa* strain PA01. These results demonstrate that exposures to sub-toxic levels of CdTe-QDs can depress cell immune-defense functions; if these occurred *in vivo*, they would likely interfere with normal neutrophil recruitment for defense against bacteria. The study reveals that pre-exposure to CdTe-QDs might impair target cell metabolism and immune responses to bacteria, which in turn could result in elevated susceptibility of hosts to manage infection. The study also provides direction for more detailed study of QD molecular and sub-cellular interactions and other toxicity biomarkers.

9.2.13. Cadmium Telluride

CdS is used as a bioorganic detector of proteins or DNA and enhances luminescence properties. It has unique properties and is interesting for photoreactivity and photocatalyst applications. CdS nanoparticles were reported to be synthesized by several microorganisms, including *E. coli*, *S. cerevisiae*, *Lactobacillus* sp., *Coriolus versiclor*, *Rhodopseudomonas palustris*, *Fusarium oxysporum*, *Schizosaccharomyces pombe*, *Clostridium thermoaceticum*, and *Klebseilla pneumoniae* (Durán and Seabra, 2012). Some of these semiconductors, such as ZnS and CdS, have been used in Mn-doped nano-crystals with different synthetic routes and surface passivations. Besides their fundamental properties, these luminescent nano-crystals are now being tested for some applications such as electroluminescent displays and biological labeling agents or biomarkers (Yang et al., 2005).

9.2.14. Iron Sulfide-greigite

It has been shown that the bacterially produced iron sulfides are adsorbents for a wide range of heavy metals and some anions, which are generally chemisorbed. Indeed, biogenically produced iron sulfides play an important role in environmental decontamination. Strongly magnetic iron sulfide materials have been produced by using sulfate-reducing bacteria and a novel bioreactor. An intracellular biomineralization of ferrimagnetic iron sulfide (single crystal), greigite (Fe_3S_4), in a multicellular magnetotactic bacterium has been reported by Durán and Seabra (2012). The magnetotactic multicellular prokaryotes (MMPs) are unique magnetotactic bacteria of the *Deltaproteobacteria* class and the first found to biomineralize the magnetic mineral greigite (Fe_3S_4); however, the unique morphology of the MMP is not restricted to marine and magnetotactic prokaryotes.

9.3. NATIVE AND ENGINEERED MICROBES FOR NANOPARTICLE SYNTHESIS

Nanoparticles are biosynthesized when the microorganisms "grab" target ions from their environment and then turn the metal ions into the element metal through enzymes generated by the cell activities. It can be classified into intracellular and extracellular synthesis, according to the location where nanoparticles are formed. The intracellular method consists of transporting ions into the microbial cell to form nanoparticles in the presence of enzymes. The extracellular synthesis of nanoparticles involves trapping the metal ions on the surface of the cells and reducing ions in the presence of enzymes (Simkiss and Wilbur, 1989; Mann, 2001). A list of microorganisms which synthesize various nanoparticles is provided in Table 9.1.

Different microorganisms have different mechanisms of forming nanoparticles. The nanoparticles are usually formed when metal ions are first trapped on the surface or inside of the microbial cells. The trapped metal ions are then reduced to nanoparticles in the presence of enzymes. The microorganisms impact the mineral formation in two distinct ways. They can modify the composition of the solution so that it becomes supersaturated or more supersaturated than it previously was with respect to a specific phase. A second method by which microorganisms can impact mineral formation is through the production of organic polymers, which can impact nucleation by favoring the stabilization of the very first mineral seeds (Benzerara et al., 2010). The formation of heavy metallic nanoparticles can be attributed to the metallophilic microorganism's developed genetic and proteomic responses to toxic environments (Reith et al., 2007).

Chen et al. (2009) reported biosynthesis of CdS nanoparticles by *E. coli*. The study revealed that Glutathione is essential for CdS nanoparticle synthesis and requires two enzymes: glutamylcysteine synthase (c-GCS) and glutathione synthetase (GS). In this study two recombinant *E.coli* ABLE C strains were constructed for over-expression of c-GCS and GS. The results indicate that the c-GCS over-expression resulted in inclusion body formation and impaired cell physiology, whereas GS over-expression yielded abundant soluble proteins and barely impeded cell growth. The maximum particle yield attained in the recombinant *E. coli* was 2.5 times that attained in the wild-type cells, and considerably exceeded that achieved in yeasts. These data implied the potential of genetic engineering approach to enhancing CdS nanoparticle biosynthesis in bacteria. Additionally, *E. coli*-based biosynthesis offers a more energy-efficient and eco-friendly method since biosynthesis is carried out at 37 °C, eliminating the need for the high-temperature toxic solvents required for chemical processes.

9.4. COMMERCIAL ASPECTS OF MICROBIAL NANOPARTICLE SYNTHESIS

The unique properties of nanoparticles mean that they find their application in various fields from medical applications to environmental science. Applications also include in fabrics and their treatments, filtration, dental materials, surface disinfectants, diesel and fuel additives, hazardous chemical neutralizers, automotive components, electronics, scientific instruments, sports equipment, flat panel displays, drug delivery systems, and pharmaceutics. Medical applications are expected to increase our quality of life through early diagnosis and treatment of diseases. Ecological applications include removal of persistent pollutants from soil and water supplies.

Several nanoparticles are known for their antimicrobial effect which includes silver, titanium oxide, fullerenes, zinc oxide, and magnesium oxide. The antimicrobial activity of zinc oxide nanoparticles is achieved by disrupting membrane permeability and being internalized by the bacteria. Silver nanoparticles have broad spectrum antimicrobial activity and their antimicrobial action is due to the destabilization of the outer membrane as well as depletion of ATP. Metallic oxides find applications in the preparation of scratch-proof lens coatings.

Nanoparticles find applications in cosmetics, replacing black soot and the mineral powders which were used in cosmetics years ago. The properties of nanoparticles which find application in the cosmetic industry include their ability to penetrate deeper into the skin, thereby acting as skin nutrients. The antioxidant properties of some nanoparticles help to maintain the youthful nature of the skin. Fullerenes are used for the preparation of cosmetics with radical scavenging properties. Alumina nanoparticles find application in the preparation of anti-wrinkle cream. Nanoparticles are also used for the preparation of personal care products such as deodorants, shampoo, soap, toothpaste, hair conditioner, and sunscreen. Titanium dioxide has been widely used in the preparation of white pigment, food colorant, sunscreens, and cosmetic creams.

In electronic equipment, the use of nano-crystalline materials greatly enhances the resolution and significantly reduces the cost. Recent developments in biomedical applications of nanoparticles include preparation of nanoscaffolds, bioseparation, drug delivery, gene transfection, and magnetic resonance imaging. The surface and bulk properties of nanoparticles make them applicable for drug delivery, tumor destruction, tissue engineering and separation, and the purification of biomolecules and cells (Li et al., 2011).

9.5. CONCLUSION

The microbial-mediated synthesis of nanoparticles is a very attractive and exciting approach. Several microbes are reported to synthesize nanomaterials during their growth in the presence of particular culture media. Microbial-mediated synthesis has several advantages over physical and chemical methods and the process is also highly economical. Active research is underway to exploit more microbes and to make the process commercially viable.

REFERENCES

Absar, A., Satyajyoti, S., Khan, M.I., Rajiv, K., Sastry, M. 2005. Extra-/intracellular biosynthesis of gold nanoparticles by an alkalotolerant fungus, *Trichothecium* sp. *Journal of Biomedical Nanotechnology* 1: 47–53.

Benzerara, K., Miot, J., Morin, G., Ona-Nguema, G., Skouri-Panet, F., Ferard, C. 2010. Significance, mechanisms and environmental implications of microbial biomineralization. *Comptes Rendus Geoscience* 343: 160–167.

Bruchez, M. Jr, Moronne, M., Gin, P., Weiss, S., Alivisatos, A.P. 1998. Semiconductor nanocrystals as fluorescent biological labels. *Science* 281: 2013–2016.

Chauhan, A., Zubair, S., Tufail, S., Sherwani, A., Sajid, M., Raman, S.C., Azam, A. 2011. Fungus-mediated biological synthesis of gold nanoparticles: potential in detection of liver cancer. *International Journal of Nanomedicine* 6: 2305–2319.

Chen, Y., Tuan, H.Y., Tien, C.W., Lo, W.H., Liang, H.C., Hu, Y.C. 2009. Augmented biosynthesis of cadmium sulfide nanoparticles by genetically engineered *Escherichia coli*. *Biotechnology Progress* 25: 1260–1266.

Coker, V.S., Bennett, J.A., Telling, N.D. et al. 2010. Microbial engineering of nanoheterostructures: biological synthesis of a magnetically recoverable palladium nanocatalyst. *ACS Nano* 4: 2577–2584.

Dhanjal, S., Cameotra, S.S. 2010. Aerobic biogenesis of selenium nanospheres by *Bacillus cereus* isolated from coalmine soil. *Microbial Cell Factories* 9: 2–11.

Dobias, J., Suvorova, E.I., Bernier-Latmani, L. 2011. Role of proteins in controlling selenium nanoparticle size. *Nanotechnology* 22(19): 1–10.

Durán, N., Seabra, A.B. 2012. Microbial syntheses of metallic sulfide nanoparticles: an overview. *Current Biotechnology* 1: 287–296.

El-Raheem, A.R., El-Shanshoury, S., Elsilk, E., Ebeid, M.E. 2012. Rapid biosynthesis of cadmium sulfide (CdS) nanoparticles using culture supernatants of *Escherichia coli* ATCC 8739, *Bacillus subtilis* ATCC 6633 and *Lactobacillus acidophilus* DSMZ 20079 T. *African Journal of Biotechnology* 11: 7957–7965.

Fesharaki, P.J., Nazari, P., Shakibaie, M., Rezaie, S., Banoee, M., Abdollahi, M., Shahverdi, A.R. 2010. Biosynthesis of selenium nanoparticles using *Klebsiella pneumoniae* and their recovery by a simple sterilization process. *Brazilian Journal of Microbiology* 41: 461–466.

Gericke, M., Pinches, A. 2006. Biological synthesis of metal nanoparticles. *Hydrometallurgy* 83: 132–140.

Gupta, P., Bajpai, M., Bajpai, S.K. 2008. Investigation of antibacterial properties of silver nanoparticle-loaded poly (acrylamide-co-itaconic acid)-grafted cotton fabric. *Journal of Cotton Science* 12: 280–286.

He, S., Guo, Z., Zhang, Y., Zhang, S., Wang, J., Gu, N. 2007. Biosynthesis of gold nanoparticles using the bacteria *Rhodopseudomonas capsulate*. *Materials Letters* 61: 3984–3987.

Hennebel, T., Nevel, S.V., Verschuere, S. et al. 2011. Palladium nanoparticles produced by fermentatively cultivated bacteria as catalyst for diatrizoate removal with biogenic hydrogen. *Applied Microbiology and Biotechnology* 91: 1435–1445.

Holmes, J.D., Smith, P.R., Evans-Gowing, R., Richardson, D.J., Russell, D.A., Sodeau, J.R. 1995. Energy dispersive X-ray analysis of the extracellular cadmium sulfide crystallites of *Klebsiella aerogenes*. *Archives of Microbiology* 163, 143–147.

Honary, S., Barabadi, H., Gharaeifathabad, E., Naghibi, F. 2012. Green synthesis of copper oxide nanoparticles using *Penicillium aurantiogriseum, Penicillium citrinum* and *Penicillium waksmanii*. *Digest Journal of Nanomaterials and Biostructures* 7: 999–1005.

Jain, N., Bhargava, A., Tarafdar, J.C., Singh, S.K., Panwar, J. 2012. A biomimetic approach towards synthesis of zinc oxide nanoparticles. *Applied Microbiology and Biotechnology* 97: 859–869.

Jiang, X., Wang, L., Wang, J., Chen, C. 2012. Gold nanomaterials: Preparation, chemical modification, biomedical applications and potential risk assessment. *Applied Biochemistry and Biotechnology* 166: 1533–1551.

Jin, T., Sun, D., Su, J.Y., Zhang, H., Sue, H.J. 2009. Antimicrobial efficacy of zinc oxide quantum dots against *Listeria monocytogenes, Salmonella enteritidis*, and *Escherichia coli* O157:H7. *Journal of Food Science* 74: 46–52.

Kalabegishvili, T.L., Kirkesali, E.I., Rcheulishvili, A.N. et al. 2012. Synthesis of gold nanoparticles by some strains of *Arthrobacter* genera. *Materials Science and Engineering* 2: 164–173.

Kang, S.H., Bozhilov, K.N., Myung, N.V., Mulchandani, A., Chen, W. 2008. Microbial synthesis of CdS nanocrystals in genetically engineered *E. coli*. *Angewandte Chemie International Edition* 47: 5186–5189.

Kowshik, M., Deshmukh, N., Vogel, W., Urban, J., Kulkarni, S.K., Paknikar, K.M. 2002a. Microbial synthesis of semiconductor CdS nanoparticles, and their use in fabrication of an ideal diode. *Biotechnology and Bioengineering* 78: 583–588.

Kowshik, M., Vogel, W., Urban, J., Kulkarni, S.K., Paknikar, K.M. 2002b. Microbial synthesis of semiconductor PbS nanocrystallites. *Advanced Materials* 14: 815–818.

Kumar, A.S., Abyaneh, M.K., Gosavi, S.W., Kulkarni, S.K., Pasricha, R. 2007. Nitrate reductase-mediated synthesis of silver nanoparticles from $AgNO_3$. *Biotechnology Letters* 29: 439–445.

Li, X., Xu, H., Chen, Z.S., Chen, G. 2011. Biosynthesis of nanoparticles by microorganisms and their applications. *Journal of Nanomaterials* 1: 1–16.

Mahendra, S., Colvin, H.V., Alvarez, P. 2008. Quantum dot weathering results in microbial toxicity. *Environmental Science and Technology* 42: 9424–9430.

Malarkodi, C., Rajeshkumar, S., Vanaja, M., Paulkumar, K., Gnanajobitha, G., Annadurai, G. 2013. Eco-friendly synthesis and characterization of gold nanoparticles using *Klebsiella pneumonia*. *Journal of Nanostructure in Chemistry* 3: 30.

Mann, S. 2001. *Biomineralization: Principles and Concepts in Bioinorganic Materials Chemistry*. Oxford University Press, Oxford, UK.

Medintz, I.I., Uyeda, H.T., Goldman, E.R., Mattoussi, H. 2005. Quantum dot bioconjugates for imaging, labeling and sensing. *Nature Materials* 4: 435–446.

Minaeian, A.R., Shahverdi, A.S., Nohi, H.R., Shahverdi, H.R. 2008. Extracellular biosynthesis of silver nanoparticles by some bacteria. *Journal of Science (Islamic Azad University)* 17: 1–4.

Mohseniazar, M., Barin, M., Zarredar, H., Alizadeh, S., Shanehbandi, D. 2011. Potential of microalgae and Lactobacilli in biosynthesis of silver nanoparticles. *BioImpacts* 1: 149–152.

Mousavi, R.A., Sepahy, A.A., Fazeli, M.R. 2012. Biosynthesis, purification and characterization of cadmium sulfide nanoparticles using Enterobacteriaceae and their application. *Proceedings of the International Conference on Nanomaterials: Applications and Properties* 1: 1862–2304.

Negahdary, M., Mohseni, G., Fazilati, M., Parsania, S., Rahimi, G., Rad, S., Rezaei-Zarchi, S. 2012. The antibacterial effect of cerium oxide nanoparticles on *Staphylococcus aureus* bacteria. *Annals of Biological Research* 3: 3671–3678.

Nguyen, K.C., Seligy, V.L., Tayabali, A.F. 2013. Cadmium telluride quantum dot nanoparticle cytotoxicity and effects on model immune responses to *Pseudomonas aeruginos*. *Nanotoxicology* 7: 202–211.

Pavani, K.V., Kumar, N.S., Sangameswaran, B.B. 2012. Synthesis of lead nanoparticles by *Aspergillus* species. *Polish Journal of Microbiology* 61: 61–63.

Rajasree, S.R., Suman, T.Y. 2012. Extracellular biosynthesis of gold nanoparticles using a gram negative bacterium *Pseudomonas fluorescens*. *Asian Pacific Journal of Tropical Diseases* 2(2): S795–S799.

Rajeshkumar, S., Malarkodi, C., Vanaja, M., Gnanajobitha, G., Paulkumar, K., Kannan, C., Annadurai, G. 2013. Antibacterial activity of algae mediated synthesis of gold nanoparticles from *Turbinaria conoides*. *Der Pharma Chemica* 5: 224–229.

Rathod, A.V., Ranganath, E. 2011. Synthesis of monodispersed silver nanoparticles by *Rhizopus stolonifer* and its antibacterial activity against MDR strains of *Pseudomonas aeruginosa* from burnt patients. *International Journal of Environmental Science* 1: 1582–1592.

Reith, F., Lengke, M.F., Falconer, D., Craw, D., Southam, G. 2007. The geomicrobiology of gold. *The ISME Journal* 1: 567–584.

Saifuddin, N., Wong, C.W., Yasumira, A.A.N. 2009. Rapid biosynthesis of silver nanoparticles using culture supernatant of bacteria with microwave irradiation. *E-Journal of Chemistry* 6: 61–70.

Saklani, V., Jain, V.K. 2012. Microbial synthesis of silver nanoparticles: a review. *Journal of Biotechnology and Biomaterials* S13: 1–3.

Salunkhe, R.B., Patil, S.V., Patil, C.D., Salunkhe, B.K. 2011. Larvicidal potential of silver nanoparticles synthesized using fungus *Cochliobolus lunatus* against *Aedes aegypti* (Linnaeus, 1762) and *Anopheles stephensi Liston* (Diptera; Culicidae). *Parasitology Research* 109: 823–831.

Sanghi, R., Verma, P. 2010. pH dependant fungal proteins in the 'green' synthesis of gold nanoparticles. *Advanced Materials Letters* 1: 193–199.

Seshadri, S., Prakash, A., Kowshik, M. 2012. Biosynthesis of silver nanoparticles by marine bacterium, *Idiomarina* sp. PR58-8. *Bulletin of Materials Science* 35: 1201–1205.

Shah, V., Shah, S., Shah, S., Rispoli, F.J., McDonnell, K.T., Workeneh, S., Karakoti, A., Kumar, A., Seal, S. 2012. Antibacterial activity of polymer coated cerium oxide nanoparticles. *Plos One* 7: 1–13.

Sharma, N., Pinnaka, A.K., Raje, M., Ashish, F.N.U., Bhattacharyya, M.S., Choudhury, A.R. 2012. Exploitation of marine bacteria for production of gold nanoparticles. *Microbial Cell Factories* 11: 86

Simkiss, K., Wilbur, K.M. 1989. *Biomineralization: Cell Biology and Mineral Deposition*. Academic Press, San Diego.

Singaravelu, G., Arockiamary, S.J., Kumar, G.V., Govindaraju, K. 2007. A novel extracellular synthesis of monodisperse gold nanoparticles using marine alga *Sargassum wightii* Greville. *Colloids and Surfaces B: Biointerfaces* 57: 97–101.

Singh, S., Bhatta, U.M., Satyam, P.V., Dhawan, A., Sastry, M., Prasad, B.L.V. 2008. Bacterial synthesis of silicon/silica nanocomposites. *Journal of Materials Chemistry* 18: 2601–2606.

Stephen, J.R., Maenaughton, S.J. 1999. Developments in terrestrial bacterial remediation of metals. *Current Opinion in Biotechnology* 10: 230–233.

Sundaramoorthi, C., Kalaivani, M., Mathews, D.M., Palanisamy, S., Kalaiselvan, V., Rajasekaran, A. 2009. Biosynthesis of silver nanoparticles from *Aspergillus niger* and evaluation of its wound healing activity in experimental rat model. *International Journal of PharmTech Research* 1: 1523–1529.

Sweeney, R.Y., Mao, C., Gao, X., Burt, J.L., Belcher, A.M., Georgiou, G., Iverson, B.L. 2004. Bacterial biosynthesis of cadmium sulfide nanocrystals. *Chemical Biology* 11: 1553–1559.

Tarafdar, J.C., Raliya, R. 2013. Rapid, low-cost, and ecofriendly approach for iron nanoparticle synthesis using *Aspergillus oryzae* TFR9. *Journal of Nanoparticles* 2013: 1–4.

Tikariha, S., Singh, S., Banerjee, S., Vidyarthi, V.S. 2012. Biosynthesis of gold nanoparticles, scope and application: A review. *International Journal of Pharmaceutical Sciences and Research* 3: 1603–1615.

Usha, R., Prabu, E., Pa, M., Venil, I.C.K., Rajendran, R. 2010. Synthesis of metal oxide nanoparticles by *Streptomyces* sp. for development of antimicrobial textiles. *Global Journal of Biotechnology and Biochemistry* 5: 153–160.

Vani, C., Sergin, G.K., Annamalai, A. 2011. A study on the effect of zinc oxide nanoparticles in *Staphylococcus aureus*. *International Journal of Pharmaceutical and Biosciences* 2: 16–22.

Varshney, R., Bhadauria, S., Gaur, M.S., Pasricha, R. 2010. Characterization of copper nanoparticles synthesized by a novel microbiological method. *Journal of the Minerals, Metals and Materials Society* 62(12): 102–104.

Xie, Y., He, Y., Irwin, P.L., Jin, T., Shi, X. 2011. Antibacterial activity and mechanism of action of zinc oxide nanoparticles against *Campylobacter jejuni*. *Applied Environmental Microbiology* 77: 2325–2331.

Yang, H., Santra, S., Holloway, P.H. 2005. Syntheses and applications of Mn-doped II-VI semiconductor nanocrystals. *Journal of Nanoscience and Nanotechnology* 5: 1364–1375.

10

SUSTAINABLE SYNTHESIS OF PALLADIUM(0) NANOCATALYSTS AND THEIR POTENTIAL FOR ORGANOHALOGEN COMPOUNDS DETOXIFICATION

Michael Bunge

Institute of Applied Microbiology, Justus Liebig University of Giessen, Giessen, Germany

Katrin Mackenzie

Department of Environmental Engineering, Helmholtz Centre for Environmental Research – UFZ, Leipzig, Germany

10.1. INTRODUCTION

Precious metal catalysts have attracted the interest of environmental chemists for some time. The unique properties which some of their representatives offer are in demand for degradation of halogenated priority pollutants in the aqueous phase. The metal is able to activate the reducing agent molecular hydrogen ($H_2 \rightarrow 2H^*$) and to

---

Bio-Nanoparticles: Biosynthesis and Sustainable Biotechnological Implications, First Edition.
Edited by Om V. Singh.
© 2015 John Wiley & Sons, Inc. Published 2015 by John Wiley & Sons, Inc.

facilitate the C–X bond cleavage. Palladium (Pd) preferably catalyzes the hydrodehalogenation and hydrogenation of olefins, whereas rhodium (Rh) also affects the hydrogenation of aromatic rings under ambient conditions. Moreover, Rh has the highest activity for hydrodefluorination of unsaturated fluorinated compounds (Baumgartner and McNeill, 2012). However, none of these catalysts are able to defluorinate saturated compounds such as perfluorinated surfactants.

Regarding water treatment, in order to reductively transform a number of priority drinking water contaminants – which are to a great extent chlorohydrocarbons – catalysis promoted by Pd is seen as one of the most promising strategies (Chaplin et al., 2012). The advantage of reductive processes such as Pd-promoted hydrodechlorination (HDC) is a high degree of selectivity towards halogenated pollutants, which all kinds of oxidative transformations cannot provide.

Due to a growing need for the development of environmentally benign technologies, the "green" biosupported synthesis of nanoparticles, including palladium nanocatalysts, is of great interest. A fundamental objective of previous research on bio-supported synthesis is the controlled production of Pd and hybrid Pd/metal nanoparticles with a narrow size distribution using microbial cells and membrane structures as nanoreaction chambers. Pd catalysts offer additional advantages when used in combination with biological bottom-up synthesis: in particular, as well as the sustainable production, size-controlled formation in natural or artificial membrane entities, and stabilization and binding to functional groups inherent to this type of synthesis, the catalysis of organohalogen compound detoxification reactions opens up further possibilities, especially when considering factors such as metal recycling and re-use. "Biosynthesized" Pd(0) nanoparticles have gained considerable attention due to their superior catalytic properties in a number of reactions in synthetic organic chemistry, such as Suzuki–Miyaura and Mizoroki–Heck reactions. In the recent past, biosynthesized palladium nanoparticles have also been used for detoxification reactions, that is, the dehalogenation of anthropogenic pollutants such as chlorobenzenes and polychlorinated biphenyls.

Chemical synthesis methods can generate palladium catalysts of high purity and catalytic activity, providing that the starting substances are equally pure. "Biogenic" pathways may represent a sustainable and environmentally sound alternative for the synthesis of precious metal nanocatalysts, especially for recycling approaches, the selective palladium recovery from multi-metal mixtures, and simultaneous nanoparticle production. Such methods may also provide novel solutions for the protection of catalysts. In this chapter we will present ideas, strategies, and the historical background concerning abiotically and biologically generated particles, their performance and how these can learn from each other.

10.2. CHEMICALLY GENERATED PALLADIUM NANOCATALYSTS FOR HYDRODECHLORINATION: CURRENT METHODS AND MATERIALS

10.2.1. Pd Catalysts

Data of a large variety of different Pd catalysts are available from Pd nanoparticles in different size regions via Pd on metals, oxidic carriers, or carbonaceous supports. The performance of a selection of catalysts is described here, among them Pd/Al_2O_3

(G-133D Pd on $\gamma\text{-Al}_2\text{O}_3$, Commercia, Germany, 63–125 µm), Pd nanoparticles (d_p =20–100 nm), Pd/nano-Fe_3O_4 (0.15 wt% Pd, d_p <50 nm; Hildebrand et al., 2009), and Pd/PDMS as membranes (1) with Pd-containing particles (SilGel® 612A/B from Wacker Chemie AG, Germany; Navon et al. 2012) and (2) with carrier-free Pd clusters generated directly in the polydimethylsiloxane (PDMS) films.

10.2.2. Data Analysis

The hydrodechlorination (HDC) reaction of chlorohydrocarbons (CHCs), commonly used as probe water pollutants (e.g., chlorobenzene, trichloroethylene or TCE, perchloroethylene or PCE) and hydrogen as reducing agent, follows a pseudo first-order kinetics. We prefer the description of the catalyst performance by the specific Pd activity ($A_{Pd,i}$) which is the equivalent to a second-order rate coefficient k'_i for the disappearance of the CHC i according to $k'_i = \ln 2 \, A_{Pd,i}$ with $dc_i/dt = -k'_i \, c_i \, c_{Pd}$. The specific Pd activity has the ability to directly show us in what time frame a certain volume of contaminated water can be treated with a certain amount of Pd (6–7 half-times are needed for complete treatment). $A_{Pd,i}$ (L g^{-1} min^{-1}) is defined:

$$A_{Pd,i} = \frac{V_{water}}{m_{Pd} \tau_{1/2}} = \frac{\ln(c_{t1}/c_{t2})}{\ln 2 c_{Pd} (t_2 - t_1)} \quad (10.1)$$

where V_{water} is the water volume to be treated, m_{Pd} and c_{Pd} are Pd mass and concentration (g L^{-1}) and $\tau_{1/2}$ is the half-life of the CHC. t_1 and t_2 are two arbitrarily chosen sampling times; c_{t1} and c_{t2} the corresponding CHC concentrations. The catalyst activities were calculated either from the CHC conversion or from the product formation rates.

10.2.3. Pd as Dehalogenation Catalyst

Metallic palladium in various forms (e.g., supported on oxidic, metallic and carbonaceous supports, as freely suspended non-supported Pd colloids and larger particles, and in combination as bimetallic clusters) has proved its ability to catalyze a broad range of reduction reactions (Chaplin et al., 2012).

Most relevant for environmental applications are HDC reactions of contaminants in the water phase. HDC refers to the conversion of a C–Cl into a C–H bond. In the course of the reaction, olefinic double bonds are hydrogenated whereas aromatic rings are usually preserved. Exceptions to this rule are phenol moieties which are also hydrogenated. By breaking the C–Cl bond, chlorinated pollutants are transformed according to the equation

$$RCl + H_2 \xrightarrow{\text{Pd catalyst}} RH + HCl \quad (10.2)$$

The dechlorinated products are commonly less toxic, environmentally more harmless substances with a better biodegradability. Pd is able to catalyze hydrodehalogenation from defluorination to deiodination, where defluorination occurs with the lowest reaction rates (Mackenzie et al. 2006).

One of the strong points of this method is that it can be efficiently carried out in the aqueous phase even at low temperatures. The naturally occurring anions such as bicarbonate, sulfate, chloride, and most abundant cations do not adversely affect the catalyst activity (Lowry and Reinhard, 2000).

10.2.4. Intrinsic Potential *vs.* Performance

Nanoscale catalyst particles are already known as excellent tools in catalytic processes and intensive research is currently optimizing their performance. As known from nano-sized metal particles, nanocatalysts have the potential of very high reaction rates due to their relatively high specific surface areas ($S \approx 10\,m^2 g^{-1}$) and low mass transfer restrictions. Nano-magnetite as a low-cost magnetic carrier for Pd has already been successfully tested and shown outstandingly high catalyst activities in clean water (Fig. 10.1; Hildebrand et al. 2009). The most active Pd-Fe_3O_4 catalyst contained only traces of Pd (0.15 wt%). In Table 10.1 some pollutant half-lives are shown, which gives an impression of the sensitivity of the highly active catalyst and the loss of activity in the various reaction media. Pd-on-magnetite was successfully tested in batch experiments for the HDC of the chlorohydrocarbons trichloroethene

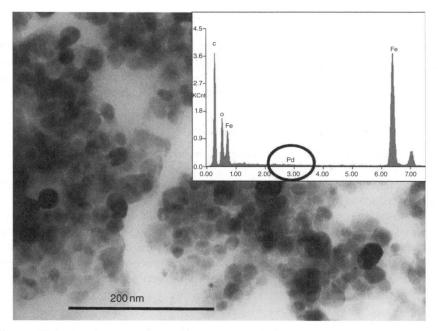

Figure 10.1. TEM image and TEM-EDX-spectrum of Pd/nano-magnetite (0.15 wt% Pd). Source: Hildebrand H, Mackenzie K & Kopinke F-D (2009). Highly Active Pd-on-Magnetite Nanocatalysts for Aqueous Phase Hydrodechlorination Reactions. *Environmental Science & Technology* 43: 3254–3259. Copyright © the American Chemical Society, 2009. *See insert for color representation of the figure.*

(TCE) and chlorobenzene. For the HDC of TCE, second-order rate coefficients of approximately $2 \times 10^4 \, L \, g^{-1} \, min^{-1}$ were measured. Such high catalyst activities have never been described before for HDC in aqueous suspensions. The ferrimagnetism of the carrier enables a separation of the nanocatalyst from the treated water by means of magnetic separation (Fig. 10.2). This allows the catalyst to be reused several times, which is an important advantage compared to other nanoscale catalytic systems such as pure Pd or Pd-on-Au colloids.

Similar to other precious metal catalysts, Pd is very sensitive towards deactivation processes. Several authors stated the poisoning effect of halogenide ions released during reaction (Urbano and Marinas, 2001). However, the heavier halogenide ions also strongly interfere when added as salt to the reaction medium (Mackenzie et al. 2006).

TABLE 10.1. Half-lives of Some Probe Compounds with Pd/nano-magnetite in Different Water Matrices ($c_{catalyst} = 100 \, mg \, L^{-1}$; $c_{Pd} = 0.15 \, mg \, L^{-1}$; $c_{0,pollutant} = 30 \, mg \, L^{-1}$).

Pollutant	Water matrix	Half-life
Trichloroethene	Deionized water	0.3 min
Chlorobenzene	Deionized water	2.7 min
Trichloroethene	Clean natural groundwater	10 min
Trichloroethene	After 5 batch cycles in clean groundwater	1 h
Trichloroethene	Deionized water +5 mg L^{-1} bisulfite	Complete deactivation

Figure 10.2. Ferrimagnetic Pd/nano-magnetite particles are easily separated in everyday laboratory life by permanent magnets. *See insert for color representation of the figure.*

At first sight, HDC using Pd seems an easy-to-use all-round method for selective removal of the often-occurring chlorinated organic water pollutants, with a spectrum of reducible substances far broader than that of other approved reducing agents (such as zero-valent iron). However, only a few working groups describe the field application of Pd catalysts for HDC in ground- or wastewater (McNab Jr and Ruiz, 1998; Schüth and Reinhard, 1998). Catalyst deactivation occurs not only due to biofouling and blocking of active sites by inorganic salts and polyelectrolytes such as dissolved humic matter causing low life-times during field application, but substances such as reduced sulfur compounds, heavy metals, or other omnipresent constituents of real waters also poison the catalyst. It was found that the catalytic system is subject to fast deactivation when used in natural water bodies. By its very nature, HDC requires reductants such as H_2 or alternative hydrogen sources such as formic acid (Kopinke et al., 2004). These act as electron donors for the intended catalytic reduction as well as for the undesired microbial sulfate reduction yielding bisulfide. This ambiguous role of the reductants can hardly be avoided. The logical consequence for catalyst design is increased effort for efficient catalyst protection against poisoning.

10.2.5. Concepts for Pd Protection

What are the technical measures for catalyst protection? In order to shield the active sites from deactivating ions, thin hydrophobic films of polydimethylsiloxane (PDMS) were applied onto commercially available Pd catalyst particles or Pd incorporated into polymer films or tubings (Fritsch et al., 2003; Kopinke et al., 2010; Navon et al. 2012). These coatings allow organic contaminants to diffuse to the reactive sites and make the catalyst resistant to inorganic salts and other ionic species including polyelectrolytes such as DOM (Fig. 10.3). However, a sufficient protection against the permeation of fairly hydrophobic sulfur compounds, such as organic sulfur compounds and H_2S ($pK_{A1}=7.05$) is much more difficult to provide. Nonetheless, these films allow a significant increase in catalyst life-time (Navon et al. 2012). The precious metal components are not subjected to direct contact with the water bulk phase, which also means that metal leaching is prevented. Although incorporation of nanoparticles into a polymer film means an increase in mass transport resistance, and therefore lower reaction rates, hydrophobic phases such as PDMS are able to enrich hydrophobic pollutants and thereby increase the rate of the catalytic reaction in return. These effects cancel out each other to some extent. The most surprising finding with PDMS as catalyst coating is the ability of HCl (the product of every dechlorination and an ionic compound) to be released from the catalyst back into the water bulk phase.

In a second amendatory protection strategy, the undesired reduced sulfur compounds are oxidized prior to contact with the catalytic system. Several oxidizing agents which are commonly used for water treatment were studied. Most of the oxidants damaged the catalyst. Only permanganate provided sulfide oxidation and at the same time kept the full catalyst activity while no interference with the HDC reaction occurred (Angeles-Wedler et al., 2008). This point can be emphasized with a clear conscience: the reductive hydrodechlorination takes place within a strongly oxidative environment!

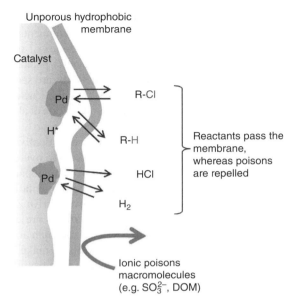

Figure 10.3. Polymer-coated catalysts particle. *See insert for color representation of the figure.*

The advantages and disadvantages of the two protections strategies are as follows. The hydrophobic protection works well against ionic poisons or unwanted precipitations (polyelectrolytes), but reduced catalyst activities have to be accepted because of higher mass transfer resistance. Furthermore, hydrogen sulfide and other non-ionic substances can penetrate the hydrophobic barrier. The oxidation is able to rapidly destroy most reduced sulfur compounds and moderate permanganate concentrations do not disturb the hydrodechlorination. As a bonus, the oxidative environment hinders the respiration of the sulfate-reducing microorganisms. However, other reduced water constituents may also consume permanganate and the unprotected catalyst cannot tolerate higher permanganate concentrations. Each protection measure alone has positive effects but has its drawbacks; the combination of the two measures may be key for a more successful protection in natural waters.

10.3. BIO-SUPPORTED SYNTHESIS OF PALLADIUM NANOCATALYSTS

10.3.1. Background

Some conventional processes for precious metal nanoparticle synthesis use hazardous chemicals as reducing agents, such as hydrazine (Choudary et al., 2002) or sodium borohydride (Artuso et al., 2003; Pittelkow et al., 2003; Wang et al., 2008), operate at elevated temperatures and pressure conditions, and can result in polydisperse particle fractions with only moderate morphology (Kimura et al., 2003; Luo et al., 2005). Despite newer convincing technologies and novel developments for chemical and physical synthesis of palladium nanoparticles in the absence of (eco)toxicologically relevant substances, it is known that unprotected ("naked") palladium nanoparticles, as synthesized by means of conventional chemical methods, are sensitive to rapid precipitation, sequestration with suspended organic matter, and deactivation by catalyst-poisoning substances such as reduced sulfur compounds (Korte et al., 2000; Mackenzie et al., 2006; Phan et al., 2006).

Under some circumstances, the synthesis of catalytically active palladium nanoparticles in the presence of microorganisms can help to overcome these limitations. For instance, synthesis of biologically produced nanoparticles operates at ambient temperature and pressure conditions and environmentally harmful chemicals are not used in the process (Narayanan and Sakthivel, 2010; Zhang et al., 2011). The organisms used for biosynthesis of Pd(0) nanoparticles are easy to handle and downstream processing appears to be less elaborate (Narayanan and Sakthivel, 2010; De Corte et al., 2012). Furthermore, it may be possible to make Pd catalysts more resistant to deactivation and agglomeration by means of biosynthesis and specific cell-molecule interactions. For example, the binding of the nanoparticles to functional groups on the cell envelope firstly provides a spacer effect, diminishing particle-particle interaction and therefore agglomeration, and secondly may provide a protective coating effect (Feldheim and Eaton, 2007; Maruyama et al., 2007; Nel et al., 2009; Virkutyte and Varma, 2011). On the other hand, some cellular components or residues may also prevent the generation of highly active catalyst surfaces (Søbjerg et al., 2011). Understanding the mechanisms and opportunities of palladium nanoparticles are one of the challenges in future Pd research. Synergies between chemical and biological approaches could bring a step forward in an efficient utilization and re-utilization of otherwise wasted precious metals.

10.4. CURRENT APPROACHES FOR SYNTHESIS OF PALLADIUM CATALYSTS IN THE PRESENCE OF MICROORGANISMS

Noble-metal nanocatalysts can be synthesized at microbial interfaces, which has been summarized in earlier reviews (Mandal et al., 2006; Theron et al., 2008; Hennebel et al., 2009, 2012; Korbekandi et al., 2009; Narayanan and Sakthivel, 2010; Alvarez and Cervantes, 2011; Li et al., 2011; Lloyd et al., 2011; Zhang et al., 2011; De Corte et al., 2012). It has been shown that such biologically produced nanoparticles can possess useful antimicrobial or catalytic properties; previous reports have had a particular emphasis towards the study of catalyzed conversions of anthropogenic environmental contaminants (e.g., Theron et al., 2008; Hennebel et al., 2009, 2012; Thakkar et al., 2010; Alvarez and Cervantes, 2011; Li et al., 2011; Zhang et al., 2011; De Corte et al., 2012; Senior et al., 2012). The formation of Pd(0) nanocatalysts at microbial interfaces and their excellent catalytic properties have also been demonstrated in various advanced reactions in synthetic organic chemistry, including Suzuki–Miyaura and Mizoroki–Heck reactions (Søbjerg et al., 2009; Bunge et al.,

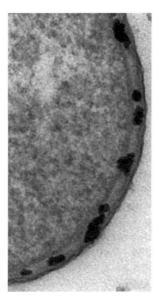

Figure 10.4. Biosynthesized Pd(0) nanoparticles in the periplasm of a *Pseudomonas* sp. Source: Modified from Bunge M, Søbjerg LS, Rotaru AE, et al. (2010) Formation of palladium(0) nanoparticles at microbial surfaces. *Biotechnology and Bioengineering* 107: 206–215. Copyright © 2010, John Wiley & Sons, Inc.

2010; Gauthier et al., 2010). Formate- or hydrogen-driven reduction in the presence of Gram-negative bacteria with no documented dissimilatory metal reduction capabilities resulted in small, well-suspended Pd(0) nanoparticles that were associated with the cells ("bioPalladium"). Whereas large and close-packed Pd(0) aggregates (200 μm) were produced in cell-free buffer solutions, nano-sized Pd(0) particles (3–30 nm) were only formed in the presence of the bacteria; particles in this size range were located in the periplasmic space (Fig; 10.4; Bunge et al., 2010). This research was further extended to the recovery of palladium, platinum, and ruthenium derived from metal leachates and industrial waste (Gauthier et al., 2010). It was shown that Gram-negative bacteria of the genus *Cupriavidus* can serve as scaffolding for the absorption and reduction of noble metal ions present in the industrial waste solutions.

10.4.1. Pd(II)-Tolerant Microorganisms for Future Biotechnological Approaches

The usual chemical synthetic route to prepare palladium nanoparticles involves the reduction of a palladium salt by various reducing agents in the presence of a stabilizer (Li and El-Sayed, 2001; Tan et al., 2003; Narayanan and El-Sayed, 2008). The reduction of Pd(II) to Pd(0) in the presence of microbial cells is the key process for sustainable and environmentally benign production of highly active catalysts using bacteria. The high cytotoxicity of Pd(II) ions usually requires the production of bio-Palladium (bioPd) to be split into separate steps for cultivation, cell harvesting, Pd(II) reduction, and Pd(0) nanoparticle formation. To the best of our knowledge, none of the existing synthesis routes employing microorganisms utilizes single-stage processes of simultaneous cell cultivation, Pd(II) reduction, and thus Pd(0) nanocatalyst formation. Biotechnological considerations may favor continuous single-stage processes rather than multi-step (spatially and temporally separated) processes for growth and reduction. Pd(II)-tolerant microorganisms capable of catalyzing the reduction of Pd(II) to Pd(0) and/or serving as scaffolding for the deposition of bioPd nanoparticles would be ideal for use in such processes.

Microorganisms from sites with a high natural background of heavy metals have been exposed to these over geological times (in contrast to those from anthropogenic contaminated areas). Such microorganisms can therefore be expected to have developed efficient metal-resistance mechanisms and should thus be optimally adapted to the elevated heavy-metal concentrations. At the same time, such sites might exhibit a yet-overlooked reservoir for the isolation of heavy-metal-tolerant bacteria useful in geobiotechnological approaches, such as for the bioremediation of industrially contaminated sites or potentially relevant for biomining via biooxidation or bioleaching of metals (Rawlings, 2002; Haferburg and Kothe, 2007).

For the synthesis of "biogenerated" nanocatalysts and the implementation of single-stage continuous processes, Schlüter et al. (2014) selectively enriched and isolated Pd(II)-tolerant strains from Alpine serpentinitic areas (high natural concentrations of heavy metals such as Cr, Ni, Co), that allowed for simultaneous culturing and synthesis of Pd(0) nanoparticles (Schlüter et al., 2014). By applying an automated high-throughput assay enabling simultaneous cultivation and kinetic analysis of heavy-metal-mediated growth inhibition (Schacht et al., 2013), the authors demonstrated that specific isolates did not only tolerate elevated Pd(II) concentrations but were also able to grow in the presence of up to 3 mM Pd(II) (Schlüter et al., 2014). Growth performance usually decreases with increase in Pd(II) concentrations, but for some Actinobacteria isolates it was demonstrated that low Pd(II) concentrations enhanced bacterial growth, most likely due to hormesis. The use of heavy-metal- and/or Pd(II)-tolerant strains would allow for production of biologically generated nanocatalysts enabling simultaneous culturing and nanoparticle synthesis on a continuous basis for large-scale production, for instance in a continuous-flow tubular reactor.

10.4.2. Controlling Size and Morphology during Bio-Synthesis

The properties of precious metal nanoparticles clearly differ from the bulk material and within the "nano-range"; the physicochemical, catalytic, biological, and toxicological properties of particles depend mainly on their size and shape (van den Tillaart et al., 1996; Le Bars et al., 1999; Jiang et al., 2008; Ju-Nam and Lead, 2008; Narayanan and El-Sayed, 2008; Nel et al., 2009). In general, many microorganisms have the ability to produce nanoscale Pd(0) particles with structural and catalytic properties equivalent to the chemically synthesized counterparts; however, different strategies to synthesize nanoparticles with controlled size, size distribution, and shape would apply (Korbekandi et al., 2009; Narayanan and Sakthivel, 2010; Zhang et al., 2011; De Corte et al., 2012). Conventional size control could be achieved by modifying different parameters including temperature, Pd salt concentration, duration of the reduction, number of reduction cycles, or pH (Beck et al., 2000; Wang et al., 2008). Cha et al. (2007) reported an approach for finely tuned size control (ca. 0.5 nm steps) in an electrochemical synthesis. The particle size decreased with increasing current density and decreasing temperature and electrolysis duration (Cha et al., 2007). Gericke and Pinches (2006) demonstrated that changes in the pH had an effect on the size distribution of noble metal nanoparticles (Gericke and Pinches, 2006). A further approach to rough control of the size of formed particles is the use of different reductants and to vary the concentration and type of the protective agents or stabilizers (Tan et al., 2003). The application of strong reducing agents usually yields bigger nanoparticles. Weak reducing agents (e.g., citrate, potassium bitartrate) commonly result in less rapid reactions and the production of smaller nanoparticles. For the biosupported generation of Pd(0) nanocatalysts, different sizes could be received by varying the Pd:biomass ratio (De Windt et al., 2006; Søbjerg et al., 2011).

A more specific strategy of controlling the size of palladium nanocatalysts would be the use of artificial and natural lipid membrane structures as nanoreaction chambers (Vriezema et al., 2005). After self-assembling of lipid molecules to liposomes in water, the size of the spherical compartments can be easily adjusted from 20 nm to more than 1 μm by application of different preparation methods (Bae et al., 2011). Fine-tuning of the nanoreaction compartment such as permeability, stiffness, surface charge, and thermal stability is controlled by lipid structure such as length and saturation of the hydrophobic tails, charge of the polar head-groups, temperature, and the incorporation of additives (e.g., cholesterol) (Philippot and Schuber, 1994). Using transmission electron microscopy of ultrathin sections of Gram-negative bacterial cells, the deposition of Pd(0) nanoparticles has been frequently observed inside the periplasmic space (De Windt et al., 2005; Creamer et al., 2008; Bunge et al., 2010). Since the distance between inner and outer membranes is typically in the range of 15–30 nm, this might be a first step in using natural membrane entities to obtain monodisperse nanoparticle fractions.

10.4.3. Putative and Documented Mechanisms of Biosynthesis of Palladium Nanoparticles

Intracellular synthesis of palladium nanoparticles, as well as extracellular formation, has been reported for a number of microorganisms (Mandal et al., 2006; Hennebel et al., 2009, 2012; Korbekandi et al., 2009; Narayanan and Sakthivel, 2010; Alvarez and Cervantes, 2011; Li et al., 2011; Lloyd et al., 2011; Zhang et al., 2011; De Corte et al., 2012).

The cell envelope plays a major role in interactions with the palladium ions, that is, the "trapping" of Pd(II) ions. This biosorption is followed by bioreduction to Pd(0) and deposition of the nanoparticles, potentially mediated by the present enzymes.

The ability of bacteria to accumulate and reduce precious metals was recognized early in the field of biosorption and biorecovery (Lovley, 1993; Yong et al., 2002; Lloyd, 2003; de Vargas et al., 2004; Mack et al., 2007; Gadd, 2009). In case of the *Desulfovibrio fructosivorans* and *E. coli* MC4100, hydrogenases have been found to be involved in the reduction of Pd(II) and the synthesis of Pd(0) nanoparticles (Mikheenko et al., 2008; Deplanche et al., 2010), a similar mechanism which is also widely accepted for the synthesis of gold nanoparticles from Au(III) (Deplanche and Macaskie, 2008). The authors proposed the involvement of periplasmic hydrogenases in reduction and nanoparticle synthesis, tentatively complemented by cytoplasmic hydrogenases and/or cytochrome c3 (Deplanche and Macaskie, 2008). A study by Bunge *et al.* (2010) with non-dissimilatory metal-reducing bacteria (DMRB) suggests that hydrogenases are not involved in Pd(0) nanoparticle formation at cellular interfaces of *Cupriavidus necator* H16 (Bunge et al., 2010). The authors also discuss the possibility that directed processes are not involved, but rather the fortuitous activity of enzymes or the role of cellular interfaces as scaffold (Bunge et al., 2010). Rotaru *et al.* (2012) even suggested an entirely cell-independent mechanism (Rotaru et al., 2012), an interesting approach using amine-functionalized abiotic surfaces which was studied later in more detail (De Corte et al., 2013).

This is in contrast to studies with gold nanoparticles which indicated that gold was only precipitated in the presence of living bacterial cells, but not in the presence of dead biomass or in chemical abiotic experiments (Lengke et al., 2007). However, the ability of the positively charged amine groups to attract anionic precious metal ions might be decreased at higher pH and should be further investigated. Furthermore, the synergistic effects resulting from sorptive enrichment of the hydrophobic pollutants in the vicinity of nanocatalyst particles covered by lipophilic cell constituents or residues may enhance the degradation of organic environmental contaminants for some catalytic reactions. In addition to the relevance as adsorbents for precious metal ions and the potential role as biosorbents for the highly lipophilic compounds, the "bioresidues" may also prevent the Pd(0) catalyst re-oxidation, or structural alterations and catalytic inactivation (Feldheim and Eaton, 2007; Maruyama et al., 2007; Nel et al., 2009; Virkutyte and Varma, 2011). Results which are contrary to this concept have been previously reported (Søbjerg et al., 2011). The exact mechanisms of microbial interfaces and biological constituents for the synthesis and protection of Pd(0) nanoparticles therefore need to be elucidated in further studies.

10.4.4. Isolation of Nanocatalysts from the Cell Matrix and Stabilization

Pd(0) nanoparticles act in a size- and shape-dependent manner (van den Tillaart et al., 1996; Le Bars et al., 1999; Jiang et al., 2008; Ju-Nam and Lead, 2008; Narayanan and El-Sayed, 2008; Nel et al., 2009). The successful use of metal catalysts for optimal activity in particular applications requires not only controlled growth in terms of particle size, particle size distribution, and shape and structure, but also the ability to disperse and stabilize the particles properly (Narayanan and El-Sayed, 2008). The same traits which result in the high catalytic performance of nanoparticles are simultaneously associated with disadvantageous effects, for instance the strong tendency to agglomerate due to the high surface energy. Stabilization is important because key properties such as activity and durability in catalytic reactions depend on the stability of the nanocatalysts (Ju-Nam and Lead, 2008; Pradeep and Anshup, 2009). In addition to related approaches to prevent the interactions, the nanocatalysts can be protected by stabilizers or capping agents, which can include organic or biological molecules or polymers acting against agglomeration by charge-stabilization or steric stabilization mechanisms (Ding and Gin, 2000; Tan et al., 2003; Narayanan and El-Sayed, 2008). In conventional synthetic techniques (e.g., the chemical or electrochemical reduction of the Pd salts), the use of stabilizers is well established in order to prevent agglomeration and allow the isolation of the nanoparticles (Tan et al., 2003; Narayanan and El-Sayed, 2008). In a study by Luo *et al.* (2005), polyethyleneglycol appeared to act as both reducing agent and stabilizer for recyclable palladium nanocatalysts for Heck reactions.

In the case of bioengineered palladium nanoparticles, methods of effectively extracting the cellular/periplasmic fraction of Pd(0) are required, thus enabling optimal catalytic access (Bunge et al., 2010). Once released from the cell boundaries, a part of the bioengineered palladium nanoparticles will aggregate to some degree despite strategies to stabilize them. Potential metal nanoparticle interactions and aggregation also involves heterogeneous material such as particulate or dissolved organic matter, biomolecules, or organic contaminants. To obtain nanoparticles with a narrow size distribution, it is usually necessary to implement post-synthetic separation, "secondary engineering", and further processing and regeneration steps. Extracts from microorganisms contain complex mixtures of biomolecules that may act both as templates and as reducing and capping agents (Feldheim and Eaton, 2007; Maruyama et al., 2007; Nel et al., 2009; Virkutyte and Varma, 2011). It is essential that the nanoparticles are characterized in detail after they have been isolated from the microbes, and that the effects of stabilizers added, or residual cellular components which may also influence aggregation behavior, are understood.

10.5. BIO-PALLADIUM(0)-NANOCATALYST MEDIATED TRANSFORMATION OF ORGANOHALOGEN POLLUTANTS

With a variety of unique properties, palladium nanoparticles can be deployed for many different applications in various fields, including synthetic organic chemistry, biology, medicine, and renewable energy production and storage (Kickelbick, 2008). Several studies have aimed to synthesize Pd- and hybrid Pd/metal nanocatalysts at microbial interfaces by means of bacteria-assisted approaches; these nanocatalysts are intended for use in the detoxification of highly persistent organic pollutants within aquatic environments.

The catalytic activity of (bio)synthesized Pd(0) (nano)catalysts has been demonstrated through the transformation of typical environmentally relevant contaminants, including the reduction of carcinogenic Cr(VI) to Cr(III) (Mabbett et al., 2004; Humphries and Macaskie, 2005; Humphries et al., 2006; Creamer et al., 2008; Macaskie et al., 2012) and the reductive/hydrodehalogenation of persistent organic pollutants (Baxter-Plant et al., 2003; De Windt et al., 2005; Mertens et al., 2007; Hennebel et al., 2010, 2012; De Corte et al., 2011; De Gusseme et al., 2011; Macaskie et al., 2012). Schlüter *et al.* (2014) showed that bioPd(0) catalyzes reductive dechlorination reactions of polychlorinated dibenzo-*p*-dioxins (PCDDs) and can therefore potentially detoxify PCDDs, which are among the most notorious environmental contaminants (Schlüter et al., 2014). It has been observed that the dechlorination of 1,2,3,4-TeCDD by means of some reductively dechlorinating bacteria results in the formation of undesired 2,3-substituted congeners (Bunge and Lechner, 2009). In contrast, the bioPd(0)-catalyzed pathway appears to circumvent this problem, as indicated by the formation of 1,2,3-TrCDD and 1,2,4-TrCDD as intermediates and 1-MCDD as putative dechlorination end product (Schlüter et al., 2014). However, the reaction pathways can vary from that commonly observed by chemically synthesized catalysts. Since the catalytic ability and the mechanism of catalyzed processes should be an inherent property of the catalyst, differences in performance should be inspected carefully in order to understand the influence of the synthesis procedure (chemical or biological) on the catalyst behavior.

10.6. CONCLUSIONS

Even though Pd catalysts bear a very high intrinsic activity, which can be seen from the performance of Pd-nanocatalysts, the potential of Pd for water treatment has not yet been fully exploited. The aim of future research efforts should therefore be to transfer the huge potential of the catalyst into the practice. Utilizing nano-sized catalysts and protecting them with thin polymer films provides a promising tool for extending the catalyst applicability, especially when metallic Pd is directly generated in the protecting film. Further research into suitable protecting materials is necessary. Such materials must allow a fast and complete reagent transfer as well as a longer catalyst life-time. If we are successful in maintaining and regaining even a fraction of the high inherent Pd activity during field application, Pd catalysts will certainly take their place as an approved tool in environmental technology.

Due to the increasing use of palladium nanoparticles in various industrial applications, there is a growing need for cost efficiency as well as conservation of resources. Implementing green approaches in the development of novel nanocatalysts is therefore of paramount importance. While conventional chemical and physical synthesis techniques can successfully produce well-defined nanoparticles with surpassing properties, these methods can be expensive, sometimes involve the use of harmful chemicals, and require defined educts and reaction conditions. During the past decade, the interaction of a number of microorganisms with Pd(II) salts and their potential for bioreduction to Pd(0) has been investigated, although the objective of previous studies was not always the biosynthesis of palladium nanoparticles. With regard to expected further increasing demand for palladium nanoparticles for industrial and environmental applications, consideration of approaches using microorganisms that minimize hazard, energy consumption, and waste will be essential. Prospective work in future nanobiotechnology should implement systematic experiments which include the development of approaches for the production of customized palladium nanoparticles of well-defined sizes and shapes.

In addition to reproducibly synthesizing nanocatalysts of a specific size and shape, the isolation of nanoparticles from the cell matrix and stabilization against unwanted agglomeration, aggregation, and deactivation (i.e., the prevention of catalyst poisoning by reduced sulfur compounds) is of major interest. It remains to be determined how the use of different cell constituents or biological residues (membranes, proteins) affects the structural and catalytical properties of the nanocatalysts, especially transformation reactions of highly lipophilic organohalogen compounds.

Better understanding of the mechanisms of palladium nanoparticle biosynthesis will enable us to achieve better control over size, size distribution, shape, and catalytic activities, and will lead to the development of sustainable techniques for tailor-made production of precious metal catalysts with precisely predictable properties and the option to recover the precious metal from otherwise wasted matrices.

ACKNOWLEDGEMENTS

We are grateful to Leonard Böhm and Andreas Bunge for their invaluable comments on the manuscript. A part of this work was funded by the German Federal Ministry of Education and Research (BMBF; FKZ: 03X3571).

REFERENCES

Alvarez, L.H., Cervantes, F.J. 2011. (Bio)nanotechnologies to enhance environmental quality and energy production. *Journal of Chemical Technology and Biotechnology* 86: 1354–1363.

Angeles-Wedler, D., Mackenzie, K., Kopinke, F.-D. 2008. Permanganate oxidation of sulfur compounds to prevent poisoning of Pd catalysts in water treatment processes. *Environmental Science & Technology* 42: 5734–5739.

Artuso, F., D'Archivio, A.A., Lora, S., Jerabek, K., Kralik, M., Corain, B. 2003. Nanomorphology of polymer frameworks and their role as templates for generating size-controlled metal nanoclusters. *Chemistry – A European Journal* 9: 5292–5296.

Bae, S.J., Jung, S., Um, S.H. 2011. Budding dynamics of the lipid membrane. *Journal of Nanoscience and Nanotechnology* 11: 6172–6176.

Baumgartner, R., McNeill, K. 2012. Hydrodefluorination and hydrogenation of fluorobenzene under mild aqueous conditions. *Environmental Science & Technology* 46: 10199–10205.

Baxter-Plant, V.S., Mikheenko, I.P., Macaskie, L.E. 2003. Sulphate-reducing bacteria, palladium and the reductive dehalogenation of chlorinated aromatic compounds. *Biodegradation* 14: 83–90.

Beck, A., Horvath, A., Szücs, A., Schay, Z., Horvath, Z.E., Zsoldos, Z., Dekany, I., Guczi, L. 2000. Pd nanoparticles prepared by 'controlled colloidal synthesis' in solid/liquid interfacial layer on silica. I. Particle size regulation by reduction time. *Catalysis Letters* 65: 33–42.

Bunge, M., Lechner, U. 2009. Anaerobic reductive dehalogenation of polychlorinated dioxins. *Applied Microbiology and Biotechnology* 84: 429–444.

Bunge, M., Søbjerg, L.S., Rotaru, A.E., Gauthier, D., Lindhardt, A.T., Hause, G., Finster, K., Kingshott, P., Skrydstrup, T., Meyer, R.L. 2010. Formation of palladium(0) nanoparticles at microbial surfaces. *Biotechnology and Bioengineering* 107: 206–215.

Cha, J.H., Kim, K.S., Choi, S., Yeon, S.H., Lee, H., Lee, C.S., Shim, J.J. 2007. Size-controlled electrochemical synthesis of palladium nanoparticles using morpholinium ionic liquid. *Korean Journal of Chemical Engineering* 24: 1089–1094.

Chaplin, B.P., Reinhard, M., Schneider, W.F., Schüth, C., Shapley, J.R., Strathmann, T.J., Werth, C.J. 2012. Critical review of Pd-based catalytic treatment of priority contaminants in water. *Environmental Science & Technology* 46: 3655–3670.

Choudary, B.M., Madhi, S., Chowdari, N.S., Kantam, M.L., Sreedhar, B. 2002. Layered double hydroxide supported nanopalladium catalyst for Heck-, Suzuki-, Sonogashira-, and Stille-type coupling reactions of chloroarenes. *Journal of the American Chemical Society* 124: 14127–14136.

Creamer, N.J., Deplanche, K., Snape, T.J., Mikheenko, I.P., Yong, P., Samyahumbi, D., Wood, J., Pollmann, K., Selenska-Pobell, S., Macaskie, L.E. 2008. A biogenic catalyst for hydrogenation, reduction and selective dehalogenation in non-aqueous solvents. *Hydrometallurgy* 94: 138–143.

De Corte, S., Hennebel, T., Fitts, J.P., Sabbe, T., Bliznuk, V., Verschuere, S., van der Lelie, D., Verstraete, W., Boon, N. 2011. Biosupported bimetallic Pd-Au nanocatalysts for dechlorination of environmental contaminants. *Environmental Science & Technology* 45: 8506–8513.

De Corte, S., Hennebel, T., De Gusseme, B., Verstraete, W., Boon, N. 2012. Bio-palladium: from metal recovery to catalytic applications. *Microbial Biotechnology* 5: 5–17.

De Corte, S., Bechstein, S., Lokanathan, A.R., Kjems, J., Boon, N., Meyer, R.L. 2013. Comparison of bacterial cells and amine-functionalized abiotic surfaces as support for Pd nanoparticle synthesis. *Colloids and Surfaces B: Biointerfaces* 102: 898–904.

De Gusseme, B., Hennebel, T., Vanhaecke, L., Soetaert, M., Desloover, J., Wille, K., Verbeken, K., Verstraete, W., Boon, N. 2011. Biogenic palladium enhances diatrizoate removal from hospital wastewater in a microbial electrolysis cell. *Environmental Science & Technology* 45: 5737–5745.

de Vargas, I., Macaskie, L.E., Guibal, E. 2004. Biosorption of palladium and platinum by sulfate-reducing bacteria. *Journal of Chemical Technology and Biotechnology* 79: 49–56.

De Windt, W., Aelterman, P., Verstraete, W. 2005. Bioreductive deposition of palladium(0) nanoparticles on *Shewanella oneidensis* with catalytic activity towards reductive dechlorination of polychlorinated biphenyls. *Environmental Microbiology* 7: 314–325.

REFERENCES

De Windt, W., Boon, N., Van den Bulcke, J., Rubberecht, L., Prata, F., Mast, J., Hennebel, T., Verstraete, W. 2006. Biological control of the size and reactivity of catalytic Pd(0) produced by *Shewanella oneidensis*. *Antonie Van Leeuwenhoek International Journal of General and Molecular Microbiology* 90: 377–389.

Deplanche, K., Macaskie, L.E. 2008. Biorecovery of gold by *Escherichia coli* and *Desulfovibrio desulfuricans*. *Biotechnology and Bioengineering* 99: 1055–1064.

Deplanche, K., Caldelari, I., Mikheenko, I.P., Sargent, F., Macaskie, L.E. 2010. Involvement of hydrogenases in the formation of highly catalytic Pd(0) nanoparticles by bioreduction of Pd(II) using *Escherichia coli* mutant strains. *Microbiology* 156: 2630–2640.

Ding, J.H., Gin, D.L. 2000. Catalytic Pd nanoparticles synthesized using a lyotropic liquid crystal polymer template. *Chemistry of Materials* 12: 22–24.

Feldheim, D.L., Eaton, B.E. 2007. Selection of biomolecules capable of mediating the formation of nanocrystals. *ACS Nano* 1: 154–159.

Fritsch, D., Kuhr, K., Mackenzie, K., Kopinke, F.-D. 2003. Hydrodechlorination of chloroorganic compounds in ground water by palladium catalysts: Part 1. Development of polymer-based catalysts and membrane reactor tests. *Catalysis Today* 82: 105–118.

Gadd, G.M. 2009. Biosorption: critical review of scientific rationale, environmental importance and significance for pollution treatment. *Journal of Chemical Technology and Biotechnology* 84: 13–28.

Gauthier, D., Søbjerg, L.S., Jensen, K.M., Lindhardt, A.T., Bunge, M., Finster, K., Meyer, R.L., Skrydstrup, T. 2010. Environmentally benign recovery and reactivation of palladium from industrial waste by using Gram-negative bacteria. *ChemSusChem* 3: 1036–1039.

Gericke, M., Pinches, A. 2006. Biological synthesis of metal nanoparticles. *Hydrometallurgy* 83: 132–140.

Haferburg, G., Kothe, E. 2007. Microbes and metals: interactions in the environment. *Journal of Basic Microbiology* 47: 453–467.

Hennebel, T., De Gusseme, B., Boon, N., Verstraete, W. 2009. Biogenic metals in advanced water treatment. *Trends in Biotechnology* 27: 90–98.

Hennebel, T., De Corte, S., Vanhaecke, L., Vanherck, K., Forrez, I., De Gusseme, B., Verhagen, P., Verbeken, K., Van der Bruggen, B., Vankelecom, I., Boon, N., Verstraete, W. 2010. Removal of diatrizoate with catalytically active membranes incorporating microbially produced palladium nanoparticles. *Water Research* 44: 1498–1506.

Hennebel, T., De Corte, S., Verstraete, W., Boon, N. 2012. Microbial production and environmental applications of Pd nanoparticles for treatment of halogenated compounds. *Current Opinion in Biotechnology* 23: 555–561.

Hildebrand, H., Mackenzie, K., Kopinke, F.-D. 2009. Highly active Pd-on-magnetite nanocatalysts for aqueous phase hydrodechlorination reactions. *Environmental Science & Technology* 43: 3254–3259.

Humphries, A.C., Macaskie, L.E. 2005. Reduction of Cr(VI) by palladized biomass of *Desulfovibrio vulgaris* NCIMB 8303. *Journal of Chemical Technology and Biotechnology* 80: 1378–1382.

Humphries, A.C., Mikheenko, I.P., Macaskie, L.E. 2006. Chromate reduction by immobilized palladized sulfate-reducing bacteria. *Biotechnology and Bioengineering* 94: 81–90.

Jiang, W., Kim, B.Y.S., Rutka, J.T., Chan, W.C.W. 2008. Nanoparticle-mediated cellular response is size-dependent. *Nature Nanotechnology* 3: 145–150.

Ju-Nam, Y., Lead, J.R. 2008. Manufactured nanoparticles: An overview of their chemistry, interactions and potential environmental implications. *Science of Total Environment* 400: 396–414.

Kickelbick, G. 2008. *Chemie für Ingenieure*. Pearson, Hallbergmoos, Germany.

Kimura, Y., Abe, D., Ohmori, T., Mizutani, M., Harada, M. 2003. Synthesis of platinum nano-particles in high-temperatures and high-pressures fluids. *Colloids and Surfaces A: Physicochemical and Engineering Aspects* 231: 131–141.

Kopinke, F.-D., Mackenzie, K., Koehler, R., Georgi, A. 2004. Alternative sources of hydrogen for hydrodechlorination of chlorinated organic compounds in water on Pd catalysts. *Applied Catalysis A: General* 271: 119–128.

Kopinke, F.-D., Angeles-Wedler, D., Fritsch, D., Mackenzie, K. 2010. Pd-catalyzed hydrodechlorination of chlorinated aromatics in contaminated waters—Effects of surfactants, organic matter and catalyst protection by silicone coating. *Applied Catalysis B: Environmental* 96: 323–328.

Korbekandi, H., Iravani, S., Abbasi, S. 2009. Production of nanoparticles using organisms. *Critical Reviews in Biotechnology* 29: 279–306.

Korte, N.E., Zutman, J.L., Schlosser, R.M., Liang, L., Gu, B., Fernando, Q. 2000. Field application of palladized iron for the dechlorination of trichloroethene. *Waste Management* 20: 687–694.

Le Bars, J., Specht, U., Bradley, J.S., Blackmond, D.G. 1999. A catalytic probe of the surface of colloidal palladium particles using Heck coupling reactions. *Langmuir* 15: 7621–7625.

Lengke, M.F., Fleet, M.E., Southarn, G. 2007. Synthesis of palladium nanoparticles by reaction of filamentous cyanobacterial biomass with a palladium(II) chloride complex. *Langmuir* 23: 8982–8987.

Li, X.Q., Xu, H.Z., Chen, Z.S., Chen, G.F. 2011. Biosynthesis of nanoparticles by microorganisms and their applications. *Journal of Nanomaterials*, doi: http://dx.doi.org/10.1155/2011/270974.

Li, Y., El-Sayed, M.A. 2001. The effect of stabilizers on the catalytic activity and stability of Pd colloidal nanoparticles in the Suzuki reactions in aqueous solution. *Journal of Physical Chemistry B* 105: 8938–8943.

Lloyd, J.R. 2003. Microbial reduction of metals and radionuclides. *FEMS Microbiology Reviews* 27: 411–425.

Lloyd, J.R., Byrne, J.M., Coker, V.S. 2011. Biotechnological synthesis of functional nanomaterials. *Current Opinion in Biotechnology* 22: 509–515.

Lovley, D.R. 1993. Dissimilatory metal reduction. *Annual Review of Microbiology* 47: 263–290.

Lowry, G.V., Reinhard, M. 2000. Pd-catalyzed TCE dechlorination in groundwater: solute effects, biological control, and oxidative catalyst regeneration. *Environmental Science & Technology* 34: 3217–3223.

Luo, C.C., Zhang, Y.H., Wang, Y.G. 2005. Palladium nanoparticles in poly(ethyleneglycol): the efficient and recyclable catalyst for Heck reaction. *Journal of Molecular Catalysis A: Chemical* 229: 7–12.

Mabbett, A.N., Yong, P., Farr, J.P.G., Macaskie, L.E. 2004. Reduction of Cr(VI) by "palladized" biomass of *Desulfovibrio desulfuricans* ATCC 29577. *Biotechnology and Bioengineering* 87: 104–109.

Macaskie, L.E., Humphries, A.C., Mikheenko, I.P., Baxter-Plant, V.S., Deplanche, K., Redwood, M.D., Bennett, J.A., Wood, J. 2012. Use of *Desulfovibrio* and *Escherichia coli* Pd-nanocatalysts in reduction of Cr(VI) and hydrogenolytic dehalogenation of polychlorinated biphenyls and used transformer oil. *Journal of Chemical Technology and Biotechnology* 87: 1430–1435.

Mack, C., Wilhelmi, B., Duncan, J.R., Burgess, J.E. 2007. Biosorption of precious metals. *Biotechnology Advances* 25: 264–271.

Mackenzie, K., Frenzel, H., Kopinke, F.D. 2006. Hydrodehalogenation of halogenated hydrocarbons in water with Pd catalysts: Reaction rates and surface competition. *Applied Catalysis B: Environmental* 63: 161–167.

Mandal, D., Bolander, M.E., Mukhopadhyay, D., Sarkar, G., Mukherjee, P. 2006. The use of microorganisms for the formation of metal nanoparticles and their application. *Applied Microbiology and Biotechnology* 69: 485–492.

Maruyama, T., Matsushita, H., Shimada, Y., Kamata, I., Hanaki, M., Sonokawa, S., Kamiya, N., Goto, M. 2007. Proteins and protein-rich biomass as environmentally friendly adsorbents selective for precious metal ions. *Environmental Science & Technology* 41: 1359–1364.

McNab, Jr W.W., Ruiz, R. 1998. Palladium-catalyzed reductive dehalogenation of dissolved chlorinated aliphatics using electrolytically-generated hydrogen. *Chemosphere* 37: 925–936.

Mertens, B., Blothe, C., Windey, K., De Windt, W., Verstraete, W. 2007. Biocatalytic dechlorination of lindane by nano-scale particles of Pd(0) deposited on *Shewanella oneidensis*. *Chemosphere* 66: 99–105.

Mikheenko, I.P., Rousset, M., Dementin, S., Macaskie, L.E. 2008. Bioaccumulation of palladium by *Desulfovibrio fructosivorans* wild-type and hydrogenase-deficient strains. *Applied and Environmental Microbiology* 74: 6144–6146.

Narayanan, K.B., Sakthivel, N. 2010. Biological synthesis of metal nanoparticles by microbes. *Advances in Colloid and Interface Science* 156: 1–13.

Narayanan, R., El-Sayed, M.A. 2008. Some aspects of colloidal nanoparticle stability, catalytic activity, and recycling potential. *Topics in Catalysis* 47: 15–21.

Navon, R., Eldad, S., Mackenzie, K., Kopinke, F.-D. 2012. Protection of palladium catalysts for hydrodechlorination of chlorinated organic compounds in wastewaters. *Applied Catalysis B: Environmental* 119–120: 241–247.

Nel, A.E., Mädler, L., Velegol, D., Xia, T., Hoek, E.M.V., Somasundaran, P., Klaessig, F., Castranova, V., Thompson, M. 2009. Understanding biophysicochemical interactions at the nano-bio interface. *Nature Materials* 8: 543–557.

Phan, N.T.S., Van Der Sluys, M., Jones, C.W. 2006. On the nature of the active species in palladium catalyzed Mizoroki-Heck and Suzuki-Miyaura couplings – Homogeneous or heterogeneous catalysis, a critical review. *Advanced Synthesis & Catalysis* 348: 609–679.

Philippot, J.R., Schuber, F. (eds) 1994. *Liposomes as Tools in Basic Research and Industry*. CRC Press Inc., Boca Raton, FL.

Pittelkow, M., Moth-Poulsen, K., Boas, U., Christensen, J.B. 2003. Poly(amidoamine)-dendrimer-stabilized Pd(0) nanoparticles as a catalyst for the Suzuki reaction. *Langmuir* 19: 7682–7684.

Pradeep, T., Anshup 2009. Noble metal nanoparticles for water purification: A critical review. *Thin Solid Films* 517: 6441–6478.

Rawlings, D.E. 2002. Heavy metal mining using microbes. *Annual Review of Microbiology* 56: 65–91.

Rotaru, A.E., Jiang, W., Finster, K., Skrydstrup, T., Meyer, R.L. 2012. Non-enzymatic palladium recovery on microbial and synthetic surfaces. *Biotechnology and Bioengineering* 109: 1889–1897.

Schacht, V.J., Neumann, L.V., Sandhi, S.K., Chen, L., Henning, T., Klar, P.J., Theophel, K., Schnell, S., Bunge, M. 2013. Effects of silver nanoparticles on microbial growth dynamics. *Journal of Applied Microbiology* 114: 25–35.

Schlüter, M., Hentzel, T., Suarez, C., Koch, M., Lorenz, W. G., Böhm, L., Düring, R.-A., Koinig, K., Bunge, M. 2014. Synthesis of novel palladium(0) nanocatalysts by microorganisms from heavy-metal-influenced high-alpine sites for dehalogenation of polychlorinated dioxins. *Chemosphere* 117, 462–470.

Schüth, C., Reinhard, M. 1998. Hydrodechlorination and hydrogenation of aromatic compounds over palladium on alumina in hydrogen-saturated water. *Applied Catalysis B: Environmental* 18: 215–221.

Senior, K., Müller, S., Schacht, V.J., Bunge, M. 2012. Antimicrobial precious-metal nanoparticles and their use in novel materials. *Recent Patents on Food, Nutrition & Agriculture* 4: 200–209.

Søbjerg, L.S., Gauthier, D., Lindhardt, A.T., Bunge, M., Finster, K., Meyer, R.L., Skrydstrup, T. 2009. Bio-supported palladium nanoparticles as a catalyst for Suzuki-Miyaura and Mizoroki-Heck reactions. *Green Chemistry* 11: 2041–2046.

Søbjerg, L.S., Lindhardt, A.T., Skrydstrup, T., Finster, K., Meyer, R.L. 2011. Size control and catalytic activity of bio-supported palladium nanoparticles. *Colloids and Surfaces B: Biointerfaces* 85: 373–378.

Tan, Y.W., Dai, X.H., Li, Y.F., Zhu, D.B. 2003. Preparation of gold, platinum, palladium and silver nanoparticles by the reduction of their salts with a weak reductant-potassium bitartrate. *Journal of Materials Chemistry* 13: 1069–1075.

Thakkar, K.N., Mhatre, S.S., Parikh, R.Y. 2010. Biological synthesis of metallic nanoparticles. *Nanomedicine: Nanotechnology, Biology and Medicine* 6: 257–262.

Theron, J., Walker, J.A., Cloete, T.E. 2008. Nanotechnology and water treatment: Applications and emerging opportunities. *Critical Reviews in Microbiology* 34: 43–69.

Urbano, F.J., Marinas, J.M. 2001. Hydrogenolysis of organohalogen compounds over palladium supported catalysts. *Journal of Molecular Catalysis A: Chemical* 173: 329–345.

van den Tillaart, J.A.A., Kuster, B.F.M., Marin, G.B. 1996. Platinum particle size effect on the oxidative dehydrogenation of aqueous ethanol. *Catalysis Letters* 36: 31–36.

Virkutyte, J., Varma, R.S. 2011. Green synthesis of metal nanoparticles: Biodegradable polymers and enzymes in stabilization and surface functionalization. *Chemical Science* 2: 837–846.

Vriezema, D.M., Aragones, M.C., Elemans, J.A.A.W., Cornelissen, J.J.L.M., Rowan, A.E., Nolte, R.J.M. 2005. Self-assembled nanoreactors. *Chemical Reviews* 105: 1445–1489.

Wang, Y., Du, M.H., Xu, J.K., Yang, P., Du, Y.K. 2008. Size-controlled synthesis of palladium nanoparticles. *Journal of Dispersion Science and Technology* 29: 891–894.

Yong, P., Rowson, N.A., Farr, J.P.G., Harris, I.R., Macaskie, L.E. 2002. Bioaccumulation of palladium by *Desulfovibrio desulfuricans*. *Journal of Chemical Technology and Biotechnology* 77: 593–601.

Zhang, X.L., Yan, S., Tyagi, R.D., Surampalli, R.Y. 2011. Synthesis of nanoparticles by microorganisms and their application in enhancing microbiological reaction rates. *Chemosphere* 82: 489–494.

11

ENVIRONMENTAL PROCESSING OF Zn CONTAINING WASTES AND GENERATION OF NANOSIZED VALUE-ADDED PRODUCTS

Abhilash and B.D. Pandey

CSIR-National Metallurgical Laboratory (NML), Jamshedpur, Jharkand, India

11.1. INTRODUCTION

Zinc is the most-used non-ferrous metal next to aluminum and copper. Apart from its major applications (about 70%) in the galvanizing of steel, other important areas of utilization are as sheets, anodes, castings, chemicals, micronutrients, paints, and dry cells. Zinc makes up about 75 ppm (0.007%) of the Earth's crust, making it the 24th most abundant element there. Soil contains 5–770 ppm of zinc with an average of 64 ppm. Seawater has only 30 ppb zinc and the atmosphere contains 0.1–$4\,\mu g\,m^{-3}$. As the usage of zinc increases, the gap between demand and supply widens. The possibilities of recycling zinc exist from many of these applications. Out of about 10 million tons per year (mtpa or Mt a^{-1}) of zinc production globally, about 30% is from secondary sources and 20% zinc is produced by secondary units (Ozberk et al., 1995). The element is normally found in association with other base metals such as copper, lead and iron in ores. Sphalerite, which is a form of zinc sulfide, is the most heavily mined zinc-containing ores and its concentrate with 60–62% zinc can be produced from these ores (USEPA, 1994). Other minerals from which zinc is extracted

---

Bio-Nanoparticles: Biosynthesis and Sustainable Biotechnological Implications, First Edition.
Edited by Om V. Singh.
© 2015 John Wiley & Sons, Inc. Published 2015 by John Wiley & Sons, Inc.

include smithsonite (zinc carbonate), hemimorphite (zinc silicate), wurtzite (another zinc sulfide), and sometimes hydrozincite (basic zinc carbonate). The world zinc resources are estimated to be about 1.8 Gt (gigatons) as in 2010–11. Nearly 200 Mt of reserves were economically viable in 2011; adding marginally economic and subeconomic reserves to that number, a total reserve base of 500 Mt has been identified. Large deposits are in Australia, China, Canada, and the United States. At the current rate of consumption, these reserves are estimated to be consumed sometimes between 2030 and 2055 unless wastes/secondary resources containing zinc and the end-of-life materials are completely recycled.

11.1.1. World Status of Zinc Production

The world's largest zinc producer is Nyrstar, a merasger of the Australian Zn Minerals and the Belgian Umicore. Commercially pure zinc is known as "Special High Grade" which the is 99.995% pure (USEPA, 1994; Ozberk, et al. 1995). Worldwide, 95% of zinc is mined from sulfidic ore deposits as sphalerite (ZnS) mineral, the main mining areas being China, Australia, and Peru. China produced over one-quarter of the global zinc output in 2006. As new sphalerite ore mines are becoming more difficult to find, new processes to produce zinc metal from oxidized zinc ores are being developed. Primary zinc production is carried out by two routes: (1) the hydrometallurgical process proposed in 1916; and (2) the imperial smelting process first introduced in Swansea (UK) in the 1960s. The hydrometallurgical process accounts for some 80% of the primary zinc production. The imperial smelting process currently represents less than 15% of the world's zinc production. Other technologies include zinc pressure leaching (Cominco Trail, Flin Flon, and Kidd Creek), atmospheric direct leaching (Outokumpu, Kokkola, Korea Zinc, Onsan), integrated sulfide–silicate process (Votorantim Zinc), and solvent extraction process (Skorpion Zinc).

The traditional hydrometallurgical production of zinc from its sulfides comprises roasting, leaching, and electrolysis (RLE). This process involves treatment of zinc sulfide concentrates obtained from the froth flotation of the ores, by roasting step to produce ZnO and sulfur dioxide at 900 °C. The calcine (ZnO) is sent to leaching reactors which is followed by an elaborate solution purification process and then to the electrolysis step, where special high-grade (SHG) zinc is produced (de Souza et al., 2007). Among the alternate technologies, the direct leaching dissolves zinc from its concentrates with a ferric iron solution (produced during the leaching step of the RLE process), and the pressure leaching (Sherritt Zinc Pressure Leach Process) uses autoclaves at 14–15 kg cm^2 oxygen pressure to extract zinc from its concentrates. The high oxygen pressure in the latter enables faster dissolution of zinc from the concentrate; about 90 min is generally the standard residence time.

11.1.2. Environmental Impact of the Process Wastes Generated

The production from sulfidic zinc ores generates large amounts of sulfur dioxide and cadmium vapor. Smelter slag and other residues of process also contain significant amounts of heavy metals (de Souza et al., 2007). The smelter in the Belgian towns of

La Calamine and Lombières has created mining dumps over the years. From these dumps a significant amount of zinc and cadmium was leached, thereby polluting the Geul River with such heavy metals (Mehta, 1992; Moreno and Neretnieks, 2006; Magyar et al., 2008). Levels of zinc in rivers flowing through industrial or mining areas can be as high as 20 ppm. Effective sewage treatment can greatly reduce this contaminant from the stream. Treatment along the Rhine, for example, has decreased zinc levels to 50 ppb. Concentrations of zinc as low as 2 ppm adversely affect the amount of oxygen that fishes can carry in their blood. The zinc works at Lutana, the largest exporter in Tasmania producing over 250 kilotons of zinc per year, is also responsible for high levels of heavy metals in the Derwent River. Soils contaminated with zinc through the mining of zinc-containing ores, refining, or where zinc-containing sludge is used as fertilizer can contain several grams of zinc per kilogram of dry soil. Levels of zinc in excess of 500 ppm in soil interfere with the ability of plants to absorb other essential metals, such as iron and manganese. Zinc levels of 2000 ppm to 180,000 ppm (18%) have been recorded in some soil samples.

11.1.3. Production Status in India

Hindustan Zinc Ltd (Vedanta Group) is the only integrated zinc producer in India with a proposed one million ton zinc-lead metal capacity by 2010 (Sahu et al., 2004). The details of its mining assets and corresponding smelter asset are listed in Tables 11.1 and 11.2. Binani Zinc Ltd is another producer operating mostly with imported zinc concentrates.

Mining and beneficiation generate large volumes of dumps and tailings, as well as the residue and slag produced from the smelters, causing damage to the ecosystem. Rampura Agucha mines possess an estimated geological reserve of 107.33 million metric tons of zinc. It is the world's third-largest zinc mine and is forecast to be the world's largest in the near future. It operates three underground mines and one opencast mine. Its expanded capacity was 5 Mtpa in 2007, and is poised for further expansion to 6 Mtpa by 2013. The mine presently has ore beneficiation plant

TABLE 11.1. Mining Assets of HZL during 2010–11

Sl. No.	Mines	Pb-Zn reserves (Mt)	Pb-Zn resources (Mt)	Zinc-grade (%)	Lead-grade (%)	Silver-grade (ppm)	Ore production capacity (Mt a^{-1})
1.	Rajpura Dariba	7.1	17.6	6.6	1.7	82	0.9
2.	Rampura Agucha	63.6	43.8	13.9	2.0	63	6.0
3.	Sindesar Khurd	1.96	35.2	5.8	3.8	215	1.5
4.	Zawar	7.2	41.9	4.4	2.3	40	1.2

TABLE 11.2. Smelting Assets of HZL during 2010–11

Sl. No.	Smelter	Pyrometallurgical capacity		Hydrometallurgical capacity		Captive power plant (MW)
		Zinc (t a^{-1})	Lead (t a^{-1})	Zinc (t a^{-1})	Lead (t a^{-1})	
1.	Chanderiya	105,000	35,000	420,000	50,000	234
2.	Dariba	–	–	210,000	100,000	160
3.	Debari	–	–	88,000	–	29
4.	Vizag	–	–	56,000	–	–

generating 3.7 Mt of tailings per year, with a stock of 24.6 million Mt as dense slurry. The average composition is 1.5–2.5% Zn, 10% Fe, and 25 ppm Ag. Mine tailings are also being used as filler in concrete, calcium silicate brick, cellular concrete, clay brick, and cement. Apart from occupying large areas, these tailings are a potential source of surface pollution around the mines due to biochemical oxidation of pyrite and metallic sulfides (Sahu et al., 2004).

11.1.4. Recent Attempts at Processing Low-Grade Ores and Tailings

Numerous proven and innovative processes for the recovery of zinc from ores and concentrates are available, including:

- neutral leach (acid leach, super hot acid leach), jarosite, goethite, and hematite process routes for solution purification, pressure acid leach, and solvent extraction;
- hydrometallurgical processing steps (e.g., processing of jarosite and goethite in washing, leaching, pressure leaching steps, and inertizing of residues);
- minerals processing steps (e.g., flotation of residues to recover Ag);
- sinter-shaft furnace processes (e.g., imperial smelter, lead shaft furnace or slag fuming, or half-shaft furnaces);
- Waelz kiln and clinker processes;
- direct reduction (e.g., of haematite);
- calcination in a rotary kiln or circulating fluidized bed reactor;
- electrothermal processes (e.g., electric arc furnace, St Joe furnace, and plasma arc furnace);
- vertical (New Jersey) retort furnace;
- flash smelting operations such as the Kivcet furnace or the flame cyclone; and
- bath smelting technologies (e.g., Isasmelt and QSL technologies).

The long-term impact of tailing deposits on the environment is noted for two types of cover: soil and water. Oxygen intrusion leads to sulfide oxidation and the generation of acid mine drainage containing toxic metals. The primary acidity generated in the tailings from the sulfide oxidation is, to a large extent, neutralized by the buffering minerals in the deposit. However, when the water leaves the deposit, more acidity is generated due to the oxidation of mainly the ferrous iron (latent acidity); the acidity is not neutralized in the interior of the deposit (Deshpande and Shekdar, 2005; Rai and Rao, 2005).

11.2. PHYSICAL/CHEMICAL/HYDROTHERMAL PROCESSING

Physical methods (flotation and altered mining) and chemical treatments (hydrometallurgy) are the common methods opted worldwide for exploiting such resources. Hydrothermal synthesis is also a process in which raw materials or residues/intermediate products are processed into a high-quality/value-added material in ambient conditions or in a high-pressure reactor to generate marketable products.

Several processes have been developed for the recycling of low-grade ores or concentrates and tailings, previously considered as waste in the zinc industry. Some of the practices being carried out internationally are described in the following sections.

11.2.1. Extraction of Pb-Zn from Tailings for Utilization and Production in China

The project involves the processing of 300 tons per day of tailings (Qingzhong, 2000), with a production line of 5000 tons per year of lead and zinc. There are nearly 250,000 tons of tailings and complex poly-metallic ores with low gold content in five major mining areas of Zhenghe County, China and more than 40% of unmanageable tailings have a lead and zinc content above 5%.

11.2.2. Vegetation Program on Pb-Zn Tailings

The experience of Canex Placer Ltd. in establishing vegetation on a Pb-Zn tailing pond is worth mentioning. The company has been producing 8 Mt of Pb-Zn (Lawrence, 1974) during 23 years of mining and milling operations.

11.2.3. Recovering Valuable Metals from Tailings and Residues

The BHP Billiton silver-lead-zinc Cannington mine is the largest single mine producer of silver in the world. The waste tailing from the unit contains small amounts of silver, lead, and zinc. Thiourea-assisted leaching was proposed for total extraction of silver (Reuter et al., 1995). Stope mining at Rosh Pinah Zinc Corporation (Pvt) Ltd, an underground mine at Rosh Pinah in southern Namibia, produces zinc and lead sulfide concentrates from 0.7 Mt a^{-1} run-off mines (ROM) containing around 6–10% zinc and

1.5–3.5% lead. Recently, a massive continuous low-grade ore body was identified with around 3% zinc. This open-stope mining from two areas has a production rate of up to 2 Mtpa with pre-beneficiated low-grade ore tails back-filled to the mined-out stopes (Fourie et al., 2007). The residue is discarded as a cake from a Waelz kiln processing Pb-Zn carbonate ores. The zinc plant residue containing 11.3% Zn, 24.6% Pb, and 8.3% Fe was blended with H_2SO_4 and subjected to a process comprising roasting, water leaching, and finally NaCl leaching. About 86% Zn and 89% Pb from the zinc residue could be recovered by leaching with $200\,g\,L^{-1}$ NaCl (Turan et al., 2004).

11.2.4. Extraction of Vanadium, Lead and Zinc from Mining Dump in Zambia

The dump consists of 8 Mt of material above the ground (Watson and Beharrell, 2006). This material is divided into particle sizes of three groups: small particles less than 20 µm in diameter, rod-like particles of the order of 10–20 µm diameter and 50 µm length, and large particles of 50–100 µm diameter. The large particles contain substantially all the iron together with almost all the Pb, V, and Zn. In the laboratory the small particles are removed with a sieve which allows the passage of <25 µm sized particles, mainly consisting of silica and calcium carbonate. The remainder consists of the rod-like calcium sulfide and the lead-containing compound, which is fairly magnetic. This material is passed through the magnetic separator. The magnetic fraction so obtained has a Pb fraction along with vanadium and zinc, and some amount of Fe-stained calcium sulfate. About 25% of the feed material is collected as the magnetic fraction, which contains 8% Pb, 6% Zn, and 0.7% V.

11.2.5. Recovery of Zinc from Blast Furnace and other Dust/Secondary Resources

Blast furnace flue dusts are a mixture of oxides expelled from the top of the blast furnace, whose major components are iron oxides. They also contain zinc, silicon, magnesium, and other minor elements as oxides in lesser amounts. The direct recycling of flue dust is not usually possible since it contains some undesirable elements, such as zinc and alkaline metals. In some cases, the dust also contains toxic elements such as zinc, cadmium, chromium, and arsenic which cannot be diverted for landfill. By selective leaching of the dust by sulfuric acid at low acid concentration and room temperature, high Zn (>80%) recovery is obtained. The metal-rich solution is subjected to purification-extraction and electro-winning of zinc (Asadi Zeydabadi et al., 1997).

The Enviroplas process developed by Mintek can treat wastes such as lead blast furnace slag, electric arc furnace dust, and neutral leach residues from the zinc industry. With a DC arc furnace and an imperial smelting process (ISP) lead splash condenser, zinc is recovered as prime western grade metal with simultaneous production of a disposable slag (Abdel-latif, 1997). Zinc was found as smithsonite and hemimorphite in the lead flotation tailings of Dandi mineral processing plant in northwestern Iran. In order to extract zinc from these tailings, agitated leaching with 2 M sulfuric acid for 2 h at 60 °C resulted in a maximum recovery of 98% Zn (Espiari et al., 2006).

11.2.6. Treatment and Recycling of Goethite Waste

Goethite (FeOOH) is an iron-rich toxic waste that is generated during the hydrometallurgical processing of zinc metal. Due to the presence of impurities in the goethite (Zn, Pb, Ni, Cd, Cu, As, etc.) and the large amount produced annually, the disposal of goethite waste is a serious environmental problem (Pelino et al., 1996). To recycle the goethite, waste production of glass ceramic material has been explored by adding other wastes. To obtain a glass with adequate properties, the goethite waste is mixed with sand, tuff, feldspar, limestone, dolomite, pumite, and glass cullet in various proportions to produce a glass-ceramic material from the molten glass using a controlled cooling treatment, yielding a 75% crystalline phase.

The leaching behavior of three Fe-containing sphalerite minerals with 0.45, 11.40, and 12.90 wt% Fe has been investigated at pH 1 in O_2-purged $HClO_4$ solutions in the temperature range 25–85 °C. For all the sphalerite samples, the leach data closely fitted a rate expression which incorporates both a shrinking sphere term and rate control via diffusion through a reacted surface layer (Weisener et al., 2004).

11.2.7. Other Hydrometallurgical Treatments of Zinc-based Industrial Wastes and Residues

Processing of nearly 1,000,000 tons of flotation tailings and 300,000 tons of slag containing considerable amounts of zinc in Turkey, abandoned almost 70 years ago, has been assessed by acid leaching (Ek, 1989; Kurama and Göktepe, 2003; Ngenda et al., 2009). The slag contains approximately 29% Fe, 13% Zn, 3% Pb, and 2% S. The acid leaching of slag in two stages shows 77.5% Zn extraction under the optimum conditions using 1.85 normality (N) and 4.07 N H_2SO_4 solutions at 1:10 (solid:liquid ratio or S/L) pulp ratio and 55 °C and 95 °C temperature in the first and second stage, respectively. The pressure leaching gives the highest extraction efficiency of 87% Zn and 80% Fe, but may have extremely high investment and operation costs.

The digestion method developed for the treatment of copper smelter slags (Kurama and Göktepe, 2003) can be applied to zinc hydro plant residues by digesting the material for 24 h in concentrated sulfuric acid (48%). The roasting of the digested material for 2 h at 750 °C selectively removes iron as Fe_2O_3 (hematite). Leaching with water at 40 °C yields nearly 98.7% Zn, 99.9% Cu, and 100% Cd with only 6.4% Fe in solution.

For the recovery of zinc and cadmium from zinc electrolyte purification waste, the material was leached out in zinc-depleted electrolyte and then cadmium was cemented out. A sulfate leach liquor containing 141 g L^{-1} Zn, 53.4 g L^{-1} Cd, 0.002 g L^{-1} Cu, 0.011 g L^{-1} Co and 0.003 g L^{-1} Ni is obtained. In the cementation stage, cadmium recovery exceeds 99.9% with a cadmium purity higher than 97 wt% (Ngenda et al., 2009). Zinc from the solution with 150 g L^{-1} Zn and 0.005 g L^{-1} Cd may be recovered by electrolysis. Solvent extraction of zinc using di-2-ethylhexyl phosphoric acid (D2EHPA) has also been investigated to recover zinc from an industrial leach residue with up to 99% extraction.

From various other sources of zinc such as slag, spent batteries, zinc ash, and zinc dross, high zinc could be extracted by roasting in the presence of 10% weight of NH_4Cl as fluxing agent for +1.25 mm size particles. The fine fraction of zinc dross, zinc ash, and the slag is treated by acid (Mackiw and Veltman, 1967; Ligiane et al., 2007; Nakamura et al., 2008; Ruşen et al., 2008; Safarzadeh et al., 2009) to yield high leaching efficiency (99%).

The process of leaching coupled with solvent extraction-electrowinning (SX-EW) and cementation is also applied for extraction of zinc from Cd–Ni plant residues (Safarzadeh et al., 2009). By leaching-solvent extraction/ion-exchange process, separation and recovery of cobalt from a zinc plant residue is achieved (Rabah and El Sayed, 1995; Vahidi et al., 2009).

A simple hydrometallurgical process has been developed to recover zinc and sulfur as zinc concentrate and silver and lead as lead-silver sponge from silver concentrate. 90% Pb, 80% Ag, 80% Zn and 87% sulphur present in the silver concentrate can be recovered. The process can be incorporated into the existing plant without any problem (Raghavan et al.,1998).

The leaching of low-grade oxide zinc ore and simultaneous integrated selective extraction of zinc have also been investigated using a small-scale leaching column and laboratory-scale box mixer-settlers. Di-2-ethythexyl phosphoric acid (D2EHPA) dissolved in kerosene is used as an extractant. The leach liquor with 32.6 g L^{-1} zinc is obtained by heap leaching and metal is recovered by solvent extraction (Qin et al., 2007).

The effect of microwave-assisted extraction of zinc from a bulk sphalerite (ZnS)/pyrrhotite (FeS) concentrate produced by flotation of tailings from the Rampura Agucha Pb-Zn mines in India has been reported (Krishnan et al., 2007). The zinc leaching in acid is found to be rapid with 90% dissolution after 6 min; further irradiation increases the zinc extraction to 96% after 16 min. Power consumption for 90% Zn recovery is calculated to be 0.36 kWh kg^{-1} concentrate.

A room-temperature cathodic electrolytic process has been developed in the laboratory to recover zinc from industrial leach residues. The various parameters affecting the electroleaching process were studied using a statistically designed experiment. Since significant amounts of iron are also present in the leach liquor, attempts were made to purify it before zinc recovery by electrowinning. Reductive dissolution and the creation of anion vacancies are found to be responsible for the dissolution of zinc ferrite present in the leach residue (Bhat and Natarajan, 1986, 1987; Reddy et al., 1990; Kori et al., 2002).

Pelletizing and alkaline leaching of powdery low-grade zinc oxide ores.
Low-grade powdery zinc oxide ores (5.2% Zn, <2 mm size) are mixed with 5 wt% cement, pelletized (5–8 mm size) and solidified. After 3 days, 92% zinc is

leached from the pellet in a column. Decreasing the solidification time can reduce the reaction time and increase the dissolution of zinc (Feng et al., 2007).

Influence of flotation reagents. Lead flotation tailings of the Dandy mineral processing plant in northwestern Iran contain oxidized zinc minerals. Zinc recovery from the tailing is reported in the presence of different flotation reagents and dispersing reagent. Among the dispersants tested, sodium hexametaphosphate gives a higher zinc grade of 40.7% with 70% recovery (Kashani and Rashchi, 2008).

Laboratory-scale investigations by Hindustan Zinc Limited (HZL), India using a 3 inch flotation column has confirmed the production of zinc concentrate of the desired grade. The column used in the cleaning circuit improves the zinc grade from 48–50% to 52–55% without any loss in the recovery (Mittal et al., 2000).

11.3. BIOHYDROMETALLURGICAL PROCESSING: INTERNATIONAL SCENARIO

Modern commercial applications of biohydrometallurgy for processing ores became a reality in the 1950s with the advent of copper bioleaching at the Kennecott Copper Bingham Mine. Bioleaching has now been extended to the commercial extraction and recovery of cobalt (Brierly and Brierly, 2001). Near-future commercial applications will likely remain focused on recoveries of copper, gold, and possibly nickel. Recent technical advances show that very refractory chalcopyrite can be successfully bioleached. Bioleaching can also be utilized to oxidize zinc sulfides. Although slower than chemical leaching, bioleaching of sphalerite (ZnS) does not require gaseous oxygen since Fe(II) oxidation is carried out by the microorganisms. In this case, air is required for bacterial respiration. This basis has been applied by numerous groups around the globe for treating high- and low-zinc-containing zinc concentrates, ores, and tailings (Viera et al., 2007).

11.3.1. Bioleaching of Zn from Copper Mining Residues by *Aspergillus niger*

In Canadian mines, large amounts of soil and rocks with low amounts of metals were exposed to wind, rainfall, and other weathering factors (e.g., temperature), producing exhaustive leachate of metals (Zn, Cu, Cd, Ni, etc.). Various groups of native microorganisms such as bacteria and fungi can accelerate the production of acid and enhance the leachability of metals to a commercially attractive concentration, thereby adding value to the mining residues. The concentration of metals (mg) per kg of mine residue is Cu: 7245; Fe: 26470; Zn: 201; and Ni: 27. *A. niger* is capable of producing gluconic and citric acids up to 15% in the presence of the mining residues, which can help in the solubilization of copper and zinc. Maximum solubilizations of 68% Cu, 46% Zn and 34% Ni are reported (Mueller et al., 1995; Castro et al., 2000; Mulligan et al., 2004; Mulligan and Galvez-Cloutier, 2005).

11.3.2. Bioleaching of Zinc from Steel Plant Waste using *Acidithiobacillus ferrooxidans*

A solid waste sample of electric arc furnace (EAF) steel-making dust from a steel plant in Turkey contained 1.23% Zn and 54.73% Fe in the dominant phases of magnetite, hematite, and calcite. Maximum extraction using *Acidithiobacillus ferrooxidans* is achieved at pH 1.3 and a solid concentration of 1% w/v to a level of 35% Zn and 37% Fe (Bayat et al., 2009).

11.3.3. Bacterial Leaching of Zinc from Chat (Chert) Pile Rock and Copper from Tailings Pond Sediment

Shake-flask experiments with *Acidithiobacillus ferrooxidans* at 26 °C leached 38% zinc from the pulverized rock in 15 days and extracted copper completely from a native sediment in 24 h. The leaching of copper from the native tailings pond sediment is relatively rapid, with complete leaching taking only 12 h in a fermentor with aeration and agitation due to small particle size. Based on the results for leaching of copper from the tailings pond sediment, two approaches that may be economically feasible for scale-up are the use of a shallow aerated pond containing a layer of sediment or the use of a continuous stirred tank reactor (CSTR) in the airlift mode.

11.3.4. Dissolution of Zn from Zinc Mine Tailings

Use of *Acidithiobacillus ferrooxidans* for the recovery of zinc metal and zinc sulfide from zinc mine tailings has been reported (Hsu and Harrison, 1995; Kaewkannetra et al., 2009). From 10,000 mg L^{-1} of initial Zn concentration, 97% zinc solubilization can be achieved from lower size fractions in 58 days. A better microbial activity is observed in case of coarser fractions. The governing attribute to the increased leaching by bacteria is because of the high ratio of Fe(III):Fe(total) encountered during the process.

11.3.5. Microbial Diversity in Zinc Mines

Total and soluble zinc is typically reported to be nearly 6% and 1% of the tailings mass, respectively (Zhang et al., 2007a). In tailings, heavy metals are unquestionably the major ecological factor responsible for the microbial distribution. However, it is also important for the surviving species to adapt to secondary ecological factors such as nutrients, pH, and oxygen concentration (Zhang et al., 2007a; Mendez et al., 2008). The results showed that the tailings had no significant gradient of heavy metals with the depths. Bacterial diversity at depths of 5–10 cm (Dy0), 100 cm (Dy1), and 200 cm (Dy2) was evaluated on the basis of their growth characteristics. The dominance of Gamma proteobacteria in layer Dy0 was possible due to the heterotrophic feature of these clones, which were mainly close to *Stenotrophomonas*, *Pseudomonas*, *Enterobacter*, and *Shigella* sp. At prolonged depths of Dy1, *Actinobacteria*, *Gemmatimonadetes*, and *Nitrospira* sp. prevailed. *Acidobacteria* also seem to prefer anaerobic activity in the presence of heavy metals, due to their increase in number with depth to Dy2.

11.3.6. Chromosomal Resistance Mechanisms of *A. ferrooxidans* on Zinc

Pulsed-field gel electrophoresis (PFGE) has been used to examine chromosomal DNA from various strains of *A. ferrooxidans* isolated from different environments and pilot plants for processing gold-bearing concentrates, and strains experimentally adapted to high-zinc and high-arsenic concentrations in growth medium. The restriction endonuclease XbaI digested *A. ferrooxidans* chromosomal DNA into a number of fragments sufficient for identification of their size and calculation of the size of the entire genome (2855f44 kb). The enhanced resistance of *A. ferrooxidans* to toxic metals is gained through increase of the copy number of resistant genes and enhanced synthesis of proteins involved in resistance (Kondratyeva et al., 1995).

11.3.7. Bioleaching of Zinc Sulfides by *Acidithiobacillus ferrooxidans*

Bioleaching of three zinc sulfides (marmatite, sphalerite, and ZnS, synthetically prepared) with *Acidithiobacillus ferrooxidans* was investigated by the electrochemical methods. The experimental results show that marmatite has the highest dissolution rate of the three zinc sulfides in the presence of bacterial strains due to the difference in physicochemical properties (Boon et al., 1998; Shi et al., 2006). Compared to synthetic ZnS, the iron is found to exist as pyrite in two concentrates and the dissolution of zinc sulfides is accelerated for galvanic cell effect. More iron existing in the crystal lattice of marmatite solid solution is released to liquid phase and is oxidized into Fe^{3+} ions by bacterial strains to accelerate the process. The bacterial leaching of zinc sulfides could become diffusion controlled when insufficient bacterial oxidation of the sulfur layer occurred, especially in the more rapid marmatite leaching reaction. The zinc sulfide bioleaching is significantly affected by mineralogical properties and leach solutions, which is supported by the electrochemical measurements based on the working electrode of zinc sulfides carbon paste electrode.

11.3.8. Bioleaching of High-Sphalerite Material

Different strategies were employed for adapting *Acidithiobacillus ferrooxidans* cells to both Zn^{2+} ions and high-grade sphalerite concentrate (Haghshenas et al. 2009a). The serial sub-culturing was found to be a very efficient strategy for adapting *A. ferrooxidans* cells to higher Zn^{2+} concentrations, as well as high-grade sphalerite concentrate, provided that a suitable protocol was employed. Adaptation of *A. ferrooxidans* cells to $30 g L^{-1}$ Zn^{2+} significantly enhanced the rate of bioleaching of Zn^{2+} from high-grade sphalerite concentrate. Pre-adaptation to Zn^{2+} ions also shortened the time required to adapt the cells to the concentrate. It is suggested that during adaptation of *A. ferrooxidans* cells to high-grade sphalerite concentrate, several events occur such as strengthening of the suspended bacterial population, a lowering in sensitivity of the cells to solids and shear, adaptation to the toxic effect of Zn^{2+} ion, and enhancement in ability of the cells to degrade the sulfur layer.

In another attempt, the feasibility of using iron- and sulfur-oxidizing bacteria for the acid leaching of a high-grade Pb-Zn ore material was tested. Three strains (ATCC 13661, NCIMB 13537, and C2-TF) of *Acidithiobacillus ferrooxidans* and two strains (MT-TH1 and MT-13) of *Sulfobacillus thermosulfidooxidans* were tested in this study. The bioleaching was monitored by measuring the dissolved metals and by X-ray diffraction analysis of leach residues. The bioleaching efficiency varied between 0.014 and 0.35. The maximum dissolution of lead was achieved with the mesophilic *A. ferrooxidans* (NCIMB 13537) at 30 °C. The maximum recovery of zinc was achieved with moderately thermopilic *S. thermosulfidooxidans* (MT-TH1) at 45 °C (Shi et al., 2006; Rehman et al., 2009).

11.3.9. Bioleaching of Low-Grade ZnS Concentrate and Complex Sulfides (Pb-Zn) using Thermophilic Species

A low-grade sphalerite concentrate was treated using native cultures of *Acidithiobacillus ferrooxidans* and *Sulfobacillus thermosulfidooxidans*. The bioleaching experiments were carried out in shake flasks at pH 1.5, 180 rpm, and temperatures of 33 °C and 60 °C for mesophilic and thermophilic bacteria, respectively. Compared to the use of laboratory reference strains or control conditions, zinc recovery from the respective concentrate was greater when native isolates were employed. The maximum zinc extraction reached was 87% after 30 days using native thermophile *S. thermosulfidooxidans* culture (Rehman et al., 2009).

The bioleaching of the complex Pb-Zn concentrate (with 43% Zn) has also been investigated using mesophilic (at 30 °C), moderate (at 50 °C), and extreme thermophilic (at 70 °C) strains of acidophilic bacteria. The results show that zinc is selectively extracted using the acidophilic strains of bacteria leaving almost completely lead (>98%) in the residue. Moderate thermophiles display superior kinetics of dissolution of zinc (97%) as compared to other two groups of bacteria in the pH regime of 1.7–1.9 (Haghshenas et al., 2009a, b).

11.3.10. Improvement of Stains for Bio-processing of Sphalerite

Three induced mutation methods by ultraviolet radiation, microwave, and multiplex factors were compared to improve the bioleaching activity of *Acidiothiobacillus caldus* in processing the sphalerite mineral. The experimental results show that the multiplex method increases the growth rate of 10% by ultraviolet radiation, and microwave is better than that of the single UV-induced (7.3%) or microwave-induced (5%) method. Moreover, there is a close relationship between the sulfate concentration and the bioleaching rate of sphalerite. The improvement of bioleaching rate by induced mutant strains is attributed to their faster growth rates and stronger sulfuric-acid-producing capabilities compared to that of the wild strains (Devecia et al., 2004; Mousavi et al., 2007; Xia et al., 2007).

11.3.11. Tank Bioleaching of ZnS and Zn Polymetallic Concentrates

Due to the lower value of zinc compared to copper and nickel, the key objective for ZnS bioleaching is to minimize costs. This cannot be achieved by a simplified hydrometallurgical circuit involving direct electrowinning of Zn. Bench-scale testing at Mintek using mesophilic and moderately thermophilic cultures for bioleaching of Petiknäs Zn concentrate has achieved zinc concentrations of $75\,g\,L^{-1}$ in the leach solution (Mousavi et al., 2006).

11.3.12. Large-Scale Development for Zinc Concentrate Bioleaching

GeoBiotics LLC has developed a proprietary heap bioleaching technology for the processing of sulfides of base metals and gold concentrates. In this process, known as GEOCOAT®, thickened flotation concentrate is contacted during heap stacking with gravel-sized support rock forming a thin adherent concentrate coating on the support rock particles (Soleimani et al., 2009). The stacked heap has an open structure and is highly permeable to the flow of solution and air. Acid solution, which contains acidophilic bacteria, is circulated through the heap to bio-oxidize and leach the contained metals. The heat produced by the exothermic oxidation reactions causes the internal temperature of the heap to rise. Heat is transferred to the percolating solution and to the air blown up through the heap. Rates of solution application and aeration are varied to control the heap temperature and maintain it within the optimum range for bacterial activity.

GeoBiotics and Kumba Resources have investigated the feasibility of applying the GEOCOAT® process to the leaching and recovery of zinc from a low-grade sphalerite concentrate produced from accumulated flotation tailings at Kumba's Rosh Pinah zinc mine in Namibia. The concentrate contained 18% Zn, 4.5% Fe, 12.8% S, and 0.2% Cu. To confirm that a GEOCOAT® heap to bioleach this concentrate would operate autothermally, a large engineering column test was conducted. The insulated stainless steel column, 6.5 m high and 1.2 m in diameter, was equipped with sample ports, airflow measurement, oxygen analysis, and thermocouples to monitor temperatures during the test. The column was filled to a 6 m height with concentrate-coated support rock which was produced using a batch-coating method. After acid stabilization to reduce the solution pH to 1.5–1.8, the column was inoculated with an adapted mixed mesophilic bacterial culture. Through control of the exothermic biooxidation reactions and the solution flow and aeration rates, the temperature in the column was increased from ambient to a maximum of 49 °C. The solution application rate and the aeration rate were adjusted to control the temperature and prevent it from rising beyond the maximum tolerable by mesophiles. The column was operated for a total of 90 days.

Final zinc dissolution after 90 days was found to be 91% with a corresponding sulfide sulfur oxidation of 89%. Concentrations of in excess of $90\,g\,L^{-1}$ Zn in the column effluent did not appear to inhibit the microbial oxidation.

11.3.13. Scale-up Studies for Bioleaching of Low-Grade Sphalerite Ore

Tests were conducted using a bench-scale column leach reactor while inoculating mesophilic (*Acidithiobacillus ferrooxidans*) and thermophilic (*Sulfobacillus*) iron-oxidizing bacteria initially isolated from the Sarcheshmeh chalcopyrite concentrate (Kerman, Iran) and Kooshk sphalerite concentrate (Yazd, Iran), respectively. In the inoculated column, jarosite and elemental sulfur were formed. The leaching rate of sphalerite increased with dissolved ferric ion concentration. Cell population of bacteria in the solution was found to be higher at lower pH. Furthermore, the effect of a decreasing particle size on the rate of zinc leaching was enhanced at low pH values. The maximum zinc recovery was achieved using a thermophilic culture. Zinc dissolution reached 72% and 85% for the mesophilic and thermophilic strains, respectively, after 120 days at the column temperatures in the range 28–42 °C.

In another attempt, the use of a draft tube fluidized bed bioreactor (DTFBB) was successfully demonstrated for bioleaching of zinc from a sphalerite-bearing low-grade ore. A strain of the thermophilic bacterium, *Sulfobacillus*, isolated from the Kooshk lead and zinc mine near the city of Yazd (Iran) was tested at 47–72 °C. This was compared to leaching with a strain of the mesophile *Acidithiobacillus ferrooxidans*, adapted to the high levels of zinc and the low-grade ore at 18–42 °C. The best conditions for leaching by mesophilic and thermophilic strains occurred at 10% w/v pulp density and pH 1.2–1.8, giving 87% and 91.4% Zn recovery, respectively, after 9 days (Sampson et al., 2005).

11.3.14. Zinc Resistance Mechanism in Bacteria

Varied mechanisms of zinc resistance are found. These mechanisms range from reduced uptake to efflux, external and internal sequestration, and, in some cases, transformation of metals to less-toxic forms. All these mechanisms are aimed at reducing the intracellular concentration of a metal in order to protect the cellular targets. Often, this may mean reducing the intracellular concentration of free metal ions as they are likely to be more toxic than the bound ions. The resistance mechanisms for essential metal ions are intricately interwoven with metal homeostasis mechanisms in order to ensure the survival of the cell, under both metal-excess and metal-depleted conditions (Bramaprakash et al., 1988; Jyothi et al., 1989; Choudhury and Srivastava, 2007).

11.4. BIOHYDROMETALLURGICAL PROCESSING: INDIAN SCENARIO

Biohydrometallurgical processing has evolved as an eco-friendly, economic and efficient technique for metal extraction from such resources. Despite the lack of commercial operations, this tool is a very viable one for tackling mine wastes and replenishing metals, that is, taking care of the environment. This section highlights some of the work carried out for Indian mine wastes using biohydrometallurgical options.

11.4.1. Electro-Bioleaching of Sphalerite Flotation Concentrate

Electrobioleaching of a sphalerite flotation concentrate at a positive potential of +0.4 V (saturated calomel electrode, SCE) was found to be more efficient than at a negative potential of −0.5 V(SCE). At positive potentials, bio-oxidation of elemental sulfur during the electro-oxidation step is favored, and the direct attack mechanism becomes predominant. The deleterious effect of various applied potentials in the positive and negative ranges can be explained based on the current generated during the electrobioleaching process. *Acidithiobacillus ferrooxidans* pre-adapted to the sphalerite concentrate was found to be more efficient in the electrobioleaching of zinc compared to the unadapted strain (Selvi et al., 1998).

11.4.2. Bioleaching of Zinc Sulfide Concentrate

Bioleaching of a zinc sulfide concentrate of Sikkim Mining Corporation containing 33.7% Zn, 16.5% Fe, 1.6% Pb, and 25.6% S, and minor amounts of Co, Cu, and Ni is also reported (Ghosh et al., 2004). A mixed consortium of acidophilic microorganisms isolated from a zinc tailing pond of Rampura Agucha mines has been used for leaching purposes in a four-stage continuous bioreactor during scale-up operations (Pani et al., 2003). Continuous bioreactor leaching is shown to keep the bacteria in a highly active log phase resulting in 83.40% zinc dissolution from the concentrate.

Bioleaching of polymetallic sulfide minerals of Gorubathan ore in Darjeeling, West Bengal was investigated using *Acidithiobacillus ferrooxidans*. The results show the optimization of various process parameters, such as pH, particle size, and time of leaching for the raw ore, as well as flotation concentrates. Leaching at low pH and at room temperature in the presence of microbes has shown influence on the metal recovery of Zn and Cu. Lead and silver are retained in the residual solid fraction. About 100% of zinc is recovered by bioleaching within 15 days of leaching of both middle-sized and coarser particles (Adhikari et al., 2007).

11.4.3. Bioleaching of Moore Cake and Sphalarite Tailings

Leaching of Moore cake using *Acidithiobacillus ferrooxidans* has also been attempted. During production of primary zinc by the roast–leach electrolysis process, Moore cake is generated as leached residue in which zinc is mostly associated with iron as zinc ferrite ($ZnO\ Fe_2O_3$). Moore cake obtained from HZL, Udaipur contains total zinc in the range 11–18% with nearly 6% Pb and up to 29% iron. Maximum zinc recovery is found to be ~40% in a percolation experiment (Gupta et al., 2003).

Bioleaching of sphalerite tailings has been successfully carried out to recover 95% zinc using *A. ferrooxidans* in the presence of a mined pyrite (Thakur et al., 1994). The presence of pyrite has helped in maintaining the acidity of the medium for bacterial growth, which in turn generates ferric sulfate for solubilization of zinc sulfide to zinc sulfate. A direct leaching mechanism is also interpreted by virtue of attachment of bacteria on the surface of sphalerite (Hossain et al., 2004).

11.5. SYNTHESIS OF NANOPARTICLES

Chemical methods of synthesis of materials play a crucial role in designing and discovering new materials and also in providing better and less cumbersome methods for preparing known materials. A large variety of inorganic solid materials has been prepared in recent years by the traditional ceramic method, which involves mixing and grinding powders of the constituent oxides, carbonates, and such compounds and heating them at high temperatures with intermediate grinding when necessary (Rao, 1993). A wide range of conditions, often bordering on the extreme conditions (e.g., very high temperatures or pressures, very low oxygen fugacities, and rapid quenching) have been employed in materials synthesis. The low-temperature chemical routes, however, are of greater interest. Noteworthy chemical methods of synthesis include the precursor method, coprecipitation, and soft-chemistry routes, the combustion method, the sol-gel method, topochemical methods, and high-pressure methods. In this section, we discuss the synthesis of inorganic solids by chemical methods with a few examples, especially oxide materials including superconducting cuprates synthesized by these means (Rao, 1993).

Various types of chemical reactions have been used for the synthesis of solid materials. Some of the common reactions employed for the synthesis of inorganic solids are as follows:

1. Decomposition of component A(s) to B and C: $A(s) \rightarrow B(s) + C(g)$
2. Combination of A(s) and B(g): $A(s) + B(g) \rightarrow C(s)$
3. Metathetic (combining (1) and (2) above) to form C(s) and D(g): $A(s) + B(g) \rightarrow C(s) + D(g)$
4. Addition of A(s) with a solid, liquid or gaseous phase: $A(s) + B(s) \rightarrow C(s)$; $A(s) + B(l) \rightarrow C(s)$; or $A(g) + B(g) \rightarrow C(s)$
5. Exchange reaction AX(s): $AX(s) + BY(s) \rightarrow AY(s) + BX(s)$; or $AX(s) + BY(g) \rightarrow AY(s) + BX(g)$

Typical examples of the above simple reactions include the following (1-5):

$$CaCO_3(s) \rightarrow CaO(s) + CO_2(g) \quad (1)$$
$$YBa_2Cu_3O_6(s) + O_2(g) \rightarrow YBa_2Cu_3O_7(s) \quad (2)$$
$$Pr_6O_{11}(s) + 2H_2(g) \rightarrow 3Pr_2O_3(s) + 2H_2O(g) \quad (3)$$
$$ZnO(s) + Fe_2O_3(s) \rightarrow ZnFe_2O_4(s) \quad (4)$$
$$ZnS(s) + CdO(s) \rightarrow CdS(s) + ZnO(s) \quad (5)$$

In the area of nanotechnology, the development of techniques for the controlled synthesis of metal nanoparticles of well-defined size, shape, and composition is a huge challenge. Metal nanoparticles exhibit unique electronic, magnetic, catalytic, and optical properties that are different from those of bulk metals. This could result in interesting new applications that could potentially be utilized in the biomedical sciences and areas such as optics and electronics. Various chemical and physical synthesis methods, aimed at controlling the physical properties of the particles, are currently employed in the production of metal nanoparticles. Most of these methods are still in the development stage and problems are often experienced with stability of the nanoparticle preparations, control of the crystal growth, and aggregation of the particles.

The use of the highly structured physical and biosynthesis activities of microbial cells for the synthesis of nanosized materials has recently emerged as a novel approach for the synthesis of metal nanoparticles. The interactions between microorganisms and metals have been well documented and the ability of microorganisms to extract and/or accumulate metals is already employed in biotechnological processes such as bioleaching and bioremediation. Many microbes are known to produce nanostructured mineral crystals and metallic nanoparticles with properties similar to chemically synthesized materials (Table 11.3), while exercising strict control over

TABLE 11.3. Synthesis of Nanoparticles by Different Microorganisms

Microorganisms	Type of nanomaterials
Bacteria	
Bacillus subtilis	Gold
Shewanella algae	Gold
Pseudomonas stutzeri	Silver
Lactobacillus	Gold, silver, Au–Ag alloy
Clostridium thermoaceticum	Cadmium sulfide
Klebsiella aerogenes	Cadmium sulfide
Escherichia coli	Cadmium sulfide
Desulfobacteriaceae	Zinc sulfide
Thermoanaerobacter ethanolicus	Magnetite
Magnetospirillium magnetotacticum	Magnetite
Thermomonospora sp.	Gold
Rhodococcus	Gold
Chlorella vulgaris	Gold
Phaeodactylum tricornutum	Cadmium sulfide
Yeast	
Candida glabrata	Cadmium sulfide
Torulopsis sp.	Lead sulfide
Schizosaccharomyces pombe	Cadmium sulfide
MKY3	Silver
Fungi	
Verticillium	Gold, silver
Fusarium oxysporum	Gold, silver, Au–Ag alloy, cadmium sulfide, Zirconia
Colletotrichum sp.	Gold

size, shape, and composition of the particles (Gericke and Pinches, 2006; Mandal et al., 2006).

Removal of heavy metals from soil and water or from waste streams at source has been a long-term challenge. Zinc is one of the metals found in effluents discharged from industries involved in galvanizing, electroplating, manufacture of batteries, and other metallurgical industries. Zinc in its metallic form has limited bioavailability and poses no ecological risk. However, zinc can react with other chemicals such as acids and oxygen to form compounds, which can be potentially toxic and can cause serious damage to biological systems (Gericke and Pinches, 2006).

Recently, the use of microorganisms for the recovery of metals from waste streams as well as the use of plants in landfill applications has generated growing interest. Many studies have demonstrated that both prokaryotes and eukaryotes have the ability to remove metals from the contaminated water or waste streams. Specific metabolic pathways leading to precipitation of heavy metals as metal sulfides, phosphates, or carbonates may be useful in biotechnological applications. Sulfate-reducing bacteria (SRB) are the important physiological group for sulfide production. Microbial reduction of sulfate in anoxic environments is the only major source of low-temperature sulfide in natural waters. SRB are commonly used for the bioremediation of metal-contaminated soil or water on a large scale. An integrated microbial process for the bioremediation of soil contaminated with toxic metals using microbially catalyzed reactions has been developed (Revina et al., 2007). In this process, bioleaching of Cd, Co, Cr, Cu, Mn, Ni, and Zn via sulfuric acid produced by sulfur-oxidizing bacteria is followed by precipitation of the leached metals as insoluble sulfides by the action of SRB.

In another scenario, the microbial diversity of a natural ZnS-producing biofilm in the Piquette Mine and the *in situ* development of the SRB population over 6 months, and potential microbial interactions necessary to sustain them, were investigated. The first aim was to determine the nature of microbial populations and interactions potentially responsible for the formation of ZnS deposits. The second aim was to understand the processes that could be mimicked for remediation of metal-contaminated aquifers in schemes that involve installation of wood or related materials after mine closure (Labrenz and Banfield, 2004). Abundant, micrometer-scale and spherical aggregates of 2-5 nm diameter sphalerite (ZnS) particles are formed within natural biofilms dominated by relatively aerotolerant sulfate-reducing bacteria of the family *Desulfobacteriaceae* sp. The biofilm zinc concentration is about 10^6 times that of associated groundwater (0.09–1.1 ppm zinc). Sphalerite also concentrates arsenic (0.01 wt%) and selenium (0.004 wt%). An almost monomineralic product is seen to result from buffering of sulfide concentrations at low values by sphalerite precipitation. These results show how microbes control metal concentrations in groundwater- and wetland-based remediation systems, and suggest biological routes for the formation of some low-temperature ZnS deposits (Labrenz et al., 2000).

Sulfate-reducing bacteria (SRB) have been used for the remediation of metal-contaminated groundwater and soil through the precipitation of metal sulfides. These metal sulfides range from 3 nm to 3 μm in diameter. Zinc sulfide nanoparticles (ZnSNPs) produced by SRBs have diameters of c. 3 nm. These ZnSNPs can be used as catalysts to grow nano-wires which, coupled with green fluorescent protein (GFP), might be used to make a cost-efficient biosensor (Labrenz et al., 2000). A novel, clean biological transformation reaction by immobilized *Rhodobacter sphaeroides* has been developed for the synthesis of zinc sulfide (ZnS) nanoparticles with an average diameter of 8 nm (Bai et al., 2006). Synthesis can be performed in a 500 mL sterile serum bottle containing 15 g immobilized *Rhodobacter sphaeroides*, 2.5 mM sterile $ZnSO_4 \cdot 7H_2O$, and then filling it with sterile culture medium. Immobilized *Rhodobacter sphaeroides* in the serum bottle are cultured at 30 °C anaerobically for 35 h. After the bio-transformation reaction is finished, the precipitate is washed several times with distilled water. The final precipitate is dried at 50 °C for 3 h in a vacuum oven. The products are obtained with about 90% yield based on Zn (Bai et al., 2006).

Studies in India have demonstrated that the fungus *Verticillium* sp. when challenged with Ag^+ and $AuCl_4^-$ ions leads to their reduction and accumulation as silver and gold nanoparticles within the fungal biomass (Sastry et al., 2003; Mukherjee et al., 2004). The acidophillic fungus *Verticillium* sp. was suspended in 100 mL of 2×10^{-4} aqueous $AgNO_3$ solution in 500 mL Erlenmeyer flasks at pH 5.5–6.0. The whole mixture was then put into a shaker at 28 °C (200 rpm) and the reaction was carried out for a period of 72 h (Sastry et al., 2003). The bio-transformation was routinely monitored by visual inspection of the biomass as well as measurement of the UV-vis spectra from the fungal cells.

When exposed to metal ions, the fungus *F. oxysporum* releases enzymes that reduce the metal ions to yield highly stable nanoparticles in solution (Mukherjee et al., 2004). Extracellular secretion of enzymes offers the advantage of obtaining large quantities in a relatively pure state, free from other cellular proteins associated with the organism, and can be easily processed by filtering of the cells and isolating the enzyme for nanoparticles synthesis from cell-free filtrate. The use of specific enzymes secreted by organisms such as fungi in the synthesis of nanoparticles is exciting for the following reasons. The process can be extended to the synthesis of nanoparticles of different chemical compositions and, indeed, different shapes and sizes by suitable identification of enzymes secreted by the fungi. Understanding the surface chemistry of the biogenic nanoparticles (i.e., nature of capping surfactants/peptides/proteins) is equally important (Mukherjee et al., 2004). This would then lead to the possibility of genetically engineering microbes to over-express specific reducing molecules and capping agents, thereby controlling the size and shape of the biogenic nanoparticles. The rational use of constrained environments within cells such as the periplasmic space and cytoplasmic vesicular compartments (e.g., magnetosomes) to modulate nanoparticle size and shape is an exciting possibility yet to be seriously explored.

The fungal- and actinomycete actinomycetes-mediated green chemistry approach towards the synthesis of nanoparticles has many advantages, such as the ease with which the process can be scaled up, economic viability, and the possibility of easily covering large surface areas by suitable growth of the mycelia. The shift from bacteria to fungi as a means of developing natural 'nano-factories' has the added advantage in that downstream processing and handling of the biomass would be much simpler. Compared to bacterial fermentations in which the process technology involves the use of sophisticated equipment for obtaining clear filtrates from the colloidal broths, fungal broths can be easily filtered by filter press of similar simple equipment, saving considerable investment costs for equipment. Fungi have been found to be extremely efficient secretors of soluble proteins and, under optimized conditions of fermentations, mutant strains secrete up to $30 g L^{-1}$ of extracellular proteins. In the strains selected for enzyme fermentations, the desired enzyme constitutes the only component or at least forms the major ingredient of the secreted protein with high specific activities. It is this trait of high-level protein secretion, besides their eukaryotic nature, that has made fungi the favorite hosts for heterologous expression of high-value mammalian protein for manufacturing by fermentation. Further, compared to bacteria, fungi and actinomycetes are known to secrete much higher amounts of proteins, thereby significantly increasing the productivity of this biosynthesis approach (Mukherjee et al., 2004).

11.6. APPLICATIONS OF ZINC-BASED VALUE-ADDED PRODUCTS/NANOMATERIALS

11.6.1. Hydro-Gel for Bio-applications

ZnO can be elected as a potential replacement for conventional dyes and toxic quantum dots in biomedical applications due to its low toxicity and high photo-stability. The major drawback of using semiconductor nanoparticles including ZnO for bio-applications is its insolubility in aqueous media. Meanwhile, hydro-gels made from Poly (N-isopropyl acrylamide) are being used in various biomedical applications due to their aqueous inner environment, biological compatibility, and feasibility for conjugating with biological molecules. A novel ZnO-hydro-gel-based fluorescent colloidal semiconductor nanomaterial system has been developed for biomedical applications such as cell imaging. UV-emitting ZnO rod particles of 200–300 nm in diameter of excellent quality have been synthesized using gas evaporation method. Since biological applications require water-soluble nanomaterials, ZnO nanoparticles are first dispersed in water by the ball-milling method, and their aqueous stability and fluorescence properties are enhanced by incorporating them in bio-compatible Poly N-isopropyl acrylamide (PNIPAM)-based hydro-gel polymer matrix. The optical properties of ZnO-hydro-gel colloidal dispersion versus ZnO-water dispersion have been analyzed. Fluorescence spectroscopy indicates an enhancement of fluorescence by a factor of ten in ZnO-hydro-gel colloidal system compared to ZnO-water system, confirming the surface modification of ZnO nanoparticles by hydro-gel polymer matrix (Xu et al., 2007).

11.6.2. Sensors

Zinc oxide (ZnO) has received considerable attention because of its unique optical, semiconducting, piezoelectric, and magnetic properties. ZnO nanostructures exhibit interesting properties including high catalytic efficiency and strong adsorption ability. Recently, interest has been focused on the application of ZnO in biosensing because of its high iso-electric point (9.5), biocompatibility, and fast electron transfer kinetics. Such features advocate the use of this exciting material as a biomimic membrane to immobilize and modify bio-molecules. This highlights the potential use of ZnO in modified electrodes and biosensing, based on the recent development at Georgia Institute, US (Xu et al., 2007).

11.6.3. Biomedical Applications

Synthetic α- and β-Hopeite, two polymorphs of zinc phosphate tetrahydrates (ZPT), have been synthesized by hydrothermal crystallization from aqueous solution at temperatures of 20 °C and 90 °C, respectively. Aside from their subtle crystallographic differences originating from a unique hydrogen bonding pattern, their thermodynamic interrelation has been investigated by means of X-ray diffraction (XRD) and differential scanning calorimetry (DSC), combined with thermogravimetry (TGA-MS). Using a new heterogeneous step-reaction approach, the kinetics of dehydration of the two forms of ZPT was studied and their corresponding transition temperature determined. Low-temperature drift, FT-Raman and ^{1}H, ^{31}P MAS-NMR (phosphorus magic-angle spinning nuclear magnetic resonance) reveal an oriented distortion of the zinc phosphate tetrahedra, due to a characteristic of the hydrogen bonding pattern, and in accordance with the molecular tetrahedral linkage scheme of the phosphate groups. Biogenic hydroxyapatite (HAP) and one of its metastable precursors, a calcium dihydrogen phosphate dihydrate (DCPD) or Brushite, were also obtained and used to underline the resulting variations of chemical reactivity in zinc phosphates (Xu et al., 2007).

11.6.4. Antibacterial Properties

The antibacterial behavior of suspensions of zinc oxide nanoparticles (ZnO nanofluids) against *E. coli* has been investigated. ZnO nanoparticles are used to formulate nanofluids. The effects of particle size, concentration, and the use of dispersants on the antibacterial behavior were examined. The results show that the ZnO nanofluids have bacteriostatic activity against *E. coli*. The antibacterial activity increases with increasing nanoparticle concentration and decreasing particle size. SEM analyses of the bacteria before and after treatment with ZnO nanofluids show that the presence of ZnO nanoparticles damages the membrane wall of the bacteria (Zhang et al., 2007b).

A method of coating paper with zinc oxide nanoparticles using ultrasound has been reported by Ghule et al. (2006). The nanoparticle-coated paper has antibacterial activity, as revealed by tests against *E. coli*. The paper could be used on hospital walls, in particular operation theatres, as well as in residential complexes. The coating approach could also be extended to textiles to generate suits with antibacterial properties to combat bioterrorism. Using ultrasound as a coating method is simpler than mechanical methods. It is also cheaper, uses less material, avoids waste, and uses water as the solvent. There are many future applications of the coated paper, from security paper for optical communications to sensors. It is also planned to develop the method to coat alumina and silica nanoparticles onto paper as an economical alternative for thin-layer chromatography (TLC) plates for chromatography. However, the disadvantage of photo-catalytically reduced volatile organic compounds generating stains on the antibacterial wallpaper requires further study.

11.6.5. Zeolites in biomedical applications

The clinoptilolite-rich rock has been evaluated to be an inorganic Zn^{2+}-releasing carrier for antibiotic *erythromycin*. It is used in the topical treatment of acne, a diffused skin pathology, given the efficacy of zinc-erythromycin combination against resistant *Propionibacterium* strains.

11.6.6. Textiles

Zinc oxide nanoparticles can be prepared by wet chemical method using zinc nitrate and sodium hydroxide as precursors and soluble starch as stabilizing agent. These nanoparticles, which have an average size of 40 nm, have been coated on the bleached cotton fabrics (plain weave, 30 s count) using acrylic binder, and the functional properties of coated fabrics were studied. On an average, 75–90% UV blocking is recorded for the cotton fabrics treated with 2% ZnO nanoparticles. Air permeability of the nano-ZnO coated fabrics is significantly higher than the control; it would therefore have increased breathability. In case of nano-ZnO coated fabric, due to its nanosize and uniform distribution, friction would be significantly lower than the bulk-ZnO-coated fabric as studied by Automated Materials Testing System. Further studies are underway to evaluate wash fastness, antimicrobial properties, abrasion properties, and fabric handle properties (Yadav et al., 2006; Becheri et al., 2008).

11.6.7. Prospects of Zinc Recovery from Tailings and Biosynthesis of Zinc-based Nano-materials

Zinc tailings are a huge global resource from which zinc can be extracted. These tailings are generated after optimal and conventionally achievable recovery as dumps. As these concentrator tailings are generated during the flotation of concentrates of zinc (sphalerite), they also contain trace levels of metals such as Co, Ni, Fe, and As. There are available technologies (solvent extraction or SX, ion exchange or IX, and electrowinning or EW) which can purify the solutions containing metal mixtures after extraction by chemical and/or biochemical treatment.

Economical processing of the zinc tailings by chemical process has yet to be realized. However, for extraction of metals from such materials, the bioprocessing route is currently being considered to produce materials from zinc sulfate solutions of high concentration for economic reasons. These raw materials could be efficiently processed by the biological or biochemical route to extract metals of the desired size and form. Of the available technologies, the investigation of microorganisms is proposed. At present, microbial methods of synthesis of nanomaterials of varying composition are extremely limited and confined to a few metals, some metal sulfides, and very few oxides that are prepared mostly from the synthetic solutions. It would be interesting to combine the bioleaching of zinc from low-grade feedstocks and tailings with the biosynthesis of nanoparticles, which could make microbial synthesis a commercially feasible proposition and help in better utilization of the discarded waste material.

11.7. CONCLUSIONS AND FUTURE DIRECTIONS

Nanotechnology is a highly promising discipline that will rewrite the future of several existing technologies, will change many aspect of our lives, and will lead to the generation of uniqueness in all streams of technology (Gericke and Pinches, 2006). The current revolution in nanoscience is a result of several advances in technology. The inherent properties of biochemical reactions occurring in microbes can be utilized to our advantage in controlled synthesis of materials from varied sources in desired forms.

When viewing the future capabilities of the area of nano-biosynthesis, the synthesis of metal nanoparticles using microbes such as bacteria, yeasts, algae, actinomycetes, and fungi is gaining momentum due to the eco-friendly nature of the organisms which reduce toxic chemicals. The interactions between microorganisms and metals have been well documented and the ability of microorganisms to extract and/or accumulate metals is already employed in biotechnological processes such as biomineralization, bioremediation, bioleaching, and microbial corrosion. The confluence of environmental biotechnology and nanotechnology will lead to the most exciting progress in the development of nano-devices having bio-capabilities in metal remediation strategies. Noble metal nanoparticle-based chemistry has immense advantages in the area of drinking water purification, as it is capable of the removal of organic compounds, heavy metals, and microorganisms. While the chemistry of noble metal nanoparticles is unique in the removal of many contaminants, their high energy surface results in systems attempting to minimize the surface energy through protection or chemical transformation or agglomeration. Nanoparticles are therefore likely to adsorb a number of other species on the surface.

Nanoparticles may exploit biological pathways to achieve payload delivery of small molecules to cellular and intracellular targets (Faraji and Wipf, 2009). Synthesis strategies, including surface, porosity, stealthing, and size modifications, can be utilized to refine the pharmacokinetic properties of nanoparticles and allow for efficient delivery.

The most important criteria in the use of nanoparticles are as follows.

- As far as possible, the use of nanoparticle supported on suitable substrates is preferred. There are many advantages of supported nanoparticle chemistry, including minimum leaching of nanoparticles into the environment, easy nanoparticle separation, and lower loss in the efficiency of nanoparticle chemistry.

- Quantitative studies on the release of nanoparticles from supports under varying environmental conditions are required.
- Appropriate mechanisms for the recovery of nanomaterials must be designed.
- Nanosystems made of materials with minimum toxicity should be used.

The highly structured physical and biosynthesis potential of the microbial cells in the production of nanosized materials should be utilized (Abhilash and Pandey, 2012). An immediate objective of further research is therefore, to use the structured activities of microbes to achieve and control manipulation of the size and shape of the particles. Issues that need to be addressed include development of a fundamental understanding of the process mechanism on a cellular and molecular level, including isolation and identification of the compounds responsible for the synthesis of desired materials.

REFERENCES

Abdel-latif, M.A. 2002. Fundamentals of zinc recovery from metallurgical wastes in the Enviroplas process. *Minerals Engineering* 15: 945–952.

Abhilash, Pandey, B.D. 2012. Synthesis of zinc-based nanomaterials: a biological perspective. *IET-Nanobiotechnology* 6(4): 144–148.

Adhikari, A., Banerjee, P.C., Bhattacharya, B., Mukherjee, S. 2007. Microbiological leaching of the polymetallic sulphides ore of Gorubathan. *Journal of the Institute of Engineers (India)* 88: 17–22.

Asadi Zeydabadi, B.D., Mowla, M.H., Shariat, J., Kalajahi, F. 1997. Zinc recovery from blast furnace flue dust. *Hydrometallurgy* 47: 113–125.

Bai, H.-J., Zhang, Z.-M., Gong, J. 2006. Biological synthesis of semiconductor zinc sulfide nanoparticles by immobilized Rhodobacter sphaeroides. *Biotechnology Letters* 28: 1135–1139.

Bayat, O., Sever, E., Bayat, B., Arslan, V., Poole, C. 2009. Bioleaching of zinc and iron from steel plant waste using *Acidithiobacillus ferrooxidans*. *Applied Biochemistry and Biotechnology* 152: 117–126.

Becheri, A., Dürr, M., Nostro, P.L., Baglioni, P., 2008. Synthesis and characterization of zinc oxide nanoparticles: application to textiles as UV-absorbers. *Journal of Nanoparticle Research* 10: 679–689.

Bhat, K.L., Natarajan, K.L. 1986. Electro-leaching of industrial zinc leach residues. *Transactions of the Indian Institute of Metal* 5: 475–480.

Bhat, K.L., Natarajan, K.A. 1987. Recovery of zinc from leach residues. Problems and development. *Transactions of the Indian Institute of Metal* 4: 361–366.

Boon, M., Snijder, M., Hansford, G.S., Heijnen, J.J. 1998. The oxidation kinetics of zinc sulphide with *Thiobacillus ferrooxidans*. *Hydrometallurgy* 48: 171–186.

Brahmaprakash, G.P., Devasia, P., Jagadish, K.S., Natarajan, K.A., Rao, G.R. Development of Thiobacillus ferrooxidans ATCC 19859 strains tolerant to copper and zinc. *Bulletin of Materials Science* 10(5): 461–465.

Brierley, J.A., Brierley, C.L. 2001. Present and future commercial applications of biohydrometallurgy. *Hydrometallurgy* 59: 233–239.

Castro, I.M., Fietto, J.L.R., Vieira, R.X., Trõpia, M.J.M., Campos, L.M.M., Paniago, E.B., Brandão, R.L. 2000. Bioleaching of zinc and nickel from silicates using *Aspergillus niger* cultures. *Hydrometallurgy* 57: 39–49.

Choudhury, R., Srivastava, S. 2007. Zinc resistance mechanisms in bacteria. *Current Science* 81(7): 768–775.

de Souza, A.D., Pina, P.S., Leão, V.A. 2007. Bioleaching and chemical leaching as an integrated process in the zinc industry. *Minerals Engineering* 20: 591–599.

Deshpande, V.P., Shekdar, A.V. 2005. Sustainable waste management in the Indian mining industry. *Waste Management Research* 23: 343–369.

Devecia, H., Akcil, A., Alp, I. 2004. Bioleaching of complex zinc sulphides using mesophilic and thermophilic bacteria: comparative importance of pH and iron. *Hydrometallurgy* 73: 293–303.

Ek, C.S. 1989. Silver recovery from zinc hydrometallurgical residues. In *Proceedings of International Symposium* sponsored by Copper, Nickel, Cobalt, and Precious Metals Committee of the Minerals, Metals and Materials Society and the International Precious Metals Institute (eds Jha, M.C. and Hill, S.D.), Las Vegas, Nevada, pp. 391–401.

Espiari, S., Rashchi, F., Sadrnezhaad, S.K. 2006. Hydrometallurgical treatment of tailings with high zinc content. *Hydrometallurgy* 82: 54–62.

Faraji, A.H., Wipf, P. 2009. Nanoparticles in cellular drug delivery. *Bioorganic and Medicinal Chemistry* 17(8): 2950–2962.

Feng, L., Yang, X., Shen, Q., Xu, M., Jin, B. 2007. Pelletizing and alkaline leaching of powdery low grade zinc oxide ores. *Hydrometallurgy* 89: 305–310.

Fourie, H., van Rooyen, P.H., Rupprecht, S., Lund, T., Vegter, N.M. 2007. Exploitation of a massive low grade zinc-lead resource at Rosh Pinah Zinc Corporation, Namibia. In *Proceedings of the Fourth Southern African Conference on Base Metals*, 109–118.

Gericke, M., Pinches, A. 2006. Biological synthesis of metal nanoparticles. *Hydrometallurgy* 83: 132–140.

Ghosh, M.K., Sukla, L.B., Misra, V. N. 2004. Cobalt and zinc extraction from Sikkim complex sulphide concentrate. *Transactions of Indian Institute of Metal* 57(6): 617–621.

Ghule, K., Ghule, A.V., Chen, B.-J., Ling, Y.-C. 2006. Preparation and characterization of ZnO nanoparticles coated paper and its antibacterial activity study. *Green Chemistry* 8: 1034–1041.

Gupta, A., Birendra, K., Mishra, R. 2003. Study on the recovery of zinc from Moore cake: a biotechnological approach. *Minerals Engineering* 16: 41–43.

Haghshenas, D.F., Alamdari, E.K., Torkmahalleh, M.A., Bonakdarpour, B., Nasernejad, B. 2009a. Adaptation of *Acidithiobacillus ferrooxidans* to high grade sphalerite concentrate. *Minerals Engineering* 22: 1299–1306.

Haghshenas, D.F., Alamdari, E.K., Bonakdarpour, B., Darvishi, D., Nasernejad, B. 2009b. Kinetics of sphalerite bioleaching by *Acidithiobacillus ferrooxidans*. *Hydrometallurgy* 99: 202–208.

Hossain, S.M., Das, M., Begum, K.M.M.S., Anantharaman, N. 2004. Bioleaching of zinc sulphide (ZnS) ore using *Thiobacillus ferrooxidans*. *Journal of the Institution of Engineers (India)* 85(1): 28–33.

Hsu, C.-H., Harrison, R.G. 1995. Bacterial leaching of zinc and copper from mining wastes. *Hydrometallurgy* 37: 169–179.

Jyothi, N., Sudha, K.N., Natarajan, K.A. 1989. Electrochemical aspects of selective bioleaching of sphalerite and chalcopyrite from mixed sulfides. *International Journal of Mineral Processing* 27(3–4): 189–203.

Kaewkannetra, P., Garcia-Garcia, F.J., Chiu, T.Y. 2009. Bioleaching of zinc from gold ores using *Acidithiobacillus ferrooxidans*. *International Journal of Minerals, Metallurgy and Materials* 16(4): 368–374.

Kashani, A.H.N., Rashchi, F. 2008. Separation of oxidized zinc minerals from tailings: Influence of flotation reagents. *Minerals Engineering* 21: 967–972.

Kondratyeva, T.F., Muntyan, L.N., Karavaiko, G. 1995. Zinc- and arsenic-resistant strains of *Thiobacillus ferrooxidans* have increased copy numbers of chromosomal resistance genes. *Microbiology* 141: 1157–1162.

Kori, S.A., Reddy, P.L.N., Bhat, K.L. 2002. Effect of process parameters for the extraction of zinc from sphalerite concentrates. *Indian Mineralogist* 36(1): 42–45.

Krishnan, K.H., Mohanty, D.B., Sharma, K.D. 2007. The effect of microwave irradiations on the leaching of zinc from bulk sulphide concentrates produced from Rampura–Agucha tailings. *Hydrometallurgy* 89: 332–336.

Kurama, H., Göktepe, F., 2003. Recovery of zinc from waste material using hydrometallurgical processes. *Environmental Progress* 22(3): 161–166.

Labrenz, M., Banfield, J. F. 2004. Sulfate-reducing bacteria-dominated biofilms that precipitate ZnS in a subsurface circumneutral-pH mine drainage system. *Microbial Ecology* 47: 205–217.

Labrenz, M., Druschel, G. K., Thomsen-ebert, T. et al. 2000. Formation of sphalerite (ZnS) deposits in natural biofilms of sulfate-reducing bacteria. *Science* 290: 1744–1747.

Lawrence, E. A. 1974. Summary of Vegetation Program on Lead-Zinc Tailings, AIME Transactions, 256: 813–814.

Ligiane, R., Gouvea, C., Morais, A. 2007. Recovery of zinc and cadmium from industrial waste by leaching/cementation. *Minerals Engineering* 20: 956–958.

Mackiw, V.N., Veltman, H. 1967. Recovery of zinc and lead from complex low grade sulphide concentrates by acid pressure leaching. *Bulletin of the Canadian Institute of Mineralogy and Metallurgy* 60(657): 80–85.

Magyar, M.I., Mitchell, V.G., Ladson, A.R., Diaper, C. 2008. Lead and other heavy metals: common contaminants of rainwater tanks in Melbourne. In *Proceedings of Water Down Under*, 14–17 April 2008 Adelaide, Australia, 409–416.

Mandal, D., Bolander, M.E., Mukhopadhyay, D., Sarkar, G., Mukherjee, P. 2006. The use of microorganisms for the formation of metal nanoparticles and their application. *Applied Microbiology and Biotechnology* 69: 485–492.

Mehta, K.D. 1992. Studies into the bioleaching of zinc from the tailings of the concentrator plant of Rajpura Dariba Mines, Hindustan Zinc Limited. M.Tech thesis, ISM Dhanbad, 1992.

Mendez, M.O., Neilson, J.W., Maier, R.M. 2008. Bacterial community characterization of an abandoned semi-arid lead-zinc mine tailings site. *Applied Environmental Microbiology*, doi: 10.1128/AEM.02883-07.

Mittal, N.K., Sen, P.K., Chopra, S.J., Jaipuri, A.A., Gaur, R.K., Jakhu, M.R. 2000. India's first column flotation installation in the zinc cleaning circuit at Rajpura Dariba Mine, India. *Minerals Engineering* 13(5): 581–584.

Moreno, L., Neretnieks, I., 2006. Long-term environmental impact of tailings deposits. *Hydrometallurgy* 83: 176–183.

Mousavi, S.M., Jafari, A., Yaghmaei, S., Vossoughi, M., Roostaazad, R. 2006. Bioleaching of low-grade sphalerite using a column reactor. *Hydrometallurgy* 82: 75–82.

Mousavi, S.M., Yaghmaei, S., Vossoughi, M., Jafari, A., Roostaazad, R., Turunen, I. 2007. Bacterial leaching of low-grade ZnS concentrate using indigenous mesophilic and thermophilic strains. *Hydrometallurgy* 85: 59–65.

Mueller, B., Burgstalles, W., Strasser, H., Zanella, A., Schiner, F. 1995. Leaching of zinc from an industrial filter dust with *Penicillium, Pseudomonas* and *Corynebacterium*: Citric acid is the leaching agent rather than amino acids. *Journal of Industrial Microbiology* 14(3–4): 208–212.

Mukherjee, P., Mandal, D., Ahmad, A., Sastry, M., Kumar, R. 2004. Process for the preparation of metal sulfide nanoparticles. US Patent No. 6,783,963.

Mulligan, C.N., Galvez-Cloutier, R. 2005. Bioleaching of copper mining residues by *Aspergillus niger*. *Water Science and Technology* 41(12): 255–262.

Mulligan, C.N., Kamali, M., Gibbs, B.F. 2004. Bioleaching of heavy metals from a low-grade mining ore using *Aspergillus niger*. *Journal of Hazardous Materials* 110: 77–84.

Nakamura, T., Shibata, E., Takasu, T., Itou, H. 2008. Basic consideration on EAF dust treatment using hydrometallurgical processes. *Resources Processing* 55: 144–148.

Ngenda, R.B., Segers, L., Kongolo, P.K. 2009. Base metals recovery from zinc hydrometallurgical plant residues by digestion method. In *Proceedings of Hydrometallurgy Conference 2009*, Southern African Institute of Mining and Metallurgy, 2009.

Ozberk, E., Jankola, W.A., Vecchiarelli, M., Krys, B.D. 1995. Commercial operations of the Sherritt zinc pressure leach process. *Hydrometallurgy* 39: 49–52.

Pani, C.K., Swain, S., Kar, R.N., Chaudhury, G.R., Sukla, L.B., Misra, V.N. 2003. Biodissolution of zinc sulfide concentrate in 160l 4-stage continuous bioreactor. *Minerals Engineering* 16: 1019–1021.

Pelino, M., Cantalini, C., Abbruzzese, C., Plescia, P. 1996. Treatment and recycling of goethite waste arising from the hydrometallurgy of zinc. *Hydrometallurgy* 40: 25–35.

Qin, W.-Q., Li, W.-Z., Lan, Z.-Y., Qiu, G.-Z. 2007. Simulated small-scale pilot plant heap leaching of low-grade oxide zinc ore with integrated selective extraction of zinc. *Minerals Engineering* 20: 694–700.

Qingzhong, B.A.I. 2000. The status and strategy of tailings comprehensive utilization in China. In *Proceedings of Third Asia-Pacific Regional Training Workshop on Hazardous Waste Management in Mining Industry*, Beijing, 4–8 September 2000, 151–159.

Rabah, M.A., El-Sayed, A.S. 1995. Recovery of zinc and some of its valuable salts from secondary resources and wastes. *Hydrometallurgy* 37: 23–32.

Raghavan, R., Mohanan, P.K., Patnaik, S.C. 1998. Innovative processing technique to produce zinc concentrate from zinc leach residue with simultaneous recovery of lead and silver. *Hydrometallurgy* 48: 225–237.

Rai, A., Rao, D.B.N. 2005. Utilisation potentials of industrial/mining rejects and tailings as building materials. *Management of Environmental Quality: An International Journal* 16(6): 605–614.

Rao, C.N.R. 1993. Chemical synthesis of solid inorganic materials. *Materials Science & Engineering: B* 18(1): 1–21.

Reddy, K.H.M., Prakash, B.V., Bhat, K.L., Natarajan, K.A. 1990. Electro-leaching of complex lead zinc copper concentrates. *Proceedings of National Symposium on Research and Practice in Hydrometallurgy*, HZL, Udaipur, Feb. 1990, 5.1.1–5.1.8.

Rehman, M., Anwar, M.A., Iqbal, M., Akhtar, K., Khalid, A.M., Ghauri, M.A. 2009. Bioleaching of high grade Pb–Zn ore by mesophilic and moderately thermophilic iron and sulphur oxidizers. *Hydrometallurgy* 97: 1–7.

Reuter, M.A., Sudholter, S., Kriger, J., Keller, S. 1995. Synthesis of processes for the production of environmentally clean zinc. *Minerals Engineering* 8(1–2): 201–219.

Revina, A., Oksentyuk, E.V., Fenin, A.A. 2007. Synthesis and properties of zinc nanoparticles: the role and prospects of radiation chemistry in the development of modern nanotechnology. *Protection of Metals* 43(6): 554–559.

Ruşen, A., Sunkar, A.S., Topkay, Y.A. 2008. Zinc and lead extraction from Çinkur leach residues by using hydrometallurgical method. *Hydrometallurgy* 93: 45–50.

Safarzadeh, M.S., Moradkhani, D., Ashtari, P. 2009. Recovery of zinc from Cd–Ni zinc plant residues. *Hydrometallurgy* 97: 67–72.

Sahu, K.K., Agrawal, A., Pandey, B.D. 2004. Recent trends and current practices for secondary processing of zinc and lead. Part II: Zinc recovery from secondary sources. *Waste Management Research* 22: 248–254.

Sampson, M.I., Van der Merwe, J.W., Harvey, T.J., Bath, M.D. 2005. Testing the ability of a low grade sphalerite concentrate to achieve autothermality during biooxidation heap leaching. *Minerals Engineering* 18: 427–437.

Sastry, M., Ahmad, A., Khan, M.I., Kumar, R. 2003. Biosynthesis of metal nanoparticles using fungi and actinomycetes. *Current Science* 85(2): 62–170.

Selvi, S.C., Modak, J.M., Natarajan, K.A. 1998. Electro-bioleaching of sphalerite flotation concentrate. *Minerals Engineering* 11(8): 783–788.

Shi, S.-Y., Fang, Z.-H., Ni, J.-R. 2006. Comparative study on the bioleaching of zinc sulphides. *Process Biochemistry* 41: 438–446.

Soleimani, M., Hosseini, S., Roostaazad, R., Petersen, J., Mousavi, S.M., Vasiri, A.K. 2009. Microbial leaching of a low-grade sphalerite ore using a draft tube fluidized bed bioreactor. *Hydrometallurgy* 99: 131–136.

Thakur, D.N., Saroj, K.K., Arya, P.K., Gupta, A., Mehta, K.D. 1994. Bioleaching of basic sphalerite tailings. *Journal of Institute of Engineers (India)* 75: 99–105.

Turan, M.D., Soner, H., Tümen, A.F. 2004. Recovery of zinc and lead from zinc plant residue. *Hydrometallurgy* 75: 169–176.

USEPA. 1994. Technical resource document on Extraction and Beneficiation of Ores and Minerals, Vol. 1, Lead-Zinc. June 1994, US Environmental Protection Agency.

Vahidi, E., Rashchi, F., Moradkhani, D. 2009. Recovery of zinc from an industrial zinc leach residue by solvent extraction using D2EHPA. *Minerals Engineering* 22: 204–206.

Viera, M., Pogliani, C., Donati, E. 2007. Recovery of zinc, nickel, cobalt and other metals by bioleaching. In *Microbial Processing of Metal Sulfides* (eds E.R. Donati, W. Sand). Springer, pp. 103–119.

Watson, J.H.P., Beharrell, P.A. 2006. Extracting values from mine dumps and tailings. *Minerals Engineering* 19: 1580–1587.

Weisener, C.G., Smart, R.S.C., Gerson, A.R. 2004. A comparison of the kinetics and mechanism of acid leaching of sphalerite containing low and high concentrations of iron. *International Journal of Mineral Processing* 74: 239–249.

Xia, L., Zeng, J., Ding, J., Yang, Y., Zhang, B., Liu, J., Qiu, G. 2007. Comparison of three induced mutation methods for *Acidiothiobacillus caldus* in processing sphalerite. *Minerals Engineering* 20: 1323–1326.

Xu, T., Zhang, N., Nichols, H.L., Shi, D., Wen, X. 2007. Modification of nanostructured materials for biomedical applications. *Materials Science and Engineering: C* 27: 579–594.

Yadav, A., Prasad, V., Kathe, A.A., Raj, S., Yadav, D., Sundaramoorthy, C., Vigneshwaran, N. 2006. Functional finishing in cotton fabrics using zinc oxide nanoparticles. *Bulletin of Materials Science* 29(6): 641–645.

Zhang, H.-B., Shi, W., Yang, M.-X., Sha, T., Zhao, Z.-Y. 2007*a*. Bacterial diversity at different depths in lead-zinc mine tailings as revealed by 16S rRNA gene libraries. *Journal of Microbiology* 45(6): 479–484.

Zhang, L., Jiang, Y., Ding, Y., Povey, M., York, D. 2007*b*. Investigation into the antibacterial behaviour of suspensions of ZnO nanoparticles (ZnO nanofluids). *Journal of Nanoparticle Research* 9: 479–489.

12

INTERACTION BETWEEN NANOPARTICLES AND PLANTS: INCREASING EVIDENCE OF PHYTOTOXICITY

Rajeshwari Sinha and S.K. Khare

Enzyme and Microbial Biochemistry Laboratory, Department of Chemistry, Indian Institute of Technology, Delhi, New Delhi, India

12.1. INTRODUCTION

The reduction of materials to nanosize alters their activity and properties to become significantly different from their bulk counterparts (Oberdörster et al., 2005). These unique properties of nanoparticles (NPs) have therefore been advantageously exploited for important applications in energy, optics, electronic, medical, remediation, food, and cosmetic sectors. Nanotechnology has recently been touted as the "next industrial revolution" (Rana and Kalaichelvan, 2013).

With increasing applications of nanomaterials, their production is predicted to increase to 58,000 tons by 2020 (Maynard et al., 2006). Consequently, an array of nanomaterials is likely to find its way into the aquatic and terrestrial environment, leading to increased exposure to living systems. The toxic effects of NPs, dubbed "nanotoxicity", have been quite evident in prokaryotic cells (Sinha et al., 2011). Similar effects causing loss of cell viability, tissue damage, and inflammatory reactions are being increasingly observed in animal systems (Oberdörster et al., 2005). Studies on nanotoxicity in plant systems are very limited.

Bio-Nanoparticles: Biosynthesis and Sustainable Biotechnological Implications, First Edition.
Edited by Om V. Singh.
© 2015 John Wiley & Sons, Inc. Published 2015 by John Wiley & Sons, Inc.

Plants are envisaged to play a critical role in the movement of NPs in the environment to various trophic levels starting from uptake, bioaccumulation, and subsequent transfer to the food chain (Monica and Cremonin, 2009). Understanding the interactions of NPs with plants, and consequent effects on physiological processes at various stages of development, is crucial for analyzing risk assessment and impact (Ma et al., 2010). Phyto-nanotoxicity is therefore an emerging area of interest. This chapter discusses the currently available information in the domain of "nanotoxicology" in plants.

12.2. PLANT–NANOPARTICLE INTERACTIONS

Plants are subjected to NP exposure from various sources such as wastewater sludge from agricultural fields, improperly treated sewage or effluents, industrial atmospheric emissions, or soil materials which have leached/eroded into groundwater (Miralles et al., 2012). The impact of NPs on a plant depends on a number of factors, namely composition, stability, concentration, size, surface area, physicochemical properties of NPs, and exposure time. The plant species, age, life-cycle stage, and anatomy are also equally important and may significantly influence the interactions.

NP-mediated phytotoxicity studies are in their early stages; very little systematic analysis is available (Judy et al., 2012; Jacob et al., 2013). It is becoming apparent that (1) nanomaterials enter into plants by penetration through the seed coat, adsorption onto roots, absorption during moisture/nutrient uptake, or internalized thorough leaves via stomatal or cuticular openings (Miralles et al., 2012); (2) once internalized, these may bioaccumulate or be further translocated to shoots through vascular system; and (3) at the cellular level, NPs enter plant cells possibly through binding onto carrier proteins, aquaporins, ion channels, creation of new pores, endocytosis, or complexing with membrane transporters.

The US Environmental Protection Agency has proposed the rate of seed germination and root/shoot elongation as standard indicators for assessing phytotoxicity. Furthermore, changes in transpiration, extent of photosynthesis, number of leaves, chlorophyll contents, microRNA or level of gene expression, and antioxidant enzyme activities in plants before and after NP exposure may form relevant endpoints for phytotoxicity assessment.

12.3. EFFECT OF NANOPARTICLES ON PLANTS

The phytotoxic effects of fullerenes, single- and multi-walled carbon nanotubes (SWCNT and MWCNTs), CdSe/ZnS quantum dots, Au, Ag, ZnO, TiO_2, SiO_2, Fe_3O_4, Al_2O_3, CeO_2 and In_2O_3 NPs have been investigated over monocots such as rice, wheat, maize, onion, and garlic and dicots such as pumpkin, cabbage, tomato, and cucumber.

12.3.1. Monocot Plants

Some of the relevant stimulatory/inhibitory effects of various NPs on the physiology and growth of monocot plants are summarized in Table 12.1. Wheat, maize, onion, and rice plants and seeds have been most commonly investigated as model systems. It is interesting to note that NP toxicity is manifested in different ways in different plants. Also, the phytotoxicity is by and large size and concentration dependent. In the case of wheat plants exposed to TiO_2, smaller TiO_2 NPs (14 nm) are more easily internalized than larger-sized 25 nm particles (Larue et al., 2012). The toxic effects of NPs in growth inhibition, decrease in plant biomass, generation of reactive oxygen species, and genotoxicity have also been recorded in the case of *Hydrilla verticillata* (Mishra et al., 2013), duckweeds, and pondweeds such as *Lemna minor*, *Spirodela polyrhiza*, *Elodea canadensis* (Gubbins et al., 2011; Jiang et al., 2012; Johnson et al., 2012) and on grasses belonging to *Lolium* spp. (Yin et al., 2011; Atha et al., 2012).

12.3.2. Dicot Plants

Crop plants such as tomato, cabbage, carrot, cucumber, lettuce, spinach, and radish have been investigated primarily because of their economical importance and routine usage. The phytotoxic effects of various nanoparticles on some representative dicot plants are summarized in Table 12.2. Although not a crop plant, *Arabidopsis thaliana* has been extensively evaluated as a model plant. Similar to monocots, the effect was dose dependent; increasing concentrations led to significant toxicity.

In general, exposure to nanomaterials leads to a decrease in root/shoot length and accumulation in cells, causing oxidative stress. Although seed germination is adversely affected, the effect on flowering and fruit production is yet to be determined. The application of NPs in agriculture is slowly catching up. Agri-nanotechnology will soon enable controlled release of agrochemicals for effective pest control, disease resistance, or efficient growth (Nair et al., 2010). A careful and considered understanding of their safety aspect is therefore quite relevant.

12.4. MECHANISMS OF NANOPARTICLE-INDUCED PHYTOTOXICITY

The primary step that brings the plant into contact with NPs is the uptake. Bioaccumulation/transformation or translocation takes place subsequently. Cell walls acts as a barrier since the pore sizes on the cell wall are in the range of 5–20 nm (Nair et al., 2010). NPs with a diameter less than 20 nm may enter directly. Nonetheless, there are other mechanisms of NP entry into plants and cells; some of these are briefly described in the following sections.

12.4.1. Endocytosis

Endocytic pathways in plants may be clathrin dependent or independent (Miralles et al., 2012). In case of clathrin-coated endocytosis, the plasma membrane forms an

TABLE 12.1. Effect of Nanoparticles on Growth, Physiology, and Germination of Monocot Plants

Type of NP	Ave. size (nm)	Concentration	Nanotoxic effects	Reference
Allium cepa				
Ag	<100	25, 20, 75, 100 ppm	NPs penetrated plant system and interfered with intracellular components; impairment of cell division, cell disintegration	Kumari et al. (2009)
CoO-ZnO	~50 nm	5, 10, 20 µg mL^{-1}	Concentration-dependent severe inhibition of root elongation; bioaccumulation of ZnO and Co adsorption responsible for phytotoxicity	Ghodake et al. (2011)
ZnO	<100	25, 50, 75, 100 µg mL^{-1}	Concentration-dependent inhibition of mitotic index and increase of chromosomal aberration; internalization and agglomeration of ZnO NPs	Kumari et al. (2011)
Zea mays				
CeO$_2$	37	10 ppm	NPs were adsorbed and incorporated in leaves	Birbaum et al. (2010)
TiO$_2$	<100	0.2, 1.0, 2.0, 4.0%	Reduction in germination rate; concentration-dependent genotoxicity	Castiglione et al. (2011)
Ag	10–30	20–100 ppm	Increased shoot/root length, leaf surface area, chlorophyll, carbohydrate and protein contents; enhanced concentrations caused inhibitory effect	Salama (2012)
CuO	20–40	100 mg L^{-1}	Inhibited the growth of maize seedlings; xylem-based transport of NPs from roots to shoots and translocation back from shoots to roots via phloem	Wang et al. (2012a)
SWCNT	1–2	20 mg L^{-1}	Accelerated root growth but inhibited root hair growth; internalization of SWCNTs into root tissues	Yan et al. (2013)
Triticum aestivum				
Magnetic carbon-coated NPs	5–50		NPs were detected easily in the xylem vessels; upward translocation following the transpiration stream	Cifuentes et al. (2010)

NP	Size (nm)	Concentration	Effect	Reference
TiO$_2$ anatase	12, 25	100 mg L^{-1}	25 nm NPs non-toxic; 12 nm NPs are taken up efficiently by plant roots and localized in the parenchymal region and vascular cylinder	Larue et al. (2011)
CuO	<50		CuO and ZnO reduced root growth significantly, but only CuO NPs impaired shoot growth; increased oxidative stress; decreased chlorophyll content; higher peroxidase and catalase activities in roots upon NP exposure	Dimkpa et al. (2012)
ZnO	<100			
Oxidized MWCNT	50–630	10–160 μg mL^{-1}	Dehydrogenase activity, faster root growth, and higher biomass production; TEM revealed internalization into the cell wall and cell elongation	Wang et al. (2012b)
Ag	10	0–5 mg kg^{-1} of sand	Dose-dependent decline in length of shoots and roots; increased branching in the roots; generation of oxidative stress in roots	Dimkpa et al. (2013)
Oryza sativa				
Ag	25	50, 500, 1000 μg mL^{-1}	Intracellular deposition of NPs inside root cells observed by TEM, damaged cell wall and vacuoles	Mazumdar and Ahmad (2011)
Ag	18.34	30, 60 mg mL^{-1}	Cell wall disorganization with increasing concentration; NP internalization, damage to cell morphology evidenced from TEM; restricted root growth.	Mirzajani et al. (2013)
CeO$_2$	~8	0–500 mg mL^{-1}	NPs were not toxic at concentrations ≤500 mg; higher concentrations, significant reduction in H$_2$O$_2$ generation in both shoots and roots, increased electrolyte leakage, and lipid peroxidation in the shoots	Rico et al. (2013)
CuO	<50	0.5, 1.0, 1.5 mM	Reduced seed germination; loss of root cell viability; severe oxidative burst.	Shaw and Hossain (2013)

TABLE 12.2. Effect of Nanoparticles on Growth, Physiology, and Germination of Dicot Plants

Type of NP	Ave. size (nm)	Concentration	Nanotoxic effects	Reference
Arabidopsis thaliana				
Al_2O_3, SiO_2, Fe_3O_4, ZnO	—	400–4000 mg mL^{-1}	Toxic effect ZnO>>Fe_3O_4>SiO_2>Al_2O_3	Lee et al. (2010)
Fullerene	100–200	0.005–0.2 mg mL^{-1}	Inhibitory effect on plant growth, decrease in root length	Liu et al. (2010)
CdSe/ZnS QDs	6.3		Generally adhered on the outside surfaces of the roots and induced oxidative stress	Navarro et al. (2012)
SiO_2	14, 50, 200	250, 1000 mg L^{-1}	Stunted growth and chlorosis in plants	Slomberg and Schoenfisch (2012)
Citrate-stabilized Ag	20, 40, 80		Concentration-dependent inhibition of root elongation, uptake and bioaccumulation in plants	Geisler-Lee et al. (2013)
CeO_2, In_2O_3		0–2000 ppm	Dose-dependent decrease in plant growth by CeO_2 while In_2O_3 NPs had no effect	Ma et al. (2013a)
Lycopersicon esculentum				
Activated carbon (AC) MWCNT	0.86–2.22 10–35	50 μg mL^{-1}	Activation of many stress-related genes; up-regulation of water channel's gene in CNT-exposed roots and leaves	Khodakovskaya et al. (2011)
Ag TiO_2	10–15 27	0–1000 mg L^{-1} 1000, 5000 mg L^{-1}	Decrease in root elongation, chlorophyll contents and less fruit productivity, higher superoxide dismutase activity. Phytotoxicity lesser than Ag	Song et al. (2013)
CeO_2	—	10 mg mL^{-1}	Second generation seedlings grown from CeO_2-NP treated seeds were smaller and weaker, developed extensive root hairs, and accumulated higher amount of ceria	Wang et al. (2013)

Nanoparticle	Size	Concentration	Effect	Reference
NiO	~24	0.025–2.0 mg mL^{-1}	Concentration-dependent retardation of root growth; increase in intracellular reactive oxygen species content leading to genotoxic effects	Faisal et al. (2013)
Cucumis sativus				
Fe$_3$O$_4$, TiO$_2$, carbon	30–50	0–5000 μg mL^{-1}	Seed germination was inhibited in the order TiO$_2$ > Fe$_3$O$_4$ > carbon NPs	Mushtaq (2011)
CuO ZnO	32–76, 500	1000 mg L^{-1}	Marked decrease in seedling biomass, generation of oxidative stress.	Kim et al. (2012)
Brassica napus				
Octahedral hexa Mo- clusters	~1	–	Inhibition of plant growth and changes in root morphology	Aubert et al. (2012)
TiO$_2$	14 or 25	10, 50, 100 mg L^{-1}	Internalization in roots, bioaccumulation and translocation up to leaves	Larue et al. (2012)
Cucurbita mixta				
Fe$_3$O$_4$	–	–	Oxidative stress in the roots; increased lipid peroxidation, SOD and catalase activities	Wang et al. (2011)
Nicotianna tabaccum				
Al$_2$O$_3$	–	0–1%	With increasing concentration, germination rate, average root length, biomass, and leaf count decreased significantly; increase in miRNA expression	Burklew et al. (2012)
Carboxy-fullerenes	100–200	0–0.144 mg mL^{-1}	Disruption of cell wall and membrane, cell growth inhibition; elevated ROS levels	Liu et al. (2013)

invagination leading to formation of clathrin-coated vesicles. SWCNTs reportedly penetrated intact cell suspensions of *Nicotianna tobaccum* cv. BY-2 through endocytic mechanism (Liu et al., 2009).

12.4.2. Transfer through Ion Channels Post-ionization

It has been proposed that ZnO and Ag become dissolved to Zn^{2+} and Ag^+ prior to cell internalization (Ma et al., 2013b; Shams et al., 2013). The ionic forms are internalized into plants through cellular ion channels. This mechanism has been observed in the case of barley plants grown in the presence of PdNPs (Battke et al., 2008). The roots of *Prosopis* sp. (mesquite) seedlings exposed to nano-$Ni(OH)_2$ contained intact NPs, whereas the aerial parts of the plant contained dissolved Ni^{2+} (Parsons et al., 2010). *P. florida*, *S. tragus* and *P. juliflora-velutina* grown in nano-ZnO suspensions also exhibited the presence of dissolved Zn^{2+} ions instead of ZnO forms (Rosa et al., 2011).

12.4.3. Aquaporin Mediated

Aquaporins are integral membrane proteins, also referred to as "water channels" that play a crucial role in controlling the water contents of cells. Their involvement in the uptake of nanoparticles, particularly multiwalled carbon nanotubes (MWCNT), was observed in *L. esculentum* plants and *N. tobaccum*, where the aquaporin gene was upregulated leading to increased water uptake along with NPs (Khodakovskaya et al., 2009, 2012).

12.4.4. Carrier Proteins Mediated

In some cases, proteins on plasma membrane enable NP entry into the plant cells.

12.4.5. Via Organic Matter

Natural organic matter (NOM) suspended fullerene C_{70} have been demonstrated to be internalized in rice plants through seeds, vascular systems, roots, stems, and leaves. NOM-C_{70} was translocated from the roots to aerial parts along with water and nutrients through the xylem vessels. Upward movement took place under the influence of plant osmotic pressure and capillary forces, while cellular internalization may have occurred through intercellular plasmodesmata (Lin et al., 2009). Humic acid, a major organic constituent of soil, has been shown to alleviate the phytotoxic effects of Cu nanoparticles and increase the accumulation of Au NP by a factor of 5.6 (Hawthorne et al., 2012).

12.4.6. Complex Formation with Root Exudates

Plant roots exude various compounds such as enzymes, sugars, phenolics, amino acids, and mucilage which may affect nanomaterial uptake and aggregation (Judy et al., 2012). Typically, a "pectin hydrogel capsule", formed by mucilage exudates from roots of *A. thaliana* seedlings, confirmed this mode by facilitating uptake of Alizarin red S-nano TiO_2 complex (Kurepa et al., 2010).

12.4.7. Foliar Uptake

NPs may gain entry though stomatal or cuticular openings or through the bases of trichomes upon application to leaf surfaces. For example, CeO_2 NPs applied as aerosol or suspension were absorbed by corn leaves but not translocated (Birbaum et al., 2010).

12.5. EFFECT ON PHYSIOLOGICAL PARAMETERS

The mechanism of NP-induced phytotoxicity is still not very clear. Nonetheless, the mere presence of active nanoparticles in a plant system will cause certain changes in the physiological activity. Some of the observed adverse effects are described in the following sections.

12.5.1. Loss of Hydraulic Conductivity

Being smaller than NPs, the pore sizes of plant cell walls obstruct their entry into the cells. This induces their accumulation on the walls, leading to the formation of cake layers on the root surface (Asli and Neumann, 2009). The cake layer restricts the transport of water from the roots to upper plant parts, causing a decrease in hydraulic conductivity of the epidermal cells. Inhibition of leaf growth and transpiration in maize seedlings by TiO_2 NPs has been attributed to this phenomenon.

12.5.2. Genotoxic Effects

Silver nanoparticles on the root tip cells of *Allium cepa* showed penetration into subcellular structures and interference with intracellular components (Kumari et al., 2009). There was a marked decrease in mitotic index, chromosomal aberrations, and irregular metaphase. Root tip cells of *Vicia narbonensis* and *Zea mays* exhibited similar genotoxic effects on exposure to TiO_2 NPs (Castiglione et al., 2010). Besides affecting cell division, NPs also damage the cellular DNA as in the case of DNA damage in radish, perennial ryegrass, and annual ryegrass by CuO nanoparticles (Atha et al., 2012) and in soybean by ZnO and CeO_2 NPs (López-Moreno et al., 2010).

12.5.3. Absorption and Accumulation

Subsequent to internalization NPs tend to accumulate in plant tissues, for example accumulation of magnetic carbon-coated NPs in leaf trichomes of wheat plants (Cifuentes et al., 2010), TiO_2 in *A. thaliana* seeds (Kurepa et al., 2010), and Ag NPs in rice seedlings (Mazumdar and Ahmad, 2011). This excess accumulation may sometimes cause cytotoxicity and growth inhibition.

The toxicity of CuO, NiO, Fe_2O_3, TiO_2, and Co_3O_4 NPs were investigated on lettuce, radish, and cucumber plants (Wu et al., 2012). NPs are primarily adsorbed on the root surface. The locally high concentration ions may have contributed to enhanced phytotoxic effects.

12.5.4. Generation of Reactive Oxygen Species (ROS)

NP-induced oxidative stress and ROS production have been amply proved (Begum et al. 2011; Wang et al., 2011; Dimkpa et al., 2012; Kim et al., 2012; Faisal et al., 2013; Hu et al., 2013; Liu et al., 2013; Rico et al., 2013). During oxidative stress, both DNA and proteins are expected to be damaged leading to stunted or wilted growth, leaf chlorosis, defoliation, and/or disruption of the cell wall or cell membrane. Sinha and Khare (2013) have reviewed the ROS-based toxicity mechanisms in NP-induced bacterial toxicity. Similar mechanisms may be functional in bringing about toxicity in plants. ROS generation and its diverse effect on plant growth and physiology are summarized in Table 12.3.

Treatment of tobacco BY-2 cell suspension culture with alumina NPs led to generation of reactive nitrogen species (RNS) (Poborilova et al., 2013). The NP-treated BY-2 cell suspension showed elevated levels of caspase-like activity, suggesting the possibilities of programmed cell death. Caspases are cysteine-aspartic proteases associated with apoptosis or cell death. Faisal et al. (2013) attributed a similar mechanism of cell death in tomato plants upon NiO treatments, implicating caspase-3 like protease activity.

Adsorption of C_{70}-carboxyfullerenes ($C_{70}(C-(COOH)_2)_{2-4}$) on tobacco BY-2 plant cells led to the disruption of cell walls and cell growth inhibition (Liu et al., 2013). AFM and confocal imaging revealed the deposition of glycosyl residue, N-acetyl glucosamine (GlcNAC), on the cell wall of treated cells. The accumulation of GlcNAC could possibly be a protection strategy of the plant cells as the increased deposition on cell surface may prevent contact with NPs, thereby ameliorating toxicity.

12.5.5. Biotransformation of NPs

There have been reports on the biotransformation of NPs into other intermediate species within the cells. Electron microscopy of cucumber root cells showed biotransformation of nano-Yb_2O_3 particles into $YbPO_4$ (Zhang et al., 2012a). The same group demonstrated the biotransformation of ceria NPs into CeO_2 and $CePO_4$ in the roots and CeO_2 and cerium carboxylates in the shoots (Zhang et al., 2012b). On similar lines, CuO and Cu(I)–sulphur complexes and Zn as Zn-phosphate have been detected in the shoots of CuO- and ZnO-treated wheat plants (Dimpka et al., 2012). Apparently, the biotransformed species may reduce or enhance the phytotoxic effect.

TABLE 12.3. ROS-Mediated Effect of Nanoparticles on Plants

Plant system	Type of NPs	ROS-mediated effect	Reference
Brassica oleracea var. capitata, Lycopersicon esculentum, Lactuca sativa, Amaranthus tricolor L., A. lividus L.	Graphene	Significant inhibition of plant growth and biomass; concentration-dependent increase in ROS and cell death; necrosis	Begum et al. (2011)
Cucurbita mixta cv. white cushaw	Fe_3O_4	NPs induced more oxidative stress than bulk particles in the roots; increased lipid peroxidation, superoxide dismutase and catalase enzyme activities	Wang et al. (2011)
C. sativus	CuO, ZnO	Marked decrease in seedling biomass, generation of oxidative stress	Kim et al. (2012)
L. esculentum	NiO	Concentration-dependent retardation of root growth, increase in intracellular ROS content leading to genotoxic effects	Faisal et al. (2013)
Salvinia natans	ZnO	Decrease in chlorophyll and carotenoid contents with increasing ZnO NP concentration; generation of oxidative stress	Hu et al. (2013)
N. tobaccum	Carboxy-fullerenes	Disruption of cell wall and membrane, cell growth inhibition; elevated levels of ROS	Liu et al. (2013)
O. sativa	CeO_2	At higher concentrations, significant reduction in H_2O_2 generation in both shoots and roots, increased electrolyte leakage and lipid peroxidation in the shoots	Rico et al. (2013)
O. sativa	CuO	Reduced seed germination; loss of root cell viability; severe oxidative stress; increased level of ROS scavenging antioxidant enzymes	Shaw and Hossain (2013)
Kiwifruit pollen	Ag	Decrease in pollen viability, membrane damage, disruption of tube elongation	Speranza et al. (2013)
T. aestivum	Zn	Increased H_2O_2 production; root lignifications and root growth inhibition	Li et al. (2012)

12.6. GENECTIC AND MOLECULAR BASIS OF NP PHYTOTOXICITY

So far, there is very little understanding about the molecular basis of phytotoxicity. Transcriptome analysis in SWCNT-treated maize root tissues indicted upregulation of root SLR1 and RTCS genes and downregulation of RTH1 and RTH3 genes (Yan et al., 2013).

Aquaporins help in the uptake of nanoparticles. These are stress-inducible genes regulated by a variety of stress factors including heavy metals. The upregulation of the NtLRX1 and water channel genes contributing to increased aquaporin production have been shown in tomato (Khodakovskaya et al., 2009) and tobacco cells (Khodakovskaya et al., 2012). Vannini et al. (2013) have very recently analyzed the proteome of *Eruca sativa* exposed to silver NPs. Proteomic profiling of nano- and bulk-Ag treated samples showed differential expression. Ag NP exposure altered some of the proteins related to the endoplasmic reticulum and vacuole, indicating these two organelles were targets of the Ag NP action.

Considerable correlation has been found between expression of microRNAs (miRs) and germination of *A. thaliana* seeds exposed to gold NP exposure (Kumar et al., 2013). Micro-RNA, namely miR164, miR167, miR395, miR398, miR408, and miR414, were downregulated upon gold NP exposure as compared to control seedlings. These results are contradictory to that reported by Burklew et al. (2012) on tobacco plants. This establishes the phytotoxic effects being species specific.

Gene expression in *Arabidopsis thaliana* roots upon treatment with ZnO, fullerene soot, and TiO_2 showed that ZnO NPs were most toxic and upregulated stress-related genes (Landa et al., 2012). Authors identified the specific roles of all upregulated and downregulated genes to postulate the mechanisms of phytotoxicity, but no clear picture has emerged so far.

Briefly, all these studies show that NPs become absorbed, transported through various mechanism, accumulated, and internalized. The cytotoxicity is then exerted at the level of gene expression, leading to growth retardation and other physiological effects. The effect varies from plant to plant. No generic process or mechanism has yet emerged.

12.7. CONCLUSIONS AND FUTURE PERSPECTIVES

The chapter has described the present level of understanding about the behavior and toxicity of different nanomaterials towards plants. Sufficient evidence exists to conclude that nanoparticles/nanomaterials exert phytotoxic effects. The effect is dependent on the type, size, and concentration of NPs. The clear mechanism of NP-mediated toxicity has yet to be understood; research in this area is at a very preliminary stage. Nonetheless, the impact of NPs on the plant system, biodiversity, and environment has huge implications and therefore needs to be addressed as a priority.

Some of the important issues for future perspectives include:

- appropriate identification of end-points as bio-indicators of nano-phytotoxcity;
- the impact on plants grown in their natural pristine habitat, in order to understand the actual nature of plant–NP interactions on the food chain;
- the possibility of "biomagnification" if NPs are incorporated within the food chain;
- the correlation between the characteristics of NPs (e.g., composition, surface area, particle size, shape, surface charge, and aggregation state) with phytotoxicity;
- assessment of the uptake kinetics and how the NP characteristics affect uptake kinetics of NPs in plant systems;
- whether NPs evoke abiotic-stress-like response and similar pathways are triggered;
- loss in yield or changes in grain quality in plants, especially during reproductive stages, and the associated safety issues;
- whether NPs are sequestered in plant tissues or are bio-available to flora, fauna, and human beings;
- minimum and maximum threshold concentrations of NPs such that standard or correct dosage for agricultural applications may be decided; and
- whether NP affected plants pass on the genotoxic effects to their next generation.

ACKNOWLEDGEMENTS

The financial grant provided by the Department of Biotechnology (Government of India) is gratefully acknowledged. Author Rajeshwari Sinha is grateful to the Council of Scientific and Industrial Research (CSIR) for her Research Fellowship.

REFERENCES

Asli, S., Neumann, P.M. 2009. Colloidal suspensions of clay or titanium dioxide nanoparticles can inhibit leaf growth and transpiration via physical effects on root water transport. *Plant, Cell & Environment* 32(5): 577–584.

Atha, D.H., Wang, H., Petersen, E.J. et al. 2012. Copper oxide nanoparticle mediated DNA damage in terrestrial plant models. *Environmental Science & Technology* 46(3): 1819–1827.

Aubert, T., Burel, A., Esnault, M.-A., Cordier, S., Grasset, F., Cabello-Hurtado, F. 2012. Root uptake and phytotoxicity of nanosized molybdenum octahedral clusters. *Journal of Hazardous Materials* 219–220: 111–118.

Battke, F., Leopold, K., Maier, M., Schmidhalter, U., Schuster, M. 2008. Palladium exposure of barley: uptake and effects. *Plant Biology* 10(2): 272–276.

Begum, P., Ikhtiari, R., Fugetsu, B. 2011. Graphene phytotoxicity in the seedling stage of cabbage, tomato, red spinach, and lettuce. *Carbon* 49(12): 3907–3919.

Birbaum, K., Brogioli, R., Schellenberg, M., Martinoia, E., Stark, W.J., Günther, D., Limbach, L.K. 2010. No evidence for cerium dioxide nanoparticle translocation in maize plants. *Environmental Science & Technology* 44(22): 8718–8723.

Burklew, C.E., Ashlock, J., Winfrey, W.B., Zhang, B. 2012. Effects of aluminum oxide nanoparticles on the growth, development, and microRNA expression of tobacco (Nicotiana tabacum). *PloS One* 7 (5): e34783.

Castligione, M.R., Giorgetti, M.L., Geri, C., Cremonini, R. 2011. The effects of nano-TiO_2 on seed germination, development and mitosis of root tip cells of *Vicia narbonensis* L. and *Zea mays* L. *Journal of Nanoparticle Research* 13(6): 2443–2449.

Cifuentes, Z., Custardoy, L., de la Fuente, J.M. et al. 2010. Absorption and translocation to the aerial part of magnetic carbon-coated nanoparticles through the root of different crop plants. *Journal of Nanobiotechnology* 8(1): 26.

Dimkpa, C.O., McLean, J.E., Latta, D.E. et al. 2012. CuO and ZnO nanoparticles: phytotoxicity, metal speciation, and induction of oxidative stress in sand-grown wheat. *Journal of Nanoparticle Research* 14(9): 1125.

Dimkpa, C.O., McLean, J.E., Martineau, N., Britt, D.W., Haverkamp, R., Anderson, A.J. 2013. Silver nanoparticles disrupt wheat (*Triticum aestivum* L.) growth in a sand matrix. *Environmental Science & Technology* 47(2): 1082–1090.

Faisal, M., Saquib, Q., Alatar, A.A. et al. 2013. Phytotoxic hazards of NiO-nanoparticles in tomato: a study on mechanism of cell death. *Journal of Hazardous Materials* 250–251: 318–332.

Geisler-Lee, J., Wang, Q., Yao, Y. et al. 2013. Phytotoxicity, accumulation and transport of silver nanoparticles by *Arabidopsis thaliana*. *Nanotoxicology* 7(3): 323–337.

Ghodake, G., Seo, Y.D., Lee, D.S. 2011. Hazardous phytotoxic nature of cobalt and zinc oxide nanoparticles assessed using *Allium cepa*. *Journal of Hazardous Materials* 186(1): 952–955.

Gubbins, E.J., Batty, L.C., Lead, J.R. 2011. Phytotoxicity of silver nanoparticles to *Lemna minor* L. *Environmental Pollution* 159(6): 1551–1559.

Hawthorne, J., Musante, C., Sinha, S.K., White, J.C. 2012. Accumulation and phytotoxicity of engineered nanoparticles to *Cucurbita pepo*. *International Journal of Phytoremediation* 14(4): 429–442.

Hu, C., Liu, X., Li, X., Zhao, Y. 2013. Evaluation of growth and biochemical indicators of *Salvinia natans* exposed to zinc oxide nanoparticles and zinc accumulation in plants. *Environmental Science and Pollution Research International*, doi:10.1007/s11356-013-1970-9.

Jacob, D.L., Borchardt, J.D., Navaratnam, L., Otte, M.L., Bezbaruah, A.N. 2013. Uptake and translocation of Ti from nanoparticles in crops and wetland plants. *International Journal of Phytoremediation* 15(2): 142–153.

Jiang, H.-S., Li, M., Chang, F.-Y.,Li, W., Yin, L.-Y. 2012. Physiological analysis of silver nanoparticles and $AgNO_3$ toxicity to *Spirodela polyrhiza*. *Environmental Toxicology and Chemistry/SETAC* 31(8): 1880–1886.

Johnson, M.E., Ostroumov, S.A., Tyson, J.F., Xing, B. 2012. Study of the interactions between *Elodea canadensis* and CuO nanoparticles. *Russian Journal of General Chemistry* 81(13): 2688–2693.

Judy, J.D., Unrine, J.M., Rao, W., Wirick, S., Bertsch, P.M. 2012. Bioavailability of gold nanomaterials to plants: importance of particle size and surface coating. *Environmental Science & Technology* 46(15): 8467–8474.

Khodakovskaya, M., Dervishi, E., Mahmood, M., Xu, Y., Li, Z., Watanabe, F., Biris, A.S. 2009. Carbon nanotubes are able to penetrate plant seed coat and dramatically affect seed germination and plant growth. *ACS Nano* 3(10): 3221–3227.

Khodakovskaya, M.V., De Silva, K., Nedosekin, D.N. et al. 2011. Complex, genetic, photothermal and photoacoustic analysis of nanoparticle-plant interactions. *Proceedings of National Academy of Sciences* 108(3): 1028–1033.

Khodakovskaya, M.V., de Silva, K., Biris, A.S., Dervishi, E., Villagarcia, H. 2012. Carbon nanotubes induce growth enhancement of tobacco cells. *ACS Nano* 6(3): 2128–2135.

Kim, S., Lee, S., Lee, I. 2012. Alteration of phytotoxicity and oxidant stress potential by metal oxide nanoparticles in *Cucumis sativus*. *Water, Air & Soil Pollution* 223(5): 2799–2806.

Kumar, V., Guleria, P., Kumar, V., Yadav, S.K. 2013. Gold nanoparticle exposure induces growth and yield enhancement in *Arabidopsis thaliana*. *The Science of the Total Environment* 461–462: 462–468.

Kumari, M., Mukherjee, A., Chandrasekaran, N. 2009. Genotoxicity of silver nanoparticles in *Allium cepa*. *The Science of the Total Environment* 407(19): 5243–5246.

Kumari, M., Khan, S.S., Pakrashi, S., Mukherjee, A., Chandrasekaran, N. 2011.Cytogenetic and genotoxic effects of zinc oxide nanoparticles on root cells of *Allium cepa*. *Journal of Hazardous Materials*, 190(1–3): 613–621.

Kurepa, J., Paunesku, T., Vogt, S. et al. 2010. Uptake and distribution of ultrasmall anatase TiO_2 Alizarin Red S nanoconjugates in *Arabidopsis thaliana*. *Nano Letters* 10(7): 2296–2302.

Landa, P., Vankova, R., Andrlova, J. et al. 2012. Nanoparticle-specific changes in *Arabidopsis thaliana* gene expression after exposure to ZnO, TiO_2, and fullerene soot. *Journal of Hazardous Materials* 241–242: 55–62.

Larue, C., Khodja, H., Herlin-Boime, N. et al. 2011. Investigation of titanium dioxide nanoparticles toxicity and uptake by plants. *Journal of Physics: Conference Series* 304: 012057.

Larue, C., Veronesi, G., Flank, A.-M., Surble, S., Herlin-Boime, N., Carrière, M. 2012. Comparative uptake and impact of TiO_2 nanoparticles in wheat and rapeseed. *Journal of Toxicology and Environmental Health, Part A* 75(13–15): 722–734.

Lee, C.W., Mahendra, S., Zodrow, K. et al. 2010. Developmental phytotoxicity of metal oxide nanoparticles to *Arabidopsis thaliana*. *Environmental Toxicology and Chemistry/SETAC* 29(3): 669–675.

Li, X., Yang, Y., Zhang, J., et al. 2012. Zinc induced phytotoxicity mechanism involved in root growth of *Triticum aestivum* L. *Ecotoxicology and Environmental Safety* 86: 198–203.

Lin, S., Reppert, J., Hu, Q. et al. 2009. Uptake, translocation, and transmission of carbon nanomaterials in rice plants. *Small* 5(10): 1128–1132.

Liu, Q., Chen, B., Wang, Q., Shi, X., Xiao, Z., Lin, Z., Fang, X. 2009. Carbon nanotubes as molecular transporters for walled plant cells. *Nano Letters* 9(3): 1007–1010.

Liu, Q., Zhao, Y., Wan, Y. et al. 2010. Study of the inhibitory effect of water-soluble fullerenes on plant growth at the cellular level. *ACS Nano* 4(10): 5743–5748.

Liu, Q., Zhang, X., Zhao, Y., Lin, J., Shu, C., Wang, C., Fang, X. 2013. Fullerene-induced increase of glycosyl residue on living plant cell wall. *Environmental Science & Technology* 47(13): 7490–7498.

López-Moreno, M.L., de la Rosa, G., Hernández-Viezcas, J.A. et al. 2010. Evidence of the differential biotransformation and genotoxicity of ZnO and CeO_2 nanoparticles on soybean (glycine max) plants. *Environmental Science & Technology* 44(19): 7315–7320.

Ma, C., Chhikara, S., Xing, B., Musante, C., White, J.C., Dhankher, O.P. 2013a. Physiological and molecular response of *Arabidopsis thaliana* (L.) to nanoparticle cerium and indium oxide exposure. *ACS Sustainable Chemistry & Engineering* 1(7): 130610093747005.

Ma, H., Williams, P.L., Diamond, S.A. 2013b. Ecotoxicity of manufactured ZnO nanoparticles: a review. *Environmental Pollution* 172: 76–85.

Ma, X., Geiser-Lee, J., Deng, Y., Kolmakov, A. 2010. Interactions between engineered nanoparticles (enps) and plants: phytotoxicity, uptake and accumulation. *The Science of the Total Environment* 408(16): 3053–3061.

Maynard, A.D., Aitken, R.J., Butz, T., Colvin, V., Donaldson, K., Oberdörster, G., Philbert, M.A. et al. 2006. Safe handling of nanotechnology. *Nature* 444(7117): 267–269.

Mazumdar, H., Ahmed, G.A. 2011. Phytotoxicity effect of silver nanoparticles on *Oryza sativa*. *International Journal of ChemTech Research* 3(3): 1494–1500.

Miralles, P., Church, T.L., Harris, A.T. 2012. Toxicity, uptake, and translocation of engineered nanomaterials in vascular plants. *Environmental Science & Technology* 46(17): 9224–9239.

Mirzajani, F., Askari, H., Hamzelou, S., Farzaneh, M., Ghassempour, A. 2013. Effect of silver nanoparticles on *Oryza sativa* L. and its rhizosphere bacteria. *Ecotoxicology and Environmental Safety* 88: 48–54.

Mishra, P., Shukla, V.K., Yadav, R.S., Pandey, A.C. 2013. Toxicity concerns of semiconducting nanostructures on aquatic plant *Hydrilla verticillata*. *Journal of Stress Physiology and Biochemistry* 9(2): 287–298.

Monica, R.C., Cremonini, R. 2009. Nanoparticles and higher plants. *Caryologia* 62(2): 161–165.

Mushtaq, Y.K. 2011. Effect of nanoscale Fe_3O_4, TiO_2 and carbon particles on cucumber seed germination. *Journal of Environmental Science and Health, Part A* 46(14): 1732–1735.

Nair, R., Varghese, S.H., Nair, B.G., Maekawa, T., Yoshida, Y., Kumar, D.S. 2010. Nanoparticulate material delivery to plants. *Plant Science* 179(3): 154–163.

Navarro, D.A., Bisson, M.A., Aga, D.S. 2012. Investigating uptake of water-dispersible CdSe/ZnS quantum dot nanoparticles by *Arabidopsis thaliana* plants. *Journal of Hazardous Materials* 211–212: 427–435.

Oberdörster, G., Oberdörster, E., Oberdörster, J. 2005. Nanotoxicology: an emerging discipline evolving from studies of ultrafine particles. *Environmental Health Perspectives* 113(7): 823–839.

Parsons, J.G., Lopez, M.L., Gonzalez, C.M., Peralta-Videa, J.R., Gardea-Torresdey, J.L. 2010. Toxicity and biotransformation of uncoated and coated nickel hydroxide nanoparticles on mesquite plants. *Environmental Toxicology and Chemistry/SETAC* 29(5): 1146–1154.

Poborilova, Z., Opatrilova, R., Babula, P. 2013. Toxicity of aluminium oxide nanoparticles demonstrated using a BY-2 plant cell suspension culture model. *Environmental and Experimental Botany* 91: 1–11.

Rana, S., Kalaichelvan, P.T. 2013. Ecotoxicity of nanoparticles. *ISRN Toxicology* 2013: 574648.

Rico, C.M., Hong, J., Morales, M.I. et al. 2013. Effect of cerium oxide nanoparticles on rice: a study involving the antioxidant defense system and in vivo fluorescence imaging. *Environmental Science & Technology* 47(11): 5635–5642.

Rosa, G. De La, Lopez Moreno, M.L., Hernandez Viezcas, J.A. et al. 2011. Toxicity and biotransformation of ZnO nanoparticles in the desert plants *Prosopis juliflora-velutina*, *Salsola tragus* and *Parkinsonia florida*. *International Journal of Nanotechnology* 8(6/7): 492.

Salama, H.M.H. 2012. Effects of silver nanoparticles in some crop plants, common bean (*Phaseolus vulgaris* L.) and corn (*Zea mays* L.). *International Research Journal of Biotechnology* 3(10): 190–197.

Shams, G., Ranjbar, M., Amiri, A. 2013. Effect of silver nanoparticles on concentration of silver heavy element and growth indexes in cucumber (*Cucumis sativus* L. Negeen). *Journal of Nanoparticle Research* 15(5): 1630.

Shaw, A.K., Hossain, Z. 2013. Impact of nano-CuO stress on rice (*Oryza sativa* L.) seedlings. *Chemosphere*, doi: 10.1016/j.chemosphere.2013.05.044

Sinha, R., Khare, S.K. 2013. Molecular basis of nanotoxicity and interaction of microbial cells with nanoparticles. *Current Biotechnology* 2(1): 64–72.

Sinha, R., Karan, R., Sinha, A., Khare, S.K. 2011. Interaction and nanotoxic effect of ZnO and Ag nanoparticles on mesophilic and halophilic bacterial cells. *Bioresource Technology* 102(2): 1516–1520.

Slomberg, D.L., Schoenfisch, M.H. 2012. Silica nanoparticle phytotoxicity to *Arabidopsis thaliana*. *Environmental Science & Technology* 46(18): 10247–10254.

Song, U., Jun, H., Waldman, B. et al. 2013. Functional analyses of nanoparticle toxicity: a comparative study of the effects of TiO_2 and Ag on tomatoes (*Lycopersicon esculentum*). *Ecotoxicology and Environmental Safety* 93: 60–67.

Speranza, A., Crinelli, R., Scoccianti, V. et al. 2013. In vitro toxicity of silver nanoparticles to kiwifruit pollen exhibits peculiar traits beyond the cause of silver ion release. *Environmental Pollution* 179: 258–267.

Vannini, C., Domingo, G., Onelli, E. et al. 2013. Morphological and proteomic responses of *Eruca sativa* exposed to silver nanoparticles or silver nitrate. *PloS One* 8(7): e68752.

Wang, H., Kou, X., Pei, Z., Xiao, J.Q., Shan, X., Xing, B. 2011. Physiological effects of magnetite (Fe_3O_4) nanoparticles on perennial ryegrass (*Lolium perenne* L.) and pumpkin (*Cucurbita mixta*) plants. *Nanotoxicology* 5(1): 30–42.

Wang, Q., Ebbs, S.D., Chen, Y., Ma, X. 2013. Trans-generational impact of cerium oxide nanoparticles on tomato plants. *Metallomics: Integrated Biometal Science* 5(6): 753–759.

Wang, X., Han, H., Liu, X., Gu, X., Chen, K., Lu, D. 2012a. Multi-walled carbon nanotubes can enhance root elongation of wheat (*Triticum aestivum*) plants. *Journal of Nanoparticle Research* 14(6): 841.

Wang, Z., Xie, X., Zhao, J., Liu, X., Feng, W., White, J.C., Xing, B. 2012b. Xylem- and phloem-based transport of CuO nanoparticles in maize (*Zea mays* L.). *Environmental Science & Technology* 46(8): 4434–4441.

Wu, S.G., Huang, L., Head, J., Chen, D., Kong, I., Tang, Y.J. 2012. Phytotoxicity of metal oxide nanoparticles is related to both dissolved metals ions and adsorption of particles on seed surfaces. *Journal of Petroleum and Environmental Biotechnology* 3(4): 2–6.

Yan, S., Zhao, L., Li, H., Zhang, Q., Tan, J., Huang, M., He, S., Li, L. 2013. Single-walled carbon nanotubes selectively influence maize root tissue development accompanied by the change in the related gene expression. *Journal of Hazardous Materials* 246–247: 110–118.

Yin, L., Cheng, Y. Espinasse, B. et al. 2011. More than the ions: the effects of silver nanoparticles on *Lolium multiflorum*. *Environmental Science & Technology* 45(6): 2360–2367.

Zhang, P., Ma, Y., Zhang, Z. et al. 2012a. Comparative toxicity of nanoparticulate/bulk Yb_2O_3 and $YbCl_3$ to cucumber (*Cucumis sativus*). *Environmental Science & Technology* 46(3): 1834–1841.

Zhang, P., Ma, Y., Zhang, Z. et al. 2012b. Biotransformation of ceria nanoparticles in cucumber plants. *ACS Nano* 6(11): 9943–9950.

13

CYTOTOXICOLOGY OF NANOCOMPOSITES

Horacio Bach

Department of Medicine, Division of Infectious Diseases,
University of British Columbia, Vancouver, BC, Canada

13.1. INTRODUCTION

Nanomedicine, or the application of nanotechnology to medicine, is an expanding field that has created new areas of research and development. Nowadays, most of the studies and research are focused on the application of nanoparticles (NPs) in fields such as imaging, drug delivery, and antimicrobial activities. Other topics of interest for application in nanomedicine are still poorly developed and include nanoelectronics, nanobotics, nanolithography, and nanoelectromechanical systems.

The fast growth in the development and application of nanosized materials needs to be tempered with the evaluation of their potentially adverse effects on biological systems. This evaluation is crucial for a safe use of nanomaterials in the human body.

NPs are characterized by having unique physico-chemical properties and are defined as particles of size range 1–100 nm. As the field of nanomedicine is developing, the need for thorough characterization of nanomaterials has become apparent. Knowledge of nanocomposites, such as particle size, morphology, aggregation tendencies, and surface charge, is necessary to substantially contribute to the knowledge about their toxicology.

For metal- and metal-oxide-based NPs that can release metal ions, one important question to ask is whether the toxicity is mediated by the particle *per se* (reactivity or characteristics) or by means of the released species. Certainly, the particle can act as a transport vehicle for delivering a "cargo" into the cells. This kind of transport

Bio-Nanoparticles: Biosynthesis and Sustainable Biotechnological Implications, First Edition.
Edited by Om V. Singh.
© 2015 John Wiley & Sons, Inc. Published 2015 by John Wiley & Sons, Inc.

mechanism, sometimes termed the "Trojan horse mechanism", may be beneficial in nanomedicine. This chapter describes the adverse effects of NPs, focusing on health implications at a cellular level when NPs are applied directly to biological systems. The toxic effects of NPs in other situations, such as inhalation, ingestion, accumulation, or transport in different organs of the body, and the effects of NP-containing devices, pollution, industrial sources, and manufacturing environments are not discussed in this chapter.

13.2. CELLULAR TOXICITY

13.2.1. Mechanisms of Cellular Toxicity

Atmospheric O_2 is an inert gas because it has two unpaired electrons, making it paramagnetic, which limits its ability to interact with organic molecules unless it undergoes activation (Cadenas, 1989). A free radical is an atom or molecule with an unpaired electron. In the case of oxygen, free radicals are referred to as reactive oxygen species (ROS). Not all ROS are radicals, but they include a univalent reduction of molecular oxygen that produces intermediates such as the superoxide radical (O_2^-), singlet oxygen ($^1O_2$), hydrogen peroxide (H_2O_2), and hydroxyl radical (HO·) (Cadenas, 1989).

All respiring cells produce ROS and their production is directly proportional to the concentration of O_2 (Jamieson et al., 1986). An oxidative stress is defined as the production and accumulation of ROS beyond the capability of the cell to quench or neutralize these species. Interestingly, ROS can also be involved in signal transduction under specific concentrations (Schrek and Baeuerle, 1991).

Cells have developed antioxidant defenses to prevent the propagation of singlet O_2 by its quenching at the site of production. In this way, the flux of reduced oxygen intermediates is lowered, preventing the production of hydroxyl radicals, the most damaging of the species (Cadenas, 1989). The most important ROS in cellular systems – singlet O_2, superoxide radicals, H_2O_2, and hydroxyl radicals – are described in the following sections (Fig. 13.1).

Singlet O_2. Singlet O_2 can be generated chemically, enzymatically, or photochemically (Krinsky, 1977). Several cellular pathways can produce singlet O_2 as a result of the energy transferred to O_2. Since this species is not a radical, it usually decays to the ground state before it has time to react with another entity (Cadenas, 1989). Although is thought that singlet oxygen will not diffuse over long distances (Moan, 1990), recent measurements revealed that a significant fraction can diffuse into the extracellular environment, reaching distances of c. 260 nm in 6 ms (Skovsen et al., 2005). It has been reported that silver NPs (AgNPs) of 13 nm covered by a pectin layer enhanced the production of singlet oxygen (de Melo et al., 2012).

Superoxide Radicals. These radicals can act as either an oxidant or a reductant in living systems. The dismutation of superoxide radicals leads to the formation of H_2O_2, which occurs spontaneously or is catalyzed by a member of the antioxidant

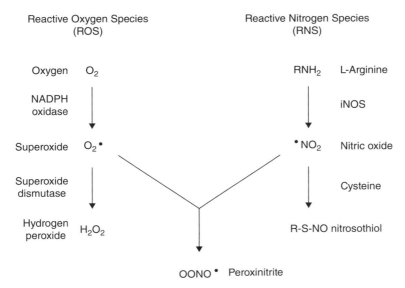

Figure 13.1. ROS and RNS generation in macrophages, showing the generation of superoxide and hydrogen peroxide by NADPH phagocyte oxidase and superoxide dismutase, respectively. The inducible nitric oxide synthase (iNOS) generates nitric oxide from L-arginine. The reaction of nitric oxide with cysteine sulphydryls results in the formation of nitrosothiols. The combination of nitric oxide and superoxide generates peroxynitrite.

defense in organisms, the enzyme superoxide dismutase (SOD; Riley and Weiss, 1994). When the superoxide radical is in its protonated state (pKa=4.8), it forms the powerful oxidant perhydroxyl radical (·OHH) (Naqui et al., 1986). Its relevance in biological systems is minor because of its low concentration at physiological pH. Within mitochondria, the superoxide radical is stable (Nohl and Jordan, 1986) and can diffuse across the membrane in a concentration-dependent manner, but at extremely slow rates (Naqui et al., 1986; Cadenas, 1989). Although SOD reduces the steady-state concentrations of the superoxide radical by several orders of magnitude in the cell, the radical still has significant and independent damaging potential (Fridovich, 1986), able to diffuse in a distance of c. 320 nm (Naqui et al., 1986; Cadenas, 1989).

H_2O_2. This uncharged compound can diffuse across biological membranes and cause significant damage because it is not restricted to its point of synthesis in the cell (Cadenas et al., 1977) and can enter into numerous other cellular reactions. H_2O_2 can damage directly many cellular biomolecules, such as DNA and enzymes, and can play an important role in apoptosis induction (Dumont et al., 1999). H_2O_2 can be further reduced and enhance the production of hydroxyl radicals, which can be facilitated by taking electrons from transition metals (such as Fe) via the Fenton reaction (Lloyd et al., 1997). The Fenton reaction comprises two consecutive steps:

1. $Fe^{2+} + H_2O_2 + H^+ \rightarrow Fe^{3+} + HO\cdot + H_2O$
2. $Fe^{3+} + H_2O_2 \rightarrow Fe^{2+} + HOO\cdot + H^+$

Recently, it has been demonstrated that AgNPs in the range of 25–69 nm can catalyze the decomposition of H_2O_2 at pH 9.5 (He et al., 2011). In this study, the authors suggested that the formation of hydroxyl radicals is unlikely (He et al., 2011) but a physiological pH was not used; the significance of this study should therefore be revised.

Hydroxyl Radical. This is the most reactive oxygen radical. It can attack any biomolecule or initiate free radical chain reactions with a mean diffusion distance of 4.5 nm. The hydroxyl radical can oxidize membrane lipids and denaturate proteins and nucleic acids (Halliwell and Gutteridge, 1984). Its production is regulated by the availability of Fe^{2+}. Any enzymatic recycling system (a reduced agent) can recycle Fe^{3+} to Fe^{2+} and maintain an ongoing Fenton reaction, leading to a concomitant generation of hydroxyl radicals.

13.2.2. Effect of Glutathione (GSH) in Oxidative Stress

Oxidative stress refers to a state in which GSH is depleted while oxidized glutathione (GSSG) accumulates. Cells are able to sense this drop in the GSH/GSSG ratio and can mount a protective or injurious response.

GSH is a tripeptide (Fig. 13.2) constituting the major endogenous antioxidant scavenger that protects cells against oxidative stress through its ability to bind to and reduce ROS. Its activity is mediated by the presence of a cysteine in the molecule. GSH is considered the primary barrier to contain ROS, protecting the cells from oxidative stress (Anderson et al., 2004; Habib et al., 2007). Preservation of the GSH-mediated antioxidant defense system is therefore critical for cell survival. Accumulation of GSSG can trigger apoptotic signaling by inhibiting mitochondrion function (Pias and Aw, 2002).

Various studies have shown that cellular levels of GSH are either increased or decreased after *in vitro* treatment with AgNPs. The increased levels of GSH in some AgNP-treated cells may involve cellular responses in order to cope with AgNP-mediated oxidative damage (Arora et al., 2009; Farkas et al., 2011). By contrast, the

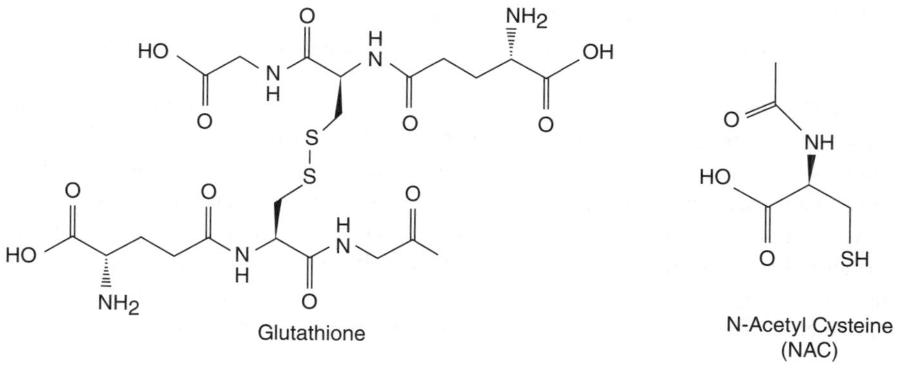

Figure 13.2. Structure of glutathione and N-acetyl cysteine.

decreased levels of GSH noted in AgNP-treated human skin carcinoma and fibrosarcoma cells suggest an inhibition of GSH-synthesizing enzymes and/or abnormally increased demand for GSH in conjugation with electrophilic molecules (Arora et al., 2008). Several studies have shown that the addition of N-acetyl cysteine (Fig. 13.2), a precursor of GSH, considerably reduced the damage produced by ROS (Hussain et al., 2005; Carlson et al., 2008).

Oxidative stress is the main event mediated by metals. An increase in ROS generation and a reduction of GSH content have been observed in cells treated with metals such as Ag, Fe, Cu, Co, and Cd (Jomova and Valko, 2011). Elevated synthesis of GSH can therefore be a key contributor to the prevention of the damage associated with ROS insults.

The induction of ROS production by NPs is well documented. Different pathways and cellular compartments can be affected by the generation of ROS by NPs. Several studies have reported the generation of ROS upon exposure of NPs to different cells, for example the exposure of murine and human fibroblasts to SiO_2NPs and AgNPs (Asharani et al., 2009a,b; Yang et al., 2009) and AgNPs exposed to HeLa and Jurkat T cells (Miura and Shinohara, 2009; Eom and Choi, 2010). Induction of oxidative stress was also reported in macrophages exposed to AgNPs in a size-dependent manner (Carlson et al., 2008).

Another group of reactive compounds involved in stress is the reactive nitrogen species (RNS; Fig. 13.1). Many cells produce physiological levels of nitric oxide or nitrogen monoxide (NO·), a molecule implicated in biological regulation and known to participate in a diverse range of cellular processes including nitrosative stress (Bredt, 1999; Ricciardolo et al., 2006). Nitric oxide is produced by the enzyme nitric oxide synthase for physiological functions, but it can also react with a superoxide radical to form the peroxynitrate anion (ONOO-), a potent oxidant (Fridovich, 1986) which is able to diffuse across biological membranes at rates 400 times greater than the superoxide radical (Marla et al., 1997).

13.2.3. Damage to Cellular Biomolecules

Oxidative and nitrosative stress can irreversibly affect biomolecules, causing lipid peroxidation, protein oxidation, enzyme inhibition, damage to nucleic acids, and activation of apoptosis, all of which will ultimately result in cell death.

Membrane Damage. Membranes can be damaged as a result of exposure to reactive molecules. A loss in the membrane fluidity can cause irreversible damage to the cell by affecting osmotic pressure and electrolyte homeostasis (e.g., a constant increase of Ca^{2+} can lead to the induction of apoptosis). For example, the hydroxyl radical can react with lipids that comprise the cellular membrane and change the saturation state of the classical lipid bilayer (Girotti, 1998), leading to lipid peroxidation (Gutteridge and Halliwell, 1990; Arora et al., 2008). Lipid peroxidation of other membranes, such as in mitochondria, can produce multiple effects on enzyme activity, energy production for the cell (ATP), and a change in the membrane potential leading to the induction of apoptosis (Green and Reed, 1998).

Proteins. Damaged proteins can exhibit site-specific amino acid modification, fragmentation of the peptide chain, aggregation of cross-linked reaction products, altered electrical charge of the protein, and increased susceptibility to removal and degradation. *De novo* carbonyl group generation in the backbone of the protein can transform functional amino acids into non-functional amino acids as in the case of histidine, lysine, proline, arginine, and serine (Stadtman and Levine, 2003).

The process of protein inactivation by a reactive attack of Ag^+ to cysteine residues was delayed by exposing the human hepatoma HepG2 cells to N-acetyl-cysteine (NAC), an antioxidant and GSH precursor. In this case, results obtained after NAC treatment were equivalent to untreated controls (Kim et al., 2009). Moreover, this NAC treatment also significantly reduced the reduction of cell viability, suggesting that an oxidative stress has been created upon exposure to AgNPs of the hepatoma cells (Kim et al., 2009).

Another toxic effect of NPs is the perturbations in the function of the endoplasmic reticulum (ER). It has been reported that SiO_2NPs at toxic levels induce an ER stress response, or the "unfolded protein response" (UPR), based on the measurement of ER stress-selective markers (Christen and Fent, 2012). When proteins are not folded properly, they deposit and accumulate in the ER which triggers the UPR (Harding et al., 2002). The UPR is an evolutionarily conserved cell-signaling pathway which, once activated, can either result in the recovery of the cell or activation of apoptosis.

To restore cellular functions and to remove the unfolded proteins from the ER, chaperons become upregulated, protein translation is inhibited and protein degradation increases. When the ER stress cannot be contained and the cell cannot restore normal ER function, apoptotic pathways will be activated (Rutkowski and Kaufman, 2004).

Since NPs can enter the cell in vesicular structures, an interaction of these vesicles with the ER can occur. As a result of ER stress, calcium is released from the ER lumen reservoirs into the cytosol (Christen et al., 2007). One of the main consequences derived from Ca increase is the phosphorylation of the transcription factor CREB, which can induce the transcription of the protein phosphatase 2A involved in a wide range of cellular functions including cell cycle regulation, cell morphology, development, signal transduction, apoptosis, and stress response (Janssens and Goris, 2001). Various studies reported the induction of intracellular calcium after exposure of AgNPs to human lung fibroblasts (Asharani et al., 2009*a,b*) or exposure of intestinal cells to TiO_2 NPs (Koeneman et al., 2010).

An induction of ER stress response genes after exposure to AgNPs has been reported in human intestinal cells. Ag^+ ions released from the NPs are probably responsible for this induction (Bouwmeester et al., 2011).

DNA. An oxidative attack can also affect the integrity of DNA. It can induce numerous unwanted changes such as deletions, mutations, and other lethal genetic effects, which can be passed onto the offspring. Both of the DNA building molecules, the sugar and base moieties, and proteins binding DNA are susceptible to oxidation, causing base degradation, single-strand breakages, and protein cross-linking (Imlay and Linn, 1988; Imlay, 2003). Interestingly, although H_2O_2 and superoxide radicals cannot by themselves cause strand fragmentation under normal physiological conditions *in vitro*, their cytotoxicity *in vivo* should not be ignored as they can serve as intermediates of hydroxyl radical production through the Fenton reaction (Imlay and Linn, 1988; Imlay, 2003). Another mechanism of DNA damage is indirect oxidative DNA damage due to inflammation as a result of the production of endogenous oxygen radicals. This damage can be exacerbated in phagocytic cells, such as neutrophils and macrophages, when they mount an immunological response (Grisham et al., 2000). These cells are able to generate ROS and nitric oxide as a response to an inflammation in inflamed tissues, causing injury to target cells and an indirect damage to the DNA (Eiserich et al., 1998; Chazotte-Aubert et al., 1999).

A decrease in the expression of the enzyme OGG1 was measured when cells were exposed to AgNPs. This enzyme is involved in DNA repair and removes 8-oxoguanine, one of the most common DNA lesions (Dizdaroglu et al., 2002). The decrease in the expression of this enzyme can be originated as: (1) a change in the gene transcription or translation; or (2) the redox state of the cell because a Nrf2 transcript factor-binding site is present in the promoter region of OGG1. As a result of the downregulation of this gene, there is then an accumulation of oxidized DNA bases (Piao et al., 2007).

DNA damage is sensed by cells and usually accompanied by cell cycle arrest. This mechanism is based on many checkpoints and dictates the fate of the cell (Fig. 13.3). If the damage can be repaired, cells will stop the cell cycle until the damage is repaired. Alternatively, if the damage is not reparable, the cell will activate its destruction by inducing apoptosis. DNA damage of NPs can be assessed depending on the cell population arrested at a specific period. For example, the sub-G1 population, which can include apoptotic cells, was found to be significantly increased in a concentration- and time-dependent manner upon exposure of

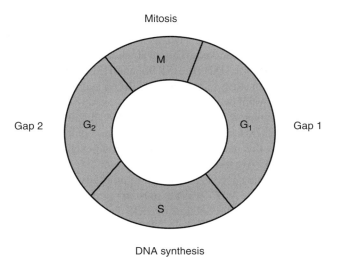

Figure 13.3. The cell cycle. In the S phase, DNA is synthesized and replication occurs. During the M phase or mitosis, nuclear chromosomes separate.

the cells to AgNPs (Chairuangkitti et al., 2013). However, a pretreatment with 10 mM NAC decreased the proportion of cells in the sub-G1 compartment. Other phases of the cell cycle, including DNA synthesis (S phase), Gap2/mitosis (G2/M phase), and Gap1 (G0/G1 phase), were analyzed. As a result of exposure to AgNPs, a significant accumulation of the cells in S phase, along with a decrease in the corresponding G1 population in a concentration-dependent manner, was observed. These results demonstrate that AgNPs could interfere with cell proliferation via induction of an S phase cycle arrest and a reduction in the population of cells in the G1 phase (Chairuangkitti et al., 2013).

The expression of the DNA damage marker protein p-H2AX was shown to be increased dramatically after 12 h of exposure of the cells to AgNPs (Eom and Choi, 2010). These results were corroborated by a Comet assay, which is a sensitive technique for quantifying DNA damage and has been widely used for the study of genotoxicity. The comet assay is conducted by electrophoresis of lysed cells and stained with ethidium bromide. Their fluorescence images resemble comets with a distinct head and tail which is composed of intact and fragmented DNA, respectively. The ratio of tail length to diameter of the head of is used to measure the degree of damaged DNA. Interestingly, a significant increase of DNA fragmentation was observed after 24 h exposure to a concentration range of AgNPs (0.1–10 µg mL^{-1}; Hackenberg et al., 2011). In this study, chromosomal aberrations consisting mainly in deletions and exchanges were observed (Hackenberg et al., 2011).

Results from a DNA microarray analysis revealed that AgNPs exposed to HepG2 cells induced the expression of a large number of genes (Kawata et al., 2009). In particular, AgNPs induced well-known stress-associated genes coding for metallothionein, such as *MT1H*, *MT1X*, and *MT2A*, and heat shock protein (*HSPA4L* and *HSPH1*), which have been reported to be induced by cellular stresses such as heavy metal and various cytotoxic agent exposures (Welch, 1992; Vasto et al., 2006). Other genes that were upregulated include those associated with cell cycle progression. In particular, checkpoint-related genes such as *BIRC5*, *BUB1B*, *CCNA2*, *CDC25B*, *CDC20*, and *CKS2* were observed. The heat shock protein (HSP) superfamily including the families HSP40, HSP 70, HSP 90, and HSP 110 families, are known to be induced upon metal ion stress (Klaasen et al., 2009) and heavy metals (Guven and De Pomerai, 1995; Mutwakil et al., 1997).

Other genes upregulated were related to DNA repair. One example is the induction of the gene *RAD51* involved in DNA damage repair (Ahamed et al., 2008). The protein encoded by this gene promotes DNA strand exchange and is involved in repair of damaged DNA (Arnaudeau et al., 1999).

Another transcriptome analysis performed in mouse lymphoma cells exposed to AgNPs showed that the gene *NCF2* was the most upregulated gene (Mei et al., 2012). This gene is involved in the production of ROS because it encodes p67phox, an essential component of the multi-protein NADPH oxidase enzyme, the major source of superoxide (Gauss et al., 2007). The *DUOX1* gene was the most downregulated gene and it is identified as H_2O_2-producing enzymes. Other upregulated genes include *XPA* and *ERCC2*, both of them related to DNA repair. Their dysregulation suggest that there is an impact of the oxidative stress induced by the AgNPs on DNA repair genes (Jaiswal, 2000).

In summary, oxidative stress that could be induced by NPs may have the following sources:

1. ROS can be generated directly on the surface of NPs when both oxidants and free radicals are present on the surface of the NP.
2. Transition metals can generate reactive ROS acting as catalysts in Fenton reactions (Risom et al., 2005). For example, the reduction of H_2O_2 with Fe^{2+} produces hydroxyl radicals that are extremely reactive, attacking biomolecules situated within their diffusion range.
3. Mitochondrial function is altered as a result of the permeability of the organelle to small NPs that produce physical damage, leading to oxidative stress (Li et al., 2003; Xia et al., 2006).
4. Inflammatory cells, such as alveolar macrophages and neutrophils, can be activated, which can phagocytose NPs leading to the generation of ROS and RNS (Long et al., 2004; Risom et al., 2005).

Taken together, NPs exposed to cellular systems can be corroded, degraded, or dissolved into their atoms, changing the chemical composition of the mixture, which can be detrimental to biomolecules and affect normal metabolic reactions or destabilize membranes necessary for maintaining cell integrity. As a result of an ionization of the metal, Ag^+ ions can then interact with protein and enzyme thiol groups such as cysteines (Elechiguerra et al., 2005; Jose et al., 2008).

13.3. NANOPARTICLE FABRICATION

Engineered NPs possess unusual physic-chemical characteristics such as their small size (surface area and size distribution), chemical composition (purity, crystallinity, electronic properties), surface structures (surface reactivity, surface groups, inorganic, or organic coating), solubility, shape, and aggregation.

NPs can be prepared and stabilized by physical and chemical methods. In the chemical method, the most common techniques include chemical reductions, which confer stability in aqueous dispersion (Wiley et al., 2005; Tao et al., 2006). Studies have shown that the size, morphology, stability, and properties (chemical and physical) of NPs are directly influenced by the experimental conditions, the kinetics of interaction of metal ions with reducing agents, and adsorption processes of stabilizing agents with NPs.

Simple reduction of metal salts by reducing agents in a controlled manner generally produces nanospheres, because spheres are the lowest-energy shape.

Commonly used reductants in NP fabrication include borohydride, citrate, ascorbate, and elemental hydrogen (Shirtcliffe et al., 1999; Sondi et al., 2003; Dahl et al., 2007). In the case of AgNPs for example, the starting material is an Ag salt, frequently $AgNO_3$, and as a result of the reduction of the cations in aqueous medium, AgNPs render with a diameter of several nanometers (Wiley et al., 2005). As a result of the formation of AgNPs with valence "zero", an agglomeration into clusters will then occur (Kapoor, 1998). To avoid this agglomeration, reductants such as borohydride can generate monodisperse suspensions of NPs, but the generation of larger particles is difficult to control (Schneider et al., 1994). On the other hand, weaker reductants such as citrate result in a slower reduction, which will narrow the size distribution of NPs (Lee and Meisel, 1982; Emory and Nie, 1997; Shirtcliffe et al., 1999).

Thermodynamic stabilization of NPs can be achieved by adding capping agents which bind to the NP surface via covalent bonds or by chemical interaction. These capping agents are essential to prevent nanoparticle aggregation and can also be used as a site for bioconjugation of the NP with other molecules of interest. Examples of capping agents include citrate (Farkas et al., 2011; Powers et al., 2011), water-soluble polymers, and polysaccharides (Ahamed et al., 2008; Schrand et al., 2008; Braydich-Stolle et al., 2010) such as chitosan (Sanpui et al., 2011), polyvinylpyrrolidone (PVP; Foldbjerg et al., 2009), carbon (Nishanth et al., 2011), hydrocarbons (Carlson et al., 2008), starch (Asharani et al., 2009a), and peptides (Haase et al., 2011). The synthesis of glycolipid-conjugated AgNPs with sophorolipids, a polysaccaharide obtained from the yeast *Candida bombicola*, and oleic acid show cytocompatibility with non-genotoxic effects to the hepatic cells HepG2 (Singh et al., 2009). The different surface chemistries imparted by these agents yield coating-specific behaviors of AgNPs in solution (Murdock et al., 2008).

Another way to prevent agglomeration is the addition of surfactants. These compounds will: (1) mediate the growth of NPs; (2) coat their surfaces; and (3) provide steric stabilization in order to prevent agglomeration and cluster formation. However, the presence of material traces used for NP fabrication must be carefully manipulated to avoid posterior damages to living systems. For instance, unwashed NPs can leave toxic agents such as formaldehyde, which has cytotoxic effects on cell cultures (Ku and Billings, 1984). Other compounds tested include 1% fetal calf serum (FCS), which stabilizes AgNP and avoids agglomeration (Foldbjerg et al., 2009). However, when AgNPs were encapsulated into PVP to enable a more effective re-dispersion and a decrease in aggregations, FCS caused adsorption of the AgNPs (Murdock et al., 2008). Another surfactant used in the preparation of NPs is the cationic detergent cetrimonium bromide (CTAB) that exhibits a significant toxicity if it is not eliminated properly from the final product (Connor et al., 2005).

13.3.1. Physico-chemical Characteristics of NPs

Zeta Potential. One of the key parameters to measure in NP fabrications is the Zeta Potential (ZP). This electrokinetic potential is a function of the surface charge developed when any material is placed in a liquid. ZP is the magnitude of an electrostatic repulsive interaction between particles and is commonly used to predict dispersion stability (Fig. 13.4).

A colloidal suspension is defined when the sizes of particles are in the range 5–200 nm in diameter; thus, NPs having a diameter <100 nm can be considered a colloidal suspension. All colloidal dispersions will eventually aggregate unless there are

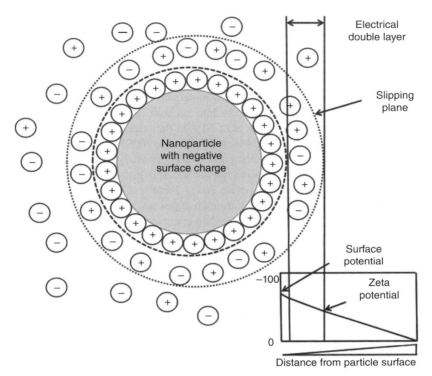

Figure 13.4 Diagram showing the zeta potential.

sufficient forces to avoid adherence of particles. For particles in water, the most common force of repulsion originates from charges on the particle surfaces. As a result of this charge in the liquid an electrical double layer is formed, consisting of the charges contributed by the particle's surface, and the concentration of counterions near the surface. An additional decrease in the ion concentrations near the surface indicates a thermally driven motion of ions. ZP is the parameter that measures the strength of the repulsion between particles, and is not a direct measure of the density of charges. ZP changes accordingly with the surface charges. In addition, the isoelectric point is the pH at which the ZP is zero; at this value the repulsive forces are zero, causing aggregation.

In cell biology, ZP has been used to diagnose the cellular interaction with charged ions or molecules. Anions will decrease the ZP at the cellular surface and cations will increase the ZP at the surface (Zhang et al., 2008). The binding of charged NPs to the cell surface will change the ZP value of the cells. As the ZP value of NPs turns negative, attachment of anionic NPs on the cellular surface will cause the ZP to become more negative (Zhang et al., 2008).

Effects of Physico-chemical Characteristics on the Toxicity of NPs. Size has a direct effect on the toxicity of NPs. Small NPs are more toxic than large NPs

(Carlson et al., 2008). Other studies reported that micro-sized particles are less toxic than their smaller counterparts (Oberdorster et al., 2005).

NPs of small size have been observed in mitochondria upon cellular internalization and induce oxidative stress and cell death via apoptosis (Xia et al., 2007). Moreover, the relatively larger surface area of NPs (per volume) can induce a greater production of ROS, which can damage intracellular biomolecules (Hussain et al., 2005; Karakoti et al., 2006). Another parameter to take into account is the nature of the charged functional groups that coat the surfaces of NPs, as they can determine whether or not they will enter the cells (Zhang and Monteiro-Riviere, 2009).

Recently, a transcriptome study comprising the analysis of gene expression changes in the human epithelial colorectal adenocarcinoma Caco-2 cells exposed to AgNPs was reported (Bouwmeester et al., 2011). Results of this study concluded that no genes were significantly regulated after 4 h exposure, suggesting that genes were not affected by the "nanoparticle effect" when compared to Ag^+ used in the study for comparison purposes (Bouwmeester et al., 2011).

The size of the NPs can directly affect the induction of oxidative stress and apoptosis in human cells. For example, hepatic cells exposed to 21 nm SiO_2 NPs induced oxidative stress and apoptosis as measured by changes of the proteins p53 and Bax. Interestingly, the same NPs but with different sizes (48 and 86 nm) did not induce these defense mechanisms (Ye et al., 2010).

13.3.2. Cellular Uptake

Once NPs enter the body, they immediately become coated in biomolecules such as proteins. It is likely that this coating can influence particle behavior and toxicity, because the binding to different proteins will change the NP surface with direct effects on the binding to different molecules. Furthermore, it is possible that the particles can indirectly alter both protein structure and function, contributing to their toxicity. For example, conclusions of previous studies related to the interaction of BSA with AgNPs determined the existence of a complex held together by van der Waals and electrostatic forces. Results from this study revealed that AgNPs increased the amount of helix and decreased the β-sheet structures, changing the conformation of the protein skeleton. As a result of this change in the structure, internal hydrophobic amino acids are exposed, leading to functional changes or inhibition of proteins (Liu et al., 2009). Moreover, it is also known that AgNPs are able to react and bind to thiol groups (present in cysteines) with high affinity within proteins, thereby promoting their denaturation.

Phagocytosis is another endocytosis pathway involved in the engulfment and breakdown of microorganisms, damaged or apoptotic cells and inert particles (Aderem and Underhill, 1999). Phagocytosis is a very complex mechanism due to the interaction of a diversity of receptors (Aderem and Underhill, 1999). The phagocytosis of particles is more effective if the particles are labeled with special molecules in the body (opsonization), such as antibodies or complement molecules, which are able to accelerate phagocytosis (Garnett and Kallinteri, 2006).

Phagocytosis comprises the following steps.

1. Specific receptors on the phagocyte membrane interact with specific molecules (ligands) on the surface of the particle. Binding of NPs to the complement protein 5a could be responsible for the chemotactic signals of NPs (Oberdorster et al., 2005), although another study suggests that the electric charge of the NPs may play a role in activating the scavenger-type receptors (Kobzik, 1995).
2. After the binding receptor-ligand, the cytoskeleton (an intracellular network of protein filaments) of the phagocyte rearranges and ultimately leads to the internalization of the particle in phagocytic vesicles termed phagosomes (Park, 2003).
3. As the phagosome internalizes, the acquisition of specific markers (maturation) through the endocytic pathways will drive the phagosome to fuse with the lysosome, an organelle containing digesting enzymes. This phenomenon is known as the phagolysosome fusion and can take from 30 min to few hours (Aderem and Underhill, 1999). Lysosomes keep an acidic pH (c. 4.5) to facilitate the activity of proteases that digest the phagosome cargo. In the case of NP, these acidic conditions can induce a chemical dissolution of the NP (Park, 2003).

Interestingly, there is also a specific distribution of some types of NPs within the cell. For example, some AgNPs appear to enter the cell by a different yet undefined mechanism (Chithrani et al., 2006). In this study, AgNPs were evenly and widely dispersed through the cells in a few minutes (12 minutes), localizing mainly in the nucleus (Vanwinkle et al., 2009). Microscopy analyses determined that the internalized NPs were smaller compared to the primary particle size and smaller than the functional diameter of the nuclear pore complex of c. 39 nm.

The mechanism of cellular uptake of NPs depends mainly on the following factors.

1. The physico-chemical properties of the NPs, chemical composition, size and shape, and agglomeration status of the NPs (ZP). For example, SiO_2–AgNPs formed loose aggregates similar to loose aggregates of AgNPs observed in cell culture medium (Farkas et al., 2011). Accumulated clusters of variable sizes of SiO_2–AgNPs were observed in murine fibroblast within vesicular structures and also as free structures in the cytosol (Christen and Fent, 2012), similar to AgNPs also found in endosomes of the same type of cells (Wei et al., 2010). Other studies reported that AgNPs and AuNPs were localized intracellularly in vacuoles, but not supermagnetic iron oxide NPs (SPION) that were found tightly bound to the cell membrane (Comfort et al., 2011). These findings strengthen the hypothesis that NPs enter the cell through an endocytotic pathway. Specific sizes of NPs could then be taken up by phagocytosis into the cells, while larger-sized NPs could not (Wei et al., 2010).

Indeed, smaller NPs can enter more easily into cells than larger NPs (Zhang et al., 2009), which may also explain the higher toxicity of smaller particles often reported. Surprisingly, a cellular size-dependent distribution of AuNPs was observed. AuNP sizes of 2.4, 5.5, 16, and 89 nm were localized in the nucleus, endosomes, cell membrane, and non-internalized aggregated or precipitated, respectively (Oh et al., 2011).

2. The experimental conditions in the *in vitro* assay which reflect the cellular microenvironment *in vivo*. For example, opsonization by serum components or modification of the NP surfaces by proteins or lipids, such as albumin and surfactants.

3. The characteristics of the targeted cells of the NPs. For example, active internalization or presence of specific surface receptors when comparing cell populations, such as in the case of monocytes vs. macrophages or "normal" vs. malignant cells.

Internalization of NPs into cells appears to be facilitated by their minute size. This internalization has been shown in animal studies where inhaled NPs were translocated into lung-associated lymph nodes and the brain (Takenaka et al., 2001; Frampton et al., 2004). This finding suggests that endothelial cells, which line the inner surface of blood vessels, must have had direct contact with the NPs. It is known that endothelial cells are able to internalize particles by different mechanisms. One of them is their unique vesicular system called the vesiculo-vacuolar organelle (VVO). The VVO together with specific invagination of the plasma membrane, called the caveolae, are involved in the regulation of the passage of particles or macromolecules (Feng et al., 2002).

Another possibility is the internalization of NPs via specific receptors or a receptor-mediated endocytosis pathway (RME). In this pathway the ligand binds onto a receptor on the cell's surface and enters the cell when the membrane invaginates. Receptors are recycled back to the membrane surface after releasing the ligand into the cell. Since cells have a specific receptor density (number of receptors per cell surface area) on the membrane, the unbound or available receptors determine the rate of internalization of a specific molecule or structure via this mechanism. For example, NPs covered with LDL (Handley et al., 1983), PDGF (Rosenfeld et al., 1984), or BSA (Geoffroy and Becker, 1984) have been shown to be internalized via endocytosis (coated pits/vesicles). An involvement of low-density lipoprotein (LDL)/LDL-receptor, platelet-derived growth factor (PDGF)/PDGF-receptor, and albumin/receptor are then expected to mediate the internalization of protein-coated NPs. AuNPs were shown to internalize via the receptor-mediated endocytosis pathway (RME) in a temperature-dependent manner (Chithrani et al., 2006).

4. Particles are digested by enzymes within the lysosome and the residues are removed by exocytosis. If the macrophage is unable to digest the phagocitosed particle, and the particle produces damage to the phagosomal membrane

(e.g., peroxidation), then interacting with the macrophage's cytoskeleton will lead to a reduce motility, impaired phagocytosis, macrophage death (Porter et al., 2006), and ultimately reduced clearance of the particles (Brown et al., 2004). If particles cannot be cleared, they can kill successive macrophages in an attempt to clear them and cause inflammation with macrophage debris accumulation (pus).

13.3.3. Factors Affecting the Internalization of NPs

Size. NP size is important for cell uptake and is a crucial factor in the design of NPs used in any application. Several parameters are dependent on the size of NPs. The size determines the mechanism and rate of NP uptake, the ability of NPs to permeate through tissues, and the adhesion of the NPs and their interaction with cells (Lee et al., 1993). For instance, particle size affects the immunological response of human dendritic cells (Vallhov et al., 2007). Experimental studies of cell uptake of NPs with uniform ZP and surface functionality are therefore needed to elucidate the effect of size on cell uptake (Zhang et al., 2009).

Endocytosis is the cellular mechanism responsible for absorbing molecules from the environment. This process is crucial for cells because of the presence of a lipid bilayer, which restrict the free movement of charged molecules towards the cell. Two types of competitive energy are involved in the endocytosis of NPs: the binding energy between ligands and receptors (first competitive energy) and the energy necessary to maintain the ligand-receptor complex (second competitive energy). The second competitive energy is the thermodynamic driving force required to drive NPs into the cells. These factors will dictate how fast and how many NPs are internalized by cells. With a diameter lesser than 40 nm, the docking of a single NP will not produce enough free energy to bind the NP onto the surface of the membrane. To be internalized, these NPs must be clustered together and therefore take a long diffusion time. For diameters greater than 80 nm, endocytosis rarely occurs.

The stability of AgNPs in solutions is also an important factor to take into account, as agglomerations can occur and disturb uptake by the cells. For instance, although AgNPs of 5–10 nm are stable in solution, they were found clustered in particles of sizes in the range 100–300 nm within the cytoplasm and nuclei of the human hepatoma HepG2 cells (Kim et al., 2009).

The size of human alveolar macrophages residing in the lungs range between 14–20 mm. Macrophages can engulf particles as big as themselves, but are significantly less effective with particles that are much larger. Experimental data show that, compared to larger particles, NPs smaller than 100–200 nm are able of evading phagocytosis by alveolar macrophages (Peters et al., 2006). This escape from macrophages allows them to enter pulmonary interstitial sites, interact with epithelial cells, and ultimately gain access to the circulatory and lymphatic systems (Takenaka et al., 2001; Oberdorster et al., 2005). *In vitro* studies have shown that NPs are engulfed by alveolar macrophages, but macrophage lavage recovery show that NP with sizes <100 nm are not efficiently phagocytosed.

The concentration of NPs has a direct effect on the extent of phagocytosis. For example, at high concentrations NPs tend to aggregate in clusters, sometimes larger than 100 nm. It is assumed that a passive uptake or adhesive interactions are the mechanisms by which these NPs are able to internalize into cells. This passive uptake does not result in the formation of vesicles and can be initiated by Van der Waals forces, electrostatic charges, steric interactions, or interfacial tension effects. Once in the cytoplasm, these NPs can then move freely inside the cell, making them very dangerous due to their direct access to different compartments, proteins, and organelles. In this regard, several studies have shown that upon non-phagocytic uptake, NPs can be found in various compartments inside the cell, including outer-cell and nuclear membranes, cytoplasm, mitochondria, lipid vesicles, and nuclei (Li et al., 2003; Garcia-Garcia et al., 2005; Xia et al., 2006). Moreover, NPs are internalized not only by professional phagocytes (Takenaka et al., 2001; Hoet et al., 2004; Xia et al., 2006) but by other types of cells, including endothelial cells, pulmonary and gastrointestinal epithelia, red blood, and nerve cells (Hopwood et al., 1995; Kreyling et al., 2002; Oberdorster et al., 2002; Garcia-Garcia et al., 2005; Gurr et al., 2005; Singal and Finkelstein, 2005; Peters et al., 2006; Rothen-Rutishauser et al., 2006).

Shape. The effect of NP shape in cellular uptake has been demonstrated by various studies. For example, the uptake of spherical and rod-shaped AuNPs nanoparticles by the same cell showed different outcomes. Cells internalized more spherical-shaped NPs than rod-shaped AuNPs (Chithrani et al., 2006). The difference in the surface chemistries between the spherical and rod-shaped AuNPs may be one of the reasons for the difference in uptake. In another study, the uptake of negatively-charged spherical and rod-shaped AuNPs by HeLa epithelial cells was studied (Chithrani et al., 2006). A preference in the uptake of spherical particles was observed and the AuNPs were contained within vesicles in the cytoplasm of the cells. Interestingly, the way NP uptake occurred suggested mediation by proteins bound to the particle surface. This is hypothesized from the fact that NP uptake increased over time until a plateau was reached, suggesting a point of saturation in the uptake process. Interestingly, 50 nm NPs were internalized more, suggesting that larger particles with a smaller surface area have less protein adsorption to the particle surface, which explains their more limited uptake. Other explanations include the mediation of the uptake by serum proteins adsorbed onto the NP surface, facilitating the uptake of these NPs via clathrin-mediated endocytosis. To confirm whether the uptake was mediated by endocytosis, particle uptake was compared at two temperatures (37 °C and 4 °C). The reason for choosing a low temperature is to decrease the level of ATP production, which is necessary for endocytosis.

The authors concluded that the spherical and rod-shaped AuNPs were coated with the serum protein transferrin after confirmation of its uptake by confocal microscopy using a fluor-tagged transferrin. The mechanism of uptake was confirmed to be endocytosis (a clathrin-mediated mechanism) as a decrease in the uptake was observed at low temperature and a dissociation of the chlatrin coat was observed when cells were pre-treated with sucrose or potassium depletion.

Charge of the NP. NPs with a ZP close to zero are internalized less in comparison with charged particles (Roser et al., 1998). Interestingly, particles with cationic or anionic charges displayed the same rate of uptake (Tabata and Ikada, 1988), but cationic NPs are more toxic than anionic NPs, probably because of the electrostatic interaction between the NP and the negative charge of the cell membrane.

13.4. IMMUNOLOGICAL RESPONSE

The existence of a hierarchical oxidative stress response based on the type of proteins expressed when cells are exposed to adverse effects has been postulated (Xiao et al, 2003). When cells are exposed to a low level of oxidative stress, the cellular response is then associated with the induction of antioxidant and detoxification enzymes (Xiao et al., 2003). In this case, the genes that encode the enzymes are under the control of the transcription factor Nrf-2. Nrf-2 activates the genes via an antioxidant response element (ARE; Xiao et al., 2003). At higher levels of oxidative stress, this protective response is overtaken by inflammation and cytotoxicity. Inflammation is initiated through the activation of pro-inflammatory signaling cascades (e.g., mitogen-activated protein kinase (MAPK) and nuclear factor κB (NF-κB) cascades), which in turn induce the expression of various pro-inflammatory genes such as COX-2, TNF-α, and IL-6. In addition, programmed cell death could result from mitochondrial perturbation and the release of proapoptotic factors. NPs can elicit an immunological response by inducing the production of cytokines (Shin et al., 2007; Martinez-Gutierrez, 2010, 2012).

13.4.1. Cytokine Production

A key event in the initiation and regulation of an immunological response such as in allergies, infections, and inflammations is the production of cytokine.

Lymphocytes are a specialized group of white cells involved in the development of an adaptive immunity. A subgroup of lymphocytes termed T helper (Th) play an important role in adaptive immunity by activating and directing other white cells. Type 1 Th (Th1) cells produce the cytokines interferon-gamma (IFN-γ), IL-2, and tumor necrosis factor beta (TNF-β). By contrast, type 2 Th (Th2) cells produce IL-4, IL-5, IL-10, and IL-13, which are responsible for strong antibody production, eosinophil activation, and partial inhibition of macrophage functions. Interestingly, NPs are able to inhibit Th1 cytokine production and can therefore inhibit inflammation associated with infections.

Upon exposure to foreign particles, alveolar macrophages are stimulated to secrete inflammatory mediators such as cytokines and chemokines to enhance an immune response. In this context, rat alveolar macrophages showed a significant increase in the levels of TNF-R, MIP-2, and IL-1, but not IL-6 when the cells were exposed to AgNPs (Carlson et al., 2008).

The induction of mitogen-activated protein kinase (MAPK) pathways linked to signaling upstream of transcription factor activation by NPs are a main focus of current research. MAPK pathways are activated by many different extracellular stimuli and cause a broad spectrum of responses.

It has been reported that AgNPs of 25 and 40 nm produced strong rat brain microvessel endothelial cell activation. This activation involves a release of proinflammatory cytokines, such as TNF-α and IL-1B, which were associated with an increase in the blood-brain barrier permeability compared to larger AgNPs of 80 nm (Trickler et al., 2010). Similarly, a higher increase in TNF-α secretion was observed when SiNPs were exposed to cells as compared to the secretion of this cytokine in larger NPs (Morishige et al., 2012). A size-dependent secretion of cytokines can then occur based on the size of the NPs.

Different immunological responses were observed when Ag-, Al-, and AuNPs were exposed to the murine macrophage cell line RAW 264.7. AgNPs and AlNPs induced the secretion of NF-κB and IL-6, whereas AuNPs did not (Nishanth et al., 2011). Similar results were observed when the secretion of Cox-2, an enzyme involved in the production of the proinflammatory prostaglandin H2, and TNF-α were examined (Nishanth et al., 2011).

Another study reported that a significant increase in p38-MAPK protein was measured as early as 4 h after exposure of Jurkat T cells to AgNPs (Eom and Choi, 2010). As p38-MAPK protein participates in signal transduction events in the cell, two downstream transcription factors substrates in this pathway corroborated its effect. In both cases, an increase in the transcription factors Nrf-2 (for oxidative stress response) and NF-κB (for inflammation response) were upregulated (Eom and Choi, 2010).

In an attempt to elucidate whether the exposure of 9 nm AgNPs modifies the epidermal growth factor (EGF)-dependent signal transduction pathways in the human epithelial cell line A-431, an analysis of the downstream proteins regulated by the EGF-receptor were studied (Comfort et al., 2011). Results showed that AgNPs were able to attenuate the production of the downstream substrates Akt and Erk by 20% (Comfort et al., 2011).

Taken together, exposing AgNPs to mammalian cells activates different signal transduction pathways that are important for the initiation of immunological responses.

13.4.2. Cytotoxicity, Necrosis, Apoptosis, and Cell Death

Apoptosis or programmed cell death is characterized by cell shrinkage, condensation of the chromatin, nuclear breakdown, cell membrane blebbing, and eventual fragmentation of the cell into apoptotic bodies that are engulfed by neighboring cells or phagocytes (Urne and Vaux, 1996). On the other hand, necrosis occurs when the cell undergoes an acute injury and is characterized by cell swelling, formation of microvesicles, and breakage of the cell membrane, which allows leakage of the cytoplasm into the intercellular space (Boulares et al., 1999).

The ability of AgNP and Ag^+ to induce apoptosis and necrosis occurs in a dose- and exposure-dependent manner. After 24 h, the decreased viability caused by both AgNP and Ag^+ was primarily detected by necrosis (Foldbjerg et al., 2009). It has been suggested that a hierarchical oxidative stress model exists in cells, and its activation depends on the level of ROS formation. Low levels of ROS then activate cellular defense mechanisms, whereas high levels of ROS activate cell death (Nel et al., 2006), although many studies have implicated intracellular ROS in signal transduction pathways leading to apoptosis (Ueda et al., 2002; Ott et al., 2007; Arora et al., 2008; Hsin et al., 2008).

Within cells, mitochondria are the organelles in charge of providing the necessary energy for vital cell functions through the formation of ATP by oxidative phosphorylation. The transfer of electrons drives this energy production across a group of respiratory complexes located within the mitochondrial inner membrane, referred to as the electron transport chain (ETC). The resulting transmembrane proton gradient is then used for the conversion of ADP$^+$ phosphate to ATP by the protein ATP synthase. Mitochondria can also dictate the fate of a cell, based on the release of cytochrome C from the outer mitochondrial as a result of an increase of the membrane permeability. Release of cytochrome C represents a key event in the induction of apoptosis via the activation of caspases, a group of apoptogenic protein-signaling cascade. Moreover, the dissipation of mitochondrial membrane potential (MMP) is an indicator of mitochondrial integrity and also an early step in apoptosis (Mammucari and Rizzuto, 2010). AgNPs caused significant dissipation of MMP with a significant increase in number of apoptotic cells.

The uptake of AgNPs into mitochondria of human lung fibroblasts has been reported (Asharani et al., 2009a,b). The effect of NP in these organelles may have various consequences for cells already discussed, for example triggering apoptosis and the impairment of ETC, which can be associated with an enhanced ROS generation from these organelles. This enhanced ROS generation has been implicated in the oxidative attack of the mitochondrial genome (Berneburg et al., 2006). Since mitochondrial DNA encodes most of genes involved in ETC as well as two subunits of ATP synthase, mitochondrial DNA mutagenesis has been connected with defective electron transfer and oxidative phosphorylation. This defect has been associated with further increases in ROS production, thereby establishing a vicious cycle characterized by gradual loss of mitochondrial function (Mandavilli et al., 2002).

Currently, several NPs have been shown to be capable of eliciting damage to nuclear DNA. The damage brings three major consequences: (1) induction and fixation of mutations; (2) induction of DNA cell cycle arrest; and (3) activation of signal transduction pathways which promote apoptosis.

AgNPs induce apoptosis by triggering mitochondria-dependent apoptosis (Hsin et al., 2008). Cell lines show different responses to the same type of NPs. For example, the fibroblast cell line NIH3T3 is more sensitive than human colon cancer cells (Hsin et al., 2008). The difference in the response of these two cell lines was assayed by measuring the activation of apoptosis and the expression of different apoptosis regulators, such as PARP cleavage, JNK phosphorylation, and p53. NIH3T3 showed activation of all of the described regulators, whereas in human colon cancer carcinoma cell line HCT116 only PARP cleavage was not activated (Hsin et al., 2008). Interestingly, a complementary experiment showed that Bcl-2 (an apoptosis regulator protein) transcript increased only in the HCT116, but not in the NIH3T3 cells. The lack of PARP cleavage indicates that apoptosis regulator caspase-3 is not activated in this cell line. However, an increase of the intracellular level of ROS was measured in both cell lines. These results suggest that different cell lines have different death mechanisms (Hsin et al., 2008).

Other studies reported a significant increase in apoptosis in Jurkat T and Dalton's lymphoma cells, in human skin carcinoma (A-431) and human fibrosarcoma (HT-1080) exposed to AgNPs (Arora et al., 2008; Eom and Choi, 2010; Sriram et al., 2010).

Taken together, understanding the molecular mechanism by which AgNPs induce apoptosis is essential for their application in nanomedicine.

13.5. FACTORS TO CONSIDER TO REDUCE THE CYTOTOXIC EFFECTS OF NP

Appropriate design of NPs is always a focus of interest to improve their *in vitro* and *in vivo* characteristics. It is desired to design NPs of low toxicity and low immunogenicity, and each component or step used for their fabrication that can affect their cytotoxicity should be considered for systematic studies to evaluate the structural modifications on immuno- and cytotoxicities.

The size of AgNPs appears to be extremely relevant to their cytotoxicity. In many studies with mammalian cells, smaller AgNPs induced more oxidative stress, DNA damage, and lethality than larger counterparts when normalized to the mass content. For example, in a comparison of hydrocarbon-coated AgNPs of various sizes, smaller-sized particles reduced the viability of macrophages and spermatogonia cells more effectively than larger particles (Carlson et al., 2008; Braydich-Stolle et al., 2010). This was also the case for polysaccharide-coated AgNPs (Braydich-Stolle et al., 2010), where both ROS expression and apoptosis were induced by only the smallest (10 nm in diameter) polysaccharide-coated particles employed in the study. Moreover, evaluation of the cytotoxicity of peptide-coated AgNPs of 20 and 40 nm in diameter indicated that the 20 nm AgNPs caused higher levels of protein carbonylation and cell death in macrophages than the 40 nm AgNPs (Haase et al., 2011). Another study reported that PVP-coated 5 nm AgNPs induced severe cell damage, DNA damage, and ROS generation compared to their counterparts of PVP-coated 20 and 50 nm AgNPs (Liu et al., 2010).

NPs should be designed to give accessibility to other molecules often used on their surface. For instance, poly(lactic-*co*-glycolic) acid or PLGA is a copolymer widely used in drug delivery and therapeutic devices because of its biocompatibility. Preparation of PLGA-NPs can then provide alternatives to reduce their cytotoxicity in living systems. Indeed, PLGA-NP functionalized with thiol groups was tested to decrease interactions with opsonins and phagocytic cells. This study found a reduction in protein adsorption, complement activation, and platelet activation (Thasneem et al., 2011). Pegylation or the covalent attachment of polyethylene glycol (PEG) to the surface of NPs is another decoration of NPs that can not only reduce immunogenicity but also reduce the clearing rate of the NPs from the circulatory system.

Encapsulation of therapeutic agents inside a nanocarrier can be an option to reduce the immunological response induced by the agent. In this context, the charge of the NPs is an important factor to avoid an exacerbated immunological response. For instance, cationic NPs are believed to be more toxic, cleared faster, and induce a higher inflammatory response than their anionic or neutral counterparts.

Taken together, polymer biodegradability and biocompatibility are essential, and a careful selection of the nanomaterials is critical in predicting the immunological response upon *in vivo* administration.

13.6. CONCLUSIONS AND FUTURE DIRECTIONS

The ability of NPs to access cells and affect biochemical functions makes them important tools at the molecular level. Advanced analyses of the physical and chemical characteristics of NPs will continue to be essential in revealing the relationship between their size, shape, composition, crystallinity, morphology, kinetics, and their reactivity in aggregation. Existing research on nanotoxicity has been concentrated on empirical evaluation of the toxicity of various NPs, with less regard given to the relationship between NP properties. In addition, studies should also include NP translocation pathways, accumulation, short- and long-term toxicity, interactions with cells, receptors, signaling pathways involved, cytotoxicity, and surface modification for effective phagocytosis. Other important research topics to pursue include NP aging, surface modification, and changes in aggregation state after interaction with bystander substances in the environment and with biomolecules and other chemicals within the organism.

The following points should be taken into consideration for future studies aimed at the evaluation of NP-induced toxicity. First, because NPs exhibit similar physicchemical characteristics compared to their parental atoms or ions, it is probably the case that NPs can interact with cells/tissues differently than atoms or ions that can cause cytotoxicity through unique pathways. Second, numerous studies have been published with regard to NP-induced cytotoxicity, reflecting a wide interest in AgNP-mediated hazards. In many of these studies, however, each research group has employed independent experimental methods. This is especially true with regard to the preparation of NPs and, in particular, procedures related to the coating and dispersion of NPs. It is therefore completely unrealistic to compare data among studies performed in different laboratories, and is critical for progress in the field to standardize experimental procedures related to the studies of NP toxicity.

In summary, toxicological screening approaches in the future should include: a well-characterized model of NPs based on the knowledge of NP properties and potential for exposure during all stages of their life cycle; knowledge about NP biokinetics; *in vitro* models that are predictive of outcomes following *in vivo* exposure; and evidence that *in vitro* outcomes are NP-specific via appropriate benchmarking. As illustrated by the variety of *in vivo* and *in vitro* models reviewed in this study, AgNPs induce cellular/tissue changes that are specific to oxidative stress, genotoxicity, and apoptosis. However, understanding of these mechanisms is still incomplete.

ACKNOWLEDGEMENTS

We thank Eviatar Bach for helpful discussions.

REFERENCES

Aderem, A., Underhill, D.M. 1999. Mechanisms of phagocytosis in macrophages. *Annual Review Immunology* 17: 593–623.

Ahamed, M., Karns, M., Goodson, M., Rowe, J., Hussain, S.M., Schlager, J.J., Hong, Y. 2008. DNA damage response to different surface chemistry of silver nanoparticles in mammalian cells. *Toxicology and Applied Pharmacology* 233(3): 404–410.

Anderson, M.F., Nilsson, M., Eriksson, P.S., Sims, N.R. 2004. Glutathione monoethyl ester provides neuroprotection in a rat model of stroke. *Neuroscience Letters* 354(2): 163–165.

Arnaudeau, C., Helleday, T., Jenssen, D. 1999. The RAD51 protein supports homologous recombination by an exchange mechanism in mammalian cells. *Journal of Molecular Biology* 289(5): 1231–1238.

Arora, S., Jain, J., Rajwade, J.M., Paknikar, K.M. 2008. Cellular responses induced by silver nanoparticles: In vitro studies. *Toxicology Letters* 179(2): 93–100.

Asharani, P.V., Low Kah Mun, G., Hande, M.P., Valiyaveeti, S. 2009a. Cytotoxicity and genotoxicity of silver nanoparticles in human cells. *ACS Nano* 3(2): 279–290.

Asharani, P.V., Hande, M.P., Valiyaveettil, S. 2009b. Anti-proliferative activity of silver nanoparticles. *BMC Cell Biology* 10: 65.

Berneburg, M., Kamenisch, Y., Krutmann, J. 2006. Repair of mitochondrial DNA in aging and carcinogenesis. *Photochemical and Photobiological Sciences* 5: 190–198.

Boulares, A.H., Yakovlev, A.G., Ivanova, V., Stoica, B.A., Wang, G., Iyer, S. 1999. Role of poly-(ADP-ribose) polymerase (PARP) cleavage in apoptosis. Caspase-3 resistant PARP mutant increases rates of apoptosis in transfected cells. *Journal of Biological Chemistry* 274(33): 22932–22940.

Bouwmeester, H., Poortman, J., Peters, R.J. et al. 2011. Characterization of translocation of silver nanoparticles and effects on whole-genome gene expression using an in vitro intestinal epithelium coculture model. *ACS Nano* 5(5): 4091–4103.

Braydich-Stolle, L.K., Benjamin, L., Schrand, A., Murdock, R.C., Lee, T., Schlager, J.J., Hussain, S.M., Hoffman M.C. 2010. Silver nanoparticles disrupt GDNF/Fyn kinase signalling in spermatogonial stem cells. *Toxicological Sciences* 116(2): 577–589.

Bredt, D.S. 1999. Endogenous nitric oxide synthesis: biological functions and pathophysiology. *Free Radical Research* 31(6): 577–596.

Brown, D.M., Donaldson, K., Stone, V. 2004. Effects of PM10 in human peripheral blood monocytes and J774 macrophages. *Respiratory Research* 5: 29.

Cadenas, E. 1989. Biochemistry of oxygen toxicity. *Annual Review of Biochemistry* 58: 79–110.

Cadenas, E., Boveris, A., Ragan, C.I., Stoppani, A.O. 1977. Production of superoxide radicals and hydrogen peroxide by NADH-ubiquinone reductase and ubiquinol-cytochrome c reductase from beef-heart mitochondria. *Archives of Biochemistry and Biophysics* 180(2): 248–257.

Carlson, C., Hussain, S.M., Schrand, A.M., Braydich-Stolle, L.K., Hess, K.L., Jones, R L., Schlager, J.J. 2008. Unique cellular interaction of silver nanoparticles: size-dependent generation of reactive oxygen species. *Journal of Physics and Chemistry B* 112(43): 13608–13619.

Chairuangkitti, P., Lawanprasert, S., Roytrakul, S. et al. 2013. Silver nanoparticles induce toxicity in A549 cells via ROS-dependent and ROS-independent pathways. *Toxicology in Vitro* 27(1): 330–338.

Chazotte-Aubert, L., Oikawa, S., Gilibert, I., Bianchini, F., Kawanishi, S., Ohshima, H. 1999. Cytotoxicity and site-specific DNA damage induced by nitroxyl anion NO(-) in the presence of hydrogen peroxide. Implications for various pathophysiological conditions. *Journal of Biological Chemistry* 274(30): 20909–20915.

Chithrani, D., Ghazani, A.A., Chan, W.C.W. 2006. Determining the size and shape dependence of gold nanoparticle uptake into mammalian cells B. *Nano Letters* 6(4): 662–668

Christen, V., Fent, K. 2012. Silica nanoparticles and silver-doped silica nanoparticles induce endoplasmatic reticulum stress response and alter cytochrome P4501A activity. *Chemosphere* 87(4): 423–434.

Christen, V., Treves, S., Duong, F.H., Heim, M.H. 2007. Activation of endoplasmic reticulum stress response by hepatitis viruses up-regulates protein phosphatase 2A. *Hepatology* 46(2): 558–565.

Comfort, K.K., Maurer, E.I., Braydich-Stolle, L.K., Hussain, S.M. 2011. Interference of silver, gold, and iron oxide nanoparticles on epidermal growth factor signal transduction in epithelial cells. *ACS Nano* 5(12): 10000–10008.

Connor, E.E., Mwamuka, J., Gole, A., Murphy, C.J., Wyatt, M.D. 2005. Gold nanoparticles are taken up by human cells but do not cause acute cytotoxicity. *Small* 1(3): 325–327.

Dahl, J.A., Maddux, B.L., Hutchison, J.E. 2007. Toward greener nanosynthesis. *Chemical Reviews* 107(6): 2228–2269.

de Melo, L.S., Gomes, A.S., Saska, S., Nigoghossian, K., Messaddeq, Y., Ribeiro, S.J., de Araujo, R.E. 2012. Singlet oxygen generation enhanced by silver-pectin nanoparticles. *Journal of Fluorescence* 22(6): 1633–1638.

Dizdaroglu, M., Jaruga, P., Birincioglu, M., Rodriguez, H. 2002. Free radical-induced damage to DNA: mechanisms and measurement. *Free Radical Biology and Medicine* 32(11): 1102–1115.

Dumont, A., Hehner, S.P., Hofmann, T.G., Ueffing, M., Droge, W., Schmitz, M.L. 1999. Hydrogen peroxide-induced apoptosis is CD95-independent, requires the release of mitochondria-derived reactive oxygen species and the activation of NF-kappaB. *Oncogene* 18: 747–757.

Eiserich, J.P., Hristova, M., Cross, C.E., Jones, A.D., Freeman, B.A., Halliwell, B., van der Vliet, A. 1998. Formation of nitric oxide-derived inflammatory oxidants by myeloperoxidase in neutrophils. *Nature* 391(6665): 393–397.

Elechiguerra, J.L., Burt, J.L., Morones, J.R., Camacho-Bragado, A., Gao, X., Lara, H.H., Yacaman, M.J. 2005. Interaction of silver nanoparticles with HIV-1. *Journal of Nanobiotechnology* 3: 6.

Emory, S., Nie, S. 1997. Near-field surface-enhanced Raman spectroscopy on single silver nanoparticles. *Analytical Chemistry* 69(14): 2631–2635.

Eom, H.J., Choi, J. 2010. p38 MAPK activation, DNA damage, cell cycle arrest and apoptosis as mechanisms of toxicity of silver nanoparticles in Jurkat T cells. *Environmental Science and Technology* 44(21): 8337–8342.

Farkas, J., Christian, P., Gallego-Urrea, J., Roose, N., Hassellov, M., Tollefsen, K.E., Thomas, K. 2011. Uptake and effects of manufactured silver nanoparticles in rainbow trout (Oncorhynchus mykiss) gill cells. *Aquatic Toxicology* 101(1): 117–125.

Feng, D., Nagy, J.A., Dvorak, H.F., Dvorak, A.M. 2002. Ultrastructural studies define soluble macromolecular, particulate, and cellular transendothelial cell pathways in venules, lymphatic vessels, and tumor-associated microvessels in man and animals. *Microscopy Research Technique* 57(5): 289–326.

Foldbjerg, R., Olesen, P., Hougaard, M., Dang, D.A., Hoffmann, H.J., Autrup, H. 2009. PVP-coated silver nanoparticles and silver ions induce reactive oxygen species, apoptosis and necrosis in THP-1 monocytes. *Toxicology Letters* 190(2): 156–162.

Frampton, M.W., Utell, M.J., Zareba, W. et al. 2004. Effects of exposure to ultrafine carbon particles in healthy subjects and subjects with asthma. *Research Report Health Effects Institute* 1–47, discussion 49–63.

Fridovich, I. 1986. Biological effects of the superoxide radical. *Archives of Biochemistry and Biophysics* 247(1): 1–11.

Garcia-Garcia, E., Andrieux, K., Gil, S., Kim, H.R., Le Doan, T., Desmaele, D., d'Angelo, J., Taran, F., Georgin, D., Couvreur, P. 2005. A methodology to study intracellular distribution of nanoparticles in brain endothelial cells. *International Journal of Pharmaceutics* 298: 310–314.

Garnett, M.C., Kallinteri, P. 2006. Nanomedicines and nanotoxicology: some physiological principles. *Occupational Medicine* 56(5): 307–311.

Gauss, K.A., Gauss, L.K., Nelson-Overton, D.W., Siemsen, Gao, Y., DeLeo, F.R., Quinn, M.T. 2007. Role of NF-B in transcriptional regulation of the phagocyte NADPH oxidase by tumor necrosis factor-a. *Journal of Leukocyte Biology* 82(3): 729–741.

Geoffroy, J.S., Becker, R.P. 1984. Endocytosis by endothelial phagocytes: uptake of bovine serum albumin-gold conjugates in bone marrow. *Journal of Ultrastructure Research* 89: 223–239.

Girotti, A.W. 1998. Lipid hydroperoxide generation, turnover, and effector action in biological systems. *Journal of Lipid Research* 39: 1529–1542.

Green, D.R., Reed, J.C. 1998. Mitochondria and apoptosis. *Science* 281(5381): 1309–1312.

Grisham, M.B., Jourd'heuil, D., Wink, D.A. 2000. Review article: chronic inflammation and reactive oxygen and nitrogen metabolism-implications in DNA damage and mutagenesis. *Alimentary Pharmacology and Therapeutics* 14(1): 3–9.

Gurr, J.R., Wang, A.S., Chen, C.H., Jan, K.Y. 2005. Ultrafine titanium dioxide particles in the absence of photoactivation can induce oxidative damage to human bronchial epithelial cells. *Toxicology* 213(1–2): 66–73.

Gutteridge, J.M., Halliwell, B. 1990. The measurement and mechanism of lipid peroxidation in biological systems. *Trends in Biochemical Sciences* 15(4): 129–135.

Guven, K., De Pomerai, D.I. 1995. Differential expression of HSP70 proteins in response to heat and cadmium in Caenorhabditis elegans. *Journal of Thermal Biology* 20(4): 355–363.

Haase, A., Tentschert, J., Jungnickel, H. et al. 2011. Toxicity of silver nanoparticles in human macrophages: uptake, intracellular distribution and cellular responses. *Journal of Physics: Conference Series* 304(1): 012030.

Habib, G.M., Shi, Z.Z., Lieberman, M.W. 2007. Glutathione protects cells against arsenite-induced toxicity. *Free Radical Biology and Medicine* 42(2): 191–201.

Hackenberg, S., Scherzed, A., Kessler, M. et al. 2011. Silver nanoparticles: evaluation of DNA damage, toxicity and functional impairment in human mesenchymal stem cells. *Toxicology Letters* 201: 27–33.

Halliwell, B., Gutteridge, J.M. 1984. Oxygen toxicity, oxygen radicals, transition metals and disease. *Biochemical Journal* 219: 1–14.

Handley, D.A., Arbeeny, C.M., Chien, S. 1983. Sinusoidal endothelial endocytosis of low density lipoprotein-gold conjugates in perfused livers of ethinyl-estradiol treated rats. *European Journal of Cellular Biology* 30(2): 266–271.

Harding, H.P., Calfon, M., Urano, F., Novoa, I., Ron, D. 2002. Transcriptional and translational control in the mammalian unfolded protein response. *Annual Review of Cell and Developmental Biology* 18: 575–599.

He, D, Jones, A.M., Garg, S., Pham, A.N., Waite, T.D. 2011. Silver nanoparticle-reactive oxygen species interactions: applications of a charging-discharging model. *Journal of Physical Chemistry* 115(13): 5461–5468.

Hoet, P.H., Bruske-Hohlfeld, I., Salata, O.V. 2004. Nanoparticles – known and unknown health risks. *Journal of Nanobiotechnology* 2(1): 12.

Hopwood, D., Spiers, E.M., Ross, P.E., Anderson, J.T., McCullough, J.B., Murray, F.E. 1995. Endocytosis of fluorescent microspheres by human oesophageal epithelial cells: comparison between normal and inflamed tissue. *Gut* 37(5): 598–602.

Hsin, Y.H., Chen, C.F., Huang, S., Shih, T.S., Lai, P.S., Chueh, P.J. 2008. The apoptotic effect of nanosilver is mediated by a ROS- and JNK-dependent mechanism involving the mitochondrial pathway in NIH3T3 cells. *Toxicology Letters* 179: 130–139.

Hussain S.M., Hess, K.L., Gearhart J.M., Geiss, K.T., Schlager, J.J. 2005. In vitro toxicity of nanoparticles in BRL 3A rat liver cells. *Toxicology In Vitro* 19(7): 975–983.

Imlay, J.A. 2003. Pathways of oxidative damage. *Annual Review of Microbiology* 57: 395–418.

Imlay, J.A., Linn, S. 1988. DNA damage and oxygen radical toxicity. *Science* 240(4857): 1302–1309.

Jaiswal, A. 2000. Regulation of genes encoding NAD(P)H:quinone oxidoreductases. *Free Radical Biology and Medicine* 29(3–4): 254–262.

Jamieson, D., Chance, B., Cadenas, E., Boveris, A. 1986. The relation of free radical production to hyperoxia. *Annual Review Physiology* 48: 703–719.

Janssens, V., Goris, J. 2001. Protein phosphatase 2A: a highly regulated family of serine/threonine phosphatases implicated in cell growth and signalling. *Biochemical Journal* 353(Pt3): 417–439.

Jomova, K., Valko, M. 2011. Advances in metal-induced oxidative stress and human disease. *Toxicology* 283(2–3): 65–87.

Jose, R., Kumar, A., Thavasi, V., Ramakrishna, S. 2008. Conversion efficiency versus sensitizer for electrospun TiO_2 nanorod electrodes in dye-sensitized solar cells. *Nanotechnology* 19: 424004.

Kapoor, S. 1998. Preparation, characterization, and surface modification of silver particles. *Langmuir* 14: 1021–1025.

Kawata, K., Osawa, M., Okabe S. 2009. In vitro toxicity of silver nanoparticles at noncytotoxic doses to HepG2 human hepatoma cells. *Environmental Science and Technology* 43(15): 6046–6051.

Karakoti, A.S., Hench, L.L., Seal, S. 2006. The potential toxicity of nanomaterials: the role of surfaces. *Journal of Materials* 58(7): 77–82.

Kim, S., Choi, J.E., Choi, J., Chung, K., Park, K., Yi, J., Ryu, D.Y. 2009. Oxidative stress-dependent toxicity of silver nanoparticles in human hepatoma cells. *Toxicology In Vitro* 23: 1076–1084.

Klaassen, C.D., Liu, J., Diwan, B.A. 2009. Metallothionein protection of cadmium toxicity. *Toxicology and Applied Pharmacology* 238(3): 215–220.

Kobzik, L. 1995. Lung macrophage uptake of unopsonized environmental particulates. Role of scavenger-type receptors. *Journal of Immunology* 155: 367–376.

Koeneman, B.A., Zhang, Y., Westerhoff, P., Chen, Y., Crittenden, J.C., Capco, D.G. 2010. Toxicity and cellular responses of intestinal cells exposed to titanium dioxide. *Cell Biology and Toxicology* 26(3): 225–238.

Kreyling, W.G., Semmler, M., Erbe, F., Mayer, P., Takenaka, S., Schulz, H., Oberdorster, G., Ziesenis, A. 2002. Translocation of ultrafine insoluble iridium particles from lung epithelium to extrapulmonary organs is size dependent but very low. *Journal of Toxicology and Environmental Health A* 65(20): 1513–1530.

Krinsky, N.I. 1977. Singlet oxygen in biological systems. *Trends Biochemical Sciences* 2(2): 35–38.

Ku, R., Billings, R. 1984. Relationships between formaldehyde metabolism and toxicity and glutathione concentrations in isolated rat hepatocytes. *Chemico-Biological Interactions* 51(1): 25–36.

Lee, K.D., Nir, S., Papahadjopoulos, D. 1993. Quantitative analysis of liposome-cell interactions in vitro: rate constants of binding and endocytosis with suspension and adherent J774 cells and human monocytes. *Biochemistry* 32(3): 889–899.

Lee, P.C., Meisel, D. 1982. Adsorption and surface-enhanced Raman of dyes on silver and gold sols. *Journal of Physical Chemistry* 86: 3391–3395.

Li, N., Sioutas, C., Cho, A. et al. 2003. Ultrafine particulate pollutants induce oxidative stress and mitochondrial damage. *Environmental Health Perspective* 111(4): 455–460.

Liu, R., Sun, F., Zhang, L., Zong, W., Zhao, X., Wang, L., Wu, R., Hao, X. 2009. Evaluation on the toxicity of nanoAg to bovine serum albumin. *Science of the Total Environment* 407: 4184–4188.

Liu, W., Wang, C., Li, H.C., Wang, T., Liao, C.Y., Cui, L., Zhou, Q.F., Yan, B., Jiang, G.B. 2010. Impact of silver nanoparticles on human cells: effect of particle size. *Nanotoxicology* 4(3): 319–330.

Lloyd, R.V., Hanna, P.M., Mason, R.P. 1997. The origin of the hydroxyl radical oxygen in the Fenton reaction. *Free Radical Biology and Medicine* 22(5): 885–888.

Long, H., Shi, T., Borm, P.J., Maatta, J., Husgafvel-Pursiainen, K., Savolainen, K., Krombach, F. 2004. ROS-mediated TNF-alpha and MIP-2 gene expression in alveolar macrophages exposed to pine dust. *Particle and Fibre Toxicology* 1(1): 3.

Mammucari, C., Rizzuto, R. 2010. Signaling pathways in mitochondrial dysfunction and aging. *Mechanisms of Ageing and Development* 131(7–8): 536–543.

Mandavilli, B.S., Santos, J.H., Van Houten, B. 2002. Mitochondrial DNA repair and aging. *Mutation Research* 509(1–2): 127–151.

Marla, S.S., Lee, J., Groves, J.T. 1997. Peroxynitrite rapidly permeates phospholipid membranes. *PNAS* 94(26): 14243–14248.

Mei, N., Zhang, Y., Chen, Y., Guo, X., Ding, W., Ali, S.F., Biris, A.S., Rice, P., Moore, M.M., Chen, T. 2012. Silver nanoparticle-induced mutations and oxidative stress in mouse lymphoma cells. *Environmental and Molecular Mutagenesis* 53(6): 409–419.

Martinez-Gutierrez, F., Olive, P.L., Banuelos, A., Orrantia, E., Nino, N., Sanchez, E.M., Ruiz, F., Bach, H., Av-Gay, Y. 2010. Synthesis, characterization, and evaluation of antimicrobial and cytotoxic effect of silver and titanium nanoparticles. *Nanomedicine* 6(5): 681–688.

Martinez-Gutierrez, F., Thi, E.P., Silverman, J.M. et al. 2012. Antibacterial activity, inflammatory response, coagulation and cytotoxicity effects of silver nanoparticles. *Nanomedicine* 8(3): 328–336.

Miura, N., Shinohara, Y. 2009. Cytotoxic effect and apoptosis induction by silver nanoparticles in HeLa cells. *Biochemical Biophysical Research Communications* 390(3): 733–737.

Moan, J. 1990. On the diffusion length of singlet oxygen in cells and tissues. *Journal of Photochemistry and Photobiology B: Biology* 6(3): 343–347.

Morishige, T., Yoshioka, Y., Inakura, H. et al. 2012. Suppression of nanosilica particle-induced inflammation by surface modification of the particles. *Archives of Toxicology* 86(8): 1297–1307.

Murdock, R.C., Braydich-Stolle, L., Schrand, A.M., Schlager, J.J., Hussain, S.M. 2008. Characterization of nanomaterial dispersion in solution prior to in vitro exposure using dynamic light scattering technique. *Toxicological Sciences* 101(2): 239–253.

Mutwakil, M.H., Reader, J.P., Holdich, D.M. et al. 1997. Use of stress-inducible transgenic nematodes as biomarkers of heavy metal pollution in water samples from an English river system. *Archives of Environmental Contamination and Toxicology* 32(2): 146–153.

Naqui, A., Chance, B., Cadenas, E. 1986. Reactive oxygen intermediates in biochemistry. *Annual Review of Biochemistry* 55: 137–166.

Nel, A., Xia, T., Madler, L., Li, N. 2006. Toxic potential of materials at the nanolevel. *Science* 311(5761): 622–627.

Nishanthl, R., Jyotsnalk, R., Schlager, J., Hussain, S., Reddana, P. 2011. Inflammatory responses of RAW 264.7 macrophages upon exposure to nanoparticles: Role of ROS-NFκB signaling pathway. *Nanotoxicology* 5(4): 502–516.

Nohl, H., Jordan, W. 1986. The mitochondrial site of superoxide formation. *Biochemical Biophysical Research Communication* 138(2): 533–539.

Oberdorster, G., Sharp, Z., Atudorei, V. et al. 2002. Extrapulmonary translocation of ultrafine carbon particles following whole-body inhalation exposure of rats. *Journal of Toxicology and Environmental Health A* 65(20): 1531–1543.

Oberdorster, G., Oberdorster, E., Oberdorster, J. 2005. Nanotoxicology: an emerging discipline evolving from studies of ultrafine particles. *Environmental Health Perspective* 113(7): 823–839.

Oh, E., Delehanty, J., Sapsford, K. et al. 2011. Cellular uptake and fate of pegylated gold nanoparticles is dependent on both cell-penetration peptides and particle size. *ACS Nano* 5(8): 6434–6448.

Ott, M., Gogvadze, V., Orrenius, S., Zhivotovsky, B. 2007. Mitochondria, oxidative stress and cell death. *Apoptosis* 12(5): 913–922.

Park, J.B. 2003. Phagocytosis induces superoxide formation and apoptosis in macrophages. *Experimental and Molecular Medicine* 35: 325–335.

Peters, A., Veronesi, B., Calderon-Garciduenas, L. et al. 2006. Translocation and potential neurological effects of fine and ultrafine particles a critical update. *Particular Fibre Toxicology* 3: 13.

Piao, M.J., Kim, K.C., Choi, J.-Y., Choi, J., Hyun, J.W. 2007. Silver nanoparticles down-regulate Nrf2-mediated 8-oxoguanine DNA glycosylase 1 through inactivation of extracellular regulated kinase and protein kinase B in human Chang liver cells. *Toxicology Letters* 207(2): 143–148.

Pias, E.K., Aw, T.Y. 2002. Apoptosis in mitotic competent undifferentiated cells is induced by cellular redox imbalance independent of reactive oxygen species production. *FASEB Journal* 16(8): 781–790.

Porter, A.E., Muller, K., Skepper, J., Midgley, P., Welland, M. 2006. Uptake of C60 by human monocyte macrophages, its localization and implications for toxicity: studied by high resolution electron microscopy and electron tomography. *Acta Biomaterialia* 2(4): 409–419.

Powers, C., Badireddy, A., Ryde, I., Seidler, F., Slotkin, T. 2011. Silver nanoparticles compromise neurodevelopment in PC12 cells: critical contributions of silver ion, particle size, coating, and composition. *Environmental Health Perspective* 119(1): 37–44.

Ricciardolo, F.L., Di Stefano, A., Sabatini, F., Folkerts, G. 2006. Reactive nitrogen species in the respiratory tract. *European Journal of Pharmacology* 533(1–3): 240–252.

Riley, D.P., Weiss, R.H. 1994. Manganese macrocyclic ligand complexes as mimics of superoxide dismutase. *Journal of American Chemical Society* 116: 387–388.

Risom, L., Moller, P., Loft, S. 2005. Oxidative stress-induced DNA damage by particulate air pollution. *Mutation Research* 592(2): 119–137.

Rosenfeld, M.E., Bowen-Pope, D.F., Ross, R. 1984. Platelet-derived growth factor: morphologic and biochemical studies of binding, internalization, and degradation. *Journal of Cell Physiology* 121(4): 263–274.

Roser, M., Fischer, D., Kissel, T. 1998. Surface-modified biodegradable albumin nano- and microspheres. II: effect of surface charges on in vitro phagocytosis and biodistribution in rats. *European Journal of Pharmaceutics and Biopharmaceutics* 46(3): 255–263.

Rothen-Rutishauser, B.M., Schurch, S., Haenni, B., Kapp, N., Gehr, P. 2006. Interaction of fine particles and nanoparticles with red blood cells visualized with advanced microscopic techniques. *Environmental Science and Technology* 40(14): 4353–4359.

Rutkowski, D.T., Kaufman, R.J. 2004. A trip to the ER: coping with stress. *Trends in Cell Biology* 1(1): 20–28.

Sanpui, P., Chattopadhyay, A., Gosh, S. 2011. Induction of apoptosis in cancer cells at low silver nanoparticle concentrations using chitosan nanocarrier. *ACS Applied Materials and Interfaces* 3(2): 218–228.

Schneider, S., Halbig, P., Grau, H., Nickel, U. 1994. Reproducible preparation of silver sols with uniform particle size for application in surface-enhanced Raman spectroscopy. *Photochemistry and Photobiology* 60(6): 605–610.

Schrand, A.M., Braydich-Stolle, L.K., Schlager, J.J., Dai, L., Hussain, S.M. 2008. Can silver nanoparticles be useful as potential biological labels? *Nanotechnology* 19: 235104.

Schrek, R., Baeuerle, P.A. 1991. A role for oxygen radicals as second messengers. *Trends in Cell Biology* 1(2–3): 39–42.

Shin, S.H., Ye, M.K., Kim, H.S., Kang, H.S. 2007. The effects of nano-silver on the proliferation and cytokine expression by peripheral blood mononuclear cells. *International Immunopharmacology* 7(13): 1813–1818.

Shirtcliffe, N., Nickel, U., Schneider, S. 1999. Reproducible preparation of silver sols with small particle size using borohydride reduction: for use as nuclei for preparation of larger particles. *Journal of Colloid and Interface Science* 211(1): 122–129.

Singh, S., Patel, P., Jaiswal, S., Prabhune, A.A., Ramana, C.V., Prasad, B.L.V. 2009. A direct method for the preparation of glycolipid–metal nanoparticle conjugates: sophorolipids as reducing and capping agents for the synthesis of water re-dispersible silver nanoparticles and their antibacterial activity. *New Journal of Chemistry* 33: 646–652.

Singal, M., Finkelstein, J.N. 2005. Amorphous silica particles promote inflammatory gene expression through the redox sensitive transcription factor, AP-1, in alveolar epithelial cells. *Experimental Lung Research* 31(6): 581–597.

Skovsen, E., Snyder, J.W., Lambert, J.D., Ogilby, P.R. 2005. Lifetime and diffusion of singlet oxygen in a cell. *Journal of Physical Chemistry B* 109: 8570–8573.

Sondi, I., Goia, D.V., Matijevic, E. 2003. Preparation of highly concentrated stable dispersions of uniform silver nanoparticles. *Journal of Colloid and Interface Science* 260(1): 75–81.

Sriram, M.I., Kanth, S.B., Kalishwaralal, K., Gurunathan, S. 2010. Antitumor activity of silver nanoparticles in Dalton's lymphoma ascites tumor model. *International Journal of Nanomedicine* 5: 753–762.

Stadtman, E.R., Levine, R.L. 2003. Free radical-mediated oxidation of free amino acids and amino acid residues in proteins. *Amino Acids* 25(3–4): 207–218.

Tabata, Y., Ikada, Y. 1988. Effect of the size and surface charge of polymer microspheres on their phagocytosis by macrophage. *Biomaterials* 9(4): 356–362.

Takenaka, S., Karg, E., Roth, C., Schulz, H., Ziesenis, A., Heinzmann, U., Schramel, P., Heyder, J. 2001. Pulmonary and systemic distribution of inhaled ultrafine silver particles in rats. *Environmental Health Perspective* 109(4): 547–551.

Tao, A., Sinsermsuksakul, P., Yang, P. 2006. Polyhedral silver nanocrystals with distinct scattering signatures. *Angewandte Chemie International Edition* 45(28): 4597–4601.

Thasneem, Y., Sajeesh, S., Sharma, C. 2011. Effect of thiol functionalization on the hemocompatibility of PLGA nanoparticles. *Journal of Biomedical Materials Research Part A* 99A(4): 607–617.

Trickler, W.J., Lantz, S.M., Murdock, R.C. et al. 2010. Silver nanoparticle induced blood-brain barrier inflammation and increased permeability in primary rat brain microvessel endothelial cells. *Toxicological Sciences* 118(1): 160–170.

Ueda, S., Masutani, H., Nakamura, H., Tanaka, T., Ueno, M., Yodoi, J. 2002. Redox control of cell death. *Antioxidants and Redox Signaling* 4(3): 405–414.

Urne A.G., Vaux D.L. 1996. Molecular and clinical aspects of apoptosis. *Pharmacology and Therapeutics* 72(1): 37–50.

Vallhov, H., Gabrielsson, S., Stromme, M., Scheynius, A., Garcia-Bennett, A.E. 2007. Mesoporous silica particles induce size dependent effects on human dendritic cells. *Nano Letters* 7: 3576–3582.

Vanwinkle, B.A., de Mesy B., Malecki, K.L. et al. 2009. Nanoparticle (NP) uptake by type I alveolar epithelial cells and their oxidant stress response. *Nanotoxicology* 3(4): 307–318.

Vasto, S., Mocchegiani, E., Candore, G. et al. 2006. Inflammation, genes and zinc in ageing and age related diseases. *Biogerontology* 7(5–6): 315–327.

Wei, L., Tang, J., Zhang, Z., Chen, Y., Zhou, G., Xi, T. 2010. Investigation of the cytotoxicity mechanism of silver nanoparticles in vitro. *Biomedical Materials* 5(4): 044103.

Welch, W.J. 1992. Mammalian stress response: cell physiology, structure/function of stress proteins, and implications for medicine and disease. *Physiological Reviews* 72(4): 1063–1081.

Wiley, B., Sun, Y., Mayers, B., Xia, Y. 2005. Shape-controlled synthesis of metal nanostructures: the case of silver. *Chemistry* 11(2): 454–463.

Xia, T., Kovochich, M., Brant, J. et al. 2006. Comparison of the abilities of ambient and manufactured nanoparticles to induce cellular toxicity according to an oxidative stress paradigm. *Nano Letters* 6(8): 1794–1807.

Xia, T., Kovochich, M., Nel, A.E. 2007. Impairment of mitochondrial function by particulate matter (PM) and their toxic components: implications for PM-induced cardiovascular and lung disease. *Frontiers in Bioscience* 12: 1238–1246.

Xiao, G., Wang, M., Li, N., Loo, J., Nel, A. 2003. Use of proteomics to demonstrate a hierarchical oxidative stress response to diesel exhaust particle chemicals in a macrophage cell line. *Journal of Biological Chemistry* 278(50): 50781–50790.

Yang, H., Liu, C., Yang, D., Zhang, H., Xi, Z. 2009. Comparative study of cytotoxicity, oxidative stress and genotoxicity induced by four typical nanomaterials: the role of particle size, shape and composition. *Journal of Applied Toxicology* 29(1): 69–78.

Ye, Y., Liu, J., Xu, J., Sun, L., Chen, M., Lan, M. 2010. Nano-SiO2 induces apoptosis via activation of p53 and bax mediated by oxidative stress in human hepatic cell line. *Toxicology In Vitro* 24(3): 751–758.

Zhang, L.W., Monteiro-Riviere, N.A. 2009. Mechanisms of quantum dot nanoparticle cellular uptake. *Toxicological Sciences* 110(1): 138–155.

Zhang, S., Li, J., Lykotrafitis, G., Bao, G., Suresh, S. 2009. Size-dependent endocytosis of nanoparticles. *Advanced Materials* 21(4): 419–424.

Zhang, Y., Yang, M., Portney, N.G. et al. 2008. Zeta potential: a surface electrical characteristic to probe the interaction of nanoparticles with normal and cancer human breast epithelial cells. *Biomedical Microdevices* 10: 321–328.

14

NANOTECHNOLOGY: OVERVIEW OF REGULATIONS AND IMPLEMENTATIONS

Om V. Singh

Division of Biological and Health Sciences, University of Pittsburgh, Bradford, PA, USA

Thomas Colonna

Center for Biotechnology Education, Zanvyl Krieger School of Arts and Sciences, The Johns Hopkins University, Rockville, MD, USA

14.1. INTRODUCTION

Nanotechnology presents great prospects for developing new products with both industrial and consumer applications in a wide range of sectors. This technology is defined as a "system of innovative methods to control and manipulate matter at near-atomic scale to produce new materials, structures, and devices" by the American National Institute of Occupational Safety and Health (NIOSH, 2007). According to the National Nanotechnology Initiative, it is the understanding and control of matter at dimensions of 1–100 nm (Fig. 14.1). Nanotechnology's primary advantages come from the new properties and functions acquired by different materials on the nanoscale (or nanomaterials), due to their higher surface-area-to-mass ratios and quantum effects (Buzea et al., 2007). The properties that change may include melting point, strength, optical, and magnetic properties, electrical conductance, and chemical reactivity (NNI, 2013a). Among the types of nanoparticles that have been produced

---

Bio-Nanoparticles: Biosynthesis and Sustainable Biotechnological Implications, First Edition.
Edited by Om V. Singh.
© 2015 John Wiley & Sons, Inc. Published 2015 by John Wiley & Sons, Inc.

are buckyballs, carbon nanotubes, liposomes, micelles, and polymeric nanospheres (Moghimi, 2012; Yamashita, 2012). Various product additives can be encapsulated on the nanoscale to enhance the taste or nutritional value of foods.

Engineered nanoscale materials (ENMs) are produced for particular effects or applications, and can be inorganic (e.g., metals and metal oxides) or organic (e.g., food additives and cosmetics ingredients). ENMs are produced in free particulate forms; they tend to stick together to form larger agglomerates due to enormous surface free energies. The chemical nature of substances used to manufacture ENMs can be inorganic or organic. To help visualize nanomaterials in context, organic life is carbon-based and the C–C bond length is about 0.15 nm. Placed in a food context, most ENMs are bigger than molecules such as lipids, are a similar size to many proteins, but are smaller than the intact cells in plant- and animal-based foods (Fig. 14.2).

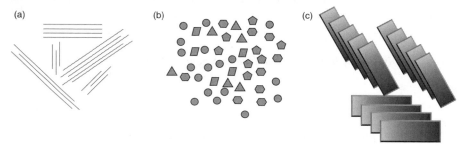

Figure 14.1. Nanomaterials in different forms as (a) rods; (b) poly-dispersal particles; and (c) layers.

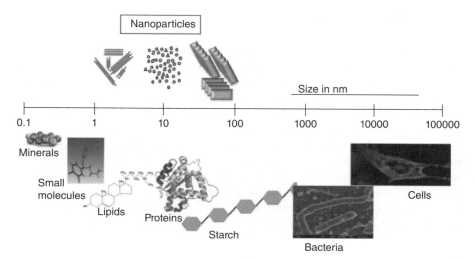

Figure 14.2. Nanomaterials and nanoparticles in context to other biological molecules. *See insert for color representation of the figure.*

Nanotechnology development is a potentially multi-billion-dollar industry with many possible applications in agriculture, medicine, and food, among other sectors. However, the science is still emerging and many of these applications, especially in agriculture and food, have yet to be realized. Furthermore, along with its advantages, nanoparticle technology presents unique health and environmental risks. Nanomaterials (i.e., nanoparticles) may cause health risks which are different from those of similar materials in micro or macro form. The precautionary risk assessment and monitoring of nanoparticles in society could be effective, however a legal framework to regulate the overuse of nanoscale material is needed that can address the unique risks involved in nanoparticles and guard against new risks that may emerge. Regulatory agencies in the United States and other countries monitor the release of nanoparticles, ensuring the safety of nanoscale materials. A legal regulatory framework is needed to control the use of nanomaterials and deal with the risks they pose. This chapter discusses the current scope of the nanotechnology field, related safety issues, and regulatory policies involving this technology, as well as how policies can be improved.

14.2. SCOPE OF NANOTECHNOLOGY

The majority of nanotechnology applications are driven by the potential to increase or improve the functionality of materials and reduce the amounts of those materials needed for particular purposes. A small quantity of an ENM can replace a much larger quantity of conventional substances. Nanomaterials and nanosensors can also be used in packaging to make food storage safer, increase shelf life, and make products more trackable. Nanomedicine is another subfield with great potential. Because they can be engineered with any variety of properties and surface modifications and are smaller than cells, nanomaterials are useful for a wide variety of medical applications, including targeted drug delivery (Moghimi, 2012). Some nanotechnological products are already on the market (Table 14.1).

The food and drug industries are constantly looking for innovations to improve their products and profit margins, as well as consumer safety. These industries are always seeking novel technologies to offer targeted products with effective means and improved safety. Consumer health concerns and tightened regulations are driving the food industry to cut down on salt, sugar, fat, and artificial additives in their products and to improve food labeling, packaging, and storage. Stricter regulations have also been imposed to address certain food-related ailments, such as obesity, diabetes, cardiovascular diseases, digestive disorders, certain types of cancer (e.g., bowel cancer), and food allergies associated with food and dietary supplements (Powell et al., 2000; FAO/WHO, 2010). The packaging and labeling of products have also changed and are regulated over time to monitor quality, safety, and security during transportation and storage (FAO/WHO, 2010). Several studies have identified potential uses for nanotechnology that address many of the issues summarized in Figure 14.3. ENMs are used in food packaging as part of polymer composites that have antimicrobial properties. Nano-sized substances are also being used in dietary supplements to

TABLE 14.1. Nanomedicine Products on the Market

Application	Composition/nanotech component	Indication	Company*
Drug delivery			
Abelcet	Amphotericin B/ lipid complex	Fungal infections	Enzon (Bridgewater, NJ, USA)
Amphotec	Amphotericin B/ lipid colloidal dispersion	Fungal infections	InterMune (Brisbane, CA, USA)
Ambisome	Liposomal Amphotericin B	Fungal infections	Gilead (Foster City, CA, USA), Fujisawa (Osaka, Japan)
DaunoXome	Liposomal daunorubicin	Kaposi sarcoma	Gilead
Doxil/ Caelyx	Liposomal doxorubicin	Cancer, Kaposi sarcoma	Ortho Biotech (Bridgewater, NJ, USA); Schering-Plough (Kenilworth, NJ, USA)
Depocyt	Liposomal cytarabine	Cancer	SkyePharma (London), Enzon
Epaxal Berna	Virosomal hepatitis vaccine	Hepatitis A	Berna Biotech (Bern, Switzerland)
Inflexal V	Berna Virosomal influenza vaccine	Influenza	Berna Biotech
Myocet	Liposomal doxorubicin	Breast cancer	Zeneus Pharma (Oxford, UK)
Visudyne	Liposomal verteporfin	Age-related macular degeneration	QLT (Vancouver, Canada), Novartis (Basel)
Estrasorb	Estradiol in micellar nanoparticles	Menopausal therapy	Novavax (Malvern, PA, USA)
Adagen	PEG-adenosine deaminase	Immunodeficiency disease	Enzon
Neulasta	PEG-G-CSF	Febrile neutropenia	Amgen (Thousand Oaks, CA, USA)
Oncaspar	PEG-asparaginase	Leukemia	Enzon
Pegasys	PEG-α-interferon 2a	Hepatitis C	Nektar (San Carlos, CA, USA), Hoffmann-La Roche (Basel)
Estrasorb	Estradiol in micellar nanoparticles	Menopausal therapy	Novavax (Malvern, PA, USA)
Adagen	PEG-adenosine deaminase	Immunodeficiency disease	Enzon
Neulasta	PEG-G-CSF	Febrile neutropenia	Amgen (Thousand Oaks, CA, USA)
Oncaspar	PEG-asparaginase	Leukemia	Enzon
Pegasys	PEG-α-interferon 2a	Hepatitis C	Nektar (San Carlos, CA, USA), Hoffmann-La Roche (Basel)

TABLE 14.1. (Continued)

Application	Composition/nanotech component	Indication	Company*
PEG-Intron	PEG-α-interferon 2b	Hepatitis C	Enzon, Schering-Plough
Macugen	Pegylated anti-VEGF aptamer	Age-related macular degeneration	OSI Pharmaceuticals (Melville, NY, USA), Pfizer (New York)
Somavert	PEG-HGH	Acromegaly	Nektar, Pfizer
Copaxone	Copolymer of alanine, lysine, glutamic acid and tyrosine	Multiple sclerosis	TEVA Pharmaceuticals (Petach Tikva, Israel)
Renagel	Crosslinked poly(allylamine) resin	Chronic kidney disease	Genzyme (Cambridge, MA, USA)
Emend	Nanocrystalline aprepitant	Antiemetic	Elan Drug Delivery (King of Prussia, PA, USA), Merck & Co. (Whitehouse Station, NJ, USA)
MegaceESc	Nanocrystalline megesterol acetate	Eating disorders	Elan Drug Delivery, Par Pharmaceutical Companies (Woodcliff Lake, NJ, USA)
Rapamune	Nanocrystalline sirolimus	Immunosuppressant	Elan Drug Delivery, Wyeth Pharmaceuticals (Collegeville, PA, USA)
Tricor	Nanocrystalline fenofibrate	Lipid regulation	Elan Drug Delivery, Abbott (Abbott Park, IL, USA)
Triglide	Nanocrystalline fenofibrate	Lipid regulation	SkyePharma, First Horizon Pharmaceuticals (Alpharetta, GA, USA)
Abraxane	Paclitaxel protein-bound nanoparticles	Cancer	Abraxis BioScience (Schaumburg, IL, USA), AstraZeneca (London)
In vivo imaging			
Resovist	Iron nanoparticles	Liver tumors	Schering (Berlin)
Feridex/ Endorem	Iron nanoparticles	Liver tumors	Advanced Magnetics (Cambridge, MA, USA), Guerbet (Roissy, France)
Gastromark/ Lumirem	Iron nanoparticles	Imaging of abdominal structures	Advanced Magnetics, Guerbet
In vitro diagnostics			
Lateral flow tests	Colloidal gold	Pregnancy, ovulation, HIV (among others)	British Biocell (Cardiff, UK), Amersham/GE (Little Chalfont, UK), Nymox (Hasbrouck Heights, NJ, USA)

(*continued*)

TABLE 14.1. (Continued)

Application	Composition/nanotech component	Indication	Company[*]
Clinical cell separation	Magnetic nanoparticles	Immunodiagnostics	Dynal/Invitrogen (Oslo, Norway), Miltenyi Biotec (Bergisch Gladbach, Germany), Immunicon (Huntingdon Valley, PA, USA)
Biomaterials			
Ceram X duo	Nanoparticle composite	Dental filling material	Dentspley (Weybridge, UK)
Filtek Supreme	Nanoparticle composite	Dental filling material	3 M Espe (Seefeld, Germany)
Mondial	Nanoparticle-containing dental prosthesis	Dental restoration	Heraeus Kulzer (Hanau, Germany)
Premise	Nanoparticle composite	Dental repair	Sybron Dental Specialties (Newport Beach, CA, USA)
Tetric EvoCeram	Nanoparticle composite	Dental repair	Ivoclar Vivadent (Schaan, Liechtenstein)
Ostim	Nano-hydroxyapatite	Bone defects	Osartis (Obernburg, Germany)
Perossal	Nano-hydroxyapatite	Bone defects	aap Implantate (Berlin)
Vitoss	Nano-hydroxyapatite	Bone defects	Orthovita (Malvern, PA, USA)
Acticoat	Silver nanoparticles	Antimicrobial wound care	Nucryst (Wakefield, MA, USA)
Active implants			
Pacemaker	Fractal electrodes	Heart failure	Biotronik (Berlin)

[*]If the product is developed by an alliance (e.g., between a drug delivery company and a pharma company) both companies are listed.
Reprinted by permission from Macmillan Publisher Ltd. (Nature Biotechnology) Wagner et al. (2006).
Abbreviations: G-CSF, granulocyte colony-stimulating factor; hGH, human growth hormone; PEG, polyethylene glycol; VEGF, vascular endothelial growth factor.

improve the bioavailability of these supplements to the body, as well as taste, texture, stability, and consistency.

By 2018, the nanomedicines market is projected to reach $96.9 billion with central nervous system nanomedicines specifically reaching $29.5 billion and anti-cancer medicines reaching $12.7 billion (www.bccresearch.com). By 2017, nanomaterials sales are expected to reach $37.3 billion and nanotools $11.4 billion (www.bccresearch.com). It is projected that the global nanotechnology applications market will be valued at $3.3 trillion by 2018 (Global Industry Analysts Inc.), with the highest growth rates in Asian countries. Today, the US is at the head of research into nanofoods, with Japan and China following; over 1200 companies are involved in

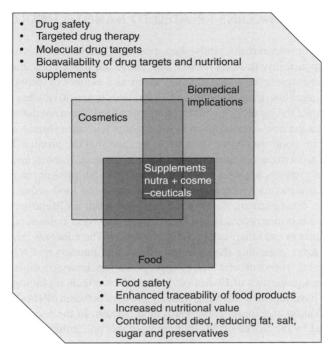

Figure 14.3. Projected benefits of nanotechnology in food- and biomedical-related sectors. *See insert for color representation of the figure.*

this research currently. The market for nanofoods is projected to reach $20.4 billion by 2020, dominated by Asian countries (http://www.hkc22.com/Nanofood.html).

The National Nanotechnology Initiative (NNI) was created by the US government in 2000 to bring together research and regulation agencies, including the Food and Drug Administration (FDA), the National Institutes of Health (NIH), the National Science Foundation (NSF), and the Environmental Protection Agency (EPA), to collaborate toward the following goals (NNI, 2013b, c):

1. advance world-class nanotechnology research and development;
2. foster the transfer of new technologies into products for commercial and public benefit;
3. develop and sustain educational resources, a skilled workforce, and the supporting infrastructure and tools to advance nanotechnology; and
4. support the responsible development of nanotechnology.

The NNI has received a total of $18 billion in funding so far, with $1.8 billion budgeted for FY2013 (NNI, 2013d). The US government is highly invested in promoting research and business opportunities in this new field, due to its great economic potential.

14.3. SAFETY CONCERNS RELATED TO NANOTECHNOLOGY

Prior to becoming successfully established, every new technology has to demonstrate a sense of responsibility towards society and, above all, to the consumer. To win the trust of consumers, new technology has to overcome a number of technological, societal, and regulatory barriers, especially when it relates to sensitive areas such as drugs and food. Despite the infancy of nanotechnology in food, there are demands for demonstrations that the new technological developments will have some real benefits for the consumer and not just for the industry alone, and that the promised benefits will outweigh any risks to the consumer and/or the environment. Nanotechnology-derived drug and food products are also new to consumers. In the present era of heightened consumer awareness, nanotechnology applications in the food sector have already sparked a new debate among the stakeholders. The rapid proliferation of nanotechnologies into consumer products, especially drug and food industries, has raised a number of concerns over their safety to the consumer. The concerns remains due to a lack of knowledge regarding the potential effects and impacts of ENMs on human health and the environment, and lack of appropriate regulatory controls.

The altered properties of ENMs could alter their effects on biological systems; some studies have reported differing toxicity profiles between ENMs and their bulk equivalents (Simon and Joner, 2008; Gatti et al., 2009). In the food industry, safety risks are posed by the use of ENMs that are insoluble and unlikely to be assimilated by the digestive tract, such as those that incorporate metal oxides; such ENMs may translocate to other tissues in the body, bind various compounds, or carry harmful substances into the blood (des Rieux et al., 2006; Arora, 2012; Yamashita, 2012). Also, because they can pass through cell membranes, ENMs may be able to enter areas of the body that larger particles cannot, even cell organelles and mitochondria (Yamashita, 2012). Evidence has been found to indicate that, in biological systems, ENMs undergo complex interactions with various other chemical entities, and that they can become coated with biomolecules that carry them to certain parts of the body (Arora, 2012; Schädlich, 2012; Yamashita, 2012). ENMs used in food products and ingested may be transformed in ways that alter their properties in a biological system; one such transformation is an increase in production of reactive oxygen species (ROS), which can cause oxidative stress, inflammation, and protein/DNA damage (Arora, 2012). The increased uptake and bioavailability of nanoscale substances also have potential health risks. Inhaled or ingested nanomaterials are easily absorbed by the body and can cross the blood–brain barrier (Arora, 2012). For example, carbon nanotubes can produce lung inflammation and fibrosis similar to the effects of long-fiber asbestos (Yamashita, 2012). Studies have also found that nanomaterials can collect in the liver, spleen, ovaries, and lymph nodes (Arora, 2012; Schädlich, 2012). Silver nanoparticles, which are released by some wound dressings, are toxic to macrophage cells and epithelial cells in the lungs and can also damage the liver (Arora, 2012).

Of course, the possible risk posed by any nanomedicine or nanofood product depends on many factors, including ENM concentration, how much of the product is consumed, and the properties of the particular ENMs in question. Significant gaps currently exist in our knowledge of how ENMs behave and react when passing through the digestive tract; much of the current evidence of harmful ENM effects has come from studies involving inhaled ENMs or *in vitro* studies. Many ENMs in food will likely bind with other food ingredients or digestive acids/enzymes, and thus be unable to translocate within the body. Some speculate that translocation risk may also be affected by gut physiology and health. Conversely, ingested microparticles are suspected to worsen the symptoms of diseases such as Crohn's disease and irritable bowel syndrome, but studies have produced conflicting results and are inconclusive (Schneider, 2007). The full extent of hazard, exposure, and risk from the ingestion of ENMs via food and drink are largely unknown. In anticipation of the likely developments in the nanofood sector however, it is imperative that the safety of nanotechnology-derived products is addressed adequately so that while the new developments bring benefits to the consumer, they are also safe to human health and the environment.

14.4. BARRIERS TO THE DESIRED REGULATORY FRAMEWORK

The recent rapid expansion in nanotechnology is challenging the currently established regulatory systems. People are questioning whether the current frameworks can be applied to nanomaterials, and there is very little nanomaterial-specific regulation thus far. Studies have highlighted the need to modify some testing methodologies to account for nanomaterials (FAO/WHO, 2010; Paradise, 2012). There are also major knowledge gaps in relation to the effects of most ENMs on human health, agreed dose units for hazard and exposure assessments, and reliable and validated methods for measurement and characterization of ENMs in complex food matrices.

The first question experts disagree on is how to even define nanotechnology. Traditionally it has been based on size, with particles between 1 and 100 nm being classified as nanoparticles (Figs 14.1, 14.2), but some molecules (e.g., carbon nanotubes) show unusual properties (e.g., longer length) when only one dimension is in this size range (Nel et al., 2006; Paradise, 2012; Yamashita, 2012). The FDA issued a guidance document defining nanotechnology based on four characteristics (FDA, 2011):

- at least one dimension between 1 and 100 nm;
- deliberate manipulation and control of particle size to produce specific properties;
- exhibits unique properties attributable to its size; and
- size range up to 1 µm.

It has been argued that regulation based on rigid definitions may be more harmful than beneficial (Maynard, 2011). One suggestion was to name nine or ten attributes with values that could "trigger" a material to be classified as a nanomaterial, with some of these being attributes other than size or surface area (Maynard, 2011).

The second major issue in regulation of ENMs is to assess the risk. The unique properties of nanomaterials led to a new sub-discipline, that of nanotoxicology (Arora, 2012). The high reactivity of ENMs could lead to toxicity in their interactions with biological systems (Arora, 2012). Due to reactive surface of nanoparticles, they may react with biological proteins, cells, and body fluids after administration, altering their properties in the body (Moghimi, 2012). In this case, *in vitro* toxicologic testing may not be very predictive of toxic effects *in vivo*. In addition, the types of interactions that nanoparticles display with cellular components such as the generation of ROS and interactions with DNA suggest possible long-term toxic effects (Arora, 2012; Schädlich, 2012). Because there are varying ENMs being used in the development of nano-drug and nano-food, each engineered specifically with certain composition, charge, and surface modifications, it is difficult to make generalizations across the discipline as to what types of toxicologic testing is necessary or relevant for each case (Arora, 2012). Current standard means of assessing toxicity for drug molecules may need to be revised in order to adequately predict toxicity of nano-drug and nano-foods *in vivo* (Moghimi, 2012).

Due to the wide variety of ENMs being produced for use in drug and food applications, it is difficult to generalize about what toxicological tests are necessary to assess them, and current assessment standards may need to be revised to accurately predict their toxicity (Moghimi, 2012). It is also difficult to measure the ADME (absorption, distribution, metabolism, and excretion) characteristics of nanomaterials, as these qualities can change *in vivo*, depending on the size, shape, charge, aggregation, and surface modifications of the nanoparticles (Nyström, 2012). Traditionally, dose response curves for drugs are calculated based on the mass of the substance administered; however, this may not apply in the same way to nanomedicines, which would have a major effect on the determination of first-in-human dosing, a factor of the NOAEL (no-observed-adverse-effect level) dose.

Furthermore, these ENMs are not homogenous: the molecules in a given batch will vary somewhat in size and surface modification, which can affect their solubility, aggregation, crystallinity, and stability and cause heterogeneity in the properties they are engineered to have (Nyström, 2012). This heterogeneity has obvious implications for goods manufacturing practice (GMP), where batch productions of drug and food must meet predetermined standards (Code of Federal Regulations: 21CFR- 210, 21CFR-211). Regardless of how they are produced, levels of public concern regarding nanomaterials vary as treatment of them has not been consistent across different regions.

14.4.1. Regulatory Framework in the United States

In the US, this regulatory framework is mainly upheld by the Food and Drug Administration (FDA), United States Department of Agriculture (USDA), and Environmental Protection Agency (EPA). The FDA's current policy is to evaluate nano-drugs in the same way that it evaluates other drugs. It views emerging technologies as neutral until evidence is gathered that they are harmful or beneficial (Holdren, 2011; Hamburg, 2012). There are several FDA centers responsible for evaluating

TABLE 14.2. FDA/s Sub-center and their Role in Biologics Evaluation (FDA, 2013b)

FDA Centers	Role and interpretation	Reference
Center for Drug Evaluation and Research (CDER)	Regulates small molecule drugs, some biologic drugs such as antibodies and proteins, and over-the-counter drugs.	CDER (2013)
Center for Biologics Evaluation and Research (CBER)	Regulates biological products such as vaccines, cellular, and gene therapy products, and tissue and blood products.	CBER (2013)
Center for Devices and Radiological Health (CDRH)	Evaluates medical devices.	CDRH (2013)
Office of Combination Products (OCP)	Responsible for developing guidance documents for products that cross two or more product classes: drug-biologics, drug-devices, or device-biologics. Assigns primary responsibility for review of these products to CDER, CBER or CDRH based on the product's primary mode of action. (For example, if a drug-device combination product's primary mode of action is attributable to the drug component, that product would be assigned to CDER for review.)	FDA (2013a)

biologics and medicines (see Table 14.2). In 2006, the FDA's Nanotechnology Task Force was created with the following goals (FDA, 2007):

- "Chair a public meeting to help FDA further its understanding of developments in nanoscale materials that pertain to FDA-regulated products;
- Assess the current state of scientific knowledge pertaining to nanoscale materials for purposes of carrying out FDA's mission;
- Evaluate the effectiveness of the agency's regulatory approaches and authorities to meet any unique challenge that may be presented by the use of nanoscale materials in FDA-regulated products;
- Explore opportunities to enable innovation using nanoscale materials to develop safe and effective drugs, biologics, and devices;
- Continue to strengthen FDA's collaborative relationships with other federal agencies, as well as with foreign government regulatory bodies, international organizations, healthcare professionals, industry, consumers, and other stakeholders;
- Consider appropriate vehicles for communicating with the public about the use of nanoscale materials in FDA-regulated products."

At that time, the findings of the task force were that the current regulatory system was sufficient to cover nanotechnological products. It did recommend that the FDA review applications as to whether labeling should be required for nanomaterials, as well as request public comment on whether the current regulations were sufficient to address complex nanomedicines (FDA, 2007). The task force also recommended issuing guidance documents to the drug industries on topics such as how to determine whether a product contains nanomaterials.

Based on these recommendations, in April 2012 the FDA issued a draft guidance document to stakeholders detailing its current positions on nanotechnology. As per the general framework for assessing the safety of nanomaterials in cosmetic products, the Federal Food, Drug, and Cosmetic Act (FDC&C Act) prohibits the marketing of adulterated or misbranded cosmetics in interstate commerce (USC 331(a); see US Code, 2014). Further, an FDA press release stated (FDA, 2012):

The food draft guidance describes the factors manufacturers should consider when determining whether changes in manufacturing processes, including those involving nanotechnology, create a significant change that may:

- affect the identity of the food substance;
- affect the safety of the use of the food substance;
- affect the regulatory status of the use of the food substance; or
- warrant a regulatory submission to FDA.

The cosmetic product draft guidance discusses the FDA's current thinking on the safety assessment of nanomaterials when used in cosmetic products. Key points include:

- The legal requirements for cosmetics manufactured using nanomaterials are the same as those for any other cosmetics. While cosmetics are not subject to premarket approval, companies and individuals who market cosmetics are legally responsible for the safety of their products and they must be properly labeled.
- To conduct safety assessments for cosmetic products containing nanomaterials, standard safety tests may need to be modified or new methods developed.

The FDA safety assessment for cosmetic products incorporating nanomaterials recommended addressing following factors:

- the physico-chemical characteristics;
- agglomeration and size distribution of nanomaterials at the toxicity testing conditions which should correspond to those of a final product;
- impurities;
- potential product exposure levels, and the potential for agglomeration of nanoparticles in the final product;
- dosimetry for *in vitro* and *in vivo* toxicology studies;
- *in vitro* and *in vivo* toxicological data on ingredients and their impurities, dermal penetration, irritation (skin and eye) and sensitization studies, mutagenicity/genotoxicity studies; and
- clinical studies to test the ingredient, or finished product, in human volunteers under controlled conditions.

The issued document is an encouraging step to ensure the safety and efficacy of nanotechnology in public perception. However, the guidance is more specific to the cosmetics rather being specific to the "food substance" concerning to the environmental, health, and safety issues. On the food substance, FDA advises that industry will have to determine on a case-by-case basis whether use of ENMs results in a "manufacturing process change", and whether industry should warrant submission of data to FDA for a formal pre-market approval. The most effective feature of guidance is that the FDA advises whether ENMs would not be covered by the existing "Generally Recognized as Safe (GRAS)" category, and that industry would have to submit product data with ENMs for a "formal pre-market review". Nanotechnology is a new and rapidly evolving field that holds great promise of advancements in the area of food and drug industries. It also poses challenges to the FDA in finding ways to apply existing regulatory pathways to the review of these materials. There are many unanswered questions that have been left behind in the guidance issued by FDA regarding nano-food and nano-drugs regulation, but it is a step forward and ensures the future of novel technology in the US.

14.4.2. Global Efforts toward Regulation of Nanotechnology

In 2006, the International Risk Governance Council (IRGC) published a white paper on nanotechnology risk governance (Grobe et al., 2008). Since then, the IRGC (along with the Austrian Federal Ministry for Transport, Innovation, and Technology and the Korean National Program for Tera-Level Nanodevices) has been working toward the following goals:

- explore the different definitions and frameworks that have been used in the debate on nanoscaled material in food and cosmetics;
- identify the applications of nanomaterials in current food items and cosmetics;
- review current risk assessments of their use;
- review current risk management and regulatory activities in different countries and continents;
- compare judgments of the acceptability of nanomaterials in food and cosmetics made by a range of actors including different countries as well as international organizations;
- identify gaps and options for global risk governance; and
- explore the possibilities of a voluntary certification program of labeling food items and cosmetics for mitigating possible risks.

The following year, the European Commission's Scientific Committee on Emerging and Newly Identified Health Risks (SCENIHR) published its opinion on "the scientific aspects of the existing and proposed definitions relating to products of nanoscience and nanotechnologies" (SCENIHR, 2007). The committee advised systematic consideration of the properties of nanoparticles and identified a number of nanotechnology-related processes, including coalescence, agglomeration, aggregation, degradation, and solubilization. The organization COmparative Challenge of NANOmaterials was created to provide data on nanotechnology risks and applications in cosmetic products (CONANO, 2007).

In 2008, the International Organization for Standardization (ISO) published a technical specification document for nanomaterials that included definitions of "nanoparticle, nanofibre, and nanoplate" (ISO, 2008).

Based on the SCENIHR recommendations, the European Food Safety Authority (EFSA) performed its own investigation into the safety of nanotechnology used in human food and animal feed products. EFSA later began requesting the following data related to nanomaterials in food and feed (EFSA, 2008):

- food and feed applications and products which contain or consist of nanomaterials or have been produced using nanotechnology;
- methods, procedures and performance criteria used to analyze nanomaterials in food and feed;
- use patterns and exposure for humans and the environment;
- risk assessments performed on nanomaterials used in food and feed;
- toxicological data on nanomaterials used in food and feed;
- environmental studies performed on nanotechnologies and nanomaterials used in food and feed; and
- other data of relevance for risk assessment of nanotechnology and nanomaterials in food and feed.

The European Union's regulations on chemical use and food/cosmetic products currently make no specific reference to nanomaterials. There is therefore no distinction between nanotechnology products and other substances, and nanotechnology is covered under the standard Registration, Evaluation, Authorization, and Restriction of Chemicals (REACH) regulations. The REACH regulation's section on operational conditions specifies the physical form in which the substance is manufactured (REACH, 2013, annex I, section 5.1.1) and states that the properties of a material have to be described in terms corresponding to the form of the application (REACH, 2014, annex IV of directive 67/548/EEC). Under REACH, manufacturers and importers have to submit a registration dossier for materials at or above one ton per year. Additionally, a Chemical Safety Report (CSR) is needed if the amount of material is at or above 10 tons per year.

The German Chemical Industry Association (VCI) goes a step beyond this and urges their members to provide data and safety data sheets even if not importing a ton of a material. They have also published guidelines on occupational health measures (BAuA/VCI, 2007) and material safety (VCI, 2008) to encourage a responsible attitude toward using and importing nanomaterials.

There have been several reports assessing nanotechnology regulation in the UK by the UK Better Regulation Taskforce (2003) and the Royal Society (2004), which vary in their conclusions. The Food Standards Agency (FSA) maintains that products based on nanotechnologies must conform to all the requirements of the EU Food

Law Regulation. In contrast, the opinions of the Food Standards Agency (FSA) (i.e., the final product of the production process is based on the nanotechnologies has to conform to all the requirements of the EU Food Law Regulation 178/2002), the UK Better Regulations Taskforce, and the report of the Department for Business, Enterprise and Regulatory Reform (BERR) suggests that "Free, engineered nanomaterials might be classed as 'hazardous' substances unless or until there is sufficient evidence of their safety in a particular context" (BERR, 2006). Both the UK and Germany currently support the idea that "soft" regulatory measures on nanotechnology such as voluntary conduct codes and public dialogue initiatives are enough to ensure protection for consumer health and the environment.

Japan has begun an array of nanotechnology regulation initiatives: for example, the Ministry of Economy, Trade and Industry (METI), together with the National Institute of Advanced Industrial Science and Technology (AIST) and National Institute of Occupational Safety and Health Japan (JNIOSH), are overseeing a five-year project on toxicity test protocols and risk assessment methods for ENMs. AIST has issued three risk assessment reports on individual nanomaterials – *Titanium Dioxide (TiO2)*, *Fullerene (C60)*, and *Carbon Nanotubes (CNTs)* – as well as a brochure on *The Principles and Basic Approach to Risk Assessment of Manufactured Nanomaterials (interim version)*. According to these publications, the Japanese government does not seem to be initiating any new regulatory changes related to nanomaterials (SAFENANO, 2013). However, the Ministry of Economy, Trade, and Industry (METI) and the Ministry of the Environment (MOE) established specific working groups dedicated to nanomaterials safety. In another attempt, the Ministry of Health, Labor, and Welfare (MHLW) published nanotechnology guidelines to workers (OECD WPMN, 2010).

Table 14.3 provides a summary of global regulatory measures for nanotechnology. There have been a number of past studies identifying gaps and inadequacies in these systems of regulation. The rule of applying existing regulations to other emerging technologies may have worked well, but nanomaterials may be an exception to this rule. ENMs are created using two types of methods: top-down physical methods, such as thermal decomposition, diffusion, irradiation, or discharge; and bottom-up chemical or biological methods, such as those involving seeded growth, polyol synthesis, electrochemical synthesis, chemical reduction, and microbe assistance. Consideration for these various manufacturing approaches should be incorporated into regulations when they are revised, in order to keep nanotechnology on the cutting edge of development.

14.5. BIOSYNTHESIS OF MICROBIAL BIO-NANOPARTICLES: AN ALTERNATIVE PRODUCTION METHOD

Nanomaterials can make technology applications in many fields cleaner and safer, but the physico-chemical methods currently used to manufacture them are not environmentally friendly, safe, or economical. Microbial bio-nanoparticles (BNPs) produced using enzymes and proteins from biological sources such as bacteria, fungi, actinomycetes, yeast, and plants present a potential solution to these problems

TABLE 14.3. Global Initiative and Key Legislations or Activities to Regulate Nanomaterials. Adopted and Modified from Grobe et al. (2008) and FAO/WHO (2010)

Region/country	Regulatory bodies/agencies	Key legislations/activities relevant to evaluate and manage the risks associated with nanomaterials and nanoproducts
Global	<ul><li>International Standards Organization (ISO) Technical Committee (TC) 229</li><li>The International Electrotechnical Commission (IEC) TC 113</li></ul>	<ul><li>Directs activities on nanotechnologies standards at the international level</li><li>ISO/TR 12885 (Technical Report): Health and safety practices in occupational settings relevant to nanotechnologies.</li><li>ISO/TS27687 (Technical Specification): Terminology and definitions for nano-objects: Nanoparticle, nanofibre and nanoplates</li><li>ISO/PRF TS 80004-3 (Technical Specification): Nanotechnologies Part 3: Carbon nano-objects</li></ul>
Australia	<ul><li>Health, Safety and Environment (HSE) Working Group</li><li>National Industrial Chemical Notification and Assessment Scheme (NICNAS)</li><li>Safework Australia, Food Standards Australian New Zealand (FSANZ)</li><li>Therapeutic Goods Administration (TGA)</li><li>Department of Environment, Water, Heritage and the Arts</li><li>Department of Innovation, Industry, Science and Research</li><li>Department of Defense</li><li>The NICNAS Nanotechnology Advisory Group (NAG)</li><li>Safe Work Australia</li><li>The Australian Pesticide and Veterinary Medicines Agency (APVMA)</li></ul>	<ul><li>A Review of Possible Impacts of Nanotechnology on Australia's Regulatory Frameworks, the Monash Report (July 2008)</li><li>The Australian Government Approach to the Responsible Management of Nanotechnology, July 2008</li><li>The National Nanotechnology Strategy (NNS) Annual Report 2007–08, Australian Office of Nanotechnology (Jan 2009)</li><li>In 2009 NICNAS introduce a regulatory reform of industrial nanomaterials</li><li>Safe Work Australia's Nanotechnology OHS Program to regulate the nanomaterials in occupational settings.</li><li>Code of Practice for Safety Data Sheets (SDS)</li></ul>
Austria	<ul><li>The Federal Ministry of Agriculture, Forestry, Environment and Water Management (BMLFUW)</li><li>Austrian Nano Initiative</li></ul>	<ul><li>The National Austrian Action Plan provides a careful analysis of risks and opportunities of nanotechnologies, focused on "particularly relevant to nanotechnology in social and political terms", health (including employee protection), environment, business, science, research, and development</li><li>Closely following the ongoing activities of EU with active participation in the REACH CASG Nano Working Group.</li></ul>

Canada	• Federal Departments of Health Canada (HC) • Environment Canada (EC) • Institut de Recherché Robert-Sauve en Santé et en Securite du Travail (IRSST) • Canadian House of Commons	• A regulatory framework for nanomaterials under the Canadian Environmental Protection Act 1999. • Best Practices Guide to Synthetic Nanoparticle Risk Management • The legislation on hazardous products (Hazardous Products Act administered by HA), a working group devoted to nanotechnology has been established within the Workplace Hazardous Materials Information System (WHMIS) • Possible amendment to the Canadian Environmental Protection Act that would introduce measures such as a pre-market review of all nanomaterials and nanoproducts and a public inventory of their use to the regulation of nanomaterials.
China	• National Steering Committee for Nanoscience and Nanotechnology (NSCNN) • Ministry of Environmental Protection (MEP) • State Administration of Work Safety (SAWS) • State Food and Drug Administration (SFDA) • Standardization Administration of China (SAC)	• Coordinate national N&N research • Inventory of Existing Chemical Substances Manufactured or Imported in China (IECSC). • 2003 regulations: Provision on the Environmental Administration of New Chemical Substances • Measures on Environmental Management of New Chemical Substances
European Union (EU)	• European Commission • Scientific Committee on Emerging and Newly Identified Health Risks (SCENIHR) • European Chemicals Agency (ECHA) • Competent Authorities Sub Group on Nanomaterials (CASG Nano) • European Medicines Agency (EMA) • European Group on Ethics in Science and New Technologies (EGE)	• Registration, Evaluation, Authorization and Restriction of Chemicals (REACH) • CLP, regulation 1272/2008 • Directive 2001/83/EC • Towards a European strategy for nanotechnology. COM(2004) 338 • Nanotechnologies Action plan COM(2005) 243 • Nanosciences and Nanotechnologies: An action plan for Europe 2005–2009. First Implementation Report 2005–2007 • Code of Conduct C(2008) 424 • Regulatory aspects of Nanomaterials, COM(2008) 366

(*continued*)

TABLE 14.3. (Continued)

Region/country	Regulatory bodies/agencies	Key legislations/activities relevant to evaluate and manage the risks associated with nanomaterials and nanoproducts
	• Working group on New and Emerging Technologies in Medical Devices (N&ET Working Group) • The European Food Safety Authority (EFSA) • EU Parliament (Committee on the Environment, Public Health and Food Safety) • The Cosmetics Directive • International Cooperation on Cosmetics Regulation (ICCR) • Restriction of Hazardous Substances (RoHS) Directive • Waste Electrical and Electronic Equipment (WEEE)	• Second Implementation Report 2007–2009 • Action Plan on Nanosciences and Nano technologies 2010–2015 • Medical Devices Directive • Report on Nanotechnology to the Medical Devices Expert Group: Findings and Recommendations • EU Novel Foods Regulation • Council Directive 76/768/EEC • Regulation (EC) 1223/2009 • EU Parliament Committee on the Environment, Public Health and Food Safety voted for a ban on nanosilver and long multi-walled carbon nanotubes (MWCNT)
France	• The French Agency for Environmental and Occupational Health Safety (AFSSET) • The French Food Safety Agency (AFSSA) • The French Health Products Safety Agency (AFSSAPS) • The French Research Ministry (Ministre de l'Enseignement Supérieur et de la Recherche)	• Following recommendations from various Government Committees and Agencies: • Évaluation des risques liés aux nanomatériaux pour la population générale et pour l'environnement, AFSSET, March 2010 • Opinion of the High Public Health Council to the French Ministry of Health: "Avis relative a la securite des travailleurs lors de l' exposition aux nanotubes de carbone" January 2009. • Assessing the toxicity of medicinal products containing nanoparticles, French Agency for the Safety of Health Products (AFSSAPS), Sept 2008. • Nanomaterials: Effects on the Environment and Human Health, French Agency for Environmental and Occupational Health Safety (AFSSET), July 2006. • "Nanotechnologies, Nanoparticles: Quels Dangers, Quels Risques?" – Ministere de l'Ecologie et du Developpement Durable, Comite De La Prevention et de la Precaution, May 2006. • Grenelle 2 Article 73, modification du Code de l'environnement • Nano-INNOV plan to develop a strategy for the innovation in the field of nanotechnologies

Germany	- The Nano-Initiative Aktionsplan 2010
- The Federal Institute for Occupational Safety and Health
- German Chemical Industry Association (VCI)
- The Federal Environmental Agency (UBA)
- The German Federal Government's Nanocommission
- The Federal Ministry for the Environment (BMU)	- Guidance for handling and use of nanomaterials at the workplace
- Health and environmental risks of nanomaterials – Research strategy
- The Actions Plan: dialogue on potential risks and opportunities of N&N
Global	- International Standards Organization (ISO) Technical Committee (TC) 229
- The International Electrotechnical Commission (IEC) TC 113	- Directs activities on nanotechnologies standards at the international level
- ISO/TR 12885 (Technical Report): Health and safety practices in occupational settings relevant to nanotechnologies.
- ISO/TS27687 (Technical Specification): Terminology and definitions for nano-objects: Nanoparticle, nanofibre and nanoplates
- ISO/PRF TS 80004-3 (Technical Specification): Nanotechnologies Part 3: Carbon nano-objects
India	- Department of Science and Technology (DST)
- Ministry of Science and Technology, National Nano Science and Technology Initiative (NSTI)
- National Institute of Pharmaceutical Education and Research (NIPER)
- The Bureau of Indian Standards (BIS)
- The Indian Council of Agricultural Research (ICAR)
- The National Environmental Protection Authority (NEPA)	- No such specific regulation to nanotechnology
- Complicated issue for regulatory activities at states (e.g., health) and national level (e.g., environment).
Italy	- National Institute of Occupational Prevention and Safety (ISPESL)
- Italian Workers' Compensation Authority (INAIL)
- National Institute of Health (The Istituto Superiore di Sanità or ISS)
- The national standardization body (UNI)
- WG Nanomaterials	- Technical Commission U22-Nanotechnologies
- White Paper on engineered nanomaterials and occupational health effects

(*continued*)

TABLE 14.3. (Continued)

Region/country	Regulatory bodies/agencies	Key legislations/activities relevant to evaluate and manage the risks associated with nanomaterials and nanoproducts
Japan	• The Ministry of Economy, Trade and Industry (METI) • The National Institute of Advanced Industrial Science and Technology (AIST) • The National Institute of Occupational Safety and Health Japan (JNIOSH) • The Ministry of Economy, Trade, and Industry (METI) • The Ministry of the Environment (MOE) • The Ministry of Health, Labour, and Welfare (MHLW)	• Under current regulatory system, the Chemical Substance Control Law obliges manufacturers to notify the government about nanomaterials • Established specific working groups dedicated to nanomaterials safety • Voluntary guidelines for the handling of nanomaterials • MHLW published nanotechnology guidelines
Norway	• The Research Council of Norway (NANOMAT) • The Climate and Pollution Agency • The Norwegian Labour Inspection Authority • The Norwegian Institute of Occupational Health • The Norwegian Institute for Agricultural and Environmental Research (BioForsk) • The Norwegian Board of Technology • The Climate and Pollution Agency	• Environmental fate and ecotoxicity of engineered nanoparticles • Nanomaterials, risk and regulation • The central register for chemical products
Switzerland	• The Federal Office for the Environment (FOEN) • The Swiss Federal Office of Public Health (SFOPH) • Swiss State Secretariat for Economic Affairs (SECO)	• Risk Assessment and Risk Management for Synthetic Nanomaterials 2006–2009 • The Precautionary Matrix • Guidelines for safe and sustainable disposal of nano wastes • Guidelines for the provision of safety information along the value chain, in particular to include information on nanomaterials in Safety Data Sheets (MSDS) • Review of the hazardous incident ordinance (HIO)

The Netherlands	• The Dutch Government • The Health Council of the Netherlands • The Risks of Nanotechnology Knowledge and Information Centre (KIR nano) • The Dutch National Institute for Public Health and the Environment (RIVM)	• Report published by the Dutch government's vision on nanotechnologies • Health significance of nanotechnologies • Coping rationally with risks, VROM, 2004 • The Dutch Action Plan on Nanotechnology • Participate in the CA subgroup of nanotechnology under the REACH authorities, OECD WPN, WPMN
Taiwan	• National Nanotechnology Programme (NNP) • Taiwan's Environmental Protection Administration (TEPA) • Safety and Health Technology Centre (SAHTECH) • Industrial Technology Research Institute (ITRI) • Taiwan Nanotechnology Industry Development Association (TANIDA) • Taiwan Nanotechnology Standards Council (TNSC) • The Bureau of Standards, Measurement and Inspection (BSMI) and ITRI	• All chemicals are regulated by the Council of Labour Affairs (CLA) • Toxic Chemical Substances Control Act • A strong lead from the EU's REACH • The Nanomark Certification System • TANIDA contributes to develop regulations to establish the Taiwan Nanotechnology Standards Council (TNSC) in conjunction with the Bureau of Standards, Measurement and Inspection (BSMI) and ITRI
UK	• Council for Science and Technology (CST) • Department for Environment, Food & Rural Affairs (Defra) • Health and Safety Executive (HSE) • Food Standards Agency (FSA) • Medicines and Healthcare Products Regulatory Agency (MHRA) • Centre for Business Relationships, Accountability, Sustainability and Society (BRASS) • National Nanotoxicology Research Centre • Nanotechnologies Leadership Group (NLG) • The British Standard Institute	• Characterizing the potential risks posed by engineered nanoparticles, a Second Government Research Report (December 2007) • HSE issued guidelines on the safe handling of carbon nanotubes • An Overview of the Framework of Current Regulations affecting Development and Marketing of Nanomaterials • Statement by the UK Government about Nanotechnologies (2008) • Supports EU initiatives; is promoting a 'case-by-case' approach to assessing the risk and suitable use of individual nanomaterials in food and food contact materials

(*continued*)

TABLE 14.3. (Continued)

Region/country	Regulatory bodies/agencies	Key legislations/activities relevant to evaluate and manage the risks associated with nanomaterials and nanoproducts
USA	• National Nanotechnology Initiative (NNI) • FDA • EPA • National Institutes of Health (NIH) • Occupational Safety and Health Administration (OSHA) • Consumer Product Safety Commission (CPSC) • National Institute for Occupational Safety and Health (NIOSH) • National Institute for Standards and Technology (NIST) • The Federal Nanotechnology Policy Coordination Group (NPCG)	• Toxic Substances Control Act (TSCA): Chemicals • Federal Insecticide, Fungicide and rodenticide Act (FIFRA): Pesticide • Clean Air Act (CAA), Clean Water Act (CWA), Safe Drinking Water Act (SDWA) • Comprehensive Environmental Response, Compensation, and Liability Act (CERCLA) • Toxics Release Inventory Program • Significant new use rules (SNUR) • EPA issued SNUR for two nanoparticles (73 FR 65743), i.e., siloxane-modified silica nanoparticles, PMN No. P-05-673 and siloxane-modified alumina nanoparticles, PMN No. P-05-687 • Pre-manufacturing notifications (PMN) for nanoscale materials • Safe Chemical Act of 2010 • FDA Nanotechnology Task Force • The Federal Food, Drug and Cosmetic Act (FFDCA) • The Public Health Service Act (PHA) • Nanomaterial Research Strategy (NRS), developed by Office of Research and Development (ORD) • In April 2012, issued draft (not for implementation): Guidance for safety of nanomaterials in cosmetic products

(Kowshik et al., 2002; He et al., 2007; Gade et al., 2008; Muckherjee et al., 2008; Nangia et al., 2009; Braunschweig et al., 2013; Manivasagan et al., 2013).

Certain molecules in biological systems are necessary for microbes to survive under different environmental conditions; the make-up of microbial cell walls is significant because they release different enzymes that can enable the synthesis of BNPs. Some recent studies indicate that nitrate reductase can be used to synthesize silver nanoparticles (Kumar et al., 2007; Gade et al., 2008; Srivastava et al., 2013). Another study found that *Rhodopseudomonas capsulate*, which secretes cofactor NADH and NADH-dependent enzymes, could be used to produce gold nanoparticles; the enzymes reduced gold ions to Au^0, converting them into nanoparticles (He et al., 2007). A later study by Nangia et al. (2009) examined the process of biosynthesizing gold nanoparticles in *Stenotrophomonas maltophilia* and stabilizing them through charge capping: Au^{3+} could be converted into Au^0 via an electron shuttle enzymatic metal reduction process using NADPH-dependent reductase enzymes. A mechanism that depended on stress protein response was suggested for CdS nanoparticles by Kowshik et al. (2002), who noted that the enzymatic complex phytochelatin-Cd was transported through the vacuolar membrane in *Schizosaccharomyces pombe* by ATP-binding cassette-type vascular proteins, and that this complex combined with vacuole sulfide to form a high-molecular-weight phytochelatin-CdS-2 complex or CdS nanocrystal. Muckherjee et al. (2008) studied nanoparticle encapsulation as a synthesis mechanism using the Raman spectrum; the hydrolysis of peptide linkages from amino acids synthesized free amino groups and carboxylate ions, which might act to encapsulate silver nanoparticles.

Intracellular and extracellular biosyntheses of BNPs have great potential as safer methods for creating nanomaterials compared to the common methods used today. However, some key mechanisms for biosynthesis have not yet been fully developed or studied. For example, the processes by which biomolecules transform ions into nanoparticles are not fully understood, and the stability of BNPs in different environments requires further study. The modulation of bio-molecules converting ions into nanoparticles under different mechanisms is yet to be researched. Other than the size and shape, the stability of BNPs under varying environmental conditions (i.e., pH, temperature, photo- and chemical-sensitivity, etc.) requires further attention. For a safer practice it is necessary to determine the role of microbial and fungal enzymes involved in bio-reduction/capping/encapsulation mechanisms in animal models prior to implementation in medical drugs and devices including food and feed industries. The *in situ* and *ex situ* roles of BNPs must be investigated prior to releasing their potential in environmental research. As for the regulatory framework, current laws can be used to assess the safety of BNPs, but they may need to be revised to accommodate widespread industrial use of these materials and distinguish BNPs from other ENMs.

14.6. CONCLUSION

The constant drive toward progress in science and technology demands improvement in our current methods for synthesizing nanomaterials, and biotechnology is a promising avenue for this improvement. Global efforts in unified legislative environment would assist corporate business, ensuring public concerns that regulations are not being compromised for the sake of profitability. Biotechnology can be used to produce great advances in agriculture, industry, environment, and health. The synthesis of BNPs in the field of nanotechnology is a promising alternative to the known and established physical and chemical methods. However, BNP synthesis requires further study before it can be applied on industrial levels. Global efforts to consistently regulate nanomaterials overall, as well as BNPs specifically, would reassure the public of their safety and help industry apply them in beneficial ways. Separate agencies or revised regulations may be necessary to control the use of nanotechnology in the future.

REFERENCES

Arora, S. 2012. Nanotoxicology and in vitro studies: The need of the hour. *Toxicology and Applied Pharmacology* 258: 151–165.

BAuA/VCI. 2007. Bundesanstalt für Arbeitsschutz und Arbeitsmedizin and Verband der Chemischen Industrie [Guidance for Handling and Use of Nanomaterials at the Workplace]. Available at: http://www.baua.de/en/Topics-from-A-to-Z/Hazardous-Substances/Nanotechnology/pdf/guidance.pdf?__blob=publicationFile (accessed 1 November 2014).

BERR (UK Department for Business, Enterprise & Regulatory Reforms) 2006. An Overview of the Framework of Current Regulation Affecting the Development and Marketing of Nanomaterials. Available at: http://orca.cf.ac.uk/44518/1/file36167.pdf (accessed 1 November 2014).

Braunschweig, J., Bosch, J., Meckenstock, R.U. 2013. Iron oxide nanoparticles in geomicrobiology: from biogeochemistry to bioremediation. *New Biotechnology* 30(6): 793–802.

Buzea, C., Pacheco Blandino, I.I., Robbie, K. 2007. Nanomaterials and nanoparticles: Sources and toxicity. *Biointerphases* 2(4): MR17–MR172.

CBER (Center for Biologics Evaluation and Research) 2013. CBER Offices and Division. Available at: http://www.fda.gov/AboutFDA/CentersOffices/OfficeofMedicalProductsandTobacco/CBER/ucm122875.htm (accessed 1 November 2014).

CDER (Center for Drug Evaluation and Research) 2013. CDER Offices and Divisions. Available at: http://www.fda.gov/AboutFDA/CentersOffices/OfficeofMedicalProductsandTobacco/CDER/ucm075128.htm (accessed 1 November 2014).

CDRH (Center for Devices and Radiological Health) 2013. CDRH Offices. Available at: http://www.fda.gov/AboutFDA/CentersOffices/OfficeofMedicalProductsandTobacco/CDRH/CDRHOffices/default.htm (accessed 1 November 2014).

CONANO. 2007. COmparative Challenge of NANOmaterials, Projektbericht, 2007, Vergleichende Nutzen-Risiko-Analysen von abbaubaren und nicht abbaubaren Nano-Delivery-Systemen sowie konventionellen Mikro-Delivery-Systemen in pharmazeutischen und kosmetischen Anwendungen. Available at: http://www.risiko-dialog.ch/Publikationen/Studien/87-Studien/311-CONANO-Dialogprojekt (accessed 20 October 2014).

des Rieux, A., Fievez, V., Garinot, M., Schneider, Y.J., Preat, V. 2006. Nanoparticles as potential oral delivery systems of proteins and vaccines: a mechanistic approach. *Journal of Controlled Release* 116(1): 1–27.

EFSA (European Food Safety Authority) 2008. Call for Scientific Data on Applications of Nanotechnology and Nanomaterials Used in Food and Feed, Deadline 28/03/2008. Available at: http://www.efsa.europa.eu/EFSA/efsa_locale-1178620753812_1178680756675.htm (accessed 20 October 2014).

FAO/WHO 2010. Expert meeting on the application of nanotechnologies in the food and agriculture sectors: potential food safety implications. Available at: http://whqlibdoc.who.int/publications/2010/9789241563932_eng.pdf (accessed 20 October 2014).

FDA. 2007 (not 2013). Nanotechnology Task Force 2007. Available at: http://www.fda.gov/ScienceResearch/SpecialTopics/Nanotechnology/UCM2006659.htm (accessed 20 October 2014).

FDA. 2011. Considering whether an FDA-regulated product involves the application of nanotechnology. Available at: http://www.fda.gov/RegulatoryInformation/Guidances/ucm257698.htm (accessed 20 October 2014).

FDA. 2012. FDA News Release April 20, 2012. FDA issues draft guidance on nanotechnology. Available at: http://www.fda.gov/NewsEvents/Newsroom/PressAnnouncements/ucm301125.htm (accessed 20 October 2014).

FDA. 2013a. Office of Combination Products. Available at: http://www.fda.gov/CombinationProducts (accessed 20 October 2014).

FDA. 2013b. Organization Charts. Available at: http://www.fda.gov/AboutFDA/CentersOffices/OrganizationCharts/ucm393155.htm (accessed 1 November 2014).

Gade, A.K., Bonde, P., Ingle, A.P., Marcato, P.D., Duran, N., Rai, M.K. 2008. Exploitation of *Aspergillus niger* for synthesis of silver nanoparticles. *Journal of Biobased Materials and Bioenergy* 2: 243–247.

Gatti, A.M., Tossini, D., Gambarelli, A., Montanari, S., Capitani, F. 2009. Investigation of the presence of inorganic micro and nanosized contaminants in bread and biscuits by environmental scanning electron microscopy. *Critical Reviews in Food Science and Nutrition* 49: 275–282.

Grobe, A., Renn, O., Jaeger, A. 2008. Risk Governance of Nanotechnology Applications in Food and Cosmetics. Report prepared by International Risk Governance Council, Switzerland.

Hamburg, M. 2012. FDA's approach to regulation of products of nanotechnology. *Science* 336: 299–300.

He, S., Guo, Z., Zhang, Y., Zhang, S., Wang, J., Gu, N. 2007. Biosynthesis of gold nanoparticles using the bacteria *Rhodopseudomonas capsulate*. *Materials Letters* 61: 3984–3987.

Holdren, J. 2011. Policy Principles for the U.S. Decision-Making Concerning Regulation and Oversight of Applications of Nanotechnology and Nanomaterials. Available at http://www.nano.gov/node/643 (accessed 20 October 2014).

ISO 2008. Technical Specification 27687, Nanotechnologies: Terminology and Definitions for Nano-objects: Nanoparticle, Nanofibre and Nanoplate. International Organization for Standardization.

Kowshik, M., Deshmukh, N., Kulkarni, S.K., Paknikar, K.M., Vogel, W., Urban, J. 2002. Microbial synthesis of semiconductor CdS nanoparticles, their characterization, and their use in fabrication of an ideal diode. *Biotechnology and Bioengineering* 78(5): 583–588.

Kumar, A.S., Abyaneh, M.K., Gosavi, S.W., Kulkarni, S.K., Pasricha, R., Ahmad, A., Khan, M.I. 2007. Nitrate reductase-mediated synthesis of silver nanoparticles from $AgNO_3$. *Biotechnology Letters* 29(3): 439–445.

Manivasagan, P., Venkatesan, J., Senthilkumar, K., Sivakumar, K., Kim, S.K. 2013. Biosynthesis, antimicrobial and cytotoxic effect of silver nanoparticles using a novel *Nocardiopsis* sp. MBRC-1. *Biomed Research International* 2013: article ID287638, doi: 10.1155/2013/287638.

Maynard, A. 2011. Don't define nanomaterials. *Nature* 475: 31.

Moghimi, S.M. 2012. Factors controlling nanoparticle pharmacokinetics: an integrated analysis and perspective. *Annual Review of Pharmacology and Toxicology* 52: 481–503.

Muckherjee, P., Roy, M., Mandal, B.P., Dey, G.K., Mukherjee, P.K., Ghatak, J., Tyagi, A.K., Kale, S.P. 2008. Green synthesis of highly stabilized nanocrystalline silver particles by a non-pathogenic and agriculturally important fungus *T. asperellum*. *Nanotechnology* 19: 103–110.

Nangia, Y., Wangoo, N., Goyal, N., Shekhawat, G., Suri, C.R. 2009. A novel bacterial isolate *Stenotrophomonas maltrophilia* as living factory for synthesis of gold nanoparticles. *Microbial Cell Factories*, doi: 101186/1475-2859-8-39.

Nel, A., Xia, T., Mädler, L., Li, N. 2006. Toxic potential of material at the nanolevel. *Science* 311: 622627.

NIOSH (National Institute for Occupational Safety and Health) 2007. Progress toward Safe Nanotechnology in the Workplace. Available at: http://www.cdc.gov/niosh/docs/2007-123/pdfs/2007-123.pdf (accessed 20 October 2014).

NNI (National Nanotechnology Initiative) 2013a. Nanotechnology 101: What's so special about the Nanoscale? Available at: http://www.nano.gov/nanotech-101/special (accessed 20 October 2014).

NNI (National Nanotechnology Initiative) 2013b. About the NNI: NSET's Participating Federal Partners. Available at: http://www.nano.gov/partners (accessed 20 October 2014).

NNI (National Nanotechnology Initiative) 2013c. About the NNI: NNI Vision, Goals and Objectives. Available at: http://www.nano.gov/about-nni/what/vision-goals (accessed 20 October 2014).

NNI (National Nanotechnology Initiative) 2013d. NNI Budget. Available at: http://nano.gov/about-nni/what/funding (accessed 20 October 2014).

Nyström, A.M. 2012. Safety assessment of nanomaterials: Implications for nanomedicine. *Journal of Controlled Release* 161: 403–408.

OECD WPMN 2010. Report on current developments/Activities in manufactured nanomaterials: Tour de Table. OECD Publications Series on the Safety of Manufactured Nanomaterials, Nanotechnology Industries Association. Paris, France.

Paradise, J. 2012. Reassessing safety for nanotechnology combination products: what do biosimilars add to regulatory challenges for the FDA. *Saint Louis University School of Law Journal* 56: 465–519.

Powell, J.J., Harvey, R.S.J., Ashwood, P., Wolstencroft, R., Gershwin, M.E., Thompson, R.P.H. 2000. Immune potentiation of ultrafine dietary particles in normal subjects and patients with inflammatory bowel disease. *Journal of Autoimmunity* 14: 99–105.

REACH (Registration, Evaluation, Authorisation and Restriction of Chemicals). 2013. Review of annexes. Available at: http://ec.europa.eu/enterprise/sectors/chemicals/documents/reach/review-annexes/index_en.htm (accessed 1 November 2014).

REACH (Registration, Evaluation, Authorisation and Restriction of Chemicals). 2014. The directive on dangerous substances. Available at: http://ec.europa.eu/environment/archives/dansub/consolidated_en.htm (accessed 1 November 2014).

Royal Society. 2004. The Royal Society and The Royal Academy of Engineering, Nanoscience and Nanotechnologies: Opportunities and Uncertainties, Royal Society Policy Document 19/04, London. Available at http://www.nanotec.org.uk/finalReport.htm (accessed 20 October 2014).

SAFENANO. 2013. Europe's Center of excellence on nanotechnology hazard and risk. Available at http://www.safenano.org/ (accessed 20 October 2014).

SCENIHR (Scientific Committee on Emerging and Newly Identified Health Risks) 2007. Opinion on the Scientific Aspects of the Existing and Proposed Definitions Relating to Products of Nanoscience and Nanotechnologies, adopted 29 November, 2007. Available at: http://ec.europa.eu/health/ph_risk/committees/04_scenihr/docs/scenihr_o_012.pdf (accessed 20 October 2014).

Schädlich, A. 2012. Accumulation of nanocarriers in the ovary: A neglected toxicity risk? *Journal of Controlled Release* 160: 105–112.

Schneider, J.C. 2007. Can microparticles contribute to inflammatory bowel disease: innocuous or inflammatory? *Experimental Biology and Medicine* 232: 1–2.

Simon, P., Joner, E. 2008. Conceivable interactions of biopersistent nanoparticles with food matrix and living systems following from their physocochemical properties. *Journal of Food and Nutrition Research* 47: 51–59.

Srivastava, P., Bragança, J., Ramanan, S.R., Kowshik, M. 2013. Synthesis of silver nanoparticles using haroarchaeal isolate Halococcus salifodinae BK3. *Extremophiles* 17(5): 821–831.

US Code. 2014. 21 US Code 331Prohibited Acts. Available at: http://www.gpo.gov/fdsys/pkg/USCODE-2010-title21/pdf/USCODE-2010-title21-chap9-subchapIII-sec331.pdf (accessed 20 October 2014).

VCI (Verband der Chemischen Industrie) 2008. Responsible Production and Use of Nanomaterials, Guidance Documents, March 2008. Available at: https://www.vci.de/Downloads/Responsible-Production-and-use-of-Nanomaterials.pdf (accessed 1 November 2014).

Wagner, V., Dullaart, A., Bock, A.-K., Zweck, A. 2006. The emerging nanomedicine landscape. *Nature Biotechnology* 24: 1211–1217.

Yamashita, T. 2012. Carbon nanomaterials: efficacy and safety for nanomedicine. Materials 5: 350–363.

NAME INDEX

Abhilash, 225–48

Bach, H., 273–93
Bansal, V., 9, 31–46, 59, 106, 112, 147, 160, 161, 170
Binod, P., 187–200
Bunge, M., 85, 205–19

Cao, B., 83–94
Colonna, T., 303–26

Dev, A., 155–79
Dixit, S., 141–9

Gudz, A., 123–35
Gupta, I., 1–19

Ingle, A., 1–19

Khare, S.K., 146, 255–67
Kriz, D.A., 123–35
Kuo, C-H., 123–35

Mackenzie, K., 205–19
Maliszewska, I., 1–19
Mishra, S., 141–9
Mohanty, A., 83–94

Ng, C.K., 83–94

Pandey, A., 187–200
Pandey, B.D., 225–48

Rai, M., 1–19, 77, 101, 115, 123
Ramanathan, R., 4–6, 31–46
Ramesh, A., 53–78

Sindhu, R., 187–200
Singh, O.V., 101–116, 303–26
Singh, S., 155–79, 282
Sinha, R., 255–67
Soni, S., 141–9
Suib, S.L., 123–35
Sundari, M.T., 53–78

Thirugnanam, P.E., 53–78

Vidyarthi, A.S., 155–79

Wanner, J., 101–16

Yadav, A., 1–19, 246

Zhang, J., 101–16

Bio-Nanoparticles: Biosynthesis and Sustainable Biotechnological Implications, First Edition.
Edited by Om V. Singh.
© 2015 John Wiley & Sons, Inc. Published 2015 by John Wiley & Sons, Inc.

SUBJECT INDEX

abiotically, 206
abiotic-stress, 267
absorption, 13, 42, 62, 64, 168–71, 178, 197, 213, 256, 263–4, 312
Acalypha indica, 149
accumulation, 2, 3, 33, 59, 61, 65, 67, 69, 71, 147, 164, 183, 191, 243, 256, 257, 262–4, 274, 276, 279, 280, 287, 293
acetate, 8, 62, 90, 112, 165, 166, 307
acetate permease (ActP), 62
Acetobacter xylinum, 130, 131
Acidiothiobacillus caldus, 236
Acidithiobacillus
  A. ferrooxidans, 234–6, 238, 239
  A. thiooxidans, 105, 106, 112
acid leaches, 93
Acidobacteria, 234
acidophilic, 236, 237, 239
acid resistance, 34
Acinetobacter sp., 4, 6, 188, 192
Acremonium, 38
Actinobacteria, 214, 234
Actinobacter sp., 4–6, 60, 160, 165
actinomycete, 2, 5, 10, 60, 108, 142, 146, 158, 161, 170–171, 243, 244, 247, 317
activated carbon, 164, 260
adaptation, 235, 236
adaptive immunity, 289
additives, 196, 199, 215, 304, 305
adenoviruses, 102
adherence, 283
adherent, 237
adhesion, 92, 287
adsorb, 247

adsorption, 57, 133, 167, 174, 193, 244, 256, 258, 264, 281, 282, 288, 292
adulterated, 314
Aedes aegypti, 166
aeration, 111, 234, 237
aerobic, 8, 70, 87, 89, 105, 111, 163, 165, 166, 192
aerobic conditions, 39, 70, 87, 89, 131, 133, 145, 163, 165, 166, 192
Aeromonas sp., 129, 130
aerotolerant, 242
aesthetics, 141
agglomerates, 217, 304
agglomeration, 54, 126, 168, 212, 217, 219, 247, 258, 281, 282, 285, 287, 314, 315
aggregate, 8, 38, 126, 129, 131–3, 135, 176, 213, 217, 242, 282, 285, 286, 288
aggregation, 13, 43, 134, 145, 176, 177, 190, 217, 219, 241, 262, 267, 273, 278, 281–3, 293, 312, 315
agitation, 167, 234
agricultural fields, 256
agriculture, 16, 18, 195, 257, 305, 312, 318, 326
agri-nanotechnology, 257
agrochemicals, 257
agucha mines, 227, 239
airflow
  measurement, 237
  rate, 115
airlift mode, 234
air permeability, 246
albumin, 286

Bio-Nanoparticles: Biosynthesis and Sustainable Biotechnological Implications, First Edition.
Edited by Om V. Singh.
© 2015 John Wiley & Sons, Inc. Published 2015 by John Wiley & Sons, Inc.

alcohol dehydrogenase propanol-preferring (AdhP), 191
alcohols, 149, 191
aldehyde, 62, 163
alfalfa, 57, 61, 67
algae, 2–4, 6, 12–16, 19, 33, 59, 60, 85, 86, 104, 106, 107, 124–6, 128, 145, 157, 159, 162, 165, 175, 189, 241, 247
alkaline medium, 146
alkaline metals, 230
alkaloids, 57
alkalothermophylic actinomycete, 5
alkalotolerent, 5
alkanes, 108
alkanethiol, 56
alkyl thiol, 129
allergies, 289, 305
Allium cepa, 258, 263
Aloe vera, 57, 143
α-and β-hopeite, 209
altered electrical charge, 278
altered mining, 229
Alteromonas putrefaciens, 4, 8
alumina, 199, 246, 264, 324
alumina nanoparticle, 199, 324
alveolar macrophages, 281, 287, 289
ambient, 40, 142, 149, 157, 158, 162, 170, 192–4, 206, 212, 229, 237
American National Institute of Occupational Safety and Health (NIOSH 2007), 303
amide group, 67
amine groups, 167, 168, 216
amino acid, 59, 65, 72, 109, 111, 126, 163, 170, 194, 262, 278, 284, 325
aminoglycosides, 90
3-aminopropyl trimethoxysilane, 54
amla, 57
ammonium sulfate, 168
amorphous gold, 67, 126
amorphous polysilic acid, 157
amorphous SeO nanospheres, 192
amorphous silica, 44–6, 158
amphiphilic, 37, 92
Anabaena, 14
anaerobic activity, 235
anaerobic condition, 39, 87, 131, 133, 145
Ananas comosus (L.), 149

anatase, 45, 259
anatomy, 256
angiogenesis, 17
angiogenic factor, 17
animals, 113, 255, 286, 304, 316, 325
animal systems, 255
anionic, 10, 35, 36, 43, 62, 147, 170, 216, 283, 289, 293
anionic hexafluorotitanate (TiF$_6^{2-}$), 36
anions, 9, 13, 54, 111, 170, 198, 208, 232, 277, 283
anisotrophic, 70
anisotropic shapes, 156
anodes, 90, 225
Anopheles subpictus, 166
anoxic zones, 89
anthraquinone pigments, 18
anthraquinones, 18, 66, 73
anthropogenic pollutants, 206
antibacterial, 12, 13, 87, 91, 92, 103, 168, 189–91, 194, 196, 197, 245–6
anti-biofilm, 90–94
antibiotic resistance, 91, 94
antibodies, 102, 189, 284, 289, 313
anticancer drugs, 189
anticancer medicines, 308
anti-hexon antibodies, 102
antimicrobial, 84, 86, 87, 90–92, 94, 126, 149, 162, 171–3, 196, 199, 212, 246, 273, 305, 308
antimony, 72, 164
antioxidant, 38, 73, 111, 196, 199, 256, 265, 274, 276, 278, 289
  antioxidant response element (ARE), 289
  enzyme, 38, 256, 265
antioxidative, 58, 73
antiseptic dressing, 190
antivirulence, 94
A7 phage, 173
Aphanocapsa sp., 13
apoptosis, 264, 275, 277–9, 284, 290–293
apoptosis regulator caspase-3, 292
apoptosis regulator protein, 291
apoptotic bodies, 290
apoptotic cells, 279, 284, 291
applications, 2, 6, 16–18, 31, 34, 36, 37, 39–41, 43, 46, 57, 58, 69, 78, 84–6, 93, 94, 101–3, 115, 116, 123, 124,

129, 134, 135, 141–9, 156, 165, 170, 178, 187, 188, 190–193, 196, 197, 199, 207, 210, 215–19, 225, 233, 237, 241, 242, 244–7, 255, 257, 263, 267, 273, 287, 292, 303–8, 310, 312, 314–17
aquaporin, 256, 262, 266
Aquaspirillum magnetotacticum, 131, 132
aquatic, 13, 61, 103, 157, 217, 255
aquatic organisms, 103
Aqueous
 dispersion, 281
 phase, 54, 205, 208
 stability, 244
aquifers, 242
Arabidopsis thaliana, 257, 260, 263, 266
Archaeoglobus fulgidus, 195
arginine, 278
aromatic and aliphatic amines, 168
aromatic rings, 206, 207
arsenate reductase, 8
arsenic, 230, 242
Arthrobacter globiformis, 188, 189
artificial additives, 305
ascorbate, 73, 74, 281
ascorbic acids, 58
aspartate, 173
Aspergillus
 A. aeneus, 160, 170, 188, 194
 A. flavus, 10, 19, 60, 67, 71, 106, 111, 113, 143, 147, 160, 168, 169
 A. foetidus, 11
 A. fumigatus, 9, 143, 148
 A. niger, 9, 10, 42, 191, 233
 A. oryzae, 188, 195
athalassohaline, 107
atomic force microscopic, 70
ATP-binding cassette-type vacuolar membrane protein, 171
ATP-dependent pumps, 59
ATP sulfurylase, 66, 109, 110
ATP synthase, 111, 291
Au(III)-chloride, 145
Au(III) ions, 145
Au(III) reductase, 124
auric chloride, 162
Austrian Federal Ministry for Transport, Innovation, and Technology, 315
Au (I)-thiosulfate, 145

autothermally, 237
Azadirachta indica, 143, 148

Bacillus
 B. cereus, 70, 86, 160, 166, 192
 B. coagulans, 42
 B. licheniformis, 4, 6, 60, 86, 106, 113, 114, 130, 143, 144, 159, 162–4
 B. megatherium, 129, 130, 159, 162
 B. selenitireducens, 4, 8, 60, 106, 107, 130, 131
 B. subtilis, 2, 4, 60, 69, 112, 124, 125, 143–5, 159, 160, 162, 166, 188–90, 193, 241
 B. thuringiensis, 166
bacterial cellulose (BC), 130–134
bacterial S-layers, 176
bactericidal, 91, 92, 94, 189, 190
bacteriophage, 92, 161, 173, 174
bacteriostatic, 245
ball milling method, 244
bandage, 190
barium carbonate ($BaCO_3$), 42
barium oxide (BaO), 40, 42
barium titanate (BT or $BaTiO_3$), 40–42, 46
bark, 149
barley plants, 262
barrier, 211, 257, 276, 290, 310–317
bath smelting technologies, 228
batteries, 232, 242
BC *see* bacterial cellulose (BC)
benzoquinones, 73
β-lactamase, 91
bicarbonate, 208
bimetallic, 5, 16, 88, 107, 148, 168, 169, 207
 alloys, 5, 107
 nanomaterials, 88
 silver and gold nanoparticles, 168
Binani Zinc Ltd, 227
binding receptor-ligand, 285
bioaccumulation, 65, 147, 256–8, 260, 261
bioavailability, 128, 242, 308, 310
biocatalysis, 84
biocatalytic processes, 59
biochemical, 6, 19, 69, 177, 194, 228, 246, 247, 293
biocidal, 87, 190
biocidal activity, 87

biocides, 84, 90, 194
biocompatibility, 19, 76, 188, 196, 245, 292, 293
biocompatible, 12, 46, 55, 74
biocomposites, 32
bioconjugate, 18
bioconjugation, 282
biocorrosion, 91
biodegradability, 207, 293
biodegradable, 142
biodiversity, 2, 266
bioengineered palladium nanoparticles, 217
biofactors, 109
biofactory, 192
biofilm, 8, 83–94, 135, 173, 197, 242
biofouling, 91, 92, 210
biogenerated, 214
biogenesis, 8, 58, 61, 70, 158, 166, 167, 169–71, 176, 177, 192, 195
biogenic, 7, 18–19, 40, 41, 65, 77, 88, 89, 158, 168, 170–173, 187, 189, 191, 194, 206, 243, 245
biogenically, 133, 198
biogenic hydroxyapatite (HAP), 245
biogenic nanoparticles, 18–19, 89, 243
biohydrometallurgy, 233–9
bioimaging, 188
bio indicators, 267
bioinorganic, 33
biokinetics, 293
bioleaching, 34, 42–6, 144, 214, 233–9, 241, 242, 247
biolipid tubules, 176
biologically controlled mineralization (BCM), 195
biologically induced mineralization (BIM), 195
biologics, 313
biomagnetite, 193
biomagnification, 267
biomarkers, 102, 197
biomass, 2, 3, 11–13, 43, 44, 57, 108, 110, 111, 114, 129, 145–7, 149, 163, 164, 167–9, 171, 172, 189, 192, 195, 215, 216, 245, 257, 259, 261, 265
biomaterials, 92, 189, 191, 308
bio-medical, 16, 78, 85, 93, 123, 187, 190, 195, 196, 199, 241, 244–6, 309
biomedicine, 36, 39, 146, 178, 188

biomembranes, 108
bio milling, 34, 44–6
biomimetic prosthetics, 103
biomimic, 245
biomineralization, 7, 11, 61–2, 124, 157, 158, 198, 247
biomining, 214
biomolecules, 32, 41, 69, 84, 148, 174, 195, 196, 199, 217, 275–81, 284, 293, 310, 325
bionanofactories, 74, 75, 147, 156
bio nanoparticle (BPN), 167, 168, 176, 177, 179, 317–26
bionanotechnology, 83–94, 115
bioorganic, 197
biooxidation, 214, 237, 239
bio oxides, 37
bioPalladium, 213, 214, 217–18
bioprocesses, 93, 94, 176, 236, 246
bioprocessing, 236, 246
bioproduction, 131
bioreactor, 62–5, 167, 176, 198, 238, 239
biorecovery, 216
bioreduction, 66–8, 76, 90, 103–5, 108, 109, 115, 126, 144, 157, 162, 164–6, 168, 171, 172, 174, 176, 177, 215, 219, 325
bioreductively, 129
bioremediation, 39, 144, 158, 194, 214, 241, 242, 247
bioresidues, 216
bio-semiconductor, 134
biosensing, 84, 188, 245
biosensors, 16, 31, 69, 85, 243
bioseparation, 199
bio-sequestration, 108
bio-silica protein, 72
biosilicates, 157
biosilicification, 36
biosorbents, 216
biosorption, 65, 104, 126, 158, 162, 167, 215, 216
biosphere, 104
biosupported generation, 215
biosynthesis, 5, 9, 10, 33–46, 53–78, 84, 87–90, 101–16, 123–35, 141–9, 158, 162–4, 168, 169, 172, 176, 177, 179, 187–200, 212, 215–16, 219, 241, 244, 246–8, 317–25

biotechnology, 3, 12, 63, 77, 84, 90, 142, 165, 193, 213, 247, 308, 325, 326
bioterrorism, 245
biotransformation, 3, 44–6, 158, 165, 166, 243, 264–5
bismuth manganese oxide (BiMnO$_3$), 45
bismuth nitrate [Bi(NO$_3$)$_3$], 37, 61
bismuth oxide (Bi$_2$O$_3$), 37, 38
bismuth oxychloride (BiOCl), 45
7-Bisphosphatase, 111
bisulfide, 210
black soot, 199
blast furnace, 230–231
blood–brain barrier, 290, 310
blood vessel, 17, 102, 286
body fluids, 312
borohydride, 55, 56, 149, 211, 281, 282
borosilicate glass, 44
bottom-up, 42, 142, 206
breathability, 246
Brevibacterium casei, 143, 160, 165
Brevis LMG 11437, 60
broad spectrum, 92, 199, 290
brookite, 36, 37, 45, 46
brushite, 245
Brust–Schiffrin, 55–7
Bryophyllum sp., 60, 73
buffering minerals, 229
building blocks, 101
bulk counterparts, 156, 255
buttermilk, 5, 107, 127, 145, 162

cabbage, 256, 257
cadmium chloride (CdCl$_2$), 107, 134, 164
cadmium–glutathione complexes, 111
cadmium sulfide (CdS), 4, 7, 15, 28, 33, 48, 60, 61, 68, 86, 106, 107, 112, 132, 134, 159–61, 164, 169, 173, 174, 180, 185, 192–4, 197–9, 240, 241, 325
cadmium telluride (CdTe), 72, 86, 197
cake layers, 263
calcination, 43–5, 228
calcite, 234
calcium, 2, 33, 157, 158, 228, 230, 245, 278
calcium carbonate, 33, 157, 230
Calothrix, 14
Campylobacter jejuni, 194
Canadian mines, 233

cancer, 17, 19, 102, 146, 188, 189, 191, 196, 291, 305–7
Candida
  C. albicans, 92, 189
  C. bombicola, 282
  C. glabrata, 60, 161, 171, 241
  C. guilliermondii, 161, 173
capillary forces, 262
capping, 11, 19, 32, 39, 42, 56, 57, 63, 65, 66, 115, 129, 142, 147, 162, 167, 168, 170, 192, 217, 243, 282, 325
Capsicum annum, 60
capsid, 161, 173, 174
carbohydrates, 108, 258
carbonaceous supports, 206, 207
carbonates, 33, 34, 59, 157, 158, 170, 208, 226, 230, 240, 242
carbonic anhydrase, 111
carbon nanotubes, 92, 256, 262, 304, 310, 311, 317, 320, 323
carbon sources, 111
carbonyl, 13, 58, 148, 278
carboxykinase, 73
carboxylate, 13, 16, 67, 109, 194, 264, 325
carboxylic, 71, 162
carcinogenic, 218
carcinoma cell line HCT116, 291
cardiovascular diseases, 305
cargo, 273, 285
carotenoids, 162, 265
carrier
  enables, 209
  proteins, 88, 256, 262
carrot, 257
carvacrol, 76
caryophyllene, 76
caspase, 264, 291, 292
castings, 225
catalysis, 13, 16, 18, 31, 37, 86, 88, 123, 190, 193, 206
catalyst, 16, 36, 55, 85, 86, 88, 165, 192, 193, 196, 205–19, 243, 281
catalytic, 17, 37, 87–9, 93, 156, 165, 206, 208–10, 212, 214, 216–19, 240, 244
catalyzed, 3, 157, 164, 167, 212, 218, 242, 274
catalyze hydrodehalogenation, 207
catechol, 74
catheters, 92

cathodic, 232
cationic detergent, 282
cationic protein, 10, 35–7, 43, 44, 46, 72, 147, 170
cations, 54, 58, 106, 163, 208, 281, 283
cation-transporting P-type ATPase, 106, 163
caves, 104
C_{70}-carboxyfullerenes ($C_{70}(C-(COOH)_2)_{2-4}$), 264
$CdCl_2$ *see* cadmium chloride ($CdCl_2$)
CdS *see* cadmium sulfide (CdS)
cell, 2, 37, 58, 66, 70, 84, 102, 124, 142, 157, 166, 190, 193, 212, 216, 235, 238, 255, 257, 274, 276, 288, 290, 291, 308
cell–cell communication, 92, 115
cell-free extract (CFE), 8, 129, 130, 166
cellular, 5, 16, 37, 59, 62–5, 110, 172, 176, 179, 190, 196, 197, 212, 216, 217, 228, 238, 243, 247, 248, 256, 262, 263, 274–81, 283–9, 291, 293, 312, 313
cement, 228, 232
cementation stage, 231
central nervous system, 17, 308
ceramic method, 240
ceramic steel, 34
ceria (cerium oxide or CeO_2), 37
cerium (Ce^{4+} and Ce^{3+}), 38
cerium carboxylates, 264
cerium nitrate [$Ce(NO_3)_3$], 38
cerium (III) nitrate hydrate ($CeN_3O_9.6H_2O$), 170
cerium oxide, 37, 170, 196
cetrimonium bromide (CTAB), 166, 282
CFE *see* cell-free extract (CFE)
21CFR-210, 312
21CFR-211, 312
Chaetoceros calcitrans, 12, 14
chalcopyrite, 233, 238
challenges, 6, 7, 9, 18, 36, 40, 42, 72, 83–94, 103, 114, 115, 143, 170–173, 176, 212, 240, 242, 243, 313, 315, 316
chaperons, 278
characteristics, 59, 101, 142, 143, 168, 170–172, 177, 189, 234, 245, 267, 273, 281–4, 286, 292, 293, 311, 312, 314

characterization, 12, 18, 133, 169, 176–7, 273, 311
charge-stabilization, 217
chat (chert) pile rock, 234
CHCs *see* Chlorohydrocarbons (CHCs)
checkpoint related genes, 280
chelation, 59
chemical precursors, 34–42, 46
chemiosmotic, 59
chemisorbed, 198
chemistry, 1, 19, 35, 46, 78, 115, 145, 156, 169, 175, 178, 189, 191, 206, 212, 217, 243, 247, 282, 288
chemo-heterotrophic, 105
chemokines, 289
chemotactic signals, 285
Chenopodium album, 61
chitosan, 282
Chlamydomonas reinhardtii, 110
chlorauric acid ($HAuCl_4$), 145
Chlorella
  C. salina, 12, 14
  C. vulgaris, 13, 191, 241
chloride, 2, 5, 112, 131, 145, 162, 208
chlorinated, 207, 210
chloroaurate ions, 11, 162
chloroauric acid ($HAuCl_4$), 12, 55, 65, 72–4, 107, 126, 145, 148, 162, 167, 169, 172
chlorobenzene, 206, 207, 209
Chlorococcum humicola, 13
chlorohydrocarbons (CHCs), 206–8
chlorohydrocarbons trichloroethene (TCE), 207–9
chlorophyll, 256, 258–60, 265
cholesterol, 215
chromatin, 290
chromatography, 171, 246
chromium, 230
chromosomal, 235, 258, 263, 280
chronic infections, 91
Cinnamomum
  C. camphora, 61, 143, 148
  C. zeylanicum, 60
circuit, 102, 103, 233, 237
circulating fluidized bed reactor, 228
circulatory system, 292
Cissus quadrangularis (CQE), 73
citrate, 54, 149, 215, 260, 281, 282

citric acid, 233
Citrobacter braakii, 188, 193
Cladosporium cladosporioides, 10, 60, 113
clarified, 115
clathrin, 257
cleaner, 317
clean waters, 208, 324
clinical studies, 314
clinoptilolite-rich rock, 246
Clostridium
  C. teurianum, 4, 8
  C. thermoaceticum, 4, 7, 61, 159, 164, 197, 241
cluster formation, 282
clusters, 108, 128, 157, 207, 261, 281, 282, 285, 288
coagulation, 57
coalmine soil, 70, 166
coarser fractions, 234
coatings, 34, 58, 92, 193, 196, 197, 199, 210, 212, 237, 245, 281, 284, 293
cobalt, 4, 7, 165, 166, 232, 233
coccal, 61
Cochliobolus lunatus, 188, 191
co-enzymes, 57
coercive force, 146
Colletotrichum sp., 11, 60, 160, 167, 241
colloidal, 11, 70, 142, 189, 244, 282, 306
colloidal gold, 5, 65, 145, 307
colloids, 5, 7, 56, 65, 69, 74, 75, 133, 169, 207, 209
column, 232, 233, 237, 238
combustion sol-gel, 240
comet assay, 280
commercial interest, 111
communities, 83, 93
Comparative Challenge of Nanomaterials (CONANO), 316
compartments, 157, 215, 243, 277, 280, 288
complement molecules, 284
complexation, 65, 108
complex interactions, 310
composition, 5, 31, 111, 133, 142, 157, 163, 177, 178, 187, 198, 228, 240, 242, 243, 247, 256, 267, 281, 285, 293, 306–8, 312
compounds, 2, 15, 57, 58, 110, 111, 115, 146, 149, 156, 162, 164, 189, 196, 205–19, 230, 240, 242, 246–8, 262, 275, 277, 282, 310
concrete, 228
condensation, 31, 104, 173, 290
confocal imaging, 264
confocal microscopy, 169, 288
conifers, 44
consortium, 165, 239
consumer, 303, 305, 310, 311, 313, 317, 324
consumer products, 310, 324
contaminants, 18, 84, 176, 206, 207, 210, 212, 216–18, 227, 247
contaminated, 207, 214, 227, 242
continents, 315
continuous stirred tank reactor (CSTR), 234
copolymer, 292, 307
copper, 6, 7, 10, 42, 105, 188, 195, 225, 231, 233, 234, 237
  aluminate, 41, 46
  oxide, 7, 41
  oxide nanoparticles (NPs), 195
coprecipitation, 146, 240
coriandrum, 57
Coriolus versicolor, 60, 113, 190
corporate business, 326
corrosion, 165, 247
Corynebacterium sp., 4, 6, 112, 125, 126, 159, 163
cosmetics, 36, 58, 187, 193, 199, 255, 304, 314–16, 320, 324
cost efficiency, 218, 243
cotton fabrics, 190, 246
covalent bonds, 164, 282
cowpea chlorotic mottle, 161, 174
C-phyotoerythrin (C-PE), 15
cristobalite polymorphic form, 43
Crohn's disease, 311
crop plant, 257
Crossandra infundibuliformis, 60
cross-linked reaction, 278
crystalline, 5, 9, 10, 35, 36, 43–5, 106, 107, 124, 148, 163, 170, 173, 174, 231
crystalline structure, 45, 107
crystallinity, 44, 45, 177, 281, 293, 312
crystallization, 6, 245
crystallographic, 191, 245

crystals, 5–7, 17, 38, 45, 54, 61, 62, 64, 124–8, 131–3, 145, 173, 198, 241
CTAB *see* cetrimonium bromide (CTAB)
C-terminus tyrosine, 65
c-type cytochrome, 8, 87, 165
cubic, 125–7, 145, 165, 173
cubic phase (δ-Bi_2O_3), 37, 40, 41
cubo-octahedral, 6, 11, 39, 160, 170
cucumber, 256, 257, 264
cultivation, 214
culture media, 163, 197, 200, 243, 285
cuprates, 240
Cupriavidus metallidurans, 105
Cupriavidus necator, 85, 216
curves, 312
cuticular, 256, 263
cyanobacteria, 13–15, 36, 67–8, 126, 127, 131, 145
Cylindrotheca fusiformis, 72
cyperquinone (type I), 73
Cyperus sp., 60, 73
cysteine, 63, 66, 68, 109, 110, 275, 276, 281, 284
  desulfhydrase, 7, 66, 68, 109, 110, 134, 164, 176
  hydrochloride, 164
  residue amines, 147
  residues, 167, 278
cysteine-aspartic proteases, 264
cytochrome C, 291
cytochrome P450 oxidases, 72
cytocompatibility, 282
cytokines, 289–90
cytoplasm, 10, 68, 72, 125, 128, 134, 159, 160, 163, 164, 166, 287, 288, 290
cytoplasmic hydrogenase, 165, 216
cytoplasmic vesicular compartments, 243
cytoskeleton, 285, 287
cytosol, 146, 171, 278, 285
cytotoxic, 171, 280, 282, 292–3
cytotoxicity, 214, 264, 266, 279, 289–93
cytotoxicology, 273–93

Dandi mineral processing plant, 231
Daphnia magna, 103
Datura metel, 60
deactivation, 209, 210, 212, 219
dechlorination, 85, 165, 210, 218
decomposers, 104

decomposition, 31, 240, 276, 317
decorative purposes, 145
deep sediment, 105
deep-tissue imaging, 196
defensins, 91
defluorination, 207
defoliation, 264
dehalogenation, 85, 193, 206–8
dehydration, 245
dehydroascorbate (DHA), 73–4
deiodination, 207
delafossite mineral, 41
deletions, 278, 280
Deltaproteobacteria, 198
denaturation, 284
dendrimers, 18, 172
dendritic cells, 287
dengue vector, 166
de novo carbonyl group, 278
dental materials, 199
deodorants, 199
Department for Business, Enterprise and Regulatory Reform (BERR) hazardous, 317
depletion, 110, 199, 288
deposits, 38, 128, 226, 228, 229, 242, 278
dermal penetration, 314
desert, 105
desiccation, 84
Desmodium trifolium, 60
desorption, 57
destabilize, 105, 124, 281
Desulfobacteriaceae sp., 132, 135, 241, 242
Desulfosporosinus sp., 4, 8
Desulfovibrio
  D. desulfuricans, 4, 6, 60, 85, 106, 107, 125, 128, 129, 146, 159, 165
  D. fructosivorans, 216
  D. magneticus, 195
Desulfuromonas acetoxidans, 195
detoxification, 87, 108, 111, 115, 124, 163, 175, 205–19, 289
devices, 40, 46, 60, 62, 64, 84, 92, 102, 104, 123, 165, 274, 292, 303, 313, 320, 325
diabetes, 305
diagnosis, 78, 102, 156, 188, 199, 283
dialkyl disulfide (RSSR), 56
dialysis, 109, 168

diameter, 3, 8–10, 35, 36, 38, 43, 44, 107, 111, 128, 131, 133, 163, 173, 174, 196, 230, 237, 242–4, 257, 280–282, 285, 287, 292
diatoms, 33, 36, 72, 157
diatrizoate dehalogenation activity, 193
dicots, 44, 256, 257, 260–261
dielectric candidate, 40
diesel, 108, 199
dietary supplements, 305
dietchequinone (type II), 73
di-2-ethylhexyl phosphoric acid (D2EHPA), 232
differential scanning calorimetry (DSC), 40, 245
diffusion, 66, 90, 93, 109, 110, 127, 231, 235, 276, 281, 287, 317
digesting enzymes, 285
3, 4-dihydroxy-L-phenylalanine (L-DOPA), 72, 73
dimensions, 2, 9, 17, 31, 44, 53, 70, 105, 141, 156, 166, 178, 187, 303, 311
Dioscorea bulbifera, 61
dipeptide, 65
discharge, 66, 242, 317
dismutation, 274
dispersion, 54, 177, 244, 281, 282, 293, 306
disposable slag, 230
dissimilatory, 8, 39, 108, 124, 165, 213, 216
dissolution, 57, 70, 226, 232–9, 285
dissolved organic matter, 217
dissolved oxygen, 89
distribution, 5, 44, 57, 87, 92, 129, 142, 176, 177, 190, 191, 195, 206, 215–17, 219, 234, 246, 281, 282, 285, 286, 312, 314
diversity, 1–19, 103, 124, 131, 187–200, 234–5, 242, 284
DNA, 58, 59, 67, 91, 133, 146, 173, 176, 193, 197, 235, 263, 264, 275, 278–81, 291, 292, 310, 312
dodecanethiol, 55
dolomite, 231
dose, 17, 197, 257, 259, 260, 290, 311, 312
dosimetry, 314
downregulated, 266, 280
downstream process, 103

downstream processing, 2, 103, 163, 164, 167, 176, 177, 179, 212, 243
draft guidance document, 314
draft tube fluidized bed bioreactor (DTFBB), 238
drink, 17, 206, 247, 311, 324
drug delivery, 17, 31, 58, 78, 84–6, 94, 102, 146, 188, 189, 199, 273, 292, 305–8
drugs, 17, 94, 189, 305, 310, 312–15, 325
dry cells, 225
dual peptide virus, 173
duckweeds, 257
Dunaliella salina, 12, 14, 191
dust, 230–231, 234
dyes, 244
dynamic light scattering (DLS), 177
dysregulation, 280

Earth, 157, 225
Eclipta
  E. alba, 74
  E. prostrate, 60
eco-friendly, 142, 148, 187, 190, 192, 193, 195, 199, 238, 247
ecological, 199, 234, 242
economic, 2, 34, 59, 115, 148, 176, 188–90, 192, 200, 220, 234, 238, 243, 246, 257, 309, 317
ecosystem, 104, 142, 227
educational resources, 309
effective drugs, 313
efficacy, 92, 179, 246, 315
efficiency, 38, 197, 218, 231, 232, 236, 244, 247
effluents, 237, 242, 256
efflux, 5, 59, 65, 108, 163, 238
electrical conductance, 303
electrical double layer, 283
electric arc, 228, 230, 234
electric charge, 285
electric circuits, 103
electrobioleaching, 239
electrochemically active biofilms (EABs), 90
electrochemical methods, 235
electrochemical synthesis, 215, 317
electrochemistry, 126
electrode, 90, 235, 245, 308
electrokinetic potential, 282

electroleaching, 232
electroluminescent, 197
electrolysis, 215, 226, 232, 239
electrolyte, 37, 196, 231, 259, 265, 277
electrolytic process, 232
electron
  acceptor, 8, 65
  carrier, 5, 70, 88
  conduit, 62
  donor, 3, 6, 8, 65, 68, 89, 90, 108, 109, 124, 129, 146, 164, 165, 193, 210
electronics, 17, 18, 31, 37, 40, 41, 85, 103, 141, 142, 156, 165, 173, 187, 190–192, 199, 240, 241, 255, 281
electronic wastes, 93
electron microscopy, 264
electron shuttle, 66, 71, 109, 143, 167, 325
electron transport chain (ETC), 63, 84, 91, 291
electrooxidation, 239
electrophilic molecules, 277
electrophoresis, 71, 169, 280
electroplating, 242
electrostatic, 15, 43, 69, 71, 72, 105, 108, 114, 167, 176, 194, 284, 288, 289
  interaction, 43, 69, 71, 72, 105, 108, 167, 289
  repulsive interaction, 282
electrowinning (EW), 232, 237, 246
elemental, 8, 61, 67, 68, 70, 72, 105, 109, 111, 124, 128, 129, 131, 135, 145, 165, 166, 192, 238, 239, 281
elements, 43, 53, 107, 131, 133, 142, 158, 176, 198, 225, 230, 289
ellipsoidal, 160, 168
Elodea canadensis, 257
Emblica officinalis, 143, 148
emerging technologies, 312, 317, 320
emodin, 73
encapsulation, 282, 293, 304, 325
endocytic, 257, 262, 285
endocytosis, 72, 256, 257, 262, 284, 286–8
endocytotic pathway, 285
endogenous oxygen radicals, 279
endophytic fungus, 11, 147, 167
endoplasmic reticulum (ER), 266, 278
  lumen, 278
  stress-selective markers, 278
endosomal membrane, 72

endosomes, 72, 285, 286
endothelial cells, 17, 286, 288, 290
energy production, 217, 277, 291
engineered microbes, 179, 198–9
engineered nanoscale materials (ENMs), 304, 305, 310–312, 315, 317, 325
engineering column, 237
ENMs *see* Engineered nanoscale materials (ENMs)
Enterobacter cloacae, 4, 6, 8, 69, 125, 128, 159, 163, 191
Enterobacteriaceae, 4, 6, 69, 109, 163
Enterobacter sp., 146
environmental conditions, 116, 196, 248, 325
Environmental Protection Agency (EPA), 103, 256, 309, 312, 324
enzymatic process, 146, 171, 187
enzymes, 5, 11, 13, 16, 34, 38, 43, 59, 69, 71–3, 76, 103, 105, 108, 109, 111, 116, 124, 126, 129, 131, 134, 142–4, 146, 147, 164, 167–9, 171, 176, 179, 194, 195, 198, 215, 216, 243, 244, 262, 265, 275, 277, 279–81, 285, 286, 289, 290, 307, 311, 317, 325
enzymes hydrogenase, 109, 165
eosinophil activation, 289
Epicoccum nigrum, 11
epidermal cells, 263
epidermal growth factor (EGF), 290
epithelial cells, 197, 287, 288, 290, 310
epithelial colorectal adenocarcinoma Caco-2 cells, 284
EPS *see* polymeric substances (EPS)
eroded, 256
Eruca sativa, 266
erythromycin, 246
Escherichia coli, 4, 6–9, 12, 13, 60, 61, 64, 67, 69, 85–8, 92, 106, 107, 125, 126, 128, 143, 159, 163, 164, 168, 188, 190–193, 196–9, 216, 241, 245
ethidium bromide, 280
ethylation, 59
eucalyptus hybrid, 60
EU Food Law Regulation, 317
eukaryotes, 149, 158, 171, 177, 242
European Food Safety Authority (EFSA), 316, 320

European Union (EU), 316, 319
evaluate, 191, 196, 234, 246, 257, 292, 312, 313, 318, 320, 322, 324
excess, 41, 56, 227, 237, 238, 264
exocytosis, 286
exogenous, 146, 165
exogenous formate, 165
exothermic, 40, 237
ex-situ, 325
extermination, 115
external energy, 57
extracellular, 2, 7, 8, 10, 11, 14–16, 32, 34, 35, 37, 39, 40, 42, 43, 59, 67–77, 83, 87–90, 104, 106, 108–10, 114, 115, 124–34, 142–8, 157, 159–65, 167–71, 173, 174, 176, 177, 189–95, 198, 215, 243, 244, 274, 290, 325
extracellularly, 8
extracellular proteins, 19, 39, 71, 169, 244
extraction, 87, 124, 164, 226, 228–34, 236, 238, 246
extreme environments, 103
extremophiles, 101–16
extremophilic actinomycete, 146, 170

fabrication, 3, 32, 103, 104, 144, 170, 171, 173, 177, 281–9, 292
fabrics, 190, 199, 246
facultative aerobic, 105
Farciminis LMG 9189, 60
fat, 305
fauna, 267
favored, 172, 239
FCC lattice, 164
FDA-regulated products, 313
fed-batch, 93
federal agencies, 313
Federal Food, Drug, and Cosmetic Act (the FDC&C Act), 314
feed, 230, 316, 325
feldspar, 231
Fenton reactions, 57, 275, 276, 279, 281
fermentation, 103, 111, 112, 114, 115, 244
fermentor, 234
Fermentum LMG 8900, 60
ferredoxin NADP+ reductase, 111
ferric ions, 61, 133, 238
ferric iron reduction, 109

ferric sulfate, 239
ferricyanide $[Fe(CN)_6]^{3+}$, 39
ferrimagnetic iron sulphide, 198
ferrimagnetic transition, 39
ferrimagnetism, 209
ferrocyanide, 39, 165, 170
ferroelectric, 40, 41
ferromagnetic alloys, 174
Ferroplasma sp., 105
ferrous/ferric ions, 61, 133, 238
fertilizer, 227
fetal calf serum (FCS), 282
fibroblast, 277, 278, 285, 291
fibrosarcoma cells, 277
fibrosis, 310
filamentous cyanobacteria, 145
filamentous fungus, 168, 169
filler, 228
film(s), 92, 147, 173, 197, 207, 210, 218
filtration, 103, 114, 199
finished product, 314
flagella, 59
flame cyclone, 228
flammable, 149
flat panel displays, 199
flavonoids, 57, 76
flexibility, 105
flora, 267
flotation, 226, 228, 229, 231–3, 237, 239, 246
flowering, 257
fluorescence, 171, 196, 244, 280
fluorescent, 54, 55, 93, 102, 196, 243, 244
fluorinated compounds, 206
fluorophores, 196
fluor-tagged transferrin, 288
foliar uptake, 263
Food and Drug Administration (FDA), 309, 311–15, 319, 324
Food Standards Agency (FSA), 316, 317, 323
foreign particles, 289
formaldehyde, 282
formate, 6, 129, 165, 213
formic acid, 210
forward osmosis (FO) membranes, 92
Fourier transform infrared spectroscopy (FTIR), 13, 18, 167, 168, 170, 171, 177

fraction, 70, 89, 115, 166, 171, 177, 212, 215, 217, 218, 230, 232, 234, 239, 274
fragmentation, 278–80, 290
fragmented DNA, 280
free radical, 38, 111, 274, 276, 281
free radical quenching, 38
freeze dried, 143, 168, 170
frictional resistance, 34
friendly, 2, 13, 32, 115, 116, 124, 142, 148, 149, 189, 190, 192, 317
froth, 226
fruit, 148, 149, 257, 260
FSA *see* Food Standards Agency (FSA)
FTIR *see* Fourier transform infrared spectroscopy (FTIR)
fuel additives, 196, 199
fuel cell, 16, 37, 90, 165, 196
fullerene (C60), 317
fungal mycelia, 109, 167
fungi, 2, 31, 59, 103, 123, 142, 158, 191, 233, 317
Fusarium
  F. acuminatum, 9, 10, 60
  F. oxysporum, 9, 11, 35–46, 60, 65, 67, 68, 71, 106, 109, 112, 113, 116, 143, 147, 148, 160, 161, 167–70, 197, 241, 243
  F. semitactum, 60
  F. solani, 9, 60

Gallionella sp., 42
galvanizing, 225, 242
Gamma proteobacteria, 234
garlic, 256
Garvieae LMG8162, 60
gastrointestinal epithelia, 288
Gelidiella acerosa, 13
Gemmatimonadetes, 234
Generally Recognized as Safe (GRAS), 315
genes, 5, 106, 108, 157, 163, 235, 260, 266, 278, 280, 284, 289, 291
genes homologue-silE, silP, and silS, 5, 106, 163
genetic engineering, 199
genetic manipulation, 88, 144, 177
genetics, 88, 115, 144, 177, 198, 199, 278
genital tract, 102
genome, 235, 291
genomics, 115

genotoxicity, 257, 258, 280, 293, 314
Geobacter sp., 87
  G. metallireducens, 4, 6, 8, 132–4
  G. sulfurreducens, 85, 188, 193
geobiotechnological, 214
GeoBiotics, LLC, 237
GEOCOAT®, 237
geological, 214, 227
geomagnetic field, 133, 157
Geothermobacterium ferrireducens, 124, 125
geraniol, 58
geranium, 11, 57, 147, 167
geranium plant, 167
German Chemical Industry Association, 316, 321
glass matrix, 156
Gleocapsa sp., 13, 14
Gliricidia sepium, 149
global nanotechnology, 308
gluconic acid, 233
Gluconoacetobacter xylinus, 132, 134
glucose, 62, 111–13, 164
glutamate, 173
glutamylcysteine synthase (c-GCS), 198
glutathione (GSH), 62, 63, 111, 147, 167, 198, 276–7
glutathione synthetase (GS), 198
glycolipid-conjugated AgNPs, 282
glycolipids, 175, 282
glycoluril, 57
glycolysis, 58
glycolytic pathway, 73
glycoproteins, 175
glycosyl residue, 264
goethite waste, 231
gold (Au), 14, 57, 102
gold (III)-chloride, 67, 145
gold nanoparticle (GNP), 2, 57, 126, 145, 156, 189, 216, 243, 325
gold nuggets, 124
gold–thiol binding, 189
gold (I) thiosulfate complex, 67, 105
goods manufacturing practice (GMP), 312
Gorubathan ore, 239
gram-negative bacteria, 92, 144, 213, 215
gram positive bacteria, 162, 194
granules, 134
grass bacillus, 124

grasses, 257
gravel, 237
green alternative, 157
green fluorescent protein (GFP), 243
green synthesis, 43, 57–8, 142, 156
green technologies, 194
greigite (Fe$_3$S$_4$), 39, 61, 146, 157, 198
greigite nanoparticles, 146
grinding, 240
groundwater, 209, 242, 256
growth, 7, 33, 54, 91, 102, 126, 144, 157, 189, 214, 234, 257, 273, 308
GSH-synthesizing enzymes, 277
guided drug delivery, 146
gut physiology, 311
gypsum, 2, 33, 157

hair conditioner, 199
half-lives, 208, 209
half-shaft furnaces, 228
Haloarchaea, 107
Halobacterium sp., 108
Halococcus salifodinae, 107, 108, 113
halogenated, 205, 206
Hansenula anomala, 161, 172
harmful, 32, 142, 192, 195, 212, 218, 310–312
HAuCl$_4$ *see* Chloroauric acid (HAuCl$_4$)
hazard, 219, 293, 311
hazardous, 17, 19, 54, 104, 142, 156, 192, 199, 211, 317, 319, 320, 322
health, 54, 187, 194, 196, 274, 303, 305, 309–11, 313, 315–24, 326
heat shock protein (HSP), 280
heat transfer, 141
heavy metal, 18, 38, 59, 105, 108, 111, 145, 146, 174, 175, 191, 194, 198, 210, 214, 226, 227, 235, 242, 247, 266, 280
height, 177, 237, 307
HeLa and Jurkat T cells, 277
helical, 61
Helminthosporum solani, 60
hematite process, 228
hemimorphite, 226, 231
hemoglobin, 126
hepatic cells, 282, 284
hepatoma cells, 278
heterogeneity, 312

heterogeneous, 44, 89, 90, 93, 173, 193, 217, 245
heterologous, 244
heterostructures, 16, 173
heterotrophic, 105, 234
hexachloroplatinic acid (H$_2$PtCl$_6$), 169
hexafluorosilicate, 36
hexafluorotitanate, 36, 45
hexafluorozirconate (ZrF$_6^{2-}$), 35
hexagonal, 5, 6, 76, 124–6, 128, 159, 160, 162, 167, 173
hexagons, 163
high resolution transmission electron microscopy, 195
high surface-to-volume ratio, 156
Hindustan Zinc Limited (HZL), 227, 228, 233, 239
histidine residue, 68
HIV-1 infections, 115
homeostasis, 238, 277
homogenous, 176, 312
hormesis, 214
Hormoconis resinae, 107, 108, 113
hormones, 189
hospital, 90, 245
hot springs, 104, 108
human
  alveolar macrophages, 287
  colon cancer cells, 291
  epithelial cell, 290
  fibroblasts, 277
  fibrosarcoma (HT-1080), 292
  health, 54, 310, 311, 320
  hepatoma, 278, 287
  lung fibroblasts, 278, 291
  urinary tract infections, 149
human food and animal feed products, 316
human skin carcinoma (A-431), 277, 292
humic acid, 262
humic matter, 210
Humicola sp, 38, 41, 42, 44, 45, 106, 161, 170
hybrids, 60, 92, 93, 206, 217
hydraulic conductivity, 263
hydrazine, 211
Hydrilla sp, 73
Hydrilla verticillata, 257
hydrocarbons, 282
hydrodechlorination (HDC), 206–11

hydrodefluorination, 206
hydrodehalogenation, 206, 207, 218
hydrogel, 64, 244, 262
hydrogel polymer, 64
hydrogen, 3, 6, 8, 73, 74, 85, 112, 124, 145,
    146, 162, 164, 165, 193, 205, 207,
    210, 211, 213, 245, 274, 275, 281
hydrogenase enzymes, 109, 165
hydrogenases, 88, 176, 216
hydrogenated, 207
hydrogenation, 206
hydrogen peroxide (H_2O_2), 274, 275
hydrogen tetrachloroaurate ($HAuCl_4$), 73,
    112
hydrolases, 35
hydrometallurgical processing, 228, 231,
    233–9
hydrometallurgy, 229
hydrophilic, 54, 92
hydrophobic, 92, 210, 211, 215, 216, 284
hydrophobicity, 93
hydrophobic microdomains, 92
hydrophytes, 60, 73, 75
hydroquinones, 143
hydrosol, 9, 65, 148
hydrothermal, 34, 196, 229, 231, 245
hydroxyl (OH^-), 13
hydroxyl radical (HO·), 274
hydrozincite, 226
hyperthermophilic microorganism, 146
Hypocotyls, 149
HZL *see* Hindustan Zinc Limited (HZL)

Idiomarina sp., 4, 6, 188, 191
IL-2, 289
IL-4, 289
IL-5, 289
IL-10, 289
IL-13, 289
imaging, 17, 39, 40, 70, 76, 83, 102, 146,
    178, 196, 197, 199, 244, 264, 273, 307
imidazole rings, 68
immobilization, 142, 164
immobilized, 59, 66, 67, 90, 93, 164, 243
immobilizing metal(loid)s, 93
immune-compromised patients, 91
immune-defense, 197
immune responses, 197, 289
immunogenicity, 292

immunological response, 279, 287, 289–93
impurities, 42, 231, 314
inclusion body, 198
increase shelf life, 305
individual cells, 197
inducer, 164
induces, 263
induction, 158, 169, 275, 277, 278, 280,
    284, 289, 291
industrial, 42, 44, 46, 84, 91, 93, 103, 134,
    192, 193, 196, 213, 214, 218, 219,
    227, 231–3, 255, 256, 274, 303, 317,
    318, 322, 323, 325, 326
industries, 242, 305, 310, 314, 315, 325
inert gas, 104, 274
infections, 91, 94, 115, 144, 149, 289, 306
inflamed tissues, 279
inflammation, 196, 279, 287, 289, 290, 310
inflammatory cells, 281
inflammatory reactions, 255
influence, 57, 76, 87, 88, 123, 157, 164,
    165, 217, 218, 233, 239, 256, 262, 284
infrared, 40
infrastructure, 309
ingestion, 274, 311
ingredient, 244, 304, 311, 314
inhalation, 274
inhibit, 17, 37, 70, 92, 102, 109, 237, 289
inhibition, 92, 96, 194, 196, 214, 257, 258,
    260, 261, 263–5, 277, 284, 289
injury, 279, 290
inner cell membranes, 127
inner membrane, 291
inorganic, 2, 7, 11, 32, 33, 46, 58, 59, 65,
    92, 108, 111, 126, 157, 176, 196, 210,
    240, 246, 281, 304
insect gut, 7, 106, 163
in situ, 54, 65, 90, 93, 190, 242, 325
installation, 242
insulated, 237
integral membrane proteins, 262
integrated sulphide–silicate process, 226
interaction, 17, 39, 67–9, 84, 90–93, 105,
    108, 115, 131, 156, 167, 170, 174,
    194, 212, 218, 255–67, 278, 281–4,
    287, 289, 293
intercellular space, 290
interfacial tension effects, 288
interferon-gamma (IFN-γ), 289

internalization, 258, 259, 261–3, 284–9
International Organization for
    Standardization (ISO), 316, 318, 321
International Risk Governance Council
    (IRGC), 315
internuclear, 53
interrelation, 245
interwoven, 238
intestinal infections, 144
intracellular, 2, 3, 5, 8–11, 13–15, 39, 43,
    59, 61, 67, 72, 107, 108, 114, 115,
    124–34, 142, 145–8, 157, 159–64,
    167, 168, 171, 174, 176, 189–92, 194,
    198, 215, 238, 247, 258, 259, 261,
    263, 265, 278, 284, 285, 291, 292, 325
intrinsic potential, 208–10
invagination, 61, 262, 286
invertebrates, 158
in vitro, 13, 39, 44, 62, 64, 65, 76, 126, 148,
    162, 196, 276, 279, 286, 287, 292,
    293, 307, 311, 312, 314
in vivo, 13, 196, 197, 279, 286, 292, 293,
    307, 312, 314
ion channels, 256, 262
ionic, 5, 36, 37, 210, 211, 262
ionization, 262, 281
ions, 3, 32, 54, 91, 103, 124, 142, 162, 187,
    209, 235, 262, 273, 325
Iran, 108, 231, 233, 238
iron-cyanide precursor, 39
iron oxides, 6, 11, 131, 148, 173, 192, 195,
    230, 285
iron-oxidizer, 42
iron pyrite (FeS_2), 61
iron-reducing bacteria, 6, 39, 133
iron sulfide, 6, 61, 67, 131, 134, 192, 198
irradiation, 31, 54, 73, 144, 145, 164, 190,
    191, 232, 317
irregular, 69, 85, 86, 101, 125, 129–32, 263
irregular metaphase, 263
irritable bowel syndrome, 311
irritation, 314
Isochrysis galbana, 12, 14
isoelectric point, 283
isolation, 214, 216–17, 219
ISP lead splash condenser, 230

Japan, 306, 317, 322
Japanese medaka, 103

jarosite, 228, 238
Jatropha curca, 60
jet fuel, 108
Jurkat T and Dalton's lymphoma cells,
    292

Kappaphycus alvarezii, 13
Kelvin probe microscopy, 40
Kennecott Copper Bingham Mine, 233
kerosene, 232
ketone, 163
kinetics, 3, 168, 179, 207, 214, 236, 245,
    267, 281, 293
Kivcet furnace, 228
Klebsiella
    K. aerogenes, 134, 188, 194, 241
    K. pneumoniae, 4, 6, 7, 132, 143, 144,
        159, 188, 189, 191–3
Kooshk sphalerite concentrate, 238
Korean National Program for Tera-Level
    Nanodevices, 315

labeling, 196, 197, 305, 314, 315
laccase, 38
lactate, 112, 129, 146, 165
lactic acid, 5, 62, 63, 162
Lactobacilli
    L. reuteri, 191
    L. strains, 4, 60
Lactobacillus
    L. acidophilus, 188, 193
    L. casei, 159, 164, 191
    L. fermentum, 4, 6
    L. fructivorans, 60
Lactobacillus sp., 4, 8, 104, 107, 125, 126,
    159, 160, 164, 165, 197
LaMer mechanism, 57
landfill, 230, 242
leaching, 43, 44, 210, 226, 228–39, 247
leach liquor, 231, 232
leach residues, 230, 232, 236
lead, 2, 35, 114, 115, 129, 135, 142, 149,
    164, 168, 177, 179, 188, 191, 192,
    194–5, 214, 216, 217, 219, 225,
    227–33, 236, 238, 239, 241, 243, 247,
    277, 287, 312, 323
legal regulatory framework, 305
legislative environment, 325
Lemna minor, 257

Leptolyngbya, 14
lethal genetic effects, 278
lettuce, 257, 264
life, 9, 36, 83, 104, 115, 141, 142, 145, 188, 199, 207, 209, 210, 218, 226, 256, 293, 304, 305
ligand, 17, 72, 104, 129, 168, 285–7
ligand-receptor complex, 287
limestone, 231
linkage, 69, 245, 325
lipid(s), 61, 83, 158, 215, 259, 261, 265, 276, 277, 286–8, 304, 306, 307
lipophilic cell, 216
lipoprotein, 61, 286
liposomes, 215, 304
liquid egg white (LEW), 197
liquid phases, 53, 54, 104, 235
liquor, 231, 232
Listeria monocytogenes, 196
liver, 307, 310
living organisms, 33, 128, 157
logarithmic phase, 162, 164
log phase, 171, 239
Lolium spp., 257
long-fiber asbestos, 310
longitudinal axis, 133
long-term toxicity, 196, 293
low-density lipoprotein (LDL), 286
luminescent, 17, 197
lung inflammation, 310
lungs, 278, 286, 287, 291, 310
Lycopersicon esculentum, 260, 262, 265
lymphatic systems, 287
lymph nodes, 286, 310
lymphocytes, 289
lymphoma cells, 280, 292
Lyngbya majuscula, 14
lysine, 278, 307
lysis buffer, 114, 176
lysosome, 285, 286

macromolecules, 108, 115, 126, 176, 286
macrophage, 197, 275, 277, 279, 281, 286, 287, 289, 290, 292, 310
maghemite, 6, 146
magnesium, 230
magnetic, 6, 7, 17, 18, 33, 39, 46, 61, 62, 64, 102, 123, 131, 133, 134, 146, 147, 157, 158, 165, 173, 193, 195, 198, 208, 209, 230, 240, 244, 245, 258, 263, 303, 307, 308
magnetic nanocrystals (NCs), 157, 165
magnetic resonance imaging (MRI), 39, 146, 199
magnetite (Fe_3O_4), 6, 39, 61, 148, 157
Magnetobacterium bavaricum, 157
magnetosome(s), 6, 7, 61, 62, 157, 158, 243
  crystals, 157
  membrane (MM), 157
Magnetospirillum
  M. gryphiswaldense, 4, 6, 157, 158, 195
  M. magneticum, 146
  M. magnetotacticum, 4, 6, 61, 132, 133, 157, 158, 195, 241
magnetotactic bacteria (MTB), 4, 6, 7, 32, 33, 39, 61–2, 133, 134, 146, 157, 158, 195, 198
magnetotactic multicellular prokaryotes (MMPs), 198
magnetotactic prokaryotes, 198
magnetotaxis, 133
magnitude, 275, 282
maize, 256–8, 263, 266
malic acid, 73
malignancy, 102
MamI, MamL, MamB, and MamQ proteins, 61
mammalian cells, 290, 292
MamM, MamN, and MamO proteins, 7, 62
Manganese, 124, 227
Manganese oxide, 38, 45
Manganese(II)sulfate, 38
mangrove plants, 149
manufacturing, 244, 274, 312, 315, 317, 324
marine alga, 189
marine microbes, 190
Marinobacter pelagius, 106, 107, 190
marmatite, 235
mass transport, 210
material, 4, 11, 16, 19, 31, 33, 35–7, 39–45, 53, 93, 101, 157, 158, 164, 188, 194, 214, 217, 229–31, 235–6, 245, 247, 281, 282, 305, 308, 311, 315, 316
matrix, 83, 89–93, 108, 126, 156, 161, 164, 174, 209, 216–17, 219, 244, 322
mechanical devices, 40, 104

mechanisms, 2, 3, 18, 53–78, 87, 93, 103–6, 108, 124, 134, 135, 145, 163, 174–6, 194, 195, 198, 212, 214–17, 219, 235, 238, 248, 257–64, 266, 274, 284, 288, 291–3, 325
medical, 17, 39, 57, 83, 84, 86, 90–92, 141–9, 156, 190, 197, 199, 255, 305, 313, 320, 325
medicine, 1, 18, 58, 101, 102, 156, 190, 191, 217, 273, 305, 308, 313, 318, 319, 323
Melia azedarach, 60
melting point, 303
membrane, 6, 59, 84, 102, 125, 146, 157, 193, 206, 245, 256, 275, 310
mercaptoethanol(ME), 142
mercury, 146
mesophiles, 114, 237, 238
mesophytes, 60, 73, 75
mesoporous, 146
metabolic pathways, 37, 39, 44, 242
metabolism, 46, 67, 76, 88, 115, 124, 174, 197, 312
metabolomics, 115
metal
  γ-glutamyl peptides, 175
  hydroxides (M(OH)$_2$), 72
  ions, 9, 10, 14, 15, 18, 32–5, 38, 54, 57, 59, 61, 65, 68, 72, 76, 105, 108–11, 124, 127, 142, 143, 145–7, 149, 167, 171, 172, 174, 176, 187, 190, 194, 195, 198, 213, 216, 238, 243, 273, 280, 281
metallic nanoparticles (MNPs), 4, 6, 13, 16, 58, 105, 109, 145–7, 189, 198, 241
metalloid, 8, 62, 131
metallophilic, 198
metallothioneins, 68, 175, 280
metallotolerant, 105
metallurgical industries, 242
metal nanoparticles, 1–19, 54–7, 59, 64, 72, 77, 102, 124–35, 142–4, 147, 148, 187, 188, 191, 194, 195, 206, 214, 215, 240, 241, 247
metal-reducing bacteria, 84, 216
metal-resistant genes, 108
metal(loid)s, 84, 85, 87–90, 93
metal thiolate complex, 111
metastable oxide phases, 46

metastable precursors, 245
metastasis, 102
metazoans, 157
methylation, 59
methylene bridges, 57
methylviologen (MV^{2+}), 87
MHLW *see* Ministry of Health, Labor and Welfare (MHLW)
micelles, 102, 304
microaerophilic, 126, 128
microalga, 13
microbial cells, 5, 66, 70, 83, 93, 109, 114, 142, 144, 148, 163, 166, 174, 175, 198, 206, 213, 241, 248, 325
microbiologists, 93
Micrococcus lactilyticus, 4, 8
Microcoleus sp., 13, 14
microconfiguration, 146
microdroplets, 62, 64
microelectromagnets, 133
microelectromechanical devices, 40
microenvironments, 89, 90, 286
microfluidic, 62–4
micrometer-scale, 242
micronutrients, 225
microorganisms, 2, 33, 58, 83, 103, 123, 142, 156, 187, 211, 233, 284
microparticles, 311
microRNAs (miRs), 256, 266
microvesicles, 290
microvessel, 290
microwave, 42, 54, 57, 73, 144, 145, 190, 191, 232, 236
microwave irradiation, 54, 73, 144, 145, 190, 191
milled, 45, 104
mine, 5, 105, 163, 170, 172, 190, 227–9, 233, 234, 237, 238, 242
mineral detoxification, 108
mining, 226, 227, 229, 230, 233–4, 239
Ministry of Economy, Trade and Industry (METI), 317, 322
Ministry of Health, Labor and Welfare (MHLW), 317, 322
Ministry of the Environment (MOE), 317, 322
mitochondrial membrane potential (MMP), 198, 291
mitochondrial perturbation, 289

mitochondrion function, 276
mitogen-activated protein kinase (MAPK), 289, 290
model plant, 257
moisture, 256
molecular, 5–7, 17, 19, 32, 35–7, 46, 53–78, 83, 84, 87, 88, 106, 110, 123, 165, 168, 170, 171, 177, 192, 194, 197, 207, 245, 248, 266, 274, 292, 293, 325
molten glass, 231
monitor, 78, 236, 237, 243, 305
monoclinic phase (α-Bi_2O_3), 37
monocots, 44, 256–8
monocytes, 286
monodisperse, 5, 18, 43, 44, 57, 104, 111, 116, 144–6, 148, 162, 170, 171, 192, 215, 282
monodispersity, 43, 103, 107, 142, 171, 191
monomineralic, 242
monooxygenases, 72
monosaccharides, 62
Moore cake, 239
Morganella sp., 5, 106, 159, 163
  M. morganii, 60
  M. psychrotolerans, 4, 6, 60
Moringa oleifera, 60
morphology, 7, 8, 11, 36, 61, 103, 104, 107, 110, 114, 124–6, 129–33, 142, 145, 156, 159–61, 163, 165, 167, 168, 170, 176, 177, 191, 198, 212, 214–15, 259, 261, 273, 278, 281, 293
motility, 287
mouse, 280
MRI *see* magnetic resonance imaging (MRI)
mucilage, 262
Mucosae, 60
multicellular forms, 61
multi-metal, 206
multi-protein, 280
multi-step processes, 214
multi-walled carbon nanotubes (MWNTs), 92, 256, 320
murine macrophage, 290
Musa paradisiaca, 60
mutagenicity, 314
mutations, 34, 236, 278, 291

mycelia, 10, 109, 167, 170, 194, 243
mycology, 9
myconanotechnology, 9

N-acetyl cysteine, 276–8
N-acetyl glucosamine (GlcNAC), 264
NADH-dependent enzymes, 5, 109, 129, 164, 325
NADH dependent nitrate reductase, 9, 10, 18
NADH-dependent reductase, 65, 70, 109, 162
$NAD^+/NADP^+$-dependent malic enzyme, 73
NADPH-dependent nitrate reductase, 9, 18, 65, 148
NADPH oxidase enzyme, 280
Namibia, 229, 237
Nannochloropsis oculata, 191
nano, 101
nanobiotechnology, 71, 73, 156, 219
nanobotics, 273
nanocarrier, 293
nanocatalyst, 205–19
nanocatalytic activity, 93
nanocomposite, 90, 92, 192, 273–93
nanocrystalline, 45, 147, 307
nanocrystalline materials, 199
nanocrystallites, 43, 46, 171
nanocrystals, 2, 7, 10, 32, 43, 54, 68, 134, 157, 163, 165, 174, 175
nanoelectromechanical, 273
nanoelectronics, 273
nanofabrication, 176
nanofactory(ies), 74, 75
nanofibre, 316, 318, 321
nanofiltration (NF), 92
nanofluids, 245
nanofood, 308–11
nanogold, 18, 108
nanolithography, 273
nanomagnetite, 193
nanomaterials, 3, 31, 57, 84, 101, 123, 142, 156, 187, 241, 255, 273, 303
nanomedicine, 101, 273, 274, 292, 305, 306, 308, 310, 312, 314
nanometer, 141, 142, 281
nanopalladium, 193
nanoparticles, 1–19, 34–45, 53–78, 87–90, 92–94, 101–116, 123–135, 141–149,

155–179, 187–200, 206, 210, 211–219,
nanoparticle-tripeptide (GNP-tripeptide), 65
nanopatterned biosilica, 157
nanoplate, 104, 316, 318, 321
nanoreaction, 206, 215
nanorobotics, 101
nanorods, 8, 18, 130, 131, 178
nanoscaffolds, 199
nanoscale, 32, 36, 44, 101, 104, 157, 169, 193, 194, 208, 209, 214, 303–5, 310, 313, 315, 324
nanosensors, 305
nanosilver, 92, 320
nanosized, 87, 123, 134, 225–48, 255, 273
nanospheres, 8, 32, 70, 107, 130, 131, 166, 178, 192, 281, 304
nanostructures, 2, 7, 8, 33, 65, 72, 73, 76, 84, 88, 89, 129, 145, 157, 190, 241, 244
nanotechnology, 1, 31, 83, 101, 123, 141, 155, 187, 240, 255, 273, 363–326
nanotization, 54, 59
nanotoxicity, 255, 256, 293
nanotoxicology, 256, 312, 323
nanotriangles, 148, 149
nanotubes, 17, 92, 173–5, 256, 262, 304, 310, 311, 317, 320, 323
nanowires, 69, 70, 103, 129, 173, 174
napthoquinones, 66
National Institute of Advanced Industrial Science and Technology (AIST), 317, 322
National Institute of Occupational Safety and Health Japan (JNIOSH), 317, 322
National Institutes of Health (NIH), 309, 324
National Nanotechnology Initiative, 303, 309, 324
National Science Foundation (NSF), 309
natural organic matter (NOM), 262
near-atomic scale, 303
necrosis, 265, 289–92
neem, 57
negative potential, 239
nerve cells, 288
Neurospora crassa, 160, 168, 169

neutralize, 111, 199, 229, 274
neutral leach, 228, 230
neutrophils, 197, 279, 281
nickel, 103, 233, 237
Nicotianna tobaccum, 262, 265
NIPAM monomers, 63
nitrate/nitrite reductase, 8, 9, 18, 62, 65, 68–9, 108, 109, 142, 144, 148, 164, 171, 176, 325
nitric oxide (NO), 92, 275, 277, 279
nitrogenase enzyme, 14, 15
nitrogen monoxide, 277
nitroreductase, 109, 130, 144, 163
nitrosative stress, 277
Nitrospira sp., 234
Nocardiopsis sp., 161, 171
non-ferrous metal, 225
non-genotoxic, 282
non-pathogenic, 84, 130, 144, 167, 168, 195
non-pathogenic bacteria, 195
nontoxic, 142
non-volatile memories, 40
No observed adverse effect level (NOAEL) dose, 312
nuclear breakdown, 290
nuclear factor κB (NF-κB), 289
nuclear histone, 111
nuclear membranes, 288
nuclear translocation, 72
nucleation, 54, 56, 57, 59, 69, 108, 126, 157, 174, 198
nuclei, 54, 69, 71, 163, 167, 287, 288
nucleic acid, 83, 102, 276, 277
nucleus, 72, 285, 286
nutrient recycling, 104
nutrients, 104, 111–13, 199, 225, 234, 256, 262
nutrient uptake, 256
nutritional value, 304

O-acetylserine, 66, 109, 110
O-acetylserine thiolyase, 109, 110
O-acetylserine through O-acetyl serine thiolyase, 66
obesity, 305
occupational health measures, 316
ocean floor, 104
Ocimum sanctum, 60

octahedral, 2, 6, 14, 39, 67, 124–7, 131, 132, 145, 159, 160, 162, 170, 261
AuNPs, 162
prism, 6, 131, 132
oil pipelines, 91
olefinic double bonds, 207
olefins, 206
oleic acid, 282
oligonucleotides, 17, 189
one billionth (10^{-9} m), 101
onion, 9, 256, 257
operation theatres, 245
opportunities, xviii, 41, 46, 83–94, 194, 212, 309, 313, 318, 321
opsonins, 292
opsonization, 284, 286
optical, 17, 87, 131, 135, 156, 165, 173, 188, 192, 196, 240, 244, 246, 303
optics, 31, 241, 255
optoelectronic(s), 37, 85, 86, 102, 123, 192
optoelectronic material, 37
Opuntia ficus, 60
ores, xix, 34, 42, 124, 225–30, 232, 233, 236, 238, 239
organelles, 59, 157, 266, 281, 285, 286, 288, 291, 310
organic, 37, 43, 55, 56, 92, 93, 102, 108, 111, 126, 129, 133, 158, 162, 176, 193, 196, 198, 206, 210, 212, 216–18, 246, 247, 262, 274, 281, 304
organisms, xviii, 5, 6, 13, 32–4, 36, 37, 42, 43, 46, 59, 84, 87, 103, 108, 109, 111, 112, 128, 143, 149, 157, 158, 192, 194, 196, 212, 243, 247, 275, 293
organohalogen, 205–19
organo-nanocomposite material, 43
organophosphorus pesticide, 18
ornamented cell walls, 157
orthodox redox, 62
orthorhombic quartz, 44
Oscillatoria sp., 13, 14
osmotic balance, 107
osmotic pressure, 262, 277
Oswald ripening stage, 54
outer-cell, 288
outer membrane, 8, 87, 88, 165, 199, 215
outer membrane c-type cytochromes (OMCs), 87, 88
ovarian cancer, 102

ovaries, 310
over express, 243
oxidants, 65, 210, 274, 275, 277, 281
oxidase, 5, 38, 63, 70, 72, 275, 280
oxidation, 17, 32, 38, 54, 57, 59, 61, 62, 65, 74, 84, 108, 124, 133, 134, 142, 165, 174, 210, 211, 214, 216, 228, 229, 233, 235, 237, 277, 278
oxidative attack, 278, 291
oxidative stress, 193, 194, 257, 259–61, 264, 265, 274, 276–8, 280, 281, 284, 289–93, 310
oxides, 2, 5, 6, 9, 27–40, 55, 59, 88, 105–15, 126, 142, 145, 146, 148, 165–9, 171, 197, 199, 205, 209–11, 233, 234, 236, 238, 244, 259, 265
oxidic, 206, 207
oxidize, 8, 65, 72, 105, 109, 233, 276
oxidoreductase, 5, 62, 63, 65, 72, 164, 176
oxidoreductive machinery, 110
oxyanions selenite (SeO_3^{2-}), 128
oxygen
gradient, 157
insensitive flavoprotein, 163
oxygenase, 72
oxygen-deficient, 133
oxygen-evolving enhancer protein, 111
oxygen-saturated medium, 89
oxyhydroxide, 134
ozone sensors, 41

Paints, 36, 225
palladium (Pd), 6, 14, 85, 107, 129, 146, 165, 188, 193, 205–19
Parabuchneri LMG 11772, 60
paramagnetic, 274
parameters, 5, 32, 54, 103, 104, 163, 164, 172, 177, 179, 190, 193, 215, 232, 239, 263–5, 282–4, 287
Parkinsonia florida, 262
Parthenium hysterophorus, 149
partial pressure, 164, 165
particle, 2, 3, 5, 9, 10, 15, 29, 30, 32–9, 45, 46, 48, 50, 52, 55–8, 88, 90, 92–4, 105–14, 122, 124, 125, 134, 135, 139, 142, 146–8, 150, 151, 161, 165–7, 170, 176, 177, 179–81, 183–5, 197, 198, 200, 203, 205–7, 210, 212, 218, 225–7, 233, 240–7, 249, 260, 265, 266

patchoulane, 76
pathogenic bacteria, 10, 173
pathogenicity, 148
pathogens, 9, 12, 94, 197
pathways, 37, 39, 44, 61, 72, 73, 88, 115, 142, 173, 206, 218, 242, 247, 257, 267, 274, 277, 278, 284–6, 289–91, 293, 315
patients, 91, 102
patterning arrays, 126
Pd nanoparticles, 88, 206, 207
pectin hydrogel capsule, 262
pectin layer, 274
Pedicoccus pentosaceus LMG 9445, 60
pedomicrobium, 105, 124
PEG *see* Polyethylene glycol (PEG)
Pelargonium graveolens, 149, 167
pelletizing, 232–3
penicillin, 91
Penicillium sp., 42, 60, 188
    P. aurantiogriseum, 195
    P. brevicompactum, 106, 160, 168
    P. citrinum, 11, 195
    P. waksmanii, 195
peptides, 7, 36, 46, 59, 65, 77, 91, 125, 126, 161, 163, 173–6, 243, 278, 282, 292, 325
perennial, 263
perfluorinated, 206
performance, 88, 92, 206–10, 214, 217, 218, 316
perhydroxyl radical (.OHH), 275
periplasm, 8, 125, 129, 146, 159, 165, 213
periplasmic nitrite reductase, 8
periplasmic proteins, 87, 163
periplasmic space, 58, 88, 124–7, 144, 159, 162, 163, 213, 215, 243
permanganate, 210, 211
permeability, 92, 194, 199, 215, 246, 281, 290, 291
permease, 62, 66, 109, 110
permeation, 102, 210
peroxynitrate anion (ONOO-), 277
Peru, 226
pest control, 257
pesticides, 17–19, 318, 324
phagocitosed, 286
phagocytes, 275, 285, 288, 290
phagocytic cells, 279, 292
phagocytic vesicles, 285

phagocytosed, 287
phagocytosis, 284, 285, 287, 288, 293
phagolysosome, 285
phagosome, 285
Phanerochaete chrysosporium, 11, 106
pharmaceutical, 141, 307, 321
pharmaceutics, 199
pharmacokinetic properties, 247
pharmacy, 18
phenolic compounds, 57
phenolics, 57, 262
phenol moieties, 207
pH modulation, 111
Phoma sp., 11, 60, 67
    P. glomerata, 11, 60, 160, 168
    P. sorghina, 18
Phormidium tenue, 15
phosphate, 59, 114, 134, 158, 162, 175, 242, 245, 291
phosphoadenosine phosphosulfate reductase, 109, 110
phosphoenol pyruvate (PEP), 73
phosphoproteins, 168
phosphorus, 133, 245
phosphorylation, 278, 291
photocatalysis, 16, 36
photocatalyst, 16, 197
photochemically, 274
photo-irradiation, 31
photoluminescence, 42, 46
photoreactivity, 197
photosynthesis, 256
photosynthetic bacterium, 134, 162, 164
photosynthetic microorganism, 13
phototoxicity, 103
Phyllanthus amarus, 61
physical, 1, 2, 15, 27, 28, 39, 40, 88–90, 95, 99, 105, 115, 121, 122, 127, 128, 150, 152, 161, 171, 181, 186, 196–9, 206, 212, 240, 250, 270, 271, 278
physical adsorption, 174
physico-chemical properties, 34, 156, 235, 256, 273, 285
physicochemical stresses, 84
physiology, 115, 198, 257, 258, 260, 264, 311
phytochelatin, 59, 65, 111, 148, 171, 175, 192, 325
phytochelatin-Cd, 325

phytochemically-derived, 149
phytochemicals, 76
phytoliths, 44
phyto-nanotoxicity, 256
phytotoxic, 256, 257, 262, 264, 266
phytotoxicity, 255–67
piezoelectric, 40, 244
pIII phage, 173
piperitone, 6, 69, 109, 130, 144, 163
piperitone (3-methyl-6-(1-methylethyl)-2-cyclohexen-1-one), 144, 163
Piper longum, 60
planktonic cells, 90
Plant
   pathogenic fungus, 35
   tissues, 57, 263, 267
Plantarum LMG 24830/LMG24832, 60
plasma membrane, 59, 162, 257, 262, 286
plasmodesmata, 262
Platelet-derived growth factor (PDGF), 286
platinum (Pt), 6, 9, 14, 16, 60, 61, 68, 93, 107, 109, 125, 128–31, 145, 146, 159, 160, 165, 213
platinum(IV) chloride, 131
platinum group metal (PGM), 128, 129, 165
platinum nanoparticles (PtNPs), 6, 9, 68, 109, 128, 145, 146, 165, 169
Plectonema boryanum, 13, 14, 61, 106, 126, 127, 130, 131
   P. boryanum UTEX485, 67
polarizations, 156
policies, xix, 305
polishing agents, 196
pollutants, 134, 199, 205–10, 216–18
pollution, 228, 274, 322
polyacrylamide gel electrophoresis, 19, 71, 168, 169
Polyalthia longifolia, 60
polychlorinated biphenyls, 165, 206
polychlorinated dibenzo-p-dioxins (PCDDs), 218
polycrystalline, 126
polydimethylsiloxane (PDMS), 207, 210
polydisperse, 108, 166, 193, 211
polydispersed, 9, 10, 167
polydispersity, 10, 57, 145, 146, 167, 194
polyelectrolyte fuel cells, 16
polyelectrolytes, 16, 210, 211

polyethylene glycol (PEG), 217, 292, 306–8
polyethylenimine-derived (PEI), 18
polymeric matrix, 164
polymeric nanospheres, 304
polymeric substances (EPS), 8, 83, 89, 90, 92
polymerization, 72
polymers, 56, 64, 103, 158, 190, 196, 198, 210, 211, 217, 218, 244, 282, 293, 305
polymetallic, 237, 239
poly-metallic ores, 229
polymorphs, 37, 245
poly N-isopropyl acrylamide (PNIPAM), 244
polyol synthesis, 317
polypeptides, 36, 59, 147
polypeptides/enzymes, 147
polysaccaharide(s), 14, 59, 83, 87, 90, 158, 175, 282, 292
polysaccharide-coated particles, 292
polystyrene, 92, 197
polysulfone, 92
polyurethane foams, 103
polyvinylpyrrolidone (PVP), 197, 282
polyvinylpyrrolidone gel, 197
Porifera, 157
porosity, 93, 167, 247
positive potential, 239
post-synthetic separation, 217
post-translational modifications, 37
Potamogeton sp., 73
potassium, 57, 58, 107, 133, 165, 215, 288
   bitartrate, 215
   ferricyanide, 165
   ferrocyanide, 165
   hexafluorotitanate, 36
   hexafluorozirconate, 35
   tetrachloroaurate (III), 57, 58
pre-adaptation, 235
precious metals, 93, 124, 193, 205, 206, 209–12, 214, 216, 219
precipitations, xvii, 3, 13, 34, 54, 59, 67, 108, 126, 128, 164, 168, 176, 193, 211, 212, 242
precursors, xviii, 7, 34–46, 53–5, 58, 63, 65, 72, 89, 127–9, 131, 134, 135, 148, 162–6, 168–70, 172, 173, 176, 177, 192, 195, 240, 245, 277, 278

pre-market approval, 314, 315
preservation, 276
pressure, xvii, xviii, 32, 36, 103, 105, 142, 149, 162, 164, 165, 167, 192, 211, 212, 226, 228, 231, 240, 262, 277
pretreatment, 280
primary and secondary amines, 168
pristine, 44, 267
proapoptotic factors, 289
profitability, 326
programmed cell death, 264, 289, 290
proinflammatory cytokines, 290
pro-inflammatory signaling cascades, 289
prokaryotes, 59, 61, 129, 158, 198, 242
prokaryotic cells, 255
proline, 278
promoter, 279
Propionibacterium, 246
proprietary, 237
Prosopis juliflora-velutina, 262
Prosopis sp., 262
prostaglandin H2, 290
protease, 5, 7, 61, 62, 264, 285
protection, xvii, 5, 59, 103, 196, 206, 210–211, 216, 247, 256, 264, 309, 312, 317–19, 321, 323
proteins, 7, 34, 59, 83, 104, 126, 147, 158, 190, 219, 235, 256, 276, 304
proteomics, 115, 198, 266
Proteus vulgaricus, 12
protocatechaldehyde, 74
protocatechuic acid, 74
protonated anionic, 62
protonated state, 275
proton gradient, 59, 291
protozoans, 104
Pseudomonas
  P. aeruginosa, 3, 4, 8, 12, 60, 67, 86, 87, 92, 112, 129, 130, 143, 145, 168, 191, 197
  P. alkaliphila, 60
  P. alkaphila, 4, 8
  P. capsulate, 60
  P. fluorescens, 189
  P. putida, 85, 86, 112, 189
  P. stutzeri, 4, 5, 8, 60, 61, 67, 106, 112, 125, 126, 128, 143, 144, 159, 163, 188–90, 195, 241
Psidium guajava, 61

psychro-tolerant, 6
PtNPs *see* platinum nanoparticles (PtNPs)
P-type ATPase, 106, 163
pulmonary interstitial sites, 287
pulsed-field gel electrophoresis (PFGE), 235
pulverized rock, 234
pumite, 231
pumpkin, 256
purity, 45, 206, 232, 281
pyrite, 61, 228, 235, 239
Pyrobaculum islandicum, 146
pyroelectric detectors, 40
pyrolysis, 31, 104
pyrrhotite (FeS), 232
pyruvate, 73

quality, xix, 43, 146, 199, 244, 267, 305
quantitative studies, 248
quantum confinement, 156
quantum dots (QDs), 7, 17, 64, 65, 86, 92, 164, 173, 196–7, 244, 256
quantum effects, xix, 303
quasi-hexagonal, 160
quasi-spherical, 36–9, 43, 147, 160, 161, 165, 170
quasispherical ZrO_2 nanoparticles, 7
quench, 274
quenching, 38, 240, 274
quinine, 66, 68, 70, 147
quinone, 9, 62, 72, 73, 148, 167

radiation, xviii, 84, 103, 105, 196, 236
radical scavenging, 199
radio-frequency, 42
radionuclides, 164, 166
radish, 257, 263, 264
Ralstonia metallidurans, 4, 5, 60
Raman spectrum, 325
Rampura Agucha Lead-Zinc mines, 232
Ramsar geothermal hot springs, 108
rat alveolar macrophages, 289
reactive nitrogen species (RNS), 264, 275, 277, 281
reactive oxygen species (ROS), 61, 62, 91, 193, 257, 261, 264, 265, 274, 276, 277, 279–81, 284, 291, 292, 310, 312
reactors, 65, 193, 214, 226, 228, 229, 234, 238

receptor-mediated endocytosis pathway (RME), 286
receptors, 17, 70, 284, 286, 287, 290, 293
recombinant, 64, 198, 199
recovery, xviii, 2, 42, 85, 93, 103, 114, 129, 163, 176, 177, 206, 213, 228, 230–234, 236–9, 242, 246–8, 278, 287
redox potential, 165, 195
redox reactions, 87
redox state, 108, 279
reducing agent, 55, 57, 62, 69, 72, 76, 110, 129, 131, 143, 147–9, 189, 205, 207, 210, 211, 213, 215, 217, 281
reductants, 65, 133, 210, 215, 274, 281, 282
reductases, 16, 19, 66, 70, 71, 74, 104, 105, 109, 168, 195
reduction, 2, 32, 54, 62, 65–6, 87, 104, 126, 142, 162, 167, 189, 195, 207, 214, 228, 255, 258, 259, 274, 317
reductive, 3, 5, 61, 65, 88, 89, 111, 124, 129–31, 148, 167, 206, 210, 218, 232
refining, 227
Registration, Evaluation, Authorization and Restriction of Chemicals (REACH), 316, 318, 319, 323
Regulation, 61, 277, 278, 286, 289, 303–26
regulatory agencies, 103, 305
remediation, 86, 192, 242, 247, 255
remirin (type III), 73
renewable materials, 142
repulsion, 114, 283
reservoir, 157, 214, 278
residues, 13, 59, 65, 68, 147, 163, 167, 174, 212, 216, 219, 226–33, 236, 239, 264, 278, 286
resins, 36, 92, 164, 192, 307
resistance, 5, 33, 34, 59, 65, 84, 90, 91, 94, 105–8, 163, 165, 190, 191, 194, 210, 211, 214, 235, 238, 257
resolution, 93, 199
resources, xix, 104, 149, 218, 226, 227, 229–31, 237, 238, 246, 309
restriction endonuclease Xbal, 235
reverse osmosis (RO), 92
revolution, 247, 255
Rhamnosus LMG 18243, 60
rhapidosomes, 176
Rhizobium selenitireducens, 4, 8
Rhizophora mucronata, 149

Rhizopus
  R. oryzae, 60, 160, 168
  R. stolonifer, 111, 191
rhizospheric soil, 170
rhodium (Rh), 129, 206
Rhodobacter
  R. capsulatus, 62, 63, 159, 162
  R. sphaeroides, 66, 67, 85, 109, 110, 114, 132, 135, 159, 164, 243
Rhodococcus sp., 4, 5, 60, 145, 146, 161, 171
Rhodopseudomonas
  R. capsulate, 3, 4, 60, 85, 104, 106, 111, 113, 129, 145, 159, 162, 189, 325
  R. palustris, 4, 7, 68, 132, 134, 159, 164, 197
rice
  husk, 42, 44–6
  mills, 44
RNS *see* reactive nitrogen species (RNS)
roasting, 226, 230–232
rocks, 189, 223, 234, 237, 246
root, 256–66
Rosh Pinah Zinc Corporation (Pvt) Ltd, 229
Rosmarinus officinalis, 149
routes, 31–40, 43, 46, 58, 65, 68, 74, 103, 108, 129, 157, 162, 176, 190, 192, 193, 197, 213, 214, 226, 228, 240, 242, 246
RTH1 and RTH3 genes, 266
RuBP carboxylase, 111
ruthenium, 213
rutile, 45
ryegrass, 263

Saccharomyces cerevisiae, 72, 147, 197
safety risks, 310
salinity, xviii, 103, 107
Salmonella
  S. enteritidis, 196
  S. typhimurium, 159, 164
  S. typhus, 67
Salsola tragus, 262
sand, 42, 43, 231, 259
Sarcheshmeh chalcopyrite concentrate, 238
Sargassumnatans, 13
Sargassum wightii, 14, 189

Scanning electron microscopy (SEM), 11, 13, 172, 177, 245
scattering, 156, 177, 178
scavenger, 276, 285
scavenger-type receptors, 285
Schizosaccharomyces pombe, 60, 106, 112, 171, 188, 192, 197, 241, 325
Scientific Committee on Emerging and Newly Identified Health Risks (SCENIHR), 315, 316, 319
secreted protein, 72, 147, 168, 195, 244
sedoheptulose-1, 111
selenate (SeO$_4^{2-}$), 70, 107, 128, 131
selenide, 107
Selenihalanaerobacter shriftii, 106, 107, 130, 131
selenite (SeO$_3^{2-}$), 69, 70, 128, 131, 166
selenium
 nanospheres, 70, 107, 166, 192
 oxyanions, 128, 131
selenium nanoparticles (SeNPs), 8, 69, 131, 166, 191, 192
self-detoxification, 105
Selnihalanaerobacter shriftii, 106, 107, 130, 131
semiconducting, 173, 244
semiconductor, 7, 8, 59, 86, 102, 107, 134, 164, 165, 168, 171, 173, 192, 193, 196, 197, 244
sensing, 31
sensitive, 209, 212, 280, 291, 310
sensitization studies, 314
sensor kinase, 107, 163
sensors, 37, 41, 107, 163, 244–6
sequester, 171
sequestration, 212, 238
serine, 66, 109, 110, 278
Serratia sp., 4, 7, 60
 S. marcescens, 61, 159, 164
serum, 243, 282, 286, 288
sewage, 144, 163, 227, 256
shampoo, 199
Shewanella
 S. algae, 3, 4, 6, 60, 85, 106, 107, 124–6, 128, 145, 162, 165, 241
 S. oneidensis, 4, 8, 60, 85–9, 106, 112, 113, 125, 129, 159, 160, 165
 S. putrefaciens, 86, 160, 165
Shigella sp., 234

shock wear, 34
shuttle quinone, 9, 148
sieve, 36, 192, 230
signaling pathways, 293
signal transduction, 274, 278, 290, 291
Sikkim Mining Corporation, 239
silaffin, 36, 59, 72
silica, 9, 17, 33, 35–8, 42–6, 72, 92, 102, 157, 158, 170, 188, 192, 230, 246, 324
silica deposited vesicles (SDV), 72
silicatein, 36, 147
silicates, 34, 36, 43, 44, 226, 228
silica transporter proteins (SIT), 72
silicic acid, 43
silicon, 2, 35, 43, 54, 55, 192, 230
silicon dioxide (SiO$_2$), 35
silver-binding protein, 106, 163, 175
silver ions, 5, 6, 9–13, 18, 19, 61, 70, 71, 73, 74, 108, 126, 144, 147, 148, 162, 163, 167, 168, 190, 191
silver nanoparticles (AgNPs), 5, 6, 9–13, 19, 61–3, 66–71, 73–7, 92, 103, 106–8, 111, 114, 126, 128, 130, 131, 144–5, 147–9, 163, 164, 167–9, 171–3, 190, 191, 199, 263, 274, 276–82, 284, 285, 287, 289–93, 308, 310, 325
silver nitrate, 11, 70, 106, 107, 144, 147, 149, 163, 164, 168, 191
silver (I) nitrate transition, 67
silver-resistant bacterium, 5, 163
silver sulfide acanthite (Ag2S), 5, 163
single-electron transistors, 102, 123
singlet oxygen (1O$_2$), 274
sinter-shaft furnace processes, 228
siRNA, 17, 102
site-specific amino acid modification, 278
size-controlled formation, 206
size distribution, 5, 57, 87, 129, 142, 177, 191, 206, 215–17, 219, 281, 282, 314
S-layer, 33, 59, 157, 176
slower reduction, 282
sludge, 227, 256
slurry, 228
small size, xviii, 91, 101, 156, 281, 284
smithsonite, 226, 230
sodium borohydride, 55, 56, 149, 211
sodium citrate, 149
sodium hexaetaphosphate(SHMP), 142
sodium hexametaphosphate, 233

sodium hydroxide, 57, 246
sodium selenite, 69
sodium sulfide (Na$_2$S), 69
soft robotics, 103
soil, 9, 10, 38, 68, 70, 71, 104, 144, 166, 170, 195, 199, 225, 227, 228, 233, 242, 256, 262
solid electrolytes, 196
solidification, 233
solids, 90, 93, 164, 196, 231, 234–6, 239, 240
solubility, 65, 196, 281, 312
solubilization, 233, 234, 239, 316
solvent extraction-electrowinning (SX-EW), 232
solvents, 32, 36, 37, 57, 104, 129, 142, 164, 196, 199, 226, 228, 232, 245, 246
solvothermal, 104
sonification, 144
sophorolipids, 282
sorption, 59, 174
sorptive enrichment, 216
spatiotemporal, 93
special high-grade (SHG) zinc, 226
species, 8, 9, 11, 12, 15, 35, 39, 44, 59–62, 91, 103, 108, 109, 128, 129, 131, 149, 157, 167, 168, 170, 191, 193–5, 210, 234, 236, 247, 256, 257, 261, 264, 266, 273, 274, 277, 310
specific receptors, 285, 286
spent batteries, 232
spermatogonia, 292
sphalerite (ZnS), 8, 225, 226, 231–3, 235–9, 242, 246
sphalerite (ZnS) crystals, 8
sphenophytes, 44
spherical, 5, 7–11, 13, 14, 67–70, 73, 76, 85, 86, 92, 101, 104, 107, 108, 111, 114, 125, 126, 129–32, 134, 135, 145, 156, 159–64, 166–71, 173, 191, 215, 242, 288
spherical aggregates, 8, 135, 242
spherical monoclinic Selenium (m-Se), 69, 70
spheroidal morphologies, 126
Spirodela polyrhiza, 257
Spirulina. sp., 13
  S. platensis, 12, 14
  S. subsalsa, 14

spleen, 310
sponges, 36, 157, 232
sports, 141, 199
spray pyrolysis, 31
stability, xix, 11, 17, 19, 71, 88, 93, 115, 148, 167–71, 188, 191, 195, 196, 215, 217, 241, 244, 256, 281, 282, 287, 308, 312, 325
stabilization, 7, 19, 37, 41, 43, 58, 111, 116, 142, 146, 149, 168, 171, 172, 198, 199, 206, 216–17, 219, 237, 282
stabilizer, 131, 162, 172, 176, 213, 215, 217
stabilizing agent, 166, 168, 170, 177, 246, 281
stakeholders, 310, 313, 314
Staphylococcus
  S. aureus, 67, 92, 126, 168, 173, 188, 194, 196
  S. epidermidis, 92
starvation, 84
stationary phase, 11, 164
stealthing, 247
stem cell sorting, 146
stems, 33, 38, 146, 262
Stenotrophomonas maltophilia, 60, 66, 125, 128, 325
steric interactions, 288
sterile, 243
stiffness, 215
Streptococcus sp., 189
Streptomyces
  S. griseus, 4, 5
  S. hygroscopicus, 4, 5
  S. viridogens, 4, 5
stress-associated genes, 280
strong alkali, 34
strontianite (SrCO$_3$), 170
sub-octahedral morphology, 124
sucrose, 113, 288
sugars, 262, 278, 305
sulfate and sulfite reductase, 66–7, 176
sulfate permease, 66, 109
sulfate-reducing bacteria (SRB), 4, 6, 68, 112, 118, 124, 125, 132, 134, 135, 165, 195, 198, 242
sulfate reductase, 6, 109, 169
sulfhydryl groups, 68
sulfide, 2, 5, 6, 33, 34, 61, 66–8, 72, 107, 109–12, 114, 127, 131–5, 163, 164,

SUBJECT INDEX

171, 173, 192, 193, 198, 210, 211, 225, 226, 228–30, 234, 235, 237, 239, 241–3, 325
sulfidic zinc, 226
sulfite reductase, 66–7, 109, 110, 176
Sulfobacillus thermosulfidooxidans, 236
sulfur dioxide, 226
sulfuric acid, 230, 231, 236, 242
Sulfurospirillum barnesii, 4, 8, 60, 106, 107, 130, 131
sulfur-oxidizing, 236
sulfurylase sulfite reductase, 66
sulphate reducing bacteria, 2, 4, 6, 8, 67, 68, 105, 112, 124, 125, 132, 134, 135, 165, 195, 198, 242
sulphate-reducing bacterium, 6, 8, 129, 146, 165
sulphate reductases, 6
sulphide oxidation, 210, 228, 229
sulphur, 232, 264
superconducting, 240
superlattices, 173
supermagnetic iron oxide, 285
superoxide dismutase (SOD), 38, 62, 111, 260, 261, 265, 275
superoxide radical, 274–8
super paramagnetic, 102, 146
supersaturated, 36, 198
supramolecular complexes, 176
surface acido-basicity, 57
surface proteins, 87, 88
surfactant, 190, 196, 206, 243, 282, 286
survival, xviii, 103, 134, 238, 276
susceptibility, 197, 278
sustainable, xix, xviii, 76, 89, 124, 205–19, 322
Suzuki-Miyaura and Mizoroki-Heck reactions, 206, 212
symmetric spherical, 156
symptoms, 102, 311
Synechococcus sp., 13
synergistic effects, 216
synthesis, 1–19, 31–46, 53, 84, 102, 124, 142, 155–79, 187, 205–19, 229, 275, 317
synthetic nanoparticles, 179, 319
synthetic strategies, 247
systematic consideration, 315
Syzygium cumini, 60

tailings, 227–34, 237, 239, 246–7
tamarind, 148
target cell, 197, 279
task force, 313, 314, 324
taste, 304, 308
tautomerization, 70, 72, 73
technology, 84, 101, 104, 141, 156, 208, 218, 237, 244, 247, 303, 305, 310, 315, 317, 321–5
Tellurite (TeO$_3^{2-}$), 8, 62, 63
tellurium, 8, 62, 63
tellurium oxyanions, 131
temperature, xvii, xviii, 8–11, 32, 34, 36, 37, 39–44, 46, 53, 57, 103–5, 107, 108, 114, 115, 131, 142, 145–7, 149, 162, 164–6, 168, 170, 171, 173–5, 190, 192, 193, 195, 199, 208, 211, 212, 215, 230–233, 236–40, 242, 245, 286, 288, 325
terpenoids, 57, 68, 147
terrestrial, 255
tertiary structure, 110
tetragonal BaTiO$_3$, 147
tetragonal phase, 37, 40, 41
tetrahedral linkage, 245
tetraoctyl ammonium bromide (TOAB), 55, 56
Tetraselmis gracilis, 12, 14
Tetrathiobacter kashmirensis, 160, 166
textiles, 245, 246
texture, 308
thalassohaline, 107
Thauera selenatis, 4, 8
therapeutic, 116, 189, 197, 292, 293, 318
therapeutic molecules, 189
thermal decomposition, 31, 317
Thermoanaerobacter ethanolicus, 134, 241
thermocouples, 237
thermodynamic driving force, 287
thermodynamic stabilization, 282
thermogravimetry (TGA-MS), 245
Thermomonospora sp., 4, 5, 10, 60, 107, 108, 146, 170, 241
thermophilic fungus, 41, 45, 170
thin-film capacitors, 40
thioglycerol (TG), 142
thioldisulfide oxidoreductase (DsbB), 62
thiol groups, 281, 284, 292
thioredoxin (Trx)-glutathione (GSH) system, 62

thiourea, 229
tissues, 32, 57, 102, 113, 196, 199, 255, 258, 263, 266, 267, 279, 287, 293, 310, 313
titania, 35–8, 40, 46, 170
titania nanoparticles, 9, 35, 37, 40, 46, 170
titanium, 8, 16, 17, 36, 72, 92, 103, 199
titanium dioxide (TiO_2), 36, 92, 199, 317
tobacco, 264, 266
tobacco mosaic virus (TMV), 161, 173
toluene, 55
tomato, 9, 256, 257, 264, 266
toothpaste, 199
top-down, 31, 42, 44, 104
top-down physical, 317
topochemical, 240
topographic, 176
Torulopsis, 188, 241
toxic, 5, 19, 32, 33, 35, 39, 57, 59, 62, 78, 84, 91, 92, 103, 104, 107, 108, 115, 128, 134, 142, 145, 146, 149, 156, 157, 163, 166, 171, 174, 187, 189–92, 194–9, 207, 229–31, 235, 236, 238, 242, 244, 247, 255, 257, 259, 260, 266, 274, 278, 282–4, 289, 293, 310, 312, 323, 324
toxic chemicals, 19, 32, 57, 84, 157, 187, 191, 195, 247, 323
toxicity, 5, 17, 19, 39, 59, 61, 65, 108, 111, 128, 163, 190, 193, 196, 197, 244, 248, 257, 264, 266, 273–84, 286, 292, 293, 310, 312, 314, 317, 320
toxicologic, 312
toxicology, 273, 314
trackable, 305
traits, 94, 217, 244
transcription factor, 278, 289, 290
transcriptome, 266, 280, 284
transferrin, 288
transformation, 69, 70, 72, 90, 217–19, 238, 243, 247, 257, 310
transition, 10, 39, 40, 59, 67, 245
transition metals, 53, 54, 64, 275, 281
translation, 278, 279
translocation, 72, 257, 258, 261, 293, 311
transmembrane proton, 291
transmission electron microscopy (TEM), 10, 11, 13, 35, 40, 73, 108, 127, 128, 133, 158, 167, 169, 175, 177, 178, 195, 215, 259

transpiration, 256, 258, 263
transportations, xvii, 59, 141, 305
transport mechanism, 62
transport vehicle, 273
trapping, 76, 108, 143, 163, 194, 198, 215
triangles, 108, 125, 163
triangular, xviii, 3, 5, 11, 101, 104, 107, 114, 125, 126, 128, 159, 160, 162, 167
Trichoderma asperellum, 11, 60, 106, 160, 167
trichomes, 263
Trichothecium sp., 11, 60, 167, 188, 189
triethanolamine, 131
tripeptide, 65, 111, 276
trisodium citrate, 54
Trojan horse mechanism, 274
trophic levels, 256
tubings, 210
tumor necrosis factor beta (TNF-β), 289
Turbinaria conoides, 188, 189

UK Better Regulation Taskforce, 316, 317
ultrasensitive detection, 102
ultra-small capacitors, 41
ultrasonication, 176
ultrasonic processor, 114
ultrasound, 245
ultraviolet ray, 236
ultraviolet-visible spectroscopy, 195
unfolded protein response (UPR), 278
United States Department of Agriculture (USDA), 312
United States Environmental Protection Agency (EPA) titanium oxide (TiO_2), 103
univalent reduction, 274
uraninite (UO_2), 8, 89
uranium, 6, 89, 165
uranium oxide (UO_2), 8, 165
Ureibacillus thermosphaericus, 107, 108
Usnea longissima, 11
usnic acid, 11

vacuolar membrane, 171, 325
vacuoles, 171, 259, 266, 285, 325
value added products, xix, xviii, 111, 225–48
vanadium, 230
Van der Waals forces, 288
vapor, xvii, 31, 104, 226

vascular system, 256, 262
vasculature, 58
vegetation, 229
vehicles, 102, 189, 273, 313
vertebrates, 158
vertical retort furnace, 228
Verticillium sp., 10, 39, 40, 60, 67, 72, 76, 143, 147, 160, 167, 170, 243
  V. luteoalbum, 60, 112
vesicles, 6, 61, 62, 67, 72, 157, 158, 262, 278, 285, 286, 288
vesiculo-vacuolar organelle (VVO), 286
viable cells, 162, 174
Vibrio cholera, 67
Vibroid, 61
Vicia narbonensis, 263
viral capsid, 161, 173, 174
viroid capsules, 176
virulence, 94
virus, 161, 173–4
viscoelastic, 93
volatilization, 59
voluntary certification, 315

Waelz kiln and clinker processes, 228
wastewater sludge, 256
water channel genes, 266
weathering, 196, 233
wetland, 242
white sand, 42, 43
wilted growth, 264
wound dressings, 310
wrinkle free cream, 199
wurtzite, 173, 175, 226

xenobiotics, 92
xerophytes, 73, 74
X-ray absorption spectroscopy (XAS), 13
X-ray diffraction (XRD), 13, 40, 177, 195, 236, 245
X-ray diffractometry, 177
X-ray photoelectron spectroscopy (XPS), 177
xygenases, 72
xylem vessels, 258, 262

Yarrowia lipolytica, 72, 73, 106, 113, 114, 161, 172
yeast, 2, 11, 33, 59, 60, 72–3, 76, 84, 104, 111–14, 146, 158, 161, 171–5, 192, 199, 241, 247, 282, 317

Zea mays, 258, 263
zeolites, 246
zero valent metal nuclei, 167
zeta potential, 114, 177, 282, 283
zinc carbonate, 226
zinc-containing ores, 225, 227
zinc-erythromycin, 246
zinc oxide quantum dots (ZnO QDs), 196, 197
zinc phosphate tetrahydrates (ZPT), 245
zirconia, 9, 34–8, 43, 60, 72, 147, 170, 241
zirconia nanoparticles, 9, 35–7, 43, 72, 170
zirconium hexafluoride (ZrF_6^{2-}), 9, 170
zirconium hexafluoride anions, 9, 170
zirconium oxide (ZrO_2), 34
zircon sand, 42, 43
ZnS deposits, 242